CLIMATE TECHNOLOGY AND LAW IN THE ANTHROPOCENE

Edited by
Leonie Reins and Alexander Zahar

BRISTOL
UNIVERSITY
PRESS

First published in Great Britain in 2025 by

Bristol University Press
University of Bristol
1–9 Old Park Hill
Bristol
BS2 8BB
UK
t: +44 (0)117 374 6645
e: bup-info@bristol.ac.uk

Details of international sales and distribution partners are available at bristoluniversitypress.co.uk

British Library Cataloguing in Publication Data
A catalogue record for this book is available from the British Library

ISBN 978-1-5292-3288-2 hardcover
ISBN 978-1-5292-3290-5 ePub
ISBN 978-1-5292-3291-2 OA PDF

Cover design: Hannah Gaskamp
Front cover image: © FAO/Steve Terrill
Bristol University Press use environmentally responsible print partners.
Printed and bound in Great Britain by CPI Group (UK) Ltd, Croydon, CR0 4YY

Bristol University Press' authorised representative in the European Union is:
Easy Access System Europe, Mustamäe tee 50, 10621 Tallinn, Estonia,
Email: gpsr.requests@easproject.com

FSC www.fsc.org MIX Paper | Supporting responsible forestry FSC® C013604

Contents

Notes on Contributors

Mirta Alessandrini is Lecturer and Researcher in the LAW Group at Wageningen University (WUR), the Netherlands. Her PhD is in food law (AGRISYSTEM) from Universitá Cattolica Sacro Cuore. She is the Academic Coordinator of the Master's Specialization 'Food Law and Regulatory Affairs' (WUR). She teaches agri-food law and is currently part of the H2020 'EU Tolerate' project, in which she addresses the legal challenges concerning the use of GM techniques to enhance the adaptation of plants to climate change. Her research addresses various aspects of agri-food law, with a focus on short food-supply chains, the EU Common Agricultural Policy, food safety, food sustainability and GMO legislation.

Justine Bell-James is Associate Professor at TC Beirne School of Law, University of Queensland, Australia. Her expertise is in environmental and climate change law. Justine's main research interest is legal mechanisms for the protection, management and restoration of coastal and marine ecosystems. She currently leads a project that considers how coastal wetland ecosystem services can be integrated into legal frameworks. Justine also has expertise in climate change litigation, and her work on opportunities for litigation under Queensland's Human Rights Act 2019 underpinned the *Waratah Coal* case decided by Queensland's Land Court in 2022. Justine's work cuts across disciplines: she is an affiliated researcher at the University of Queensland's Centre for Biodiversity and Conservation Science.

Kerryn Brent is Senior Lecturer at University of Adelaide Law School. She holds a PhD from the University of Tasmania. Kerryn teaches climate law, environmental law, and law of the sea. Her research is in climate law, focusing on the governance of solar geoengineering and carbon-dioxide-removal technologies. She is Deputy Director of the Australian Forum for Climate Intervention Governance.

Gareth Davies is Professor of European Law at Free University Amsterdam. He has been a barrister in London, an advisor on internal market law to the European Parliament, European Commission and other organizations, and a research fellow at NYU law school and the EUI in Florence. He writes and researches on European integration, European citizenship, free movement law and the internal market, the European Court of Justice, and climate engineering. He teaches EU internal market law, EU anti-discrimination law and technology law.

Felicity Deane is Associate Professor at Queensland University of Technology, Australia. She has published extensively in areas of the intersection of economics and law. For over a decade, Felicity has been researching and teaching the law of the World Trade Organization, and has extensive knowledge of international trade law and practice. Her first book, *Emissions Trading and WTO Law: A Global Analysis* (Edward Elgar, 2015), examined the impact of trade rules on market-based instruments related to climate change. In recent years, Felicity has collaborated with industry partners on projects that focus on land-use practices, export markets and the impacts of development on natural resources.

Gunnar S. Eskeland is Professor of Resource and Environmental Economics at the Norwegian School of Economics (NHH), Bergen. He has worked in senior research positions with the World Bank and Centre for Climate Change, Oslo. Topics have included environmental protection, transport and air quality, energy economics, shipping markets, and climate policy. Recent work on climate change has included energy-sector analysis, European policy review and research on technological change and an R&D treaty. Professor Eskeland teaches in NHH's Master's programme on Energy, Natural Resources and the Environment, and is the director of the Energy and Natural Resources and Environment Centre, ENE, at NHH/Samfunns- og næringslivsforskning (SNF).

Ruven Fleming is an expert on hydrogen regulation. He is Assistant Professor of Energy Law at the University of Groningen, where he is also Scientific Coordinator of the Groningen Centre of Energy Law and Sustainability. He is working as Project Manager on hydrogen projects at the German Fuel Institute (DBI), TU Freiberg, Germany. He is Visiting Associate Professor at TU Berlin, Germany. His research interest is energy law, with a particular focus on hydrogen, 'unconventional' forms of energy production, renewable energy and energy investment and trade law.

Floor Fleurke is Associate Professor of European Environmental Law in the Department of Public Law and Governance at Tilburg Law School. Her research focuses on European environmental law and nature protection law with an interest in the relationship between law, science and technology. She is part of the research project *Constitutionalizing in the Anthropocene.*

Morag Goodwin holds the Chair in Global Law and Development at Tilburg Law School. Morag's research focuses on questions of exclusion, inclusion and participation, and the role of law or regulatory phenomena (for example, indicators or discourses) in creating and structuring exclusionary dynamics. She is particularly interested in the role that global (development) narratives play in such dynamics. Morag is also co-director of the LL.B. Global Law programme at Tilburg, series editor of the CUP Global Law book series, and director of the European Joint Doctorate in Law and Development (EDOLAD). She has been a visiting professor at FGV Direito, São Paulo; Osgoode Hall Law School, Toronto; and, in 2022, IDC School of Law, Tel Aviv.

Ignacio Herrera Anchustegui is Associate Professor at the University of Bergen's Faculty of Law. He leads the Research Group for Climate, Energy and Environmental Law at the University of Bergen, and is a member of the Bergen Offshore Wind Centre and the Bergen Centre for Competition Law and Economics. He obtained his PhD in Competition Law and Economics at the University of Bergen with a thesis entitled *Buyer Power in EU Competition Law*. Ignacio's research focuses on regulating energy markets and new-energy technologies, such as offshore wind. In addition to economic perspectives, his work combines energy law with competition and public procurement law. In 2018, he received the University of Bergen's Meltzer Award for Young Researchers. He is the author of more than 40 publications and co-editor of the *Routledge Handbook of Energy Law*.

Kateryna Holzer is Senior Researcher at the Centre for Climate, Energy and Environmental Law at University of Eastern Finland Law School, where she leads a project on regulatory cooperation on carbon standards and teaches a course on international economic law and green transitions. She holds a PhD in Law from the University of Bern and a PhD in Economics from Ukraine and has many years of experience in climate change and energy research, teaching and consultancy, with a focus on trade rules and measures supporting sustainable development.

Kaisa Huhta is Associate Professor of European Law at UEF Law School and at the Centre for Climate Change, Energy and Environmental Law, where her research focuses on EU energy law. Her research on the nexus of EU law and the energy sector has produced publications on state-aid law, free-movement law, competence and institutional issues in EU law, the regulation of new-energy technologies and energy investment, security and consumers in EU law. Her broader research objective is understanding the roles of law in governing the energy sector and the low-carbon energy transition in the EU. Kaisa is the Principal Investigator in an EUJUST project (2021–24), funded by the Academy of Finland, that studies EU law as an instrument for energy justice. She is also a leader in the RELIEF project (2023–26), funded by the Strategic Research Council, which studies the role of law in sustainability transformations.

Laura Kaschny is a PhD Researcher specializing in EU and energy law at Tilburg Law School. Her research focuses on how to regulate and accelerate a just energy transition, with a particular emphasis on the concept of energy justice. Her research places importance on understanding the institutional framework and interplay between the EU, national governments and national regulatory authorities in addressing regulatory trade-offs while considering public interests.

Bettina Lange is Associate Professor in Law and Regulation at the Centre for Socio-Legal Studies, University of Oxford. She has published widely in the field of environmental regulation, and is the author of *Implementing EU Pollution Control: Law and Integration* (Cambridge University Press, 2008), and co-author (with E. Fisher and E. Scotford) of *Environmental Law: Text, Cases and Materials* (Oxford University Press, 2019). She is a joint winner of the *Journal of Environmental Law*'s Richard Macrory 'Best Article' Prize for her article (with M. Shepheard) 'Changing Conceptions of Rights to Water? An Eco-Socio-Legal Perspective' (2014).

Jan McDonald is Professor of Environmental and Climate Law at the University of Tasmania's Faculty of Law and Centre for Marine Socioecology, and New Star Professor of Environmental Law at the University of Tasmania. Her current research and teaching spans the legal and policy dimensions of climate change adaptation, including coastal management, biodiversity conservation and species range shifts, as well as the emerging field of geoengineering governance. Jan is a former president and current board member of the National Environmental Law Association and a member of the Australian Panel of Experts on Environmental Law. She is also a Deputy Director of the Australian Forum for Climate Intervention Governance.

Jeffrey McGee is Associate Professor of Climate Change, Marine and Antarctic Law at the Institute for Marine and Antarctic Studies and the Faculty of Law, University of Tasmania. He is Director of the Australian Forum for Climate Intervention Governance, assistant editor of *Carbon and Climate Law Review*, and editorial board member for the journals *Review of European, Comparative & International Environmental Law*, *International Environmental Agreements: Politics, Law and Economics*, and *Australian Journal of Maritime and Ocean Affairs*.

Phillip Paiement is Professor in Law and Governance in the Anthropocene at Tilburg Law School where he conducts research in transnational legal theory and socio-legal studies. He is engaged in a collaborative project on *Constitutionalizing in the Anthropocene*, in which he evaluates the role of legal institutions in both facilitating and confronting the exploitative relations between and within human and nonhuman communities that have given rise to the Anthropocene. Since September 2022, he is also the Principal Investigator in the ERC Starting Grant project TransLitigate, which combines doctrinal and qualitative socio-legal methods to study the transnational collaborative networks of strategic litigators' work on climate change, pollution from the extractive industries, biodiversity conservation and land conflicts. Phillip has also published widely on transnational private sustainability governance, with a focus on transnational product standards and the auditing practices that underpin them. He serves as co-Convening Editor of *Transnational Legal Theory*.

Kelsey Pailman is an energy law PhD Researcher at the University of Groningen focusing on the regulation of renewable and low-carbon hydrogen in the promotion of sustainable mobility. Her research forms part of the EU Marie Skłodowska-Curie Actions 'Training for a Hydrogen-Economy-Based Renewable Energy Society in the Anthropocene' (THERESA) Project. Prior to joining the University of Groningen, she served as an energy lawyer in South Africa, advising global clients in the renewables, oil and gas and mining sectors. She also holds an LLM in energy law from the University of Cape Town in South Africa, having also completed an international exchange at Bucerius Law School in Hamburg, Germany, on EU and German energy law.

Alberto Quintavalla is Assistant Professor at the Department of Law and Markets, Erasmus University Rotterdam. He received his doctoral degree at the Rotterdam Institute of Law and Economics and the Department of International and European Union Law, Erasmus University, in 2020 for research on water governance. He has been a Visiting Researcher at the Hebrew University of Jerusalem (recipient of the Guidetti Prize) and the European University Institute (recipient of Erasmus + funding). His research interests are at the intersection of environmental governance, human rights and digital technologies. He has published in leading journals and has served as a guest editor of the *European Journal of Risk Regulation*. He has been awarded the prize for the best 'new voices' 2020/2021 published in the *European Journal of Legal Studies*. He is admitted to the Italian Bar.

Leonie Reins is Professor for Public Law and Sustainability at Erasmus School of Law (ESL). Prior to joining ESL she worked as an Assistant Professor at the Tilburg Institute for Law, Technology and Society at Tilburg University. She was awarded an FWO-financed PhD on *The Coherent Regulation on Energy and Environment – Using Shale Gas as a Case Study* from KU Leuven, where she subsequently worked as a Post-Doctoral Researcher. In addition Leonie worked as Legal Advisor at a Brussels-based environmental law and policy consultancy, where she was involved in projects on environmental, energy and climate change law and policy. Leonie also holds an LLM in International, European and Comparative Energy and Environmental Law.

Matthew Rimmer (BA/LLB ANU, PhD UNSW) is Professor in Intellectual Property and Innovation Law at the Faculty of Business and Law at Queensland University of Technology (QUT). He is a chief investigator at the QUT Centre for Behavioural Economics, Society, and Technology and at the QUT Australian Centre for Health Law Research. He is a member of the NHMRC Centre of Research Excellence on Achieving the Tobacco Endgame (CREATE).

Daria Shapovalova is Senior Lecturer in Energy Law at the University of Aberdeen, where she is also Director of the Centre for Energy Law and Coordinator of the Just Transition Lab. Prior to obtaining a PhD from the University of Aberdeen, Daria completed legal education in Ukraine and the Netherlands. Her main research interests are in energy sector regulation, with a focus on environmental protection, human rights and just transition. She has published extensively on the intersections of environmental and energy law, Arctic governance and just transition.

Manon Simon is Lecturer at the University of Tasmania Faculty of Law. She completed her PhD on *Governing Cloud Seeding in Australia and the United States: Lessons for Regional Solar Radiation Management*. She is currently working as a Research Fellow on an Australian Research Council Discovery Project entitled 'Governing Solar Radiation Management Research, Development and Deployment' with the Australian Forum for Climate Intervention Governance.

Josephine van Zeben is a full-time professor and Chair of Transnational Law at the Florence School of Transnational Governance. She also serves as the Academic Director for the Master in Transnational Governance. Josephine's research focuses on the regulation of environmental issues by public and private actors across jurisdictions, with particular attention on polycentric governance theory and developments related to the European Union. Josephine is co-editor-in-chief of *Transnational Environmental Law* and series editor of the Transnational Environmental Law book series. Prior to her current position, she was Professor and Chair of the LAW group at

Wageningen University and a Law Fellow at Worcester College, University of Oxford. She has held visiting professorships at Notre Dame Law School, Bocconi University, and La Trobe Law School. Between 2012 and 2022, she taught Environmental Law and Policy at ETH Zürich. She holds a PhD in Law and Economics from the University of Amsterdam, LLM degrees from Harvard University and the University of Amsterdam (European Private Law), an LLB in Scots Law from the University of Edinburgh and a BA in Social Sciences from University College Utrecht.

Jonathan Verschuuren is Professor of International and European Environmental Law at Tilburg University and is on the faculty of the Tilburg Sustainability Centre and the Centre for Transboundary Legal Development. His research mainly focuses on the role of international and EU environmental law in legal practice. His current research focuses on climate change. As of 2015, Jonathan primarily researches the impact of climate change on agriculture (and vice versa), particularly with a view to developing a regulatory framework for climate-smart agriculture. In 2015, he was awarded a two-year Marie Skłodowska Curie Global Fellowship on this topic.

Seita Vesa holds the position of Professor of Environmental Law with a specialization in sustainability transition at the University of Eastern Finland Law School. Additionally, she is a Research Professor within the Climate Solutions Unit of the Finnish Environment Institute. Her publications span a wide range of topics in international and EU environmental and climate law. With extensive international collaboration experience in research, education and project consortia, Seita also holds several external expert positions and offers consultancy services in environmental law.

Yulia Yamineva is Associate Professor in Climate Law and Policy at the Centre for Climate, Energy and Environmental Law, University of Eastern Finland Law School. Her research addresses various aspects of international climate law and governance, as well as the science-policy interface. This work has been published in *Transnational Environmental Law*, *Environmental Science and Policy* and *Nature Climate Change*, among other journals. Prior to academia, she was with the UNFCCC Secretariat and the International Institute for Sustainable Development. She holds a PhD in International Studies from the University of Cambridge.

Alexander Zahar is a solicitor in natural resources and climate law at the law firm Ashurst.

Hao Zhang is Associate Professor at the Faculty of Law, The Chinese University of Hong Kong, where he specializes in teaching energy law and Chinese law. His primary research interests lie in the fields of energy law and climate law, with a particular focus on China. His monograph, *In Pursuit of Carbon Neutrality: Energy Law and Regulation in China* (Cambridge University Press, 2023), critically examines the essential energy laws and regulations pertaining to the strategies and pathways to achieve China's goal of carbon neutrality.

Acknowledgements

This publication has been made Open Access thanks to the generous support of the Erasmus Initiative Dynamics of Inclusive Prosperity and the EUR Open Access fund.

1

Introduction

Alexander Zahar and Leonie Reins

1.1 Aim of this book

The justification for this book starts with climate change. The problem of climate change has been steadily worsening. Thirty years after the United Nations Framework Convention on Climate Change (UNFCCC) was signed, only a few countries have substantially improved their practices. Most have not. Many plan to emit more greenhouse gases per year in the next decade (or decades) than they have in the past.[1] Collective emissions continue to rise. Yet, in the official literature of the community of states, there is still hope. Despite critical deadlines being declared and missed,[2] it would seem it is never too late to solve the problem of climate change.

This apparent paradox – of a problem constantly worsening while remaining solvable – avoids becoming an actual paradox because of a conviction that technologies exist that will solve the problem of climate change eventually.[3] As time passes, more and more of the solution relies

[1] This is apparent from the Nationally Determined Contributions (NDCs) of many developing countries: Alexander Zahar, 'Climate Change Mitigation Law and Policy in the Asia-Pacific', in *Research Handbook on Climate Change Mitigation Law*, ed. Leonie Reins and Jonathan Verschuuren, 155–77 (Cheltenham, UK: Edward Elgar, 2021).

[2] For example, United Nations Development Programme, *Human Development Report 2007/2008: Fighting Climate Change: Human Solidarity in a Divided World* (New York: UNDP, 2007), 1 ('climate change … is still a preventable crisis – but only just. The world has less than a decade to change course').

[3] The foundational paper that popularized a new way of thinking about mitigation technologies is Stephen Pacala and Robert Socolow, 'Stabilization Wedges: Solving the Climate Problem for the Next 50 Years with Current Technologies', 305 *Science* 968 (2004).

on these technologies intervening and less and less of it on the adoption of climate-friendly lifestyles, such as energy saving, reduced travel, plant-based diets and avoidance of products linked with deforestation. The prevailing attitude seems to be that, while pricing of greenhouse gas emissions (in the few places where pricing is practised) may encourage some lifestyle change, we can keep doing almost everything we are used to doing, and even more of it, because houses will be powered by wind and solar, transport will be electrified, and in the background there will be an army of contraptions working to offset or capture past emissions or reduce the impact of solar radiation on Earth.[4] Such is the confidence of states in a technological solution which installs technology around us without demanding any significant behavioural change that dozens of states have announced net-zero targets for 2050, 2060 or 2070, or for unspecified dates in the second half of the century.[5] Vague and incomplete though these pledges are,[6] each of them presumes that an unprecedented technological transition will become a reality.

As the critical warming threshold of 2 °C approaches,[7] the technological toolbox that will power the transition away from fossil fuels is not becoming larger. It has remained unchanged or is becoming smaller,[8] because the narrative of the realistically achievable has been given priority, as time runs out, over that of the merely promising or possible.[9] The technologies of the future that we now count on number about a dozen: enhanced energy efficiency, scaled-up renewables (modern bioenergy,[10] solar photovoltaics (PV), wind power, hydropower, biofuels and geothermal energy), electrification to replace most direct fuel uses, extensive battery storage, and

4 Studies of mitigation pathways generally do not assume lifestyle change; for example, James H. Williams et al, 'The Technology Path to Deep Greenhouse Gas Emissions Cuts by 2050: The Pivotal Role of Electricity', 335 (6 January) *Science* 53 (2012), 58 ('Assuming plausible technological advances, we find that it is possible for California to achieve deep GHG reductions by 2050 with little change in life-style').

5 See UNFCCC website, 'Communication of Long-Term Strategies' <https://unfccc.int/process/the-paris-agreement/long-term-strategies>.

6 See Alexander Zahar, 'Emission Neutrality in Asia in the Paris Agreement Context', available from SSRN <https://papers.ssrn.com/sol3/papers.cfm?abstract_id=3809690>.

7 Olivier Dessens, Gabrial Anandarajah and Ajay Gambhir, 'Limiting Global Warming to 2°C: What Do the Latest Mitigation Studies Tell Us About Costs, Technologies and Other Impacts?', 13–14 *Energy Strategy Reviews* 67 (2019), 70 ('According to the latest reviews, equilibrium climate sensitivity is likely in the range 2.1 °C–4.7 °C').

8 Pacala and Socolow considered *15* technologies: (n 3) 970.

9 Pete Smith et al, 'Biophysical and Economic Limits to Negative CO_2 Emissions', 6 (January) *Nature Climate Change* 42 (2016), 42–3 (on why technologies such as ocean iron fertilization and soil carbon and biochar storage are not being considered in modelling).

10 To be distinguished from traditional biomass uses.

more nuclear power and natural sinks (forests), as well as the novelties of hydrogen fuels, negative-emission technologies (NETs) including carbon-dioxide removal (CDR), and carbon capture and storage (CCS). The extent to which these novelties can actively be relied upon to function at scale is currently unknown.[11] Some fossil-fuel technologies will need to be retained at a much-reduced scale – natural gas for flexibility in power generation, coal for iron and steel production, oil for aviation and shipping – and some albedo-modification techniques, such as stratospheric aerosol injection will need to be readied commercially as a backup plan, in case the aggregate of other efforts proves insufficient.[12] At a regional scale, for example at the EU level, the key technology set reduces to about eight types.[13] At a state level it can be reduced to five or fewer, depending on each state's resource endowment.[14]

We are becoming increasingly dependent for a solution on this fixed or diminishing set of tried or known technologies, which means that the scale and speed of their deployment increases with every business-as-usual-emissions day that goes by.[15] This was never so clearly stated as at the 28th United Nations Climate Change Conference (COP 28), in 2023:

> The Conference of the Parties … recognizes the need for deep, rapid and sustained reductions in greenhouse gas emissions in line with 1.5°C pathways and calls on Parties to contribute to the following global efforts, in a nationally determined manner, taking into account the Paris Agreement and their different national circumstances, pathways and approaches:

---

[11] See, for example, Motoko Rich and Hikari Hida, 'Japan Says It Can Make Coal Cleaner. Critics Say Its Plan Is "Almost Impossible"', *The New York Times*, 22 July 2023 <www.nytimes.com/2023/07/22/world/asia/japan-coal-renewable-energy-ammonia.html> ('Japan … stands alone in insisting it can make coal less damaging to the planet').

[12] On PV: Felix Creutzig et al, 'The Underestimated Potential of Solar Energy to Mitigate Climate Change', 2 *Nature Energy* 1 (17140) (25 August 2017); on BECCS and CCS: Matteo Muratori et al, 'Global Economic Consequences of Deploying Bioenergy with Carbon Capture and Storage (BECCS)', 11 (095004) *Environmental Research Letters* 1 (2016), 4; and on BECCS and other NETs: Smith et al (n 9) 43. On hydrogen, direct air capture, CCS, and in general on the limited number of technology options: Dessens, Anandarajah and Gambhir (n 7) 74; and International Energy Agency, *Energy Technology Perspectives 2020* (Paris: 2020), 66, 80–1 and 104–6.

[13] International Energy Agency, *World Energy Outlook 2020* (Paris: 2020), 157.

[14] For California, see Williams et al (n 4) 53–4.

[15] Varun Rai, Kaye Schultz and Erik Funkhouser, 'International Low Carbon Technology Transfer: Do Intellectual Property Regimes Matter?', 24 *Global Environmental Change* 60 (2014), 60.

(a) Tripling renewable energy capacity globally and doubling the global average annual rate of energy efficiency improvements by 2030;
(b) Accelerating efforts towards the phase-down of unabated coal power;
(c) Accelerating efforts globally towards net zero emission energy systems, utilizing zero- and low-carbon fuels well before or by around mid-century;
(d) Transitioning away from fossil fuels in energy systems, in a just, orderly and equitable manner, accelerating action in this critical decade, so as to achieve net zero by 2050 in keeping with the science;
(e) Accelerating zero- and low-emission technologies, including, inter alia, renewables, nuclear, abatement and removal technologies such as carbon capture and utilization and storage, particularly in hard-to-abate sectors, and low-carbon hydrogen production;
(f) Accelerating and substantially reducing non-carbon-dioxide emissions globally, including in particular methane emissions by 2030;
(g) Accelerating the reduction of emissions from road transport on a range of pathways, including through development of infrastructure and rapid deployment of zero-and low-emission vehicles;
(h) Phasing out inefficient fossil fuel subsidies that do not address energy poverty or just transitions, as soon as possible.

This decision to 'transition away' from fossil fuels affirms the need to ramp up alternatives rapidly. Simon Stiell, the UNFCCC Executive Secretary, in his closing speech to COP 28, referred to this decision as 'the beginning of the end' of the fossil-fuel era and appealed to 'all governments and businesses … to turn these pledges into real-economy outcomes, without delay'.[16] The terms that stand out in the decision are 'accelerating', 'doubling', 'tripling', 'by 2030'.

The two concepts – of scale and speed – are inseparable: 'speed at scale' is the more accurate expression.

Such other expressions as 'scaled up rapidly' and 'ramped up significantly in the next decade' colour the climate technology literature. Here is a quantitative illustration of speed at scale under a modestly ambitious mitigation scenario (the SDS scenario, with net-zero carbon dioxide (CO_2) emissions by 2070):

[16] UN Climate Press Release, 'COP28 Agreement Signals "Beginning of the End" of the Fossil Fuel Era', 13 December 2023 <https://unfccc.int/news/cop28-agreement-signals-beginning-of-the-end-of-the-fossil-fuel-era>.

> By 2070, about 90% of all the CO_2 emitted globally in cement and around 75% in chemicals and iron and steel is captured. For this to happen, an average of almost 40 cement plants, over 15 chemicals plants and over 10 steel plants operating with CO_2 capture need to be built every single year through to 2070.[17]

The more low-key the technology, the more mind-boggling the quantity over time.[18] Speed at scale affects the shuttering of the old as much as the rolling out of the new. Under an ambitious, 1.5 °C, mitigation scenario, and given that there is not enough time to scale up CCS by 2030, 'The majority of emissions reductions [needed by 2030] from coal-fired power plants … comes from reducing operations and closing facilities',[19] risking 'large job losses in coal mining and coal-fired power plants'.[20] For the same scenario, the average annual increase in global renewable electricity generation is around 1,100 TWh from 2019 through to 2030; for reference, the largest ever single-year increase so far has been about 440 TWh – in 2018.[21]

The future will, we now know officially, courtesy of COP 28, be oriented toward deploying and maintaining this specific set of technologies.[22] 'The pathways indicate that the energy transformations need to be initiated without delay, gain momentum rapidly, and be sustained for decades.'[23] Their fundamental and unprecedented purpose will be 'climate control'.[24] When the Biden Administration announced its 2030 mitigation target, in April 2021, learned commentary instantly hit upon its main challenge: the carrying out of a technological overhaul of the United States at breakneck speed.[25] Precisely what will have to give way from the point of view of

[17] IEA (n 11) 106.

[18] See Cécile A. J. Girardin et al, 'Nature-Based Solutions Can Help Cool the Planet – If We Act Now', 593 *Nature* 191 (13 May 2021), 192–3.

[19] IEA (n 12) 132–3.

[20] Ibid, 130.

[21] Ibid, 133.

[22] Dessens, Anandarajah and Gambhir (n 7) 72 (on the necessary large-scale deployment of CCS).

[23] Ibid, 71.

[24] The earliest extensive use of the concept of 'climate control' was by Alan Carlin, 'Global Climate Change Control: Is There a Better Strategy Than Reducing Greenhouse Gas Emissions?', 155(6) *University of Pennsylvania Law Review* 1401 (2007). For a recent use, see Donald Zillman et al, 'Conclusion', in *Innovation in Energy Law and Technology: Dynamic Solutions for Energy Transitions*, ed. Donald Zillman et al, 412–35 (Oxford: Oxford University Press, 2018).

[25] Brad Plumer, 'Biden Wants to Slash Emissions. Success Would Mean a Very Different America', *The New York Times*, 22 April 2021 <www.nytimes.com/2021/04/22/climate/biden-emissions-target-economy.html>.

settled law and values to allow for the speed at scale of this never-before-seen technological rescue is debatable. The logic of speed at scale fixes our interest in this transformation and supplies the core rationale for this book.

Our aim is to systematically review the legal issues that arise from these premises. The essays in this book sharpen up the issues, often presenting them as clashing positions, inviting the reader to decide them. Some of the issues are broadly known because they are already being debated in public policy. For example, the desirability of, and conditions of experimentation with, albedo-modification techniques are questions that are currently generating controversy.[26] Any suggested use of genetic modification as a mitigation technology is also guaranteed to give rise to starkly opposing views.[27] Assisted migration of species as an adaptation technology, or gene editing for adaptation ('assisted evolution') with or without assisted migration, are increasingly debated in public fora.[28] However, most of the topics discussed in this book have not (yet) been popularized, or even examined much by scholars, and may thus be considered 'new' – and ripe for further development. Their connection with technology makes them important, because a technological revolution to rival the Industrial Revolution is inevitable for the purposes of climate control – as affirmed at COP 28. But they are also important for a more subtle reason.

A body of climate law has emerged at the international level, assembled from elements of treaty and customary law. Scholars have made efforts to strengthen it and fill in its gaps, but the result still does not amount to very much.[29] How could it, when the states are so divided about who is obliged to do what about a problem that is both expensive to fix and collective and intergenerational? At the domestic level, only a few countries outside the EU could claim to have developed a substantial body of climate law. The chapters in this book show that climate law can also be conceptualized in

[26] See, for example, Jesse L. Reynolds, 'Is Solar Geoengineering Ungovernable? A Critical Assessment of Governance Challenges Identified by the Intergovernmental Panel on Climate Change', 12 *WIREs Climate Change* 1 (e690) (2021); Paul Voosen, 'U.S. Needs Solar Geoengineering Research Program, Report Says', 372 (6537) *Science* 19 (2021); and Editorial, 'Give Research Into Solar Geoengineering a Chance', 593 *Nature* 167 (13 May 2021).

[27] See, for example, the controversy brewing over the EU's position on GMOs: Sarah Garland, 'EU Policy Must Change to Reflect the Potential of Gene Editing for Addressing Climate Change', 28 *Global Food Security* 1 (100496) (2021).

[28] For example, Jedediah F. Brodie et al, 'Global Policy for Assisted Colonization of Species', 372 (6541) *Science* (2021). This 'technology' may need to extend to people themselves: J. B. Ruhl and Robin Kundis Craig, '4°C', 106 *Minnesota Law Review* 191 (2021).

[29] See Benoit Mayer and Alexander Zahar (eds), *Debating Climate Law* (Cambridge, UK: Cambridge University Press, 2021), for a snapshot of this deeply unsettled area of the law.

contrast with existing law, which, being for the most part *not* climate law,[30] will need to change in ways that will allow for more ambitious action to be taken to keep global warming under control. Much of the climate law to come will need to inhabit a space that is now occupied by other law.[31]

---

[30] For the idea of an 'anti-law' that supports harmful acts against nature, see J. B. Ruhl, 'Farms, Their Environmental Harms, and Environmental Law', 27(2) *Ecology Law Quarterly* 263 (2000), 263.

[31] As already recognized, although taking a broader policy analysis and limiting it to a particular state jurisdiction, in Michael B. Gerrard and John C. Dernbach (eds), *Legal Pathways to Deep Decarbonization in the United States* (Washington, DC: Environmental Law Institute, 2019); see also Alastair R. Lucas and Chidinma B. Thompson, 'Transition to a Low-Carbon Energy Economy: The Legal Agenda', in *Innovation in Energy Law and Technology: Dynamic Solutions for Energy Transitions*, ed. Donald Zillman et al, 38–54 (Oxford: Oxford University Press, 2018) (on 'the legal norms that constrain the development of a consistent low carbon transition' in Canada); and Sara Traerup, Lucy Ellen Gregersen and Vladimir Hecl, *Summary of Country Priorities: Technology Needs Assessments, 2015–2018* (Copenhagen and Bonn: UNEP DTU Partnership and UNFCCC Secretariat, 2018), 23 (finding that, for the scaling up of mitigation technologies in developing countries, 'the most commonly identified barriers related to legal and regulatory issues (73%)'). Several recent news reports have observed that wind power, solar power, and the cabling infrastructure needed to supply them will change the face of the United States in ways that will be resisted, both for economic reasons (the huge expense involved) and by current nature-conservation arrangements: for example, Veronica Penney, 'Where Wind and Solar Power Need to Grow for America to Meet Its Goals', *The New York Times*, 28 May 2021 <www.nytimes.com/interactive/2021/05/28/climate/climate-wind-solar-energy-map.html>; and Ivan Penn and Clifford Krauss, 'More Power Lines or Rooftop Solar Panels: The Fight Over Energy's Future', *The New York Times*, 11 July 2021 <www.nytimes.com/2021/07/11/business/energy-environment/biden-climate-transmission-lines.html>; Vernon C. Bleich et al, 'Green Energy Projects at Odds with Conservation', 380(6649) *Science* 1021 ('two green energy projects … are planned between proximate, albeit separate, protected areas, posing threats to the species the protected areas were intended to benefit'); Patrick McGeehan, 'Wind Power, Key to Democrats' Climate-Change Goals, Faces a Crisis', *The New York Times*, 2 November 2023, <www.nytimes.com/2023/11/02/nyregion/wind-power-climate-change-democrats.html>. To further illustrate, CCS, CDR, next-generation nuclear, and emission-reducing GMOs will all need to push back against certain bodies of law to be enabled: see, respectively, Richard Macrory, 'Carbon Capture and Storage', in *The Oxford Handbook of Law, Regulation and Technology*, ed. Roger Brownsword, Eloise Scotford and Karen Yeung, 1232–51 (Oxford: Oxford University Press, 2017), 1235; Alexandra Deprez et al, 'Sustainability Limits Needed for CO_2 Removal', 383(6682) *Science* 484 (2024) ('CDR deployments … pose major economic, technological, and social feasibility challenges; threaten food security and human rights; and risk overstepping multiple planetary boundaries, with potentially irreversible consequences'); Daniel F. Stenger, Amy C. Roma and Sachin Desai, 'Innovation in Nuclear Power: How We Got Here and How to Move Forward', in *Innovation in Energy Law and Technology: Dynamic Solutions for Energy Transitions*, ed. Donald Zillman et al, 117–36 (Oxford: Oxford University Press, 2018), 124–6; and David Zilberman, Tim G. Holland and Itai Trilnick, 'Agricultural

While this observation acknowledges that climate law is a future body of law, rather poorly developed at present, it also acknowledges climate law's ongoing development as a subdiscipline of law, despite its current sketchiness. It is unfortunate, indeed, that the term 'Law and Technology' has become almost synonymous with a branch of intellectual property law ('Law and *Information* Technology'). This book demonstrates that the term stands for a much richer set of issues, containing the key for the emergence of a disciplinary identity for climate law.

Another service of this book is to indicate areas of legal research that will become important. As noted earlier, there is the question of what to do with existing law, which tends to prevent or limit certain technological deployments. 'New technologies create novel issues that press at the extremes of the established laws.'[32] There is also the question of how technological deployments, when not constrained by existing law, should be incentivized in order that they be scaled up globally at speed; there is the related question of opportunity costs, as incentivization is almost synonymous with subsidization, and subsidies mean that money is taken away from one area and put into another, causing resentment and resistance; there is the question of which laws will need to be changed to accommodate the expansion of climate-friendly technologies and with what potential effect on such values as equity or human rights; and there is the question, introduced earlier, of whether such deployments will, themselves, soon need to be regulated, not only due to their potential negative impacts but also because the need to control global warming (climate control) implies international coordination.

At the domestic level, there are indeed many questions to be answered. In a world successful in deploying the critical set of technologies at adequate scale and speed for stabilization at below 2 °C (not to mention 1.5 °C), each country will be contributing a different mix. To some extent, it is already known what a given country can or cannot contribute. Australia can contribute wind and PV but (physically) not hydro, for example. It can move away from energy-related CO_2 but (economically) not from husbandry-related methane. It can deploy CCS but (politically) not nuclear power, although it is being discussed. The current domestic laws – but also their supporting policies, mindsets and cultures – that stand in the way of such deployments need to be understood and debated in advance.

GMOs – What We Know and Where Scientists Disagree', 10 *Sustainability* 1514 (2018), 3. An example often in the news concerns the environmental effects of lithium mining for battery production: Ivan Penn and Eric Lipton, 'The Lithium Gold Rush: Inside the Race to Power Electric Vehicles', *The New York Times*, 6 May 2021 <www.nytimes.com/2021/05/06/business/lithium-mining-race.html>.

32 Michael Guihot, 'Coherence in Technology Law', 11(2) *Law, Innovation and Technology* 311 (2019), 323.

At the international level, multilateral environmental agreements, trade and intellectual property treaties, and treaties protecting human rights, among other types of agreement, will present barriers for climate control at below 2 °C achieved and maintained through massively scaled-up mitigation and adaptation technologies. For example, the rapid transition to new forms of energy, and the abandonment of old forms, will disadvantage people throughout the world and will impact on traditional ways of life (including, possibly, methane-intensive rice cultivation and cattle raising) regarded as cultural heritage. Of some concern also is that the international order, weakened as it currently is by multiple armed conflicts, will break down under the strain of a revolution about the control of greenhouse gases, a revolution few are excited about but nevertheless necessary.

It is clear from the foregoing that the completion of the climate/energy transition, its precise form, and its cost are all very much up in the air at the time of writing. However, it is also clear that the seemingly transitory decisions about law and climate technology that are being made today will preordain the course and outcome of this transition. It is best, therefore, to try to make those decisions in full knowledge of the facts and in full awareness of the ideological currents that carry them. This is the aim uniting the 19 chapters that form the core of this book.

1.2 Chapter overview

This book is divided into two parts. Part 1 is dedicated to the discussion of individual climate technologies and related (regulatory) perspectives and challenges. Part 2 deals with the broader normative context in the Anthropocene in which these technologies are embedded.

1.2.1 Climate technologies

In Chapter 2, Floor Fleurke and Phillip Paiement study the interplay between energy, conservation and law. The specific question that animates their chapter is whether renewable-energy technologies are compatible with the conservation of biodiversity. The governments of the United States and the EU have pushed policy packages that have ramped up the production of renewable energy as their chief purpose. Those packages would result in infrastructural transformations of a kind not seen in two generations. The various trade-offs between the construction of facilities for renewable energy and conservation have been documented extensively in the literature. Fleurke and Paiement review those trade-offs within the frameworks of energy and environmental law, which have distinct regulatory toolkits and orientations. The chapter seeks to show how the trade-offs are struck within the US Green New Deal and the EU Green Deal. The analysis reveals that, on the whole, the two packages mostly advance an energy-first principle. The conclusion is that conservation will be compromised in the course of the forthcoming rapid expansion of renewable-energy capacity.

In Chapter 3, Alexander Zahar investigates dam-building for climate change mitigation. Hydropower expansion is an active mitigation policy in many countries, with dam-building programmes of often unprecedented size adumbrated in Nationally Determined Contributions (NDCs) under the Paris Agreement or in associated instruments under the UNFCCC. In most cases it is unclear how well thought out these plans are, including in relation to their expected mitigation benefits. The incorporation of hydropower into NDCs may have the effect of both legitimizing hydropower as a leading mitigation technology and bestowing an appearance of greater ambition in NDCs in light of the technology's supposedly gargantuan mitigation potential. But an ambitious-looking NDC, no matter how well-intentioned, is not an end in itself, despite the emphasis currently placed on ramping up ambition in NDCs. One must also ask what are the side-effects of an NDC's mitigation plans, as well as how realistic they are. This is especially pertinent for NDCs with a significant hydropower component. This chapter argues that the paradigm-shifting mitigation gains from hydropower's expansion promised in the early rounds of NDCs are unlikely to materialize, due to the financial and environmental uncertainties of hydropower and the disadvantage faced by technologies with long lead times.

In Chapter 4, Kaisa Huhta explores the nuclear debate from the vantage point of EU law. Nuclear reactors enable the low-carbon production of electricity, but their eschatological connotations continue to divide public opinion. Huhta is agnostic on the arguments for and against nuclear energy. She seeks neither to discredit one side nor to support another, and she takes no legal, social or economic – let alone personal – stand on any specific viewpoint. Instead, she reviews the facts relevant to the technology and uses the EU as a case study to show how a sensitive and politically charged debate has imprinted itself on the law and the legal governance of the transition to low-carbon energy.

Chapter 5, by Seita Vesa, discusses critical raw materials (CRMs). In a zero-carbon world, materials such as nickel, lithium and cobalt, which are used in batteries, would become as critical as oil and gas are in our fossil-fuel-dependent world. The global race for these new critical materials is becoming much more frantic as governments accelerate their pursuit of a net-zero future. It is widely thought that the evolving circular economy will satisfy demand for materials; however, in reality, supplies will still have to come from primary mining. The extraction and processing of critical materials in this fashion are associated with innumerable environmental and human-rights concerns. The emerging governance around CRMs showcases the tense connection between the demand for rapid production and deployment, on the one hand, and the difficulty of sustainably managing limited resources, on the other. That tension raises several legal questions, which Vesa examines in detail. In the final analysis, it emerges that the most pressing issue for lawyers is that of finding holistic solutions.

Ruven Fleming and Kelsey Pailman examine the legal implications of the deployment of hydrogen at speed and scale in Chapter 6. They start from the perspective that since the 1970s there has been a constant stream of news that hydrogen will be 'the new thing in energy'. Even though these ambitions have not materialized, the authors argue that two factors are different in the current debate: first, governments seem to have understood that green hydrogen can be a tool to 'green' the industrial sector; second, they are willing to invest in the technology. Fleming and Pailman look at the case for hydrogen, particularly in the EU, and discuss the potential tension between rapid technological expansion and the law's general tendency of making normative compromises. Three central regulatory 'battlegrounds' for hydrogen are used to illustrate their points.

In Chapter 7, Mirta Alessandrini and Josephine van Zeben examine the regulation of genetically modified organisms. Typically, debates in this domain revolve around the perceived trade-off between the safety of using such organisms and the sustainability of the food supply. Alessandrini and van Zeben adopt a legal perspective on the various genetic-engineering and gene-editing techniques around which these debates revolve. They delve into the intricacies of the legal definition of genetically modified organisms, the optimal level of precaution, and the functions that genetically modified organisms can discharge in improving food security and sustainability.

Closely related to the previous topic, Chapter 8 concerns cultured meat and dairy. Here, technology is expected to induce a profound transformation in agriculture. As Jonathan Verschuuren explains, growing meat and dairy in labs can eliminate greenhouse gas emissions in one fell swoop while also mitigating nitrogen and phosphate overload and transforming landscapes by freeing up space currently taken up by pastures and cattle-feed crops. Law can smooth this transition. Verschuuren contemplates three (related) regulatory interventions: measures that constrict the production of and curb demand for conventional meat and dairy; measures that stimulate the large-scale retail and consumption of cultured meat and dairy; and regulations that guarantee the safety, fairness and sustainability of these new products.

The next two chapters are on the regulation of geoengineering. In Chapter 9, Gareth Davies suggests that the ideological commitments of an epistemic community of climate change experts may be causing them to overlook the promise of geoengineering. There is tentative evidence for the proposition that climate change may be managed effectively through geoengineering, but more research is needed to determine its effectiveness and safety. However, many climate change experts are hostile to the idea.

A significant proportion of those experts is also opposed to the conduct of further research. This tendency may appear bizarre in light of the urgency of the climate crisis. Davies explores the origins of resistance to geoengineering, a movement that has always emphasized moral-hazard arguments. He shows that those arguments are often put forth without any reference to the underlying evidence, and he draws an analogy with the treatment of lifestyle-caused diseases, which are commonly approached in an equally dogmatic manner. He also argues that the path-dependent self-interest of certain expert communities may explain their hostility to geoengineering more convincingly than policy arguments or evidence.

In Chapter 10, four authors – Manon Simon, Kerryn Brent, Jeffrey McGee and Jan McDonald – argue that research on technologies for the management of solar radiation should intensify in view of the fact that the risks that those technologies entail can be managed. CDR has been mooted as a means of meeting shortfalls in state-set targets. However, many removal techniques are not yet fully developed, and their effectiveness at scale remains unproven. Earth's albedo can be altered to reflect some solar radiation away from the planet. Stratospheric aerosol injection is one technique for achieving this result; another is to thin cirrus clouds in order to enable additional heat energy to escape the atmosphere. The two proposals are often called 'solar radiation management', or 'SRM'. SRM is not free of controversy – for instance, stratospheric aerosol injections can damage parts of the planet that no nation controls, such as the ozone layer, and the availability of SRM might dampen politicians' incentives to reduce emissions at source. Unilateral or minilateral actions initiating SRM could aggravate geopolitical conflicts. The authors, after elaborating on these concerns, contend that the technology should nevertheless be developed, up to the point at which its deployment becomes feasible – because only then will decision-makers grasp its full value and be in a position to compare it with the harm it can inflict. In this chapter's view, the most acute risk at present is that an SRM programme will be initiated without its consequences being fully understood and without appropriate governance mechanisms.

The capture, use and storage of carbon have been particularly prominent in climate change discourse in recent years, and they supply the topic of the next two chapters. In Chapter 11, Hao Zhang compares the extent to which robust legal measures can promote carbon capture, use and storage in the EU and China. Technologies of this kind provide a glimmer of hope to those anxious about the attainability of carbon-neutrality goals, especially in industries such as power generation, petrochemicals, steel and cement production. Zhang's chapter is about the legal and regulatory obstructions to the progress of capture-use-and-storage projects. These have to do with matters such as site characterization, leaks, operator liability and the cross-border transport and sequestration of CO_2. Zhang introduces the issues as they manifest in the EU and China before teasing out their implications for climate law and technology in the Anthropocene.

Bettina Lange (Chapter 12) examines the economic evidence for the cost-effectiveness of carbon capture, use and storage (CCUS). These technologies require significant outlays of financial resources. Should states intervene in private business activity, not just to subsidize research and CCUS pilot plants, but also their commercial application? This question forms part of a lively debate in a number of jurisdictions. Lange focuses on the UK. The first two parts of the UK Energy Act 2023 are meant to promote the deployment of CCUS at speed and scale. The chapter examines arguments for and against this economic regulatory framework. Building on science and technology studies, Lange argues that this framework is a 'boundary object' around which the limited economic evidence about CCUS may be arranged and by virtue of which actors with divergent interests would be able to coordinate their plans for the technology. Thus, even though the economic data remain ambiguous, the UK Energy Act creates law that is capable of reflecting the links between science and policy.

Chapter 13 is on nature-based climate solutions and the rights of nature. Felicity Deane and Justine Bell-James argue that nature-based solutions, which are centred on the active nurturing of particular ecosystem functions, can be double edged. Human interventions that have the narrow objective of supporting carbon sequestration can affect biodiversity in unexpected ways that are misaligned with the rights-of-nature paradigm. The authors, while not minimizing this problem, contend that the benefits of human 'use' can be balanced against the intrinsic value of the environment. Given how pressing the climate crisis is, deploying a wide range of solutions is vital. Finding a balanced means of implementing nature-based solutions should, therefore, be a matter of priority for policy-makers and scholars.

Chapter 14, by Daria Shapovalova, wraps up the discussion of individual climate technologies. Shapovalova discusses the rationale behind, and the legality of, restrictions on the production, sale and use of coal. As this is the fossil fuel that releases the most CO_2 per unit of energy through its combustion, coal is usually at the very top of the phase-out agenda. Most experts are of the view that coal use should fall dramatically soon, and that, in the long term, it should only be permitted when appropriate carbon-capture technologies are available. At the same time, coal is currently used to meet a quarter of global demand for energy. It has a substantial share in the energy mix, even in advanced economies. Some countries have implemented restrictions. For instance, the United Kingdom has banned the sale of household coal and Germany has passed a law to phase out coal use for energy. However, it is undeniable that a ban on coal, whatever its environmental benefits, could imperil mining communities. Such a ban would also have economy-wide ramifications – coal is a staple source of base-load power and remains the only feasible means of ensuring a secure supply of power in some areas. A ban could not be imposed on unwilling states, as this would infringe against the long-standing principle that states enjoy untrammelled sovereignty over their natural resources. Shapovalova's chapter examines existing and potential policies and laws on restrictions of coal use and production, providing a critical outlook in light of concerns about the climate emergency and the need for a just transition.

1.2.2 The broader normative context

Part 2 of the book continues with the broader normative context in which the individual technologies are embedded. It starts with a discussion of the question of whether the owners of carbon-intensive assets should be compensated for the energy transition. In Chapter 15, Ignacio Herrera-Anchustegui and Gunnar S. Eskeland propose a law-and-economics answer to this question. According to them, the security of property rights and contractual entitlements is essential to law. It follows that expropriated owners and those whose pre-existing rights are curbed unilaterally ought to be compensated. However, the state is also free to pursue its policies, provided that it does not discriminate and it complies with the law. From an economic standpoint, secure property rights give owners confidence and are thus conducive to investment and the accumulation of assets. Requiring the government to pay compensation would serve two ends: it would discourage the government from changing its policies abruptly and it would mitigate the damage that expropriations otherwise inflict on investor confidence and the accumulation of assets. Conversely, uncompensated expropriations would sensitize investors to changes in policy and lower the expectation of compensation. Therefore, compensation has both advantages and disadvantages. Much hinges on the notion of the public interest, because, according to the chapter authors, compensation should be due for unduly discriminatory or sudden policies, even if they are otherwise lawful, but not when the government seeks to attain its objectives in an incremental and nondiscriminatory manner.

In Chapter 16, Laura Kaschny examines the intersection of climate mitigation and energy justice. Her view is that the notion of 'energy justice' can, if it is adopted as a guiding principle, ensure that the costs and the benefits of the transition are distributed equitably. The same principle can also serve as a foundation for representative and inclusive decision-making in the energy sector. Kaschny explores the meaning of energy justice in the Anthropocene and evaluates its role in the implementation of the substantial climate-mitigation measures that are now required. Her analysis of energy justice as a guiding principle points to a comprehensive approach to the economic transformation that should yield fairer and more democratic legislation and jurisprudence. At the same time, Kaschny shows how the trade-offs that inhere in the notion of energy justice can restrict its efficacy as a principle and, potentially, produce injustice.

Chapter 17 is a treatment of intellectual property and climate change in the Anthropocene, by Matthew Rimmer. Rimmer positions the debate about intellectual property, innovation policy and climate change within the fields of geopolitics and international relations. The tension between the intellectual property regime and climate change may be traced to the development of the UNFCCC. Traditionally, developed nations have found common cause with industry in lobbying for the robust protection and strict enforcement of intellectual property rights. Conversely, the BRICS/BASIC group, has called for more rapid and expansive transfers of technology. The developing countries that form the Group of 77 have likewise voiced their concern about slow progress on climate finance and have sought further concessions on intellectual property. Small island states, least developed countries, nations vulnerable to climate change, and climate justice activists have all called for clean technologies to be treated as global public goods, to be subjected to the same regulatory treatment as other commons. Rimmer enquires whether the procedural and substantive deadlocks that previously typified the interplay between intellectual property and climate change were overcome at the climate change summits at COP 26, COP 27 and COP 28, and whether a new consensus has emerged. The chapter also scrutinizes UN Secretary-General António Guterres' call for renewable-energy technologies to be treated as global public goods and thus exempted from the strictures of the intellectual property regime.

Chapter 18, by Kateryna Holzer and Yulia Yamineva, is about climate technology and international trade law. As noted previously, the Paris Agreement implicitly calls for a technological revolution. International trade law must accommodate the need to deploy and diffuse climate technologies. Holzer and Yamineva begin with a brief description of the race for industrial-decarbonization subsidies in the major economies. Thereafter, they enquire whether the rules of the World Trade Organization on subsidies, non-discrimination and intellectual property rights enable climate technologies to be promoted effectively. The possibility of accelerating the diffusion of such technologies by eliminating tariff and non-tariff trade barriers for environmental goods and services is also discussed, as are various means of leveraging synergies between the international climate and trade regimes.

In Chapter 19, Alberto Quintavalla and Leonie Reins seek to determine whether the architecture of international climate and environmental law can accommodate the particularities of climate-related technologies. Accordingly, they evaluate four initiatives that are directed at improving this accommodation, namely the Earth Charter, the Global Pact for the Environment, Earth System Law and the proposal for a Framework

Convention for the Regulation of New Technologies. It transpires that most of these efforts overlook climate-related technologies; only the Paris Agreement is exempt from this tendency. Consequently, that Agreement may be said to contain foundational principles for the regulatory treatment of climate-related technology. However, its full potential is yet to be harnessed, partially because international climate and environmental law continues to revolve around the principle of state sovereignty.

In Chapter 20, Morag Goodwin adumbrates an approach to technology transfer that goes beyond development. Decolonization and a departure from the development paradigm are both necessary if climate change is to be addressed. Even if it is accepted that technological development will suffice to attain that end, a functioning technology-transfer regime would be indispensable for the adaptation-and-mitigation effort. The development paradigm is among the principal mechanisms that structure international relations. The manner in which development is conceptualized and implemented as a regulatory practice, therefore, determines how obligations and duties are understood, both among individual states and in the dialogue between the Global North and the Global South. Consequently, development conditions the operation of technology-sharing obligations.

1.3 Conclusion

It would be difficult to downplay the great variety of themes that this volume explores – we are asking the reader to think of the coal in the ground, the ozone in the stratosphere, and almost everything that stands between them. The reasons for doing so should be obvious from the short synopses in the preceding section. There is no one-size-fits-all solution to climate change; the dramas of the Anthropocene are about the specific, not about the general. Furthermore, nearly all of the debates that the authors cover eventually harden into strict dilemmas. For the climate transition to unfold at the speed and scale that our times require, policy makers will soon have to make decisions that their predecessors chose to postpone. Our hope in preparing this volume is that they will take stock of the many original ideas that run through the following pages.

PART 1

Climate Technologies

2

Are Renewable Energy Technologies Compatible with Biodiversity Conservation? The Energy–Conservation Legal Nexus

Floor Fleurke and Phillip Paiement

2.1 Introduction

In an effort to mitigate the harmful consequences of a changing climate, states around the world are engaged in the rapid development and expansion of renewable energy infrastructure, constituting the largest set of infrastructural development projects since the period immediately following the Second World War. While these projects could help ensure the long-term viability of habitats essential for biodiversity conservation, they also pose considerable risks to nature conservation and protection in the immediate term. In this chapter, we review the Green Deal and Green New Deal legislative developments facilitating the rapid expansion of renewable energy (RE) in the European Union (EU) and the United States (US), respectively, in order to assess whether and how they confront and respond to the green-green trade-offs between RE growth and biodiversity conservation. While the EU has been more proactive in constructing a comprehensive framework for tackling these trade-offs, developments in both legal systems show signs of structurally prioritizing RE growth over biodiversity conservation concerns. While there are clear examples of legislative support for win-win outcomes, such as incentivizing RE development in degraded sites or the built environment, the general tenor of these developments is the rise of an 'energy first' principle for resolving trade-offs with biodiversity conservation.

In Section 2.2 we review the state of the art on the green-green trade-offs associated with the five most prominent sources of RE: solar, wind, hydropower, bioenergy and geothermal. By identifying the scale of projected development for each type of energy source, according to intergovernmental mitigation pathway studies, and the major conservation risks associated with the development, operation and decommissioning of each type, we seek to provide an overview understanding of the potential impacts that the energy transition poses for biodiversity conservation. Overall, we find that all forms of RE pose substantial threats to conservation, but some are highly problematic (for example, bioenergy), others are relatively unproblematic (geothermal), and all of them are considerably dependent on the particular siting and operational contexts of the specific energy facilities. In Section 2.3, we discuss the divergent frameworks of energy law and environmental law and use them to problematize the RE and conservation trade-offs that arise at their intersection. We compare the market-oriented, enabling logic of energy law focused on the mitigation of long-term climate-related harm with the restrictive logic of environmental law and its focus on protecting species and individual specimens from immediate risks. Finally, in Section 2.4 we assess the main legislative components of the Green Deal and Green New Deal, reviewing how they affect nature-conservation protections in the planning, development and operation of RE facilities. While the EU has developed a comprehensive framework for integrating nature conservation and RE considerations in the development process, the legislative developments of both legal systems are found to largely tilt the scales in favour of 'energy first', reducing, streamlining and deprioritizing conservation concerns that may pose obstacles to a rapid energy transition. While some opportunities for win-win outcomes have been incentivized, on the whole the RE/biodiversity conversation trade-off will remain a problematic source of conflict for the coming decades under these legislative frameworks.

2.2 The RE/biodiversity conservation trade-off

The trade-off between the expansion of RE production and biodiversity conservation has been very well studied by environmental scientists through diverse formats of research, including project-specific, energy-specific, regional and global spatial analyses, and other approaches.[1] Here, we present

[1] Helpful general overviews of this trade-off as assessed in the environmental sciences include: Jose A Rehbein et al, 'Renewable energy development threatens many globally important biodiversity areas' (2020) 26 *Global Change Biology* 3040; Andrea Santangeli et al, 'Global change synergies and trade-offs between renewable energy and biodiversity' (2016) 8 *GCB Bioenergy* 941; Alexandros Gasparatos et al, 'Renewable energy and

general findings that characterize the state of the art pertaining to this trade-off by drawing on the insights from systematic literature reviews on the subjects as well as illustrative case studies. Overall, our review of this literature finds that all sources of RE pose potential risks to biodiversity, but some pose greater risks than others. These risks arise from both the direct and indirect consequences of RE production, which manifest in the construction, operation and decommissioning of RE facilities. Likewise, these risks are often context-dependent, posing risks to biodiversity in particular contexts (for example, proximity to critical habitats) while not in others (degraded land or the built environment). For this reason, general statements about the risks posed by RE production are rather unhelpful law and policy discussions. In this section we present scientific findings about the risks of each of the following five types of RE: solar, wind, hydropower, bioenergy and geothermal. While the following paragraphs outline major trade-offs associated with each RE type, it should be noted that there are diverse opportunities for win-win scenarios in which RE development simultaneously supports biodiversity, suggesting that REs and biodiversity are not categorical trade-offs.[2]

Solar energy production consists of large- and small-scale photovoltaics (PV), which are the most prominent forms of solar energy production; and concentrated solar power generation, which uses mirrors or lenses to concentrate solar energy in large-scale power stations. As of 2019, solar contributed 4 EJs (exajoules) to the global energy system (of 585 EJ in total), and was projected to rise to as much as 261 EJs per year by 2060 in the most RE-reliant mitigation pathway (IMP-REN-2.0) of the Intergovernmental Panel on Climate Change (IPCC).[3] Here, we focus only on biodiversity concerns relating to solar PV, as it is the most common form of solar energy production.[4] In general, solar PV production poses the least risk to biodiversity among RE types, mainly because it can be integrated into existing infrastructure (for example, solar panels on buildings).[5] However, when facilities are located on previously undeveloped land, solar power

biodiversity: Implications for transitioning to a Green Economy' (2017) 70 *Renewable and Sustainable Energy Reviews* 161.

2 See, for instance: MC Ashley, SC Mangi and LD Rodwell, 'The potential of offshore windfarms to act as marine protected areas: A systematic review of current evidence' (2014) 45 *Marine Policy* 301.

3 IPCC AR6 Chapter 6, page 617 (Scenario IMP-REN-2.0).

4 Concentrated solar power, the other major form of solar energy production, uses lenses in a large-scale facility to concentrate the solar energy over the site to a single point where a receiver converts that energy into heat and, subsequently, into electricity. This form of energy production has been far less popular than the rapidly expanding use of solar PV.

5 Santi Pratiwi and Nataly Juerges, 'Review of the impact of renewable energy development on the environment and nature conservation in Southeast Asia' (2020) 5 *Energy, Ecology*

poses risks to habitat disruption and fragmentation, as large plots of land are required. In addition, chemical applications commonly used for suppressing plant growth and keeping sunlight unobstructed pose risks of water and soil pollution.[6] These risks are minimized when solar PV is developed on degraded lands or on lands with marginal conservation value.[7]

Wind energy, produced by increasingly large wind turbines, has long been publicly accused of posing a substantial threat to birds.[8] In 2019, it produced 5 EJs globally, and the IPCC projects growth to as much as 104 EJs by 2060.[9] Many studies have confirmed that birds and bats routinely collide with turbines, resulting in casualties or barotrauma.[10] While the habitat loss associated with a wind turbine is relatively minimal due to its small physical footprint, turbines have been found to cause disruptions in migratory pathways.[11] Depending on the species involved and the siting context, the *construction* of a wind-turbine site can be as, or more, disruptive than its subsequent operation.[12] Offshore wind turbines pose unique risks to the marine environment, particularly during construction and maintenance periods when there is substantial disruption to surrounding marine habitats.[13]

and Environment 221; Rehbein et al, 'Renewable energy development threatens many globally important biodiversity areas'; Santangeli et al (n 1).

6 Pratiwi and Juerges (n 5) 227; Gasparatos et al (n 1) 163.

7 Kara A Moore-O'Leary et al, 'Sustainability of utility-scale solar energy: Critical ecological concepts' (2017) 15 *Frontiers in Ecology and the Environment* 385, 391; Gasparatos et al (n 1) 163.

8 See, for example, Oliver Milman, 'America's big shift to green energy has a woolly mammoth problem', *The Guardian* (London, 22 May 2023) <www.theguardian.com/environment/2023/may/22/america-green-energy-obstacles-fossils> last accessed 30 November 2023; Nikolaus J Kurmayer, 'Berlin's new wind power goals seen clashing with EU bird protection laws', *EURACTIV* (Brussels, 14 January 2022) <www.euractiv.com/section/energy/news/berlins-new-wind-power-goals-seen-clashing-with-eu-bird-protection-laws/> last accessed 30 November 2023.

9 L Clarke et al, 'Energy Systems' in IPCC, *Climate Change 2022: Mitigation of Climate Change. Contribution of Working Group III to the Sixth Assessment Report of the Intergovernmental Panel on Climate Change* (Cambridge University Press 2022) 617.

10 Gasparatos et al (n 1) 164; Pratiwi and Juerges (n 5) 227; Filipa Peste et al, 'How to mitigate impacts of wind farms on bats? A review of potential conservation measures in the European context' (2015) 51 *Environmental Impact Assessment Review* 10.

11 R Villegas-Patraca et al, 'Bird-community shifts in relation to wind farms: A case study comparing a wind farm, croplands, and secondary forests in southern Mexico' (2012) 114 *The Condor* 711.

12 JW Pearce-Higgins et al, 'Greater impacts of wind farms on bird populations during construction than subsequent operation: Results of a multi-site and multi-species analysis' (2012) 49 *Journal of Applied Ecology* 386.

13 DYC Leung and Y Yang, 'Wind energy development and its environmental impact: A review' (2012) 16 *Renewable and Sustainable Energy Reviews* 1031.

Hydropower is produced in a variety of forms, though the most common facilities consist of reservoir-fed dams and run-of-river plants. In addition, tidal power energy generation is expected to expand considerably in coming decades. In 2019, hydropower produced 15 EJs in the global energy system, and the IPCC projects that it could grow to 24 EJs by 2060.[14] The construction of reservoir-fed hydropower poses large-scale negative consequences to biodiversity due to habitat destruction and fragmentation, sediment loading, flooding and disruption of natural seasonal water flow and its impacts on aquatic food chains.[15] The creation of upstream reservoirs is also known to cause high levels of carbon dioxide and methane as vegetation in flooded areas decomposes.[16] The environmental consequences of large-scale hydropower facilities have long been the subject of environmental law, as hydroelectric dams were the first large-scale RE production facilities.[17] However, the increasing use of smaller-scale run-of-river facilities is also associated with negative risks for biodiversity, such as their impacts on groundwater aquatic fauna.[18] Tidal power generation, a technology that transforms the energy of tidal movements into electricity, also poses consequences for biodiversity, such as sediment loading, salinization and habitat destruction in tidal estuaries, as well as species entrapment risks (for example, of whales and other large aquatic fauna) and sound pollution during construction.[19]

Bioenergy has perhaps the worst reputation for the risks it poses to biodiversity. It consists of various forms of energy production, including biomass for heating and liquid biofuels for transportation. In 2019, bioenergy contributed 57 EJs to the global energy system. The IPCC projects this to nearly double by 2060 or triple by 2070 under various mitigation pathways.[20] Bioenergy can be produced through a wide variety of materials

14 Clarke et al (n 9) 617.

15 Pratiwi and Juerges (n 5) 228. See also Chapter 3 of this book.

16 Gasparatos et al (n 1) 165.

17 For example, the US Supreme Court's first decision concerning the Endangered Species Act pertained to the construction of the Tallico Dam. *Tennessee Valley Authority v. Hill, et al.* 437 U.S. 153 (1978).

18 C Fišer et al, 'The European Green Deal misses Europe's subterranean biodiversity hotspots' (2022) 6 *Nature Ecology & Evolution* 1403.

19 Gasparatos et al (n 1) 169.

20 See scenarios IMP-REN-2.0 and IMP-NEG-2.0 in Clarke et al (n 9) 617. The category 'biomass' used in the IPCC projection scenarios includes solid, liquid and gas forms of bioenergy derived from biomass. The scope of biomass used by the IPCC is derived from the International Energy Agency (IEA). In its latest modelling publication, the IEA's definition of bioenergy does not differentiate bioenergy derived from re-growth processes from the whole of bioenergy supply. IEA, 'Global Energy and Climate Model: Documentation 2023' (2023) 103. As such, the growth projects indicated here do not distinguish between non-renewable and renewable forms of bioenergy.

including woodchips, soybeans, oil palm, sugar cane, rice husks, maize, rapeseed, eucalyptus, and miscanthus or switchgrass. Not all bioenergy can be categorized as RE, however, as only that which is derived from circular harvest and regrowth practices will result in a carbon-neutral energy source; without being linked to regrowth, bioenergy still produces a net contribution to greenhouse gas (GHG) emissions as it emits carbon without replacing it in the soil.[21] Given the wide variety of production methods, the potential risks posed by this type of RE vary considerably. However, in general, all forms of bioenergy pose significant risks of deforestation and large-scale land-use conversion, including potential indirect deforestation. These land transitions risk habitat destruction and fragmentation.[22] Establishment of monoculture plantations for bioenergy production often entails large-scale deforestation and chemical applications (fertilizers, pesticides and insecticides) that risk soil and water pollution.[23] A global spatial analysis of RE production sites and areas of biodiversity importance found bioenergy to be the biggest RE threat to biodiversity, with more than half of its potential area of global production being located in top biodiversity areas.[24] Other studies have found that the risks posed by bioenergy production are greatest in Southeast Asia and South America, where 45 per cent and 31 per cent, respectively, of existing oil palm plantations are established on lands that were forested in 1989.[25] Notably, the land-use transitions associated with oil palm plantations, as well as other crops that serve as biomass sources for bioenergy (such as rice, soybeans, maize, and rapeseed), are the result of compounded increasing demands on these raw commodities for human consumption (vegetables and seed oils) and livestock feed, in addition to their demand for bioenergy production. As such, it is challenging to distinguish which land-use transitions correspond to which demand factors (food, feed, bioenergy). A civil society report on oil palms found that bioenergy accounts for half of oil palm consumption in the EU in 2018, and it is the fastest growing category of consumption since 2013.[26]

---

21 Michael Hoel, 'The rise and fall of bioenergy' (2020) 101 *Journal of Environmental Economics and Management* 102314, 4.

22 J Popp et al, 'The effect of bioenergy expansion: Food, energy, and environment' (2014) 32 *Renewable and Sustainable Energy Reviews* 559.

23 Gasparatos et al (n 1) 166–8; Pratiwi and Juerges (n 5) 227–8.

24 Santangeli et al (n 1) 947.

25 V Vijay et al, 'The Impacts of Oil Palm on Recent Deforestation and Biodiversity Loss' (2016) 11 *PLOS ONE* e0159668; V Vijay et al, 'Deforestation risks posed by oil palm expansion in the Peruvian Amazon' (2018) 13 *Environmental Research Letters* 114010.

26 Transport & Environment, 'The trend worsens: More oil for energy, less for food' (2019) <www.transportenvironment.org/discover/trend-worsens-more-palm-oil-energy-less-food/>. For an in-depth, though slightly dated, analysis of the land-use transitions associated with EU bioenergy consumption, see Hugo Valin et al, 'The land use change

Finally, geothermal energy is produced by capturing subterranean heat and transforming it into electricity. In 2019, it contributed just 1 EJ to the global energy system and is not expected to rise to more than 10 EJ by 2060 according to IPCC mitigation pathways.[27] As such, it is the smallest RE type covered in this chapter and also the type with the least projected growth. Given the relatively small size of geothermal facilities, the immediate biodiversity consequences are also limited. The construction and operation of geothermal facilities, including supporting infrastructure (roads, electricity lines) risks habitat destruction and fragmentation and the introduction of invasive species.[28] The operation of geothermal energy facilities has also been reported to contribute to local air pollution due to the discharge of subterranean hydrogen sulfide gas and increased heat.[29]

In summary, all types of RE production are associated with potential risks to biodiversity. At the same time, some offer far more attractive pathways for expansion that have neutral impacts on biodiversity, such as the expansion of solar PV in existing infrastructure, or even positive impacts, such as solar PV or wind turbine facilities that incorporate compensatory habitat conservation of higher conservation value than the site itself.[30]

The risks posed to biodiversity are complex, location- and technology-dependent, and they arise across all phases of RE development (construction, operation and decommissioning).[31] RE facilities can have 'zones of impact' on biodiversity that extend well beyond the physical sites of RE production.[32] Some scholars have used the variety of biodiversity concerns connected to the expansion of RE as a basis to argue for nuclear energy production to be given a greater role in the decarbonization process, despite the environmental risks posed by nuclear facilities and the long-term storage of nuclear waste.[33] Furthermore, this review has not included the impact that mining operations

impact of biofuels consumed in the EU: Quantification of area and greenhouse gas impacts' (*ECOFYS*, 27 August 2015) <https://pure.iiasa.ac.at/id/eprint/12310/1/Final%20Report_GLOBIOM_publication.pdf> last accessed 30 November 2023.

27 Clarke et al (n 9) 617.

28 Gasparatos et al (n 1) 170.

29 Hrefna Kristmannsdóttir and Halldór Ármannsson, 'Environmental aspects of geothermal energy utilization' (2003) 23 *Geothermics* 451, 456. Pratiwi and Juerges (n 5) 228–9.

30 This expanded use of compensatory habitat conservation as a mitigation mechanism for RE sites has been recommended in M Gerrard, 'Legal pathways for a massive increase in utility-scale renewable generation capacity' (2017) 47 *Environmental Law Reporter* 10591, 10613.

31 Rehbein et al (n 1) 3047.

32 BB Niebuhr et al, 'Renewable energy infrastructure impacts biodiversity beyond the area it occupies' (2022) 119 *Proceedings of the National Academy of Sciences* e2208815119.

33 P Mateo-Tomás and JV López-Bao, 'A nuclear future for biodiversity conservation?' (2022) 270 *Biological Conservation* 109559.

for rare materials – elements that are essential for the electrification of energy systems – will have on biodiversity.[34] Overall, the complexity of these potential RE/biodiversity conflicts in the green transition illustrate why existing legal frameworks for environmental impact assessment ought to be further supported through more robust legal frameworks that entrench biodiversity concerns into RE legislation and policy.[35]

2.3 The nexus of energy law and conservation law

The trade-offs between biodiversity protection and RE development pose a challenge to legal systems in that they arise at the nexus of two legal fields: energy law and environmental law. This nexus is made problematic by the divergent characteristics of these fields, described later. Rather than resolving these trade-offs, the two legal fields often reinforce the conflicting interests caught in the trade-offs. As RE expansion has developed over the course of the previous decades, legislative and administrative actions in the US and EU have sought to harmonize conflicts between the two fields. In the EU, the integration principle enshrined in Article 11 of the Treaty on the Functioning of the EU (TFEU) seeks to harmonize environmental protection with other interests: 'Environmental protection requirements must be integrated into the definition and implementation of the Union policies and activities, in particular with a view to promoting sustainable development.' This principle obliges the EU institutions to take stock of environmental objectives when proposing, amending or adopting non-environmental policies, including energy.[36] However, literature has showed that the principle has had little effect on the integration of environmental requirements, as EU institutions enjoy a wide margin of discretion in weighing different interests in the context of the objective of 'sustainable development'.[37] The principle has played a modest role in the case law of the EU courts when interpreting EU secondary legislation in fields other than environmental law. A notable case is *Preussen Electra*, in which the court

[34] LJ Sonter et al, 'Renewable energy production will exacerbate mining threats to biodiversity' (2020) 11 *Nature Communications* 4174.

[35] Pratiwi and Juerges (n 5) 229.

[36] N de Sadeleer, *Environmental Principles: From Political Slogans to Legal Rules* (2nd edn, Oxford University Press 2020) 472–83, at 473–4; E Scotford 'Environmental Rights and Principles: Investigating Article 37 of the EU Charter of Fundamental Rights' in S Bogojević and R Rayfuse (eds), *Environmental Rights in Europe and Beyond: Swedish Studies in European Law* (Hart Publishing 2018) 133, 144.

[37] NML Dhondt, *Integration of Environmental Protection into other EC Policies: Legal Theory and Practice* (Europa Law Publishing 2003); G van Calster and L Reins, *EU Environmental Law* (Edward Elgar 2017); de Sadeleer (n 36) 472–83; Scotford (n 36) 133.

justified a discriminatory scheme by invoking the integration principle in the context of the objective of promoting the use of RE sources in accordance with international climate change obligations.[38] In addition, TFEU Article 194(1), which is the legal basis of EU energy policy, states the need to preserve and improve the environment within the context of that policy.[39]

Despite these attempts, energy law and environmental law (specifically natural-resources or nature-conservation law) operate as two distinct fields of law, with their own goals, modes of regulation and temporal horizons.[40] Energy law is largely oriented to the enabling of energy markets and their associated commodities, working with instruments of market regulation, and – in the context of the energy transition – seeking to protect ecosystems and biodiversity writ large in the long term from the hazardous implications of climate change.[41] It is a field focused on national, regional and global scales of production, distribution and consumption of energy. In contrast, environmental law, and particularly its sub-field of nature conservation and biodiversity protection, is a restrictive field of law aimed at prohibiting conduct that creates a risk of harm for rare, threatened or endangered species and their habitats. It is directed in the first instance at the local scale, oriented toward the protection of specific species, communities and their habitats, in the present. To achieve this, it uses tools that slow down or restrain activities, such as environmental impact assessment and activity- and site-licensing/permitting procedures. Furthermore, these two fields of law are overseen by different administrative or legislative entities, such as, in the US, the Fish and Wildlife Service (FWS), located within the Department of the Interior, for environmental law, as opposed to the US Department of Energy and its

---

[38] Case C-379/98 *Preussen Elektra* ECLI:EU:C:2001:160 paras 73–6. See also case C-573/12, *Ålands Vindkraft* ECLI:EU:C:2014:2037 paras 77–80.

[39] Scotford (n 36) 129.

[40] Davies portrays the cleavage between the two fields in stark terms: 'Although there certainly are exceptions, the general trajectory of energy and environmental law should thus be clear. The fields work in separate spheres. They promote different objectives and, even where they share commonalities, such as the trend toward market-based regulation, fail to regulate in a coordinated, holistic manner.' LL Davies, 'Alternative energy and the energy-environment disconnect' (2010) 46 *Idaho Law Review* 473, 499. See also DR Hodas, 'Biodiversity and Climate Change Laws: A Failure to Communicate?' in MI Jeffery, J Firestone and K Bubna-Litic (eds), *Biodiversity, Conservation, Law and Livelihoods: Bridging the North-South Divide* (Cambridge University Press 2008) 383–99; ALR Jackson, 'Renewable energy vs. biodiversity: Policy conflicts and the future of nature conservation' (2011) 21 *Global Environmental Change* 1195; CL Brunette, J Byrne and CK Williams, 'Resolving conflicts between renewable energy and wildlife by promoting a paradigm shift from commodity to commons-based policy' (2013) 16 *Journal of International Wildlife Law & Policy* 375.

[41] On the growing autonomy of energy law see K Huhta, 'The coming of age of energy jurisprudence' (2021) 39 *Journal of Energy & Natural Resources Law* 199.

Federal Energy Regulatory Commission, for energy law, or, in the EU, the Directorate-General Environment versus the Directorate-General Energy.

The distinct forms of functioning of these two legal fields further aggravate the RE and biodiversity trade-offs. For context, it is important to note that the rise of RE amounts to infrastructural development on a scale that most industrialized nations have not experienced since the years immediately following the Second World War, when many modern highway systems were constructed. Thus, it is also one of the largest infrastructural development projects since the rise of modern environmental law. In the US context, Gerrard identifies the potential obstacles that environmental law poses to a rapid development of RE: siting requirements, the National Environmental Policy Act (NEPA), state and local-government approval, and species-protection laws.[42] Given the large spatial scale of most commercial RE projects (with the exception of geothermal), many are likely to be located on leased public land. In particular, US federal land (national forests, Bureau of Land Management land, Tribal land managed by the Bureau of Indian Affairs, and so on), offshore areas, and disturbed land (for example, land with contaminated soil) offer attractive sites for RE development while also being subject to extensive procedural requirements related to environmental protection. There is a risk that these obstacles will prevent projects from being developed. For example, the Cape Wind project near Cape Cod underwent nearly a decade of licensing-approval processes at the state and federal levels only to have its approval by the Bureau of Ocean Energy Management ultimately overturned by a federal appeals court decision due to insufficient data on the project's impact on the seafloor, as required by NEPA.[43] Similar experiences have occurred in EU member states, where the permitting process can take up to ten years as a result of substantive and procedural requirements for environmental legal protection.[44]

In relation to obstacles posed by species-protection law, RE projects in the US must comply with a number of environmental legislative acts, including the Endangered Species Act (ESA),[45] the Migratory Bird Treaty

[42] Gerrard (n 30).

[43] J Firestone et al, 'Wind in the sails or choppy seas? People-place relations, aesthetics and public support for the United States' first offshore wind project' (2018) 40 *Energy Research & Social Science* 232; L Moosa, 'The energy capital of the east coast: Lessons Virginia can learn from Cape Wind failure and European success in offshore wind energy' (2014) 39 *William & Mary Environmental Law & Policy Review* 713.

[44] European Commission, Directorate-General for Energy, J Tallat-Kelpšaitė et al, *Technical support for RES policy development and implementation – Simplification of permission and administrative procedures for RES installations (RES Simplify) – Interim report*, Publications Office of the European Union, 2022 <https://data.europa.eu/doi/10.2833/239077>.

[45] Endangered Species Act, 16 U.S.C. §§1531–44.

Act,[46] the Bald and Golden Eagle Protection Act,[47] as well as NEPA.[48] These obstacles have prevented some development of RE projects, such as a proposed wind farm on an Ohio National Guard base near Lake Erie, which was cancelled after environmental non-governmental organizations challenged the proposed project's impact on protected bird and bat species. Section 9 of the ESA prohibits the 'taking' of any endangered species, intentional or incidental, and this prohibition applies to RE facilities.[49] There have been instances when wind-turbine operators have been fined for unintentionally taking golden eagles and migratory birds that are protected under the Migratory Bird Treaty Act and the Bald and Golden Eagle Protection Act.[50] To avoid such liabilities, RE facility operators would need to request incidental-take permits from the US Secretary of Interior. These can only be issued on a project-by-project basis rather than as a general exception for RE facilities. As Ruhl poignantly states, 'Giving wind power such a green pass through administrative reform would violate both the letter and the spirit of the statute.'[51] This obstacle offers a stark illustration of the green-green trade-off between a restrictive, short-term and locally focused environmental law and an energy law framework that is enabling, long-term and globally focused and seeks to support the development of RE projects.

With these obstacles in mind, agencies responsible for the implementation of environmental law have taken numerous substantial measures to adapt to the demands of a rapid RE transition. One approach is the use of programmatic environmental impact studies (EIS) for entire regions, which then allow for RE projects within the scope of the programmatic EIS to skip the environmental impact assessment requirement in the licensing process. In 2012, the Bureau of Land Management (BLM) conducted a programmatic EIS for solar energy development in six states in the south-west of the US. The EIS is currently in the process of being updated.[52] The California Desert Renewable Energy Conservation Plan, developed in 2015 by a number of state and federal agencies working across energy and environmental

46 Migratory Bird Treaty Act, 16 U.S.C. §§703–11.

47 Bald and Golden Eagle Protection Act, 16 U.S.C. §§668–668d.

48 National Environmental Policy Act, 42 U.S.C. §§4321–70h.

49 Endangered Species Act, 16 U.S.C. §1538(a)(1)(B). JB Ruhl, 'Harmonizing commercial wind power and the Endangered Species Act through administrative reform' (2012) 65 *Vanderbilt Law Review* 1769, 1791–2.

50 The incidents pertain to a 2013 fine issued against Duke Energy Renewables, Inc. for taking 14 golden eagles and 149 migratory birds, and a 2014 fine issued against PacifiCorp Energy for taking 38 golden eagles and 336 migratory birds. Gerrard (n 30) 10611.

51 Ruhl (n 49) 1789.

52 Bureau of Land Management, '2023 Solar Programmatic Environmental Impact Statement' <www.blm.gov/2023-solar-programmatic-environmental-impact-statement> last accessed 30 November 2023.

law, offers another example of streamlining RE development focused on federal/state coordination. It established a co-management land-use plan for public lands that simplifies RE development while maintaining conservation objectives.[53] This kind of streamlining has been said to be necessary to ensure a rapid uptake of RE.[54] While the BLM has been a leader among US federal agencies in prioritizing RE interests in its policy making, other federal agencies, such as the FWS and the National Forest Service (NFS), have taken less proactive measures, despite there being evident, actionable measures that would facilitate RE development: creating clear guidelines for the issuance of incidental-take permits for protected species (FWS) and developing programmatic EIS along the lines of the BLM's Western Solar Plan for public lands managed as national forests (NFS).

Ironically, environmental law has historically offered exemptions to fossil-fuel extraction activities, and these often remain more extensive than those that have been made for RE projects.[55] For example, oil and gas wells and pipelines are exempted from some hazardous air-pollution controls established under the US Clean Air Act.[56] RE proponents are now looking to environmental law to make comparable accommodations to help fast-track RE development and place it on an even regulatory footing with carbon-based energy. As Ruhl and Salzman write, 'the Green New Deal needs New Green Laws, or perhaps a New Green Law, that somehow balances the infrastructure deployment goals and timelines with the ideals of environmental protection, distributive justice and public participation'.[57]

Compared to the US, EU species-protection law has done little to streamline RE projects. The Habitats Directive (HD) requires member states to prohibit the deliberate disturbance and killing of animals belonging to Annex IV species;[58] and in accordance with Article 5 of the Birds Directive (BD), member states must prohibit the deliberate disturbance of *all* wild bird species and their deliberate killing by any method.[59] Both Directives

---

53 Bureau of Land Management, 'Desert Renewable Energy Conservation Plan' <www.blm.gov/programs/planning-and-nepa/plans-in-development/california/desert-renewable-energy-conservation-plan> last accessed 30 November 2023.

54 Gerrard (n 30).

55 For an extensive list of accommodations made for fossil-fuel energy activities within environmental law, see U Outka, 'Environmental law and fossil fuels: Barriers to renewable energy' (2012) 65 *Vanderbilt Law Review* 1679, 1704–11.

56 Clean Air Act § 112(n)(4)(a), 42 U.S.C. § 7412(n)(4)(A).

57 JB Ruhl and J Salzman, 'What happens when the Green New Deal meets the old green laws?' (2020) 44 *Vermont Journal of Environmental Law* 693, 720.

58 Directive 92/43 on the conservation of natural habitats and of wild fauna and flora, OJ 1992, L 206, p 7 (Habitats Directive), Article 12(1) (b).

59 Directive 2009/147 on the conservation of wild birds, OJ 2010, L 20, p 7 (Birds Directive), Articles 5 and 1, 9(1) and 13.

are applicable to economic activities, including RE projects.[60] The Court of Justice of the European Union has ruled that the prohibitions of HD Article 12 must be applied at the level of individual specimens. It is not limited to showing that a given activity causes a risk of adverse impact on the conservation status of the animal species concerned.[61] The term 'deliberate' has been interpreted as also encompassing 'accepting the possibility' of harm, regardless of intention.[62] Concerning the HD, an activity (for example, wind power) that entails deliberate killing or disturbance of one or more Annex IV animals requires an exemption.[63] For RE projects, authorities may grant such a derogation in the interests of public health and safety or for other imperative reasons of overriding public interest, including those of a social or economic nature or having beneficial consequences of primary importance for the environment.[64] In such a case, there must be no satisfactory alternatives for the disturbing activity, such as alternative sites or operating practices. Lastly, the disturbance caused must not be detrimental to the achievement or maintenance of the population concerned at a favourable conservation status. This safeguard clause is to be applied restrictively, requiring authorities to provide proof based on the best scientific and technical evidence while applying the precautionary principle and refusing the derogation in case of doubt.[65] As for the BD, it also allows for derogation for a number of interests, such as public health and safety, although an overriding public interest is not listed as a ground that justifies derogation.

The Directives include a system for the strict protection of sites. The HD requires the designation of Special Areas of Conservation (SACs) for the habitats and species listed in its Annexes I and II, and the BD requires the designation of Special Protection Areas (SPAs) for the species of wild birds listed in its Annex I.[66] Article 6(3) of the HD requires member states to carry out an 'appropriate assessment' of any plans or projects likely to have a significant effect on SACs established under the HD, or SPAs established

---

[60] For example, CJEU, joined cases C-473/19 and C-474/19, *Föreningen Skydda Skogen et al*, ECLI:EU:C:2021:166, paras 51–3.

[61] Ibid, paras 57 and 58. See also H Schoukens, 'Common hamsters *in* and *outside* the city: Some reflections on urban biodiversity, species recovery and the EU Habitats Directive' (2022) 19 *Journal of European Environmental and Planning Law* 180, 206.

[62] For example, CJEU, case C-221/04, *Commission v Spain*, ECLI:EU:C:2006:329, para 71.

[63] Habitats Directive, Article 16(1)(a)–(c).

[64] Ibid, Article 16(1)(c).

[65] For example, CJEU, case C-674/17, *Luonnonsuojeluyhdistys Tapiola Phojois-Savo*, ECLI:EU:C:2019:851, para 30, 66. Similar to the Habitats Directive, Article 9 of the Birds Directive includes a safeguard clause that provides for the possibility to derogate from the obligations of protection of wild birds included in Articles 5–8 of the BD when there is no other satisfactory solution and in accordance with a closed list of reasons.

[66] Habitats Directive, Articles 3.1 and 4; Birds Directive, Article 4.1.

under the BD. Therefore, RE projects are subject to an 'appropriate assessment' if they pose a risk to the conservation status of the designated site, once again taking into account the precautionary principle.[67] Authorities may approve the plan or project only after having ascertained that it will not adversely affect the integrity of the site concerned. However, HD Article 6(4) (which also applies to the BD) offers an escape route that allows for carrying out projects for 'imperative reasons of overriding public interest' on the condition that compensatory measures are implemented to ensure the integrity of the designated sites.[68] Although this regime of strict protection is diversely implemented in the member states, it is clear that it has a huge impact on the development, authorization and operation of RE projects.[69]

The accommodations made in the green-green trade-off between energy and environmental law have mostly come from one side of the conflict: environmental law. Authorities charged with the implementation of environmental legislation – air-, water- and soil-quality laws, species and habitat conservation, and so on – have taken measures that seek to overcome this trade-off, or are criticized for failing to do so. In doing so, they have reflected a technology-driven 'good Anthropocene' vision of the energy–conservation issue, prioritizing technological solutions in RE production rather than the reduction of energy production in order to remain within key planetary boundaries.[70] In the legal scholarship on this issue, there has

---

[67] Case C-127/02, *Waddenvereniging and Vogelbeschermingsvereniging*, ECLI:EU:C:2004:482 para 54.

[68] This regime applies also outside the protected areas if they have effects within such areas.

[69] C Soria-Rodríguez, 'The European environmental regulation of marine renewable energies' (2020) 29 *Review of European, Comparative and International Environmental Law* 95–106; C Backes and S Akerboom (eds), *Renewable Energy Projects and Species Law: A Comparison into the Application of the EU Species Protection Regulation with Respect to Renewable Energy Projects in the Netherlands, United Kingdom, Belgium, Denmark and Germany* (Utrecht Center for Water, Oceans and Sustainability Law, 28 May 2018) <https://dspace.library.uu.nl/handle/1874/366324> last accessed 30 November 2023.

[70] Ellis is frequently ascribed with developing the notion of the 'good Anthropocene', the notion that humans must benevolently deploy technologies to become stewards of the planet rather than defer to the natural limits set by planetary boundaries: 'It is not natural limits that will determine whether this planet will sustain a robust measure of its evolutionary inheritance into the future. Our powers may yet exceed our ability to manage them, but there is no alternative except to shoulder the mantle of planetary stewardship. A good, or at least a better, Anthropocene is within our grasp. Creating that future will mean going beyond fears of transgressing natural limits and nostalgic hopes of returning to some pastoral or pristine era.' Erle Ellis, 'The planet of no return' (Fall 2011) *Breakthrough Journal* 2. For a critique of this perspective see Clive Hamilton, 'The theodicy of the "good Anthropocene"' (2016) 7(1) *Environmental Humanities* 233–8. On the topic of planetary boundaries, see J Rockström et al, 'Planetary boundaries: Exploring the safe operating space for humanity' (2009) 14(2) *Ecology & Society* 32.

been little critique of the lack of action on the side of energy law and of the agencies tasked with the implementation of energy law to take measures that would limit green-green trade-offs in RE development and incentivize 'win-win' RE projects. We now turn to evaluate recent legislation within the ambit of the US Green New Deal and the EU Green Deal to identify whether and how they seek to prevent green-green trade-offs from manifesting and steer RE development into win-win and low-conservation-conflict scenarios.

2.4 The RE-conservation trade-off in the Green New Deal and Green Deal

Since gaining widespread attention and political support in 2018–19, the Green New Deal and Green Deal policy programmes have impacted the development of numerous new pieces of legislation geared toward tackling climate change, including the social, environmental and economic consequences of the energy transition. In the US context, the most notable federal legislative measures focusing on the energy transition include the Consolidated Appropriations Act of 2021 (including the Energy Act of 2020), the Infrastructure Investment and Jobs Act of 2021, and the Inflation Reduction Act of 2022. In the EU context, the most notable legislative development is the revised Renewable Energy Directive (2018/2001/EU), which includes provisions that simplify permitting processes. Next to this, the Commission has proposed a series of measures in its so-called 'REPowerEU' policy as a response to rising energy prices and in order to break the EU's dependence on Russian fossil fuels. They include a proposal for a targeted amendment of the Renewable Energy Directive, as well as a temporary emergency Regulation, adopted by the Council on 22 December 2022, to tackle the energy-market turmoil and to accelerate the clean energy transition.

In the following, we will present the facets of these pieces of legislation that aim at facilitating an update of RE. We will also evaluate whether and how they seek to address the green-green trade-off between environmental conservation and RE development.

2.4.1 United States

Signed into law on 27 December 2020 by President Trump, the Consolidated Appropriations Act of 2021 included the Energy Act of 2020, proclaimed as a US$35 billion investment into updating US energy policy for the first time in over a decade. It affected issues as diverse as energy storage, carbon capture technology, nuclear energy and grid modernization. Title III, devoted to Renewable Energy and Storage, touches on a number of issues affecting the green-green trade-off discussed here. It instructs the Secretary of Energy to:

- Carry out research aimed at improving the environmental impact of water power, hydropower, marine energy, geothermal, and wind technologies.[71]
- Prioritize the development of wind-energy projects in 'economically distressed areas and areas disproportionately impacted by pollution' and projects that are carried out 'in collaboration with Tribal energy development organizations, Indian Tribes, Tribal organizations, Native Hawaiian community-based organizations, minority-serving institutions, or territories or freely associated States'.[72]
- Specifically for solar power, conduct research and development with the aim 'to reduce and mitigate potential life cycle negative impacts of solar energy technologies on human communities, wildlife and wildlife habitats'.[73]
- Specifically for solar power, reduce market barriers to the adoption of solar energy which arise from their impacts on 'local communities, with special consideration given to economically distressed areas, previously disturbed lands such as landfills and former mines, and other areas disproportionately impacted by environmental pollution' and 'wildlife and wildlife habitats'.[74]
- Establish a national 'Renewable Energy Coordination Office' to improve coordination in the Federal permit process for RE projects on lands that are not excluded from RE uses under their respective land use plans.[75]
- Facilitate the co-production of geothermal energy from wells leased for oil and gas production.[76]

While these are only a few of the many reforms introduced in the 2020 Energy Act, they illustrate a small-yet-growing awareness of the green-green trade-offs posed by the rapid transition to RE. In particular, the Renewable Energy Coordination Office, along with associated requirements to establish national targets for RE development on federal lands (including at least 25 gigawatts of electricity by 2025),[77] illustrates how Congress seeks to minimize the procedural burdens that slow down the development of RE projects on federal lands. As the first major legislative push in the Green New Deal era, it exemplifies the growing attention to environmental protection via

[71] Consolidated Appropriations Act, 2021, Public Law 116–260, 27 December 2020, § 633(2), 634, 635(a)(10), 614(b), 3003(b)(iv).

[72] Ibid, § 3003(b)(2)(C)(i)(II)(a) and 3003(b)(2)(C)(i)(V).

[73] Ibid, § 3004(b)(1)(B)(vii).

[74] Ibid, § 3004(b)(2)(B)(viii)(II).

[75] Ibid, § 3102(a).

[76] Ibid, § 3105.

[77] Ibid, § 3104.

technological progress and the increasing demand to put federal lands – much of them sites of nature conservation – to use for RE development.

Less than a year later, on 15 November 2021, President Biden signed into law the Infrastructure Investment and Jobs Act (commonly referred to as the Infrastructure Bill) which had featured as a main plank of his electoral campaign and partially served as an economic stimulus in the second year of the COVID-19 pandemic. It constituted a massive federal investment in RE, including US$73 billion allocated to updating the nation's energy policy and infrastructure as well as more than US$50 billion allocated to public transport and electric vehicles.[78]

The Infrastructure Investment and Jobs Act introduced a few measures directly pertaining to the conservation/RE trade-off. First, it instructed the Secretary of Energy to assess the viability of RE projects on current and former mine sites.[79] Second, it allocated a considerable budget to the Secretary of Energy for funding research and development of RE technologies with, among other goals, an emphasis on improving their environmental impacts: US$106 million for hydropower and marine energy,[80] US$84 million for geothermal,[81] US$100 million for wind power[82] and US$40 million for solar power.[83] Through these investments, the Act principally sought to make resources available to better achieve the policy goals established in the earlier Energy Act. Despite a Title devoted to 'Natural Resources-Related Infrastructure, Wildfire Management, and Ecosystem Restoration' in the energy portion of the Act (Division D), it offered no further substantive solutions for the RE/ conservation trade-off.

On 16 August 2022, President Biden signed into law the Inflation Reduction Act of 2022, proclaimed by some to be 'the largest bill to address climate change' in US history and potentially capable of achieving a 42 per cent reduction in emissions by 2050.[84] Estimations of the Act's total spending on energy and climate-related matters range from US$700–900

78 For a helpful overview of the bill's spending allocations, see Clark Hill, 'Infrastructure Investment and Jobs Act (IIJA)' <www.clarkhill.com/infrastructure-investment-and-jobs-act-iija/> last accessed 6 January 2025.

79 Infrastructure Investment and Jobs Act, 2021, Public Law 117–58, 15 November 2021, § 40341-2.

80 Ibid, § 41006(a-b).

81 Ibid, § 41007(a).

82 Ibid, § 41007(b).

83 Ibid, § 41007(c)(1).

84 Princeton University Zero Lab, 'Preliminary Report: The Climate and Energy Impacts of the Inflation Reduction Act of 2022' (August 2022) <https://repeatproject.org/docs/REPEAT_IRA_Prelminary_Report_2022-08-12.pdf> last accessed 30 November 2023.

billion.[85] It includes subsidies and tax credits aimed at incentivizing climate-friendly residential heating (for example, heat pumps), clean-electricity production, electric vehicles and clean-hydrogen production.[86] Despite the big price tag, the Act offers few substantive changes to the RE/conservation policy trade-off. It allows the Secretary of Energy to lease federal offshore sites on the outer continental shelf for the development of wind power, reversing a previous Presidential Determination instating a ten-year ban on such development, issued by President Trump in 2020, during the final months of his presidency.[87] Additionally, it funds the 'Rural Energy for America Program', which offers financial assistance to farmers and rural small businesses who seek to develop RE systems.[88] It also allocates US$500 million to funding projects aimed at increasing the sale and use of biofuels.[89] The Act also increases royalties on various federal land leases for oil and gas production, thereby making RE sources more competitive and reducing indirect federal support for carbon-based energy sources.[90]

In sum, these three key legislative components of the US Green New Deal exemplify the regulatory mode of energy law insofar as they allocate enormous sums of money to influence the position and development of RE in national and global energy markets. Nonetheless, the RE-conservation trade-offs identified above emerge in myriad ways throughout these legislative initiatives. Notably, the Energy Act began to address one of the key requirements identified by scholars working on this trade-off: the coordination of federal licensing processes for RE projects through the establishment of the Renewable Energy Coordination Office. The utility of this coordination lies in the Office's actions toward minimizing the obstacles that conservation law poses to RE development. A number of other measures seek to prioritize potential 'win-win' RE opportunities in which negative conservation impacts are minimal. These measures include the prioritization and funding for RE projects on polluted lands, landfills, closed mines and closed wells originally built for oil and gas production. Similarly, measures that

85 See for instance, Paul Ciampoli, 'Ultimate Public Climate Spending Spurred by Inflation Reduction Act Could Be Over $800 Billion: Credit Suisse' (American Public Power Association, 19 October 2022) <www.publicpower.org/periodical/article/ultimate-public-climate-spending-spurred-inflation-reduction-act-could-be-over-800-billion-credit> last accessed 30 November 2023.

86 Justin Badlam et al, 'The Inflation Reduction Act: Here's what's in it' (McKinsey & Company, 24 October 2022) <www.mckinsey.com/industries/public-sector/our-insights/the-inflation-reduction-act-heres-whats-in-it> last accessed 30 November 2023.

87 Inflation Reduction Act, Public Law 117–169, 16 August 2022, § 502521.

88 Ibid, § 22002.

89 Ibid, § 22003.

90 Ibid, § 50261–3.

fund research and development of RE technologies with lesser environmental impacts offer support for future RE development that does not come at the expense of biodiversity conservation. Yet, other measures risk potentially aggravating the RE/conservation trade-off by facilitating or incentivizing RE projects with increased chances of harming protected species or habitats. For example, the re-authorization of leasing federal offshore sites for wind-power development potentially exposes nearby marine habitats as well as marine bird species to risk during construction and operation. Likewise, the Inflation Reduction Act's US$500 million investment in biofuels is a continuation of federal support for the RE type that poses the greatest global risk to biodiversity.

2.4.2 European Union

The European Green Deal (EGD) establishes the objective of making the EU climate neutral in 2050.[91] The European Climate Law implements the EGD. It adopts an emission-reduction target of at least 55 per cent below the 1990 level by 2030.[92] The 'Fit for 55' policy package seeks to strengthen eight existing pieces of legislation and presents five new initiatives, across a range of policy areas and economic sectors, including energy, climate and biodiversity. Energy use accounts for 75 per cent of the EU's emissions, so the transformation of the energy sector is central to the EU's climate ambitions.[93] Regarding the implementation of the EGD, while the Integration Principle of Article 11 of the TFEU is not mentioned, the need for 'mainstreaming sustainability in all EU policies' is considered.[94] In this context, the 'no harm principle' is introduced as a new principle according to which an explanatory memorandum is attached to each legislative proposal or executive Act explaining how the aim of no (significant) harm is to be achieved.[95] The meaning of the principle is still evolving, but it implies an *ex ante* examination at an early stage to avoid (significant) harm.[96]

---

91 Communication from the Commission COM/2019/640 final, The European Green Deal.

92 Regulation (EU) 2021/1119 of the European Parliament and of the Council of 30 June 2021 establishing the framework for achieving climate neutrality and amending Regulations (EC) No 401/2009 and (EU) 2018/1999 (European Climate Law), [2021] OJ L 243/1, pp 1–17.

93 Ibid, at p 9.

94 Communication COM(2019) 640 final, p 15.

95 Ibid, 19(2)(2)(5). See also M Onida, 'The "Do Not (Significantly) Harm" Principle' in K de Graaf et al (eds), *Grensoverstijgende rechtsbeoefening*, p 45; E Chiti, 'Managing the Ecological Transition of the EU: The European Green Deal as a regulatory process' (2022) 59(1) *Common Market Law Review* 19–48, at 36–7.

96 The principle has found some traction in other EU legal instruments: Regulation (EU) 2020/852 of the European Parliament and of the Council of 18 June 2020 on the

2.4.2.1 *RE targets*

The Renewable Energy Directive (RED) forms the main instrument in the transition to the use of energy from renewable sources for the purpose of reducing emissions. Since the introduction of the first RED,[97] the share of RE sources in EU energy consumption has increased from 12.5 per cent in 2010 to 21.8 per cent in 2021. Sweden had the highest share of renewables in its consumption (62.6 per cent), ahead of Finland (43.1 per cent) and Latvia (42.1 per cent).[98] The first phase of the programme went into effect in 2009 with the goal of achieving a 20 per cent usage of RE by 2020, particularly in the transportation sector. In 2018, the targets for 2021 to 2030 were set, and were called RED II. Directive (EU) 2018/2001 has since set a binding EU target to reach a share of at least 32 per cent of energy from renewable sources in the EU's gross final consumption of energy by 2030.[99] Under the Climate Target Plan, the share of RE in gross final energy consumption would need to increase to 40 per cent by 2030 in order to achieve the EU's emission-reduction target.[100] The Commission proposed in July 2021 to double the share of RE in the energy mix in 2030 compared to 2020, so that it reaches at least 40 per cent. As part of the 'REPowerEU' proposal, the European Commission has proposed to raise the 2030 target for energy from renewable energy sources (RES) further, to 45 per cent of final energy consumption.[101] This would bring the total RE generation capacity to 1,236 GW by 2030, in comparison to 1,067 GW by 2030 envisaged under 'Fit for 55'. According to the proposal for RED III,

establishment of a framework to facilitate sustainable investment and amending Regulation 2019/1088; Regulation (EU) 2021/241 of the European Parliament and of the Council of 12 February 2021 establishing the Recovery and Resilience Facility; Regulation (EU) 2021/1229 of the European Parliament and of the Council of 14 July 2021 on the public sector loan facility under the Just Transition Mechanism.

97 Directive 2009/28/EC of the European Parliament and of the Council of 23 April 2009 on the promotion of the use of energy from renewable sources and amending and subsequently repealing Directives 2001/77/EC and 2003/30/EC (text with EEA relevance) OJ L 140/16.

98 Eurostat, Shedding Light on Energy – 2023 Edition <https://ec.europa.eu/eurostat/web/interactive-publications/energy-2023#about-publication>.

99 Directive (EU) 2018/2001 of the European Parliament and of the Council of 11 December 2018 on the promotion of the use of energy from renewable sources [2018] OJ L 328/28, p 82.

100 Communication from the Commission COM(2020) 562 final, 'Stepping up Europe's 2030 climate ambition: Investing in a climate-neutral future for the benefit of our people' <https://eur-lex.europa.eu/legal-content/EN/TXT/HTML/?uri=CELEX:52020SC0176>.

101 COM 2022, 222 final, Article 3(1). See Paris Agreement Compatible Scenario for Energy Infrastructure (PAC scenario) <www.pac-scenarios.eu/pac-scenario/scenario-development.html>.

the compulsory target for RE in total energy consumption is now to be 42.5 per cent by 2030, more than 10 per cent more than in RED II. In addition, there is an indicative target of a further 2.5 per cent. Specific targets are set for solar photovoltaic,[102] heat pumps,[103] acceleration of renewable hydrogen[104] and scaling up sustainable biomethane.[105]

2.4.2.2 Shifting the nexus: from 'climate' to 'security'

When considering the nexus between RE and nature protection, the successive REDs display a clear shift toward more elaborate streamlining of processes for approving RE projects and an emphasis on RE as a security concern. The first RED (2001/77/EC), required member states to evaluate the existing regulatory and administrative frameworks on authorization procedures to reduce the regulatory and non-regulatory barriers to the increase in electricity production from RE sources and streamline and expedite procedures at the appropriate administrative level, but it left it to the member states to decide how to achieve this streamlining.[106] This was followed by Directive 2009/28/EC, which introduced requirements to simplify the administrative procedures for RE developers. The subsequent version of the RED (EU) 2018/2001 states in recital 51:

> Lengthy administrative procedures constitute a major administrative barrier and are costly. The simplification of administrative permit

[102] For solar photovoltaics as the fastest growing renewable sector, REPowerEU sets a target of over 320 GW of solar photovoltaic newly installed by 2025, over twice today's level, and almost 600 GW by 2030. As part of this the Commission presents the EU Solar Strategy and introduces the European Solar Rooftop Initiative anchored around a legally binding EU solar rooftop obligation for certain categories of buildings; see EU Solar Energy Strategy, COM(2022) 221 (18 May 2022) <https://eur-lex.europa.eu/legal-content/EN/TXT/HTML/?uri=CELEX:52022DC0221>.

[103] For heat pumps, the EU aims at doubling the current deployment rate of individual heat pumps, resulting in a cumulative 10 million units over the next five years. See also, Commission Staff Working Document: 'Implementing the REPowerEU Action Plan: Investment Needs, Hydrogen Accelerator and Biomethane Targets' <https://eur-lex.europa.eu/legal-content/EN/TXT/PDF/?uri=CELEX:52022SC0230>.

[104] REPowerEU sets a target of 10 million tonnes of domestic renewable hydrogen production and 10 million tonnes of renewable hydrogen imports by 2030.

[105] Production is to be scaled up to 35 billion cubic metres by 2030 as a cost-efficient path to achieve a reduction of imports of natural gas from Russia; see Biomethane Action Plan in the accompanying staff working document <https://eur-lex.europa.eu/legal-content/EN/TXT/PDF/?uri=CELEX:52022SC0230>.

[106] Directive 2001/77/EC of the European Parliament and of the Council of 27 September 2001 on the promotion of electricity produced from renewable energy sources in the internal electricity market [2001] OJ L 283/33, Article 6.

> granting processes, and clear time-limits for decisions to be taken by the authorities competent for issuing the authorization for the electricity generation installation on the basis of a completed application, should stimulate a more efficient handling of procedures, thereby reducing administrative costs. … In order to respond to the increasing need for the repowering of existing renewable energy plants, streamlined permit-granting procedures should be provided for.

Article 15(1) of Directive (EU) 2018/2001 therefore sets rules for the administrative permitting procedures and obliges member states to ensure that national regulations on the authorization, certification and licensing for generating electricity from RE are proportionate and necessary and contribute to the implementation of the energy-efficiency-first principle.[107] This includes setting timeframes, considering the particularities of individual RE technologies, and simplifying procedures for decentralized devices and the production and storage of energy from RE. However, this Directive still considers that the provisions on the organization and duration of the administrative permit-granting process, in particular, 'should apply without prejudice to international and Union law, including provisions to protect the environment and human health'.[108]

In March 2022, EU leaders agreed in the European Council to phase out Europe's dependency on Russian energy imports as soon as possible.[109] According to the Commission, the purpose of REPowerEU is 'about rapidly reducing our dependence on Russian fossil fuels by fast forwarding the clean transition and joining forces to achieve a more resilient energy system and a true Energy Union'.[110] The need to fast-track the deployment of RE is therefore to be viewed in the context of the EU's plan to become independent from Russian fossil fuels – a narrative that has sharply shifted from 'climate' to 'security'.[111] In the proposed REPowerEU Plan it is emphasized that 'slow and complex permitting processes are a key obstacle to unleashing the

[107] Directive (EU) 2018/2001 of the European Parliament and of the Council of 11 December 2018 on the promotion of the use of energy from renewable sources [2018] OJ L 328/82.

[108] Ibid, recital 51. See also European Commission, Directorate-General for Environment, *Guidance Document on Wind Energy Developments and EU Nature Legislation* (Publications Office of the European Union, 2020).

[109] European Council Conclusions (24 and 25 March 2022).

[110] Ibid, p 1.

[111] See J van Zeben, L de Almeida and M Alessandrini, 'Stress testing the European Green Deal: The "securitization" of energy, food and climate' (2022) *EU Law Live Weekend Edition* <https://issuu.com/eulawlive/docs/weekend_edition_106> last accessed 14 December 2023.

renewables revolution and for the competitiveness of the renewable energy industry'. The revised proposal operationalizes the principle of RE as an 'overriding public interest' and introduces the designation of 'go-to' areas and other ways to shorten and simplify permitting, while also minimizing potential risks and negative impacts on the environment.[112]

The need to accelerate RE development has been further underlined by the Commission's recommendation on speeding up permitting procedures for RE projects and facilitating Power Purchase Agreements,[113] the adoption of Council Regulation (EU) 2022/2577 laying down a framework to accelerate RE deployment,[114] and the Provisional Agreement on the Renewable Energy Directive (that also includes accelerated permitting procedures for RE projects).[115]

The shift toward RE development as a security priority is also evident from the use of Article 122 of the TFEU as the legal basis for the adoption of Council Regulation 2022/2577. This competence allows the Council to decide – without involvement of the European Parliament – on a proposal from the Commission and, in a spirit of solidarity between the member states, upon the measures appropriate for the economic situation, in particular if severe difficulties arise in the supply of certain products, notably in the area of energy. According to the Commission, in light of political tensions and the high risk of a complete halt in Russian gas supplies, combined with the uncertain outlook for alternatives, there is a significant threat of disruption of energy supplies, justifying the deployment of the emergency competence of Article 122 of the TFEU.[116]

2.4.2.3 RED III

As part of REPower EU, the Commission issued a proposal to amend the existing Directive (EU) 2018/2001 on the promotion of the use of energy from renewable sources (RED III).[117] The revised Directive was adopted

[112] Communication from the Commission to the European Parliament, the European Council, The European Economic and Social Committee of the Regions REPowerEU Plan, COM/2022/230 final, p 11.

[113] COM/2022/3219 final.

[114] [2022] OJ L 335/36–44.

[115] Council of the European Union, 19 June 2023, Interinstitutional File: 2021/0218(COD).

[116] Reg. 2022/2257, recital 22.

[117] Proposal for a Directive of the European Parliament and of the Council amending Directive (EU) 2018/2001 of the European Parliament and of the Council, Regulation (EU) 2018/1999 of the European Parliament and of the Council and Directive 98/70/EC of the European Parliament and of the Council as regards the promotion of energy from renewable sources, and repealing Council Directive (EU) 2015/652. COM/2021/557 final.

in 2023 to raise the share of RE in the EU's overall energy consumption to 42.5 per cent by 2030 with an additional 2.5 per cent indicative top-up to allow the target of 45 per cent to be achieved.[118] As mentioned above, central to the amendment is the aim to accelerate the energy transition and tackle 'key barriers' to investment in renewables and their related infrastructure. Slow and complex permitting processes are considered to be the key obstacle for RE projects, as it has been reported that permitting can take up to nine years for wind projects, and up to four and a half years for ground-mounted solar projects. Administrative and grid issues make up about 46 per cent of all identified barriers and this is expected to rise in the future.[119] Conflicting public goods are assessed by the Commission as a main source of obstacles for the deployment of RE installations.[120] This is particularly the case for wind power, geothermal power, hydropower and solar PV. The most prominent among them concern environmental protection, in particular biodiversity and protection of endangered species and bodies of water.[121] In response, the amendment adopts a further simplification and shortening of the administrative permit-granting processes in a coordinated and harmonized manner, while still acknowledging the relevance of the 'do no harm' principle.[122] Importantly, the amendment introduces an overall 'presumption of overriding public interest' as a central feature to the energy law framework by stating in Article 16(f) of the revised Directive:

> By 21 February 2024, until climate neutrality is achieved, Member States shall ensure that, in the permit-granting process, the planning, construction and operation of plants for the production of energy from renewable sources, their connection to the grid and the related

[118] Directive (EU) 2023/2413 of the European Parliament and of the Council of 18 October 2023 amending Directive (EU) 2018/2001, Regulation (EU) 2018/1999 and Directive 98/70/EC as regards the promotion of energy from renewable sources, and repealing Council Directive (EU) 2015/652. OJ L, 2023/2413, 31.10.2023.

[119] Technical support for RES policy development and implementation – Simplification of permission and administrative procedures for RES installations ('RES Simplify') <https://data.europa.eu/doi/10.2833/239077>.

[120] Ibid, pp 14–16.

[121] This has been confirmed by an open public consultation, held 18 January–22 April. See <https://ec.europa.eu/info/law/better-regulation/have-your-say/initiatives/13334-Renewable-energy-projects-permit-granting-processes-power-purchase-agreements_en>.

[122] Directive (EU) 2023/2413 recitals 22 and 39. In line with the European Green Deal objectives, research and innovation activities should comply with the 'do no significant harm' principle, according to which the research and innovation activities should not be supporting or carrying out activities that cause significant harm to any of the six environmental objectives, within the meaning of Article 17 on the establishment of a framework to facilitate sustainable investment (EU Taxonomy Regulation).

> grid itself and storage assets are presumed as being in the overriding public interest and serving public health and safety when balancing legal interests in the individual cases for the purposes of Articles 6(4) and 16(1)(c) of Directive 92/43/EEC, Article 4(7) of Directive 2000/60/EC and Article 9(1) (a) of Directive 2009/147/EC.

The revision therefore has a direct impact on existing EU environmental law, particularly on the BD and HD, the Water Framework Directive, Marine Strategy Framework Directive and the Maritime Spatial Planning Directive.[123]

2.4.2.4 Permit-granting process in 'renewables acceleration areas'

RED III, and particularly its inclusion of the 'do no harm' principle, establishes that member states should designate 'renewables acceleration areas' as areas that are particularly suitable to develop RE projects where the deployment of the specific type of RES is not expected to have a significant environmental impact.[124] In the designation of go-to areas for renewables, member states should avoid protected areas to the extent possible.[125] In order to avoid adverse environmental impact or, where that is not possible, to significantly reduce it, effective mitigation measures should be adopted.[126]

The Directive defines a renewables acceleration area as 'a specific location or area, whether on land, sea or inland waters, which a Member State designated as particularly suitable for the installation of renewable energy plants'.[127] In order to designate go-to areas for renewables, member states should prepare a plan or plans encompassing the identification of areas and the applicable rules and mitigation measures for projects located in each go-to area. In doing so, Article 15(c)(a)(i) instructs member states to give priority to artificial and built surfaces, such as rooftops, transport infrastructure areas, parking areas, waste sites, industrial sites, mines, artificial inland water bodies, lakes or reservoirs, and, where appropriate, urban wastewater treatment sites, as well as degraded land not usable for agriculture. Natura 2000 sites and nature parks and reserves, migratory bird routes, as well as other areas identified based on sensitivity maps are to be excluded, and all appropriate tools and datasets must be used to identify the areas where the

[123] The amendment withdraws the member states' discretion under Article 2, para 4 of the EIA Directive, and contradicts Articles 6(2) and (3) and 12(1) of Directive 92/43/EEC and Article 5 of Directive 2009/147/EEC.

[124] Article 15c.

[125] Article 15c(a)(ii).

[126] Article 15c(b).

[127] Article 2(9a).

RE plants would not have a significant environmental impact, including wildlife sensitivity mapping.

Each plan should be subject to an environmental assessment carried out in accordance with the conditions set out in Directive 2001/42/EC in order to assess the impacts of each renewable technology on the relevant areas designated in such a plan.[128] According to the Commission, this will ensure a more integrated and efficient approach to planning and take environmental considerations into account at an early phase of the planning process at a strategic level. Spatial planning is hence perceived as an important tool to identify and steer synergies for land- and sea-use at an early stage. Following this, in the designated renewables acceleration areas RE projects that comply with the rules and measures identified in the plan or plans prepared by member states will benefit from a presumption of not having significant effects on the environment. Therefore, they are exempted from the need to carry out an environmental impact assessment (EIA) at the project level.[129] Moreover, projects located in renewables acceleration areas should benefit from clearly delimited deadlines and legal certainty on the outcome of the procedure, such as a one-year in-principle limit.[130]

Following the application for projects in a renewables acceleration area, member states should carry out a fast screening of such applications with the aim to identify if any such project is highly likely to give rise to significant unforeseen adverse effects in view of the environmental sensitivity of the geographic area where it is located that were not identified during the environmental assessment of the plan or plans designating renewables acceleration areas.[131] All projects to be located in renewables acceleration areas should be deemed approved at the end of such a screening process.[132] This also results in a derogation from Article 6(3) of the HD (assessment of the implications of such projects for Natura 2000 sites).[133] Only if, following

[128] Directive 2001/42/EC of the European Parliament and of the Council of 27 June 2001 on the assessment of the effects of certain plans and programmes on the environment [2001] OJ L 197/30.

[129] It is worth noting that Article 2.4 of the EIA Directive already allows member states to derogate in 'exceptional cases' a specific project from the provisions laid down in that Directive. Projects which are likely to have significant effects on the environment in another member state, or where a member state likely to be significantly affected so requests, are excluded from the presumption, as the obligations under the UNECE Espoo Convention on environmental impact assessment in a transboundary context of 25 February 1991 remain applicable; see Article 16a(2).

[130] Directive (EU) 2023/2413, Articles 16a, 16b and 16c.

[131] Ibid, Article 16a(4). The screening process relating to applications for new renewable energy plants is to be finalized within 45 days of the date of submission of sufficient information necessary for that purpose.

[132] Ibid, Article 16a(5).

[133] Ibid, Article 16a(3).

the screening process, the competent authority adopts an administrative decision, 'duly motivated' and based on 'clear evidence', that a specific project is 'highly likely' to give rise to significant unforeseen adverse effects in view of the environmental sensitivity of the geographic area where it is proposed to be located that cannot be mitigated by the measures, should a full environmental impact assessment be carried out in accordance with Article 6(3) of the HD within six months following the screening decision.[134] Ultimately, the practical result of this proposed amendment is a reversal of the protection regime established under the HD, as discussed earlier. Going forward, RE projects within renewables acceleration areas would be presumed to be compatible with nature conservation and protection, unless demonstrated otherwise.

It is moreover noted that if the construction and operation of RE plants results in the occasional killing or disturbance of birds and other protected species under the HD or BD, such killing or disturbance is not to be considered 'deliberate' in the sense of these Directives. In such a case, appropriate mitigation measures to avoid collisions or prevent disturbance need to be taken, and 'proper monitoring' needs to be carried out to assess the effectiveness of such measures and to ensure no significant negative impact on the population of the species concerned.[135] Critically also, no list of approved mitigation measures is included in the Directive or in the Commission Recommendation.[136]

2.4.2.5 Permit-granting process outside renewables acceleration areas

According to the Directive, the identification of renewables acceleration areas should not prevent the ongoing and future installation of RE projects in *all* areas available for RE deployment.[137] For these projects, a project EIA under Directive 2011/92/EU or Directive 92/43/EEC is still required, but it is noted that 'permit-granting procedures applicable to projects outside renewables acceleration areas should be simplified and streamlined with the introduction of clear maximum deadlines for all steps of the permit-granting procedure'.[138] This means, for example, that, in principle, the permit-granting process for these projects is not to exceed two years.[139] The

[134] Ibid, Article 16a(5).
[135] Ibid, Article 16b(2).
[136] COM/2022/3219 final.
[137] Ibid, recital 32.
[138] Ibid, recital 36.
[139] Ibid, Article 16b(1). The permit-granting process for the repowering of projects and for new installations with an electrical capacity of less than 150 kW, co-located storage facilities as well as their grid connection, located outside renewables go-to areas is not to exceed one year including environmental assessments: Article 16b(2).

EIA is to be carried out in a single procedure that combines all relevant assessments for a given project.[140] Similar to the projects within go-to areas, any killing or disturbance of the species protected under HD Article 12(1) and BD Article 5 is not to be considered deliberate, on the condition that appropriate mitigation measures have been adopted.[141]

2.4.2.6 Regulation 2022/2577 to accelerate the European transition toward RE

The link between RE capable of causing significant environmental harm and biodiversity protection has been further altered by the EU legislature by adopting 'emergency' Regulation 2022/2577.[142] The Regulation establishes temporary rules of an emergency nature to accelerate the permit-granting process applicable to the production of energy from RE sources, with a particular focus on RE technologies or types of projects that are capable of achieving a short-term acceleration of the pace of deployment of renewables in the EU.[143] Member states can apply the Regulation to ongoing permit-granting processes that have not resulted in a final decision before 30 December 2022, provided that pre-existing third-party legal rights are preserved.[144] The Regulation applies for a period of 18 months from its entry into force.[145] However, the Commission can propose to prolong its validity in view of the development of the security of supply and energy prices and the need to further accelerate the deployment of RE.[146]

The Regulation introduces a rebuttable presumption that RE projects are of overriding public interest for the purposes of the relevant environmental legislation, limits the scope of certain environmental Directives, and simplifies the permit-granting framework for the repowering of RE power plants. Article 3 lays down this presumption:

1. The planning, construction and operation of plants and installations for the production of energy from renewable sources, and their connection to the grid, the related grid itself and storage assets shall be presumed as being in the overriding public interest and serving public health and safety

---

[140] Ibid, Article 16b(2).

[141] Ibid.

[142] Regulation 2022/2577 laying down a framework to accelerate the deployment of renewable energy [2022] OJ L 335/36.

[143] Ibid, Article 1.

[144] Ibid, Article 2.

[145] Ibid, Article 10.

[146] Ibid, Article 9.

when balancing legal interests in the individual case, for the purposes of Article 6(4) and Article 16(1)(c) of Council Directive 92/43/EEC,[147] Article 4(7) of Directive 2000/60/EC[148] and Article 9(1)(a) of Directive 2009/147/EC.[149] …

2. Member States shall ensure, at least for projects which are recognised as being of overriding public interest, that in the planning and permit-granting process, the construction and operation of plants and installations for the production of energy from renewable sources and the related grid infrastructure development are given priority when balancing legal interests in the individual case. Concerning species protection, the preceding sentence shall only apply if and to the extent that appropriate species conservation measures contributing to the maintenance or restoration of the populations of the species at a favourable conservation status are undertaken and sufficient financial resources as well as areas are made available for that purpose.

In order for those exemptions to be introduced, two conditions should be met, namely that the project is located in a dedicated renewable or grid area and that such area should have been subject to a strategic environmental assessment.[150] To minimize adverse impacts on the species concerned, the competent authority must ensure that, on the basis of existing data, appropriate and proportionate mitigation measures are applied in order to ensure compliance with HD Article 12(1) or BD Article 5, again without adding a list of approved measures or conditions. If mitigation measures are not available, monetary compensation for species-protection programmes must be paid by the operator in order to secure or improve the conservation status of the affected species.[151] In addition, the presumption of overriding public interest is only rebutted when there is 'clear evidence' that a given project has 'major adverse effects on the environment which cannot be mitigated or compensated for'.[152] This is clearly sidestepping the case law of the Court of Justice of the European Union (CJEU) that has consistently emphasized that the burden of proof of no significant harm lies with the

[147] Council Directive 92/43/EEC of 21 May 1992 on the conservation of natural habitats and of wild fauna and flora [1992] OJ L 206/7.

[148] Directive 2000/60/EC of the European Parliament and of the Council of 23 October 2000 establishing a framework for Community action in the field of water policy [2000] OJ L 327/1.

[149] Directive 2009/147/EC of the European Parliament and of the Council of 30 November 2009 on the conservation of wild birds [2010] OJ L 20/7.

[150] Ibid, Article 6.

[151] Ibid.

[152] Ibid, Preamble, para 8.

national authorities, must be based on the best scientific evidence, and must be read in the light of the precautionary principle.[153]

For the installation of solar energy equipment and co-located energy-storage assets, Article 4 states that the permit-granting process is not to exceed three months, provided that the primary aim of such structures is not solar energy production.[154] Installations of solar energy equipment are to be exempted from being subjected to a determination of whether the project requires an EIA, or from the requirement to carry out a dedicated EIA.[155] According to Article 5 of the Regulation, the permitting process of repowering projects is not to exceed six months, including EIA where required by relevant legislation. Prior determination or environmental impact assessment here is limited to the potential significant impacts stemming from the change or extension compared to the original project, and if the repowering of solar installations does not entail the use of additional space and complies with the applicable environmental mitigation measures established for the original installation, the project is to be exempted from EIA pursuant to Article 4 of Directive 2011/92/EU. Concerns have been raised about the mismatch between the set timelines and the permitting procedures and the production of knowledge that is needed to assess their (cumulative) impacts and potential significant harm, particularly concerning cumulative effects.[156]

2.4.2.7 Nature Restoration Law

Finally, the EU Nature Restoration Law should be mentioned here, as this proposed legislation was issued almost simultaneously with the REPowerEU proposal. As part of the EGD, the EU Biodiversity Strategy announced that the EU legal framework for nature restoration needed to be strengthened to fill 'significant gaps' in (the implementation of) existing legislation.[157] This

[153] For example, CJEU, Case C-404/09, *Commission v Spain*, ECLI:EU:C:2011:768; CJEU, Case C-674/17, *Luonnonsuojeluyhdistys Tapiola Phojois-Savo*, ECLI:EU:C:2019:851.

[154] Reg. 2022/2577, Article 4.

[155] This means a derogation from Article 4(2) of Directive 2011/92/EU, and Annex II, points 3(a) and (b), read alone or in conjunction with point 13(a) of Annex II to that Directive.

[156] M Ferrer, M de Lucas, GF Janss, E Casado, AR Munoz, MJ Bechard et al, 'Weak relationship between risk assessment studies and recorded mortality in wind farms' (2012) *Journal of Applied Ecology* 49(1), 38; CJ Durá-Alemañ, M Moleón, JM Pérez-García, D Serrano and JA Sánchez-Zapata, 'Climate change and energy crisis drive an unprecedented EU environmental law regression' (2023) *Conservation Letters* 16, 12958; European Court of Auditors, *Offshore renewable energy in the EU: Ambitious plans for growth but sustainability remains a challenge*, (2022(23)) <www.eca.europa.eu/ECAPublications/SR-2023-22/SR-2023-22_EN.pdf> last accessed 14 December 2023.

[157] EU Biodiversity Strategy, para 2.2.1.

Strategy states that at least 30 per cent of the EU's land area and 30 per cent of its maritime area must be legally protected. Currently, for the entire EU, this is 26 per cent for land (including inland waters) and 11 per cent for sea, adding up all designated Natura 2000 areas and other (internationally or nationally) protected areas.[158] The European Commission has since proposed a Regulation setting binding restoration targets for specific habitats and species.[159] It aimed to restore 30 per cent of the EU's degraded areas, and ultimately all ecosystems in need of restoration by 2050. Nature restoration at sea is very urgent, partly due to the radical transformation in the coming years called for by the energy transition. The Regulation was adopted, after an unprecedentedly tense legislative process, on 24 June 2024.[160] The Regulation sets out a general legally binding target that is quantified and time-bound: member states are to put in place effective and area-based restoration measures which together are to cover, by 2030, at least 20% of the Union's land and sea areas and, by 2050, all ecosystems in need of restoration.[161] Member states have discretion to determine in which areas restoration measures are taken. However, the Regulation does require that the most suitable areas for recovery measures be determined on the basis of the best available knowledge and the most recent scientific findings on the status of the habitat types included in Annex I and on the quality and quantity of the species' habitats.[162]

In the political debate on the Nature Restoration Law, it was pointed out that there exists an apparent contradiction between the obligations to restore in the draft Nature Restoration Law and the REPowerEU proposal.[163] The Regulation calls for coordination and 'synergy' between the development of national restoration plans and the designation of renewables acceleration areas for full energy.[164] Indeed, existing wind-energy areas offer opportunities for taking nature restoration measures (for example, restoration of suitable oyster banks), as bottom-disturbing fishing can no longer occur where wind farms are placed. However, it seems unavoidable that a green-green trade-off

[158] Ibid, para 2.1.

[159] Proposal for a Regulation of the European Parliament and of the Council on nature restoration, COM(2022) 304 final.

[160] Regulation (EU) 2024/1991 of the European Parliament and of the Council of 24 June 2024 on nature restoration and amending Regulation (EU) 2022/869 (Text with EEA relevance) OJ L, 2024/1991, 29.7.2024, EL.

[161] Ibid, Article 1(2). This general restoration objective has been further specified in targets and obligations for nature restoration in six specific ecosystems, including terrestrial, coastal and freshwater ecosystems.

[162] Ibid, Article 4(8).

[163] BNC-fiche Verordening Natuurherstel, p 12.

[164] Regulation (EU) 2024/1991, recital 68 and Article 14(12).

between environmental conservation and RE development will occur to some extent due to conflict of use, especially at sea.[165]

2.5 Conclusion

Under the umbrellas of the US New Green Deal and the EU Green Deal, the rise of RE amounts to infrastructural development on a scale that is unprecedented. It is inevitable that these sources of RE pose potential risks to biodiversity arising from both the direct and indirect consequences of RE development and production, which manifest in the construction, operation and decommissioning of RE facilities. The trade-offs between biodiversity protection and RE development are a challenge to legal systems in that they arise at a nexus of two separate legal fields: energy law and environmental law. Where as energy policy uses law to accelerate the energy transition, with an eye toward mitigating the long-term hazardous effects of climate change, environmental law uses legal tools that slow down and restrain activities, such as EIA and activity- and site-licensing/permitting procedures. As RE expansion has developed over the course of recent decades, legislative and administrative actions in the US and the EU have sought to harmonize conflicts between the two fields. Underneath these conflicting legal fields lies a more fundamental conflict between competing visions of environmental governance in the Anthropocene. One vision emphasizes the duty of humans to benevolently deploy technologies in their newfound positions as planetary stewards of the 'good Anthropocene', without any historic 'natural boundaries'.[166] In contrast, the other vision emphasizes the importance of ensuring biosphere integrity as a key planetary boundary and prioritizes the reduction of energy consumption alongside the transition to renewable production.[167]

In this chapter, we have analysed key pieces of legislation within the ambit of the Green New Deal and Green Deal to identify whether and how they seek to prevent these green-green trade-offs from manifesting and steer RE development into win-win and low-conservation-conflict scenarios. We found that, in both jurisdictions, RE/conservation trade-offs remain pervasive in myriad ways throughout these legislative initiatives. In both jurisdictions attempts are made to address this trade-off, but, generally, the scale tips toward the rapid development of RE.

---

[165] EA Willsteed, S Jude, AB Gill, SNR Birchenough, 'Obligations and aspirations: A critical evaluation of offshore wind farm cumulative impact assessments' (2018) 82 *Renewable and Sustainable Energy Reviews* 3, 2332.

[166] Ellis (n 70) 2.

[167] G Chapron et al, 'Bolster legal boundaries to stay within planetary boundaries' (2017) 1 *Nature Ecology & Evolution* 1.

In the US context, the Consolidated Appropriations Act of 2021 (including the Energy Act of 2020), the Infrastructure Investment and Jobs Act of 2021 and the Inflation Reduction Act of 2022 all allocate enormous sums of money to develop RE energy production and markets. A number of measures have been included that seek to prioritize potential 'win-win' RE opportunities, such as the prioritization and funding for RE projects on polluted lands, landfills, closed mines and closed wells initially built for oil and gas production, measures that fund research and development of RE technologies with lesser environmental impacts, and the establishment of the Renewable Energy Coordination Office. Yet, other measures risk potentially aggravating the RE/conservation trade-off by facilitating or incentivizing RE projects with increased chances of harming protected species or habitats. Most importantly, the legislative instruments have failed to develop any comprehensive framework to mediate the green-green trade-offs.

By contrast, in the EU, such a comprehensive framework has been developed, and further revised. As part of REPowerEU, the European Commission has proposed a series of measures, including a revision of the RED 2018/2001/EU, a proposal for a targeted amendment of the RED, and a temporary emergency Regulation adopted by the Council. These instruments are very much geared toward fast-tracking the deployment of RE to break the EU's dependence on Russian carbon-based energy after Russia's invasion of Ukraine, and must be viewed more in the context of security concerns than of climate change. The measures operationalize the principle of RE as an 'overriding public interest' and introduce measures to shorten and simplify permitting while also aiming to minimize potential risks and negative impacts on the environment by, for example, the designation of renewables acceleration areas, the requirement of a Strategic Impact Assessment, and the need to take mitigation measures. However, RE development outside the renewables acceleration zones is not excluded, existing EU environmental laws (as well as the case law of the CJEU) are sidelined, and the imbalance between the rapid roll-out of RE and the understanding of its (cumulative) impacts is not addressed. Notwithstanding attempts to reconcile the green-green trade-offs and the integration of the 'do no harm' principle in energy laws, it is clear that the balance has dramatically tilted toward the 'energy first' principle.

Funding acknowledgment

The research for this publication was supported in part by the ERC 2021 Starting Grant project TransLitigate (101039648), funded by the EU. Views and opinions expressed are those of the authors alone and do not necessarily reflect those of the EU or the European Research Council. Neither the EU nor the European Research Council can be held responsible for them.

3

Ambition in Nationally Determined Contributions: The Case of Hydropower

Alexander Zahar

3.1 Introduction

The damming of rivers for electric power is, in principle, an important replacement technology for energy derived from fossil fuels. It rivals mineral combustion for flexibility and reliability.[1] The technology's current portfolio includes around 12,000 large dams (wall height of 41.5 metres or more).[2] Although the technology has been in use for more than a century, in several countries it has not been deployed to anything close to its maximum potential extent.[3] Hundreds of large dams could yet be built, along with thousands of smaller ones.[4] Because damming a river significantly degrades nature over an area that extends well beyond the dam itself,[5] domestic law

1 International Energy Agency (IEA), *Hydropower Special Market Report: Analysis and Forecast to 2030* (Paris: IEA, 2021), 11.

2 Ottmar Edenhofer et al, *Renewable Energy Sources and Climate Change Mitigation: Special Report of the Intergovernmental Panel on Climate Change* (New York: Cambridge University Press, 2012), 441. See also Atif Ansar et al, 'Should We Build More Large Dams? The Actual Costs of Hydropower Megaproject Development', 69 *Energy Policy* 43 (2014), 45.

3 Edenhofer et al (n 2) 441, 484–7 (worldwide technical potential for hydropower is 3,721,000 MW of installed capacity, about four times more than the 2009 standing total of 926,000 MW).

4 Luis Berga, 'The Role of Hydropower in Climate Change Mitigation and Adaptation: A Review', 2 *Engineering* 313 (2016), 317.

5 N. LeRoy Poff et al, 'The Natural Flow Regime', 47(11) *BioScience* 769 (1997), 772, 777; and Antti Iho et al, 'Rivers under Pressure: Interdisciplinary Feasibility Analysis of Sustainable Hydropower', 33(2) *Environmental Policy and Governance* 191 (2022), 193.

in several countries has sought to constrain further hydropower growth, on environmental (and environmental-rights) grounds.[6] A few countries have even gone to the trouble of dismantling decommissioned plants to reconnect segmented rivers.[7]

However, this does not reflect the general situation. International law does not directly regulate what countries can do with rivers, although it might have that effect indirectly where such action affects the interests of another state (for example, by denying or polluting water) or harms a treaty-protected feature within a state. Clearly this has been a very marginal constraint on hydropower's expansion in practice, given the number of dams in existence and the number recently built.[8] In the domestic law of most countries, especially developing ones, constraints on the damming of rivers, whether for power-generation or other purposes, is relatively weak.[9] As a result, rivers that are free-flowing over their entire length are exceptional today, with many having been reduced to chains of reservoirs.[10]

The problem of climate change characterizes the Anthropocene. A management plan for this global problem is being developed through the Paris Agreement and its subordinate decisions. The Agreement puts pressure on states to identify and adopt mitigation measures to lay the foundations for emission neutrality in the second half of the century.[11] Dam-building programmes slated to break ground in the years through to 2030 and beyond have been announced by numerous states in their official communications under the Paris Agreement or under earlier frameworks of the international climate change regime. In reality, at this point, significant hydropower growth at a scale that would count as a meaningful global response to climate change is possible in only a few countries. These have the necessary physical attributes of being sloped or hilly, with large rivers and unexploited hydropower locations; and they generally sit lower on the scale

---

6 Edenhofer et al (n 2) 468–70; Iho et al (n 5) 196–8; and Xiao Zhang et al, 'Impacts of Climate Change, Policy and Water-Energy-Food Nexus on Hydropower Development', 116 *Renewable Energy* 827 (2018), 829.

7 United States, *Eighth National Communication and Fifth Biennial Report of the United States to the UN Framework Convention on Climate Change* (Washington, DC: Department of State, 2022), 261; and Beatrice Wagner, Christoph Hauer and Helmut Habersack, 'Current Hydropower Developments in Europe', 37 *Current Opinion in Environmental Sustainability* 41 (2019), 43.

8 G. Grill et al, 'Mapping the World's Free-Flowing Rivers', 569 *Nature* 215 (2019), 216 (on the 'boom' since 2000).

9 See, for example, Philip M. Fearnside, 'Brazil's Madeira River Dams: A Setback for Environmental Policy in Amazonian Development', 7(1) *Water Alternatives* 256 (2014); and IEA (n 1) 25.

10 Grill et al (n 8) 216; Wagner et al (n 7) 44; and Iho, et al (n 5) 192.

11 Paris Agreement, Art 4(1).

of economic development.[12] They number hardly more than 20.[13] However, their combined population size is huge (they include China and India); and hydropower expansion decisions in these countries are affected to some extent by mitigation policy in countries neighbouring them (thus slightly increasing the number of interested players), as hydroelectricity is at times (or could be) produced in excess of local needs in countries with sizeable unexploited hydropower endowments and sold to adjacent jurisdictions, displacing fossil-fuel use there.

If we focus on the expansion plans that are large in absolute terms and are scheduled for the near term, the number of key actors is, as I have indicated, small. In most of the 20 or so countries with globally significant hydropower growth potential and corresponding plans for the technology, environmental regulation is not strong.[14] Large dams can be built there with little legal hindrance, if finance for them can be found domestically or can be made available for this purpose from international sources.

In this chapter, I examine the form that the incorporation of hydropower expansion takes in the Paris Agreement's Nationally Determined Contributions (NDCs). The Paris Agreement implicitly encourages hydropower development wherever possible, no less than it encourages other mitigation measures.[15] I discuss, in particular, the tendency of NDCs

---

12 Many developed countries that were early movers on hydropower have no new plans for the technology except to maintain their existing installations at more or less current levels of output. See, for example, Australia, *Eighth National Communication on Climate Change* (Canberra: Ministry for Climate Change and Energy, 2022), 134 (increases will be in pumped hydropower technology, not in the river-reservoir type); Canada, which has the fourth-largest hydropower installed capacity in the world, foresees very slight increases in hydroelectric production through to 2035 (385 TWh in 2025, 386 TWh in 2030, and 396 TWh in 2035): Canada, *Eighth National Communication on Climate Change and Fifth Biennial Report* (Gatineau: Environment and Climate Change Canada, 2022), 23 and 311; and New Zealand, *First Nationally Determined Contribution (Updated 4 November 2021)* (Wellington: Ministry for the Environment, 2021), 10. Switzerland is in the same general position: *Information Necessary for Clarity, Transparency and Understanding in Accordance with Decision 1/CP.21 of its Updated and Enhanced Nationally Determined Contribution under the Paris Agreement (2021–2030)* (2021), 5–6 (hydropower production by 2035 is expected to be 37,400 GWh, only slightly up from 36,137 GWh in 2019). Countries in this category (that is, highly developed countries) also plan to turn some existing *non-powered* water reservoirs or conduits into hydropower facilities by adding turbines, thus producing clean power at a low environmental cost. See, for example, United States (n 7) Annex 2–14; and Zhang et al (n 6) 829.

13 IEA (n 1) 13 ('less than thirty' countries). See also Berga (n 4) 314.

14 See the country-by-country discussion to follow.

15 The decision adopting the Paris Agreement calls for 'the enhanced deployment of renewable energy': Decision 1/CP.21, *Adoption of the Paris Agreement*, FCCC/CP/2015/10/Add.1 (2015), preamble. Hydropower was explicitly brought into the fold of the climate treaties by the Kyoto Protocol, when it was designated as a source of tradable emission

to say almost nothing about the implicit costs – financial, environmental, social – of hydroelectricity produced from river-fed reservoirs. My argument is not that more large dams should not be built. If the Paris Agreement's limit of 2°C of global warming is to be observed, there may be no rational alternative than to build many more dams quickly and at scale – large and small ones alike. My argument, rather, is that states take a cavalier approach to dam-building as a mitigation measure, paying little attention to the obstacles and uncertainties involved. Dam-building proposals have thus come to round off many NDCs. But the enthusiasm for the measure can be misleading. In particular, NDCs with ambitious hydropower components do not address the key question of whether the planned power dams can be built and put into operation in physical locations and in time frames that are conducive to meaningful mitigation outcomes.

To illustrate the problem, lowland reservoir installations, especially in tropical and subtropical countries, which are planned in several NDCs, may deliver no mitigation benefit at all over relevant time frames (the first 20 years of their operation, which coincides with the critical period 2030–50),[16] due to high greenhouse gas emissions (mainly carbon dioxide) from the construction materials used, methane emissions and carbon dioxide from the reservoir's surface (through ebullition and diffusion of gases produced in the reservoir bed and in the water column), methane emissions from the depressurized water exiting the turbines from the depths of the reservoir (outlet degassing), and additional release of methane downstream from the run-off of the dam.[17] Deforestation emissions from access routes servicing remote dams (including those cleared for transmission cables) can also be significant when considered over relevant time frames.[18] Emissions per unit (megawatt hours) of hydropower can easily exceed emissions per unit of fossil-fuel power, and even coal power, in the medium-term (that is, through to 2050), depending on the dam's location and surface-area-to-installed-capacity

reductions under the Clean Development Mechanism (CDM); see *CDM Methodology Booklet*, 14th ed. (Bonn: UNFCCC, 2022), 145 (methodology 'ACM0002: Grid-Connected Electricity Generation from Renewable Sources').

16 Rafael M. Almeida et al, 'Reducing Greenhouse Gas Emissions of Amazon Hydropower with Strategic Dam Planning', 10 *Nature Communications* 1 (4281) (2019), 2.

17 For an explanation of the physiological processes leading to these emissions, see Felipe A. M. de Faria et al, 'Estimating Greenhouse Gas Emissions from Future Amazonian Hydroelectric Reservoirs', 10 *Environmental Research Letters* 1 (124019) (2015), 2. See also Edenhofer et al (n 2) 472.

18 Philip M. Fearnside, 'Tropical Hydropower in the Clean Development Mechanism: Brazil's Santo Antônio Dam as an Example of the Need for Change', 131(4) *Climatic Change* 575 (2015), 583.

ratio.[19] Hydropower policy incorporated into NDCs that does not take into account these emission factors and fails to weigh them against the mitigation benefits of dams may justly be challenged for using the cover of a dire need for new mitigation actions to promote the development of hydropower technology for profit, energy independence, economic growth, or another consideration unrelated to the mitigation of emissions.

My focus in this chapter, then, is on the intersection of the climate change regime and hydropower. The international regime, through its NDC system, identifies and encourages the technology as an important part of the solution to the problem of climate change. What evidence is there that NDCs actually offer such a solution when they incorporate hydropower?

Section 3.2 of the chapter reviews the hydropower plans of key countries. Section 3.3 concludes that NDC hydropower plans must be taken with a grain of salt, because hydropower is either unlikely to grow as much as NDCs suggest it will, or, if it does grow strongly, it is unlikely to deliver the promised mitigation outcomes.

3.2 Power dams in NDCs

In this section, examples of state practice will be used to unpack the issues.

3.2.1 Preliminary observations

Angola is a relatively small country (population 36 million), but its hydropower ambitions are sizeable. In its 2021 NDC, mitigation targets for 2025 include an unconditional 700 MW of new large hydropower, as well as 100 MW of new 'mini hydro', for a total cost of US$1.25 billion, to be funded from the national budget.[20] With international financial support, Angola expresses a willingness to increase installed capacity to 2,050 MW and 150 MW, respectively.[21] These enhancements would deliver 22 per cent of the country's mitigation target for its unconditional NDC scenario for 2025 and 37 per cent for its maximal scenario (unconditional plus conditional) for the same year. By any measure, hydropower's proportional contribution to Angola's mitigation ambition is significant. Angola's NDC also outlines a slightly more ambitious set of hydropower growth targets for 2030.[22] The

[19] De Faria et al (n 17) 10–11; and Philip M. Fearnside, 'Emissions from Tropical Hydropower and the IPCC', 50 *Environmental Science and Policy* 225 (2015), 226 (finding this to be true whether a 20-year or a 100-year global warming potential is used for the gases).

[20] Angola, *Nationally Determined Contribution* (Luanda: Ministry of Culture, Tourism and Environment, 2021), 46.

[21] Ibid, 48.

[22] Ibid, 46, 48.

country's ambition is not large in absolute terms, but it is still significant relative to the country's circumstances.

NDCs in general do not go into much detail about their mitigation targets and corresponding mitigation measures (irrespective of the underlying technology).[23] Angola's NDC is typical in this respect. The physical location and other characteristics of the planned hydropower plants in both the unconditional and conditional scenarios are not discussed. Their environmental impact is not discussed either. The NDC does not mention whether Angola's new dams will undergo environmental impact assessment (EIA) or what will happen in the case of a negative EIA. And the NDC does not mention where the US$1.25 billion for the unconditional hydropower-building measure will be found in the national budget. Has this expenditure already been budgeted for, or does it involve an appropriation subject to parliamentary approval?

Most problematic of all in Angola's case is the time line: it is hardly possible in five years (2021–25) to build a relatively large dam of 700 MW and put it into operation. These plants take on average 8.6 years to complete – not counting lead time for preparation, including geological investigation, economic feasibility analysis and social and environmental impact studies.[24]

The evasion of detail is not particular to hydropower, as I have indicated. NDCs from the start have settled on a narrative of asserted future achievements in emission reductions that is almost never qualified by a concern for negative impacts or implementation obstacles. What makes them assertions is that they are generally unelaborated and undefended. This could mean several things; it could mean that NDCs are *implicitly* qualified by various assumptions about contingencies, and are therefore more uncertain (less predictive of outcomes) than their matter-of-fact style suggests. It could also mean that elements of NDCs may have the status of proposals, yet to be firmed up, or that they represent mere good intentions – after all, NDCs are not legally binding.[25] It may even mean that governments, rightly or wrongly, presume that the

---

23 The rules on 'Information to facilitate clarity, transparency and understanding of nationally determined contributions' are set out in Annex I to Decision 4/CMA.1, *Further Guidance in Relation to the Mitigation Section of Decision 1/CP.21*, UN Doc. FCCC/PA/CMA/2018/3/Add.1 (December 2018). They do not require clarification of any of the matters I later refer to, which seem to me vital to a basic appreciation of the proposed measures. On the laconic approach of NDCs in general, see Alexander Zahar, 'Climate Change Mitigation Law and Policy in the Asia-Pacific', in *Research Handbook on Climate Change Mitigation Law*, edited by Leonie Reins and Jonathan Verschuuren, 155–77 (Cheltenham, UK: Edward Elgar, 2022, 2nd ed.).

24 Ansar et al (n 2) 49, 52.

25 Alexander Zahar, 'The Paris Agreement's "Nationally Determined Contribution": Law and Practice', in *Research Handbook on the Law of the Paris Agreement*, edited by Alexander Zahar, 143–66 (Cheltenham, UK: Edward Elgar, 2024).

measures listed in NDCs are assured smooth passage through permitting procedures to realization because they represent a country's ambition in a climate change context, and ambition is always a good thing. In relation to this last point, it is possible that some governments are hoping that the interest groups most likely to object to environmentally invasive measures such as river damming, or costly public works that increase budget deficits, might refrain from criticism of unprecedented roll-outs of mitigation technology, in order to support the country's mitigation ambition expressed in its NDC. If such self-muting is actually occurring, which would be a difficult thing to prove, NDCs would be having a legitimizing effect for the technologies that are listed in them, including hydropower – not by deliberate design, but incidentally and de facto.[26]

Yet other explanations are possible. Whatever the reason, there is a general silence in NDCs on the economic, technical and nature-altering challenges of the measures they contain.[27] With some very modest exceptions, discussed later, this is also true of hydropower. In several cases, significant hydropower growth targets over short time frames – far more ambitious than Angola's in absolute terms – are postulated. For a technology normally so controversial, the general silence on costs, impacts, uncertainties and other obstacles to its future growth, not only of the infrastructure itself but of its supposed mitigation benefits, is an invitation for us to consider the very nature of NDCs – are they really what they purport to be, that is, literal emission-reduction plans?

Angola does not indicate whether it has the interest or the potential to continue substantial growth of its hydropower infrastructure beyond 2030. Possibly the 2,200 MW of its 2025 maximal scenario is close to the country's overall (hydro-geological) limit for this technology. But other countries do have both an interest in, and a claimed potential for, growing installed hydropower almost indefinitely, and some of the figures mentioned are remarkable for being an order or two of magnitude larger than Angola's.

Information on these countries' intentions is not always found in NDCs themselves, however. One is sometimes obliged to look for it in countries' periodic reports under the broader climate change regime of the United Nations Framework Convention on Climate Change (UNFCCC), namely the instruments known as National Communications, Biennial Reports

[26] This point is also made by Akarath Soukhaphon, Ian G. Baird and Zeb S. Hogan, 'The Impacts of Hydropower Dams in the Mekong River Basin: A Review', 13 *Water* 1 (265) (2021), 12.

[27] The solution would be to more strictly regulate NDC content. (For the current state of regulation, see n 23.) But the Paris Agreement's credo of 'national determination' runs against stricter NDC regulation.

(BRs) and Biennial Update Reports (BURs).[28] The fact that information to disambiguate NDCs must be sought elsewhere than in the NDCs themselves heightens concern about how well the Paris Agreement system is performing procedurally.

The legal status of the three UNFCCC communication instruments I have just named differs from that of NDCs, which report mitigation 'contributions' pursuant to a dedicated Paris Agreement article.[29] Moreover, BRs and BURs, while they have certainly influenced the design of NDCs,[30] are, as I have indicated, instruments of the UNFCCC, not the Paris Agreement.

To reiterate, the main drawback of an NDC which lacks detail about how it is going to be realized is that its feasibility and overall impacts are not apparent from the face of the instrument. The Paris Agreement has produced a remarkably weak result on this point.

Cambodia's case may be used as an illustration. Although situated along the course of the Mekong River, Cambodia does not indicate this source of power at all in its 2020 NDC. By contrast, in its BUR of the same year, the country mentions that its hydroelectricity production in 2015 was 63 times higher than in 2010.[31] This remarkable growth rate was continued over the following years, and electric power from harnessing the Mekong River is due to keep growing in the country, as evidenced by Cambodia's 2022 National Communication, according to which electricity from river installations is due to grow uninterruptedly through to 2050, this being the country's target year for a planned elimination of coal from energy production.[32] Water power will help to drive coal out of Cambodia's energy profile under the country's plan, perhaps; but the technological specifications that will enable this to happen are absent from the NDC. Logically, however, the NDC is where one expects to find them. Specifications are also absent from the other communication instruments Cambodia files under the international climate

---

28 BRs are submitted by developed countries and BURs by developing countries; see Decision 1/CP.16, *'The Cancun Agreements': Outcome of the Work of the Ad Hoc Working Group on Long-Term Cooperative Action under the Convention*, UN Doc. FCCC/CP/2010/7/Add.1 (2010).

29 Paris Agreement, Art 3.

30 These two instruments perform functions under the UNFCCC that overlap with those of the NDC under the Paris Agreement, especially in relation to the declaration of mitigation measures. See Decision 2/CP.17, *Outcome of the Work of the Ad Hoc Working Group on Long-Term Cooperative Action under the Convention*, UN Doc. FCCC/CP/2011/9/Add.1 (Bonn: UNFCCC, 2011), Annexes I and III.

31 Cambodia, *First Biennial Update Report to the United Nations Framework Convention on Climate Change* (Phnom Penh: General Secretariat of the National Council for Sustainable Development/Ministry of Environment, 2020), 11.

32 Cambodia, *Third National Communication under the United Nations Framework Convention on Climate Change* (Ministry of Environment, 2022), 143.

change regime. For the pace of hydropower development in the Mekong Basin, present and future, one must look to sources external to the regime in an effort to discern what the country's plans in this respect actually are.[33]

3.2.2 The major players

I will now discuss the NDCs and associated reporting instruments of the countries with the largest claimed hydropower potential.

China is in the behemoth category of dam-building countries. It towers above the rest in its aggregate installation of hydropower. It has the world's single largest plant, the Three Gorges Dam, at 22,500 MW.[34]

In 2016, installed hydropower capacity in China was 332,000 MW.[35] To this, the country announced that it would add another 68,000 MW by 2020 (a remarkable 17,000 MW per year).[36] China's NDC is nearly silent on hydropower expansion plans through to 2030 (and beyond), stating only that 'hydro power in accordance with local condition [*sic*] will be developed'.[37]

This manages to sound so secretive that it could mean anything. It could be good news from a mitigation point of view, if China's roll-out of the technology continues at past levels; however, it is impossible to know what the net benefits will be without seeing the specifications. The country's NDC presents no information on installation plans and locations, let alone lifecycle emissions[38]

33 According to the Mekong River Commission, as recently as 2001 there were just 17 hydropower plants in operation in the Mekong Basin, with an installed capacity of less than 1,400 MW in total. Forty new plants were added between 2002 and 2015, with a total capacity of 6,442 MW. As of 2021, 88 plants were operating in the Basin, with a combined capacity of around 12,600 MW. Another 15 dams with a total capacity of 1,600 MW were under construction at the time. The Commission estimates that, by 2040, hydropower capacity in the Basin will have grown to more than 30,000 MW. See Mekong River Commission, *Sustainable Hydropower Development Strategy* (Vientiane: MRC Secretariat, 2022), 20. For other studies of the region, see Xinlin Li, 'Hydropower in the Mekong River Basin: A Balancing Test', 24(1) *Environmental Claims Journal* 51 (2012); Zhang et al (n 6) 832; and May Tan-Mullins, Frauke Urban and Grace Mang, 'Evaluating the Behaviour of Chinese Stakeholders Engaged in Large Hydropower Projects in Asia and Africa', 230 *The China Quarterly* 464 (2017), 479 (on the Kamchay Dam, in Cambodia, built by Sinohydro in Bokor National Park).

34 IEA (n 1) 16.

35 China, *Third National Communication on Climate Change* (Beijing: Ministry of Ecology and Environment, 2018), 12.

36 Ibid, 128.

37 China, *Achievements, New Goals and New Measures for Nationally Determined Contributions* (Beijing: Ministry of Ecology and Environment, 2021), 34.

38 For a breakdown of hydropower lifecycle greenhouse gas emissions, see Edenhofer et al (n 2) 470–1.

from the planned plants or any negative impacts. In the past, the negative impacts of hydropower in China have been significant.[39] Because few free-flowing rivers still exist in China,[40] the losses from the future expansion of hydropower in the country will not be in the category of river connectivity. The main issues from a strictly mitigation point of view are where the new installations will be placed (highland or lowland), what their short-term emissions will be, and when they will begin operations. China's documentation under the climate change regime is silent on these matters. To further cloud matters, China's NDC seeks to limit the growth of carbon dioxide only,[41] not methane, which is a significant output of reservoir installations. One is only given a small part of an already small part of the mitigation picture.

While China's installed capacity for hydroelectricity, whether or not it increases by much, will probably always remain unrivalled in size,[42] the hydropower-building programmes of greatest concern in the early rounds of NDCs and in related reports under the UNFCCC are those of tropical and subtropical countries. An inescapable implication of these gigantic building programmes is that vast areas of dense vegetation will be submerged.

The Democratic Republic of the Congo (DRC) sees 'enormous potential' for hydropower development in the country. Its NDC claims that 110,000 MW of new capacity exists ('equivalent to 30 million tons of oil per year'), with 44 per cent of it 'concentrated in the Inga site alone'.[43] Although the DRC's NDC (at 102 pages) is not short, it is silent on the country's plans to harness the power of the Congo River, whether 'at the Inga site' or elsewhere. It is as if, by making a mere reference to it in its NDC, the DRC's entire future dam-building programme is given the imprimatur (and credit) of a mitigation project under the country's mitigation policy, supported by the Paris Agreement.

This apparent deeming exercise masks a complex reality. The DRC's Inga region already serves two dams (2,132 MW capacity in total), built years before the first climate treaty came into existence.[44] Expansion at Inga has been under discussion since 2011.[45] Its purpose is economic development – of

[39] International Rivers (IR), *Advancing Ecological Civilization? Chinese Hydropower Giants and Their Biodiversity Footprints* (Oakland, CA: IR, 2021).

[40] Grill et al (n 8) 218.

[41] China (n 37), 2, 39.

[42] IEA (n 1) 9.

[43] Democratic Republic of the Congo, *Contribution Déterminée À L'échelle Nationale Révisée* (Kinshasa: Ministère de l'Environnement et Développement Durable, 2021), 34 (original in French).

[44] Environmental Justice Atlas, 'Inga 3 and Gran Inga complex hydropower project on Congo river, DRC', 13 October 2020, <https://ejatlas.org/conflict/gran-inga-hydropower-project-on-congo-river-drc>.

[45] Zhang et al (n 6) 829.

the DRC as much as of neighbouring countries to which some of the produced electricity will be sold.[46] The mitigation benefit is, in historical perspective, an afterthought.

This is understandable. Hydropower growth is, or can be, *accelerated* by mitigation policy. But the reasons favouring it continue to exist independently of that policy.[47] What should be apparent from the DRC's NDC, but is not, is whether the Inga plant, potentially as large as 39,000 MW,[48] will deliver any mitigation benefit at all over the relevant term (that is, through to 2050). A verdant valley of 22,000 hectares will be flooded for a new reservoir. Inga is low on the river's course, hence the vast surface of the reservoir. Sediment flow to the sea, which creates an enormous plume in the Atlantic Ocean which acts as a carbon sink, will be interrupted (dams capture all but the finest particles of sediment carried by a river).[49] Transmission lines leading out of the area (including to South Africa, thousands of kilometres away) will clear swathes of forest.[50] And because each megawatt of installed capacity for a large dam costs, on average, more than US$3 million,[51] the DRC's lack of clarification about the sources of finance for this colossal project (potentially costing in the region of US$100 billion) is a source of concern.

The 'contribution' of the DRC's NDC incorporates an aspiration for hydropower growth that clearly exudes ambition. At the same time, it *excludes* critical information relating to how the ambition will be realized.

Africa's other contender for hydropower supremacy is Ethiopia, on the upper Nile. Its NDC does not mention the technology at all. On the other hand, its National Communication affirms that hydropower is a component of the country's mitigation strategy ('its bid to become a major … green economy in East Africa').[52] Ethiopia identifies 45,000 MW of untapped hydropower potential, *all* of which it says it will exploit, inter alia to boost hard-currency receipts from cross-border energy sales, domestic industrial development, and other exports.[53] A flagship project, already well advanced,

46 For a history, see International Rivers, 'Grand Inga Hydroelectric Project: An Overview', no date, <https://archive.internationalrivers.org/resources/grand-inga-hydroelectric-project-an-overview-3356>.

47 Edenhofer et al (n 2), 442, 486.

48 This size would make it the largest hydropower facility on Earth by far: Zhang et al (n 6) 829.

49 Poff et al (n 5) 773.

50 Ibid.

51 Ansar et al (n 2) 48.

52 Ethiopia, *Third National Communication to the United Nations Framework Convention on Climate Change* (Addis Ababa: Ministry of Planning and Development, 2022), 3.

53 Ibid, 6. At another point in the text, Ethiopia claims that its untapped potential is 48,030 MW: ibid, 311.

is the Grand Ethiopian Renaissance Dam (designed for 6,450 MW).[54] It is the largest dam currently under construction on the continent.[55] Inundation of 187,400 hectares for the reservoir, which includes forested areas, was under way at the time of writing.[56] Other, unspecified, installations will be located 'in eight major river basins and their innumerable tributaries'.[57]

There is a coercive factor in Ethiopia's case, which may be relevant to understanding how a major expansion of mitigation technology, not necessarily limited to hydropower, may be realized in certain countries, despite the inevitable conflict of a high-impact roll-out on other norms (including, in this case, with the water needs of Egypt, downriver). Hydropower's mitigation credentials relative to the shrinking time lines for action are, moreover, debatable. The coercive factor is that authoritarian regimes in developing countries (Ethiopia's being one of them) may simply push through such measures, depending on the package of advantages they offer. In the case of hydroelectric dams, the advantages go well beyond electric power, of course. Electricity is just one of several benefits that go with the control of a vast reservoir of water. The technologies used for other types of renewables (photovoltaic panels, wind-driven generators) may cause less friction when it comes to their physical expansion, but compared with hydropower they are entirely one-dimensional. They do one thing only. River damming, by contrast, offers water storage for dry seasons, year-round irrigation, flood control, recreation and tourism, subsistence fishing, and so on, as added benefits.[58] In Ethiopia's case, it can also be a trump card against Egypt. All of this, plus its monumental image, ensure that the technology will continue to spread relatively easily in certain places. The mitigation imperative's effect might be to support coercion.

Asia is where the greatest concentration of dam-building aspirants is found today, at least if NDCs are read literally. Pakistan's NDC outlines the country's plan to phase out coal and replace much of it with hydropower. In fact, Pakistan's headline-friendly self-imposed 'moratorium' on coal as of 2020 affects only imported coal and new coal-fired power plants. Use of indigenous coal in existing plants continues unabated.[59] Nevertheless, the

54 Zhang et al (n 6) 829.

55 IEA (n 1) 10.

56 NASA Earth Observatory, 'A Grand New Dam on the Nile', 2021, <https://earthobservatory.nasa.gov/images/149691/a-grand-new-dam-on-the-nile>. On the political history of this dam, see Harry Verhoeven, 'The Grand Ethiopian Renaissance Dam: Africa's Water Tower, Environmental Justice and Infrastructural Power', 150(4) *Daedalus* 159 (2021).

57 Ethiopia (n 52) 311.

58 IEA (n 1) 13.

59 Pakistan, *Updated Nationally Determined Contributions* (Islamabad: Ministry of Climate Change, 2021), 14–15.

level of Pakistan's reliance on the Indus River and its tributaries to provide the additional power generation the country needs for its development is considerable. Pakistan calculates its hydropower potential to be around 51,600 MW.[60] The estimated cost of realizing it is not specified, but it is in the order of US$80 billion through to 2040. Pakistan's NDC assumes that, if Pakistan is to meet its mitigation targets under the Paris Agreement, some of this money (an unspecified portion) will be sourced from international climate finance.[61]

Pakistan's dam-building programme has a dual role for its planners, as it does for many other hydropower (aspiring) countries. It involves another of the technology's many advantages. Hydroelectricity is generally assigned responsibility for a fixed slice of day-to-day energy generation (which may include a peaking-power service, when electricity prices are high).[62] On top of this, it is generally made (or can be) responsible for compensating instantly for output falls in renewable energy of the variable kind (wind and solar), where this technology is integrated into the national grid. Sheer storage of water is behind these conveniences. Hence, Pakistan, which also plans to develop its wind and solar, describes hydropower as 'critical' to its mitigation objectives.[63] The reality is that a country in Pakistan's circumstances cannot plan to phase out coal and phase in wind and solar without expanding hydropower. It is part of the equation. The mitigation contribution of hydropower, per se, may not be the point. It will nevertheless be listed as a mitigation measure in the NDC.

Pakistan's NDC refers to 12,000 MW of hydropower under construction.[64] While no further information is given, it presumably includes the Diamer-Bhasha dam on the Indus River, with a planned installed capacity of 4,500 MW. Construction was launched, as a joint Pakistan-China venture, in 2020. It will be the largest dam in Pakistan.[65] As with other elements of the country's dam-building programme, information about the Diamer-Bhasha dam must be sought from sources external to Pakistan's NDC and the other instruments of the climate change regime. A 200-square-kilometre reservoir will flood 100 kilometres of the Karakoram Highway, inundating

60 Ibid, 27. Around 10,000 MW is already installed: ibid, 28.

61 Ibid, 16–17.

62 IEA (n 1) 27.

63 Pakistan (n 59) 28.

64 Ibid.

65 Sana Jamal, 'Pakistan Begins Construction of Diamer Bhasha Dam', *Gulf News*, 15 July 2020, <https://gulfnews.com/world/asia/pakistan/pakistan-begins-construction-of-diamer-bhasha-dam-1.72607867#>; and Kaswar Klasra, 'Pakistan Gets China's Help in Building Dam', *China Daily*, 2 July 2020, <https://global.chinadaily.com.cn/a/202007/20/WS5f14f269a31083481725a98f.html>.

villages along it housing more than 35,000 people.[66] The dam's wall requires 12 million tons of steel and 32 million tons of concrete (with emissions of around 2.5 tons of carbon dioxide per ton of steel and 1.2 tons of carbon dioxide per ton of cement). Mangrove forests in the Indus delta will be affected by the large amount of water held back to fill the reservoir.[67] It is notable that this NDC, whose mitigation contribution revolves around the expansion of hydropower at an unprecedented rate (for Pakistan), does not include even passing reference to any likely socio-legal obstructions to, or any negative environmental side-effects of, the technology. Given that the dam is officially due to be completed in 2028, but in practice probably no earlier than 2035 due to schedule overruns,[68] its greenhouse gas emissions over the period to 2050 may exceed the emissions it displaces from fossil-fuel power. An NDC dealing with the substance of the issue might be expected to note these matters. However, the pattern we are seeing take hold is that the NDC/Paris Agreement narrative is frequently limited to indicia of ambition, not a demonstration of how ambition will be realized.

Nepal (population 31 million) is a much smaller country than Pakistan (236 million), a fact that makes its planned acceleration of 'clean energy generation' all the more noteworthy: from 1,400 MW in 2020 to 15,000 MW in 2030, with as much as 95 per cent of this coming from large hydropower.[69] It amounts to more than three of Pakistan's Diamer-Bhasha dams added to this small country in ten years. Because all Nepalese dams are of the highland variety, with a low reservoir-surface-to-installed-capacity ratio, greenhouse gas emissions from the reservoirs are not the primary concern. Rather, the issue is how a relatively poor developing country can deliver this scale of technological change within a meaningful time frame, even with substantial international support.[70] Nepal's NDC does not say. About half the world's large dams suffer cost overruns so large that the costs end up exceeding the benefits, even assuming that their planned benefits are realized.[71] For large

---

[66] International Rivers, 'Fast Track Power Generation: Will a Large Dam Increase Access to Electricity in Pakistan?', 2008, <https://archive.internationalrivers.org/resources/fast-track-power-generation-1834>.

[67] Hassan Abbas and Asghar Hussain, 'Opinion: Pakistan's Diamer Bhasha Dam is Neither Green Nor Cheap', *The Third Pole*, 5 November 2021, <www.thethirdpole.net/en/energy/pakistans-diamer-bhasha-dam-neither-green-nor-cheap/>.

[68] Ansar et al (n 2) 54.

[69] Nepal, *Second Nationally Determined Contribution* (Kathmandu: Ministry of Forests and Environment, 2020), 2. The current installed capacity of 1,400 MW is mainly hydropower: ibid, 3–4.

[70] Nepal's NDC clarifies that only 5,000 MW of the 13,600 MW it hopes to install will be self-funded: ibid, 2. Nevertheless, this is still a considerable amount of new hydropower for a small country.

[71] Ansar et al (n 2) 48.

dam projects, the mean cost overrun is 96 per cent (compared with just 6 per cent for thermal power plants).[72]

Undeterred, Nepal's Third National Communication, which postdates its NDC by a year, presents an even more ambitious plan: 'The estimated hydropower potential of Nepal is 83,000 MW of which 114 projects having 45,610 MW [total capacity] have been identified as commercially feasible.'[73] The largest of these is the Chisapani Karnali dam, whose installed capacity would be an astounding 10,800 MW (about half the size of China's Three Gorges Dam).[74] Nepal does not need all this power, of course. Much of it is planned for export to India to boost hard-currency earnings.[75] It would appear that India's mitigation plans under the Paris Agreement will be partly realized through hydropower expansion in Nepal (although this is not acknowledged by India, as I discuss later). Bhutan is another small impoverished country in the same region that plans to invest heavily in hydropower dams to sell electricity to India as part of an export-enhancement strategy.[76]

The risk in these cases is that the cost and schedule overruns that will almost certainly be the upshot of the hydropower-based mitigation plans of Nepal and Bhutan may leave these countries in debt to Chinese state-owned enterprises, in particular Sinohydro, which dominates the hydropower-construction business in developing countries.[77] An NDC whose ambition derives from hydropower could very well lead a poor country to financial dependency or ruin. If international climate finance were plentifully available (contrary to what is actually the case), vulnerable developing countries could avoid this prospect. But by using their own resources to invest in expensive mitigation technologies as a development strategy, they could end up trapped under a mountain of debt.

India's NDC reveals nothing about its plans for hydropower, with the exception of this one sentence: 'With a vast potential of more than 100 GW, a number of policy initiatives and actions are being undertaken to aggressively pursue development of country's [*sic*] vast hydro potential.'[78] There is an expectation also in non-government circles that an immense roll-out of

---

[72] Ibid, 54.

[73] Nepal, *Third National Communication to the United Nations Framework Convention on Climate Change* (Kathmandu: Ministry of Forests and Environment, 2021), 20.

[74] Ibid.

[75] Ibid.

[76] Bhutan, *First Biennial Update Report to the UNFCCC* (Thimphu: National Environment Commission, 2022), 67.

[77] Tan-Mullins et al (n 33) 465; and IEA (n 1) 10.

[78] India, *Intended Nationally Determined Contribution: Working Towards Climate Justice* (New Delhi: Ministry of Environment, Forest and Climate Change, 2016), 9–10. It is not clear whether the 100,000 MW of potential hydropower includes or excludes currently installed hydropower totalling 46,100 MW (ibid). India's NDC from 2016 was updated

the technology is indeed likely in India.[79] A National Communication of India from an earlier date raised a point that is relevant to an understanding of the likelihood of India's 100 GW expansion claim being realized. The Communication remarked that India's thermal power generation is expanding more rapidly than hydropower 'due to … increasing opposition to large hydro projects on social and environmental grounds, and higher political risks of hydro projects due to inter-state disputes on power and water sharing'.[80]

This is a rare concession in this context. Sweden is likewise quite direct about hydropower's problems in its communications under the UNFCCC.[81] A tiny number of other countries merely hint at these issues.[82] However, although India raises the spectre of opposition to hydropower in its National Communication, it does not revisit it in its NDC. A later BUR, in fact, omits any discussion of the technology's future growth potential in the country. It reports only on past growth of 4,000 MW of added hydropower over a period of about five years (2016–20) – minuscule in comparison with the technology's growth potential claimed in India's NDC.[83] As I indicated earlier, there is no mention in any of these instruments of a plan to cooperate with Nepal or Bhutan on greenhouse gas mitigation policy through purchases of hydropower from those countries.

The NDC of Laos has half a page of information on its hydropower plans, which makes it relatively detailed compared to other NDCs. The instrument specifies that Laos had 4,500 MW of installed hydroelectric capacity in 2018. It claims that these plants were offsetting the equivalent per year of 26 per cent of national emissions in the year 2000.[84] This sounds impressive, and it amounts to an argument for hydropower expansion in the country.

in 2020 with a very short text (less than three pages long) that made some revisions that are not pertinent here; it otherwise left the earlier NDC unchanged and in force.

79 See IEA (n 1) 9.

80 India, *Second National Communication to the United Nations Framework Convention on Climate Change* (New Delhi: Ministry of Environment and Forests, 2012), 25.

81 Sweden, *Eighth National Communication on Climate Change* (Stockholm: Ministry of Climate and Enterprise, 2023), 219.

82 Cambodia identifies 'hydropower development throughout the Mekong Basin' as a 'non-climate stressor': Cambodia (n 32) 117. Brazil implies that it will wind down its large-dam programme (while increasing construction of smaller dams) for environmental reasons: Brazil, *Fourth National Communication to the UNFCCC* (Brasilia: Ministry of Science, Technology and Innovation, 2020), 225. Myanmar's NDC notes 'national sensitivities around social and environmental safeguards' concerning the technology: Myanmar, *Nationally Determined Contributions* (Rangoon: Ministry of Natural Resources and Environmental Conservation, 2021), 7–8.

83 India, *Third Biennial Update Report to the United Nations Framework Convention on Climate Change* (New Delhi: Ministry of Environment, Forest and Climate Change, 2021), 11.

84 Laos, *Nationally Determined Contribution* (Vientiane: Ministry of Natural Resources and the Environment, 2021), 3–4.

Underlying it, however, is a simplistic calculation that does not factor in the greenhouse gas emissions from the installations themselves. In Laos, the dams are of the lowland variety, located in the Mekong River basin, and therefore likely to have high methane emissions.

Laos further states in its NDC that installed hydropower capacity in the country had, by 2020, increased by 1,000 MW, and that it would be further increased by 7,500 MW by 2030 without reliance on international climate finance.[85] How such a massive growth in the technology (costing more than US$20 billion based on the average per-megawatt cost cited earlier) would be realized in Laos, which is a Least Developed Country, without international support, is not explained.[86] Nor is it explained how so much construction could be completed in so short a period of time in light of the years-long delays that typically mar reservoir projects.[87] There is an indication that some of the hydropower Laos plans to generate will be surplus and will be exported (to Vietnam?),[88] but no further information is provided on this point.

Laos's NDC refers in passing to an incident of 'dam collapse' in 2018, caused by storms and heavy rains, leading to 'huge losses and damages'.[89] This is, of course, a pertinent point, which, however, is not further detailed in Laos's NDC or its UNFCCC communications. As with the mirror-image threat of the impacts *of* dams on the world around them, NDCs and related instruments produced by states pursuant to the climate change regime choose to avoid discussion of the vulnerabilities of hydropower itself under a warming climate. Sturdy though dams are,[90] they are not immune to the impacts of climate change. Storms and droughts can lead to their closure.[91] In the Laos dam-collapse incident of 2018, at least 71 people were killed according to two non-governmental organizations, Inclusive Development International and International Rivers. The two organizations have alleged shoddy construction as the main cause – not severe weather, which is the

[85] Ibid, 5. Significantly higher figures for installed capacity are given by Soukhaphon et al (n 26) 3.

[86] The country's BUR dating to the year before the NDC states that 20,000 MW of hydropower is planned by the country for the period after 2020: Laos, *First Biennial Update Report* (Vientiane: Ministry of Natural Resources and Environment, 2020), 26–7. No further information is given on this plan.

[87] Ansar et al (n 2) 49.

[88] Laos (n 84) 5; see also IEA (n 1) 9–10.

[89] Laos (n 84) 13.

[90] The structural elements of a power dam have a projected life of up to 100 years: Edenhofer et al (n 2) 474.

[91] Ying Meng et al, 'Undermined Co-Benefits of Hydropower and Irrigation under Climate Change', 167 *Resources, Conservation and Recycling* 1 (105375) (2021), 1.

explanation given in the Laos NDC.[92] An acceleration of dam-building for mitigation purposes in developing countries must take account of this kind of risk as well, which is less well understood than the more conventional concerns underlying critiques of hydropower growth. The track record of large, complex engineering feats undertaken by poor developing countries is unenviable.[93] The risk of 'huge losses and damages' from them, especially if a multitude of them were to be fast-tracked in parallel in the same country, is higher in these circumstances.

On the periphery of Europe, Turkey is the country with the largest hydropower growth potential and corresponding ambition. 'Tapping the full hydroelectric potential' is the country's plan, according to its NDC.[94] In 2016, hydroelectricity held a 24.5 per cent share of total electricity generation in Turkey.[95] Installed capacity was 26,678 MW, an amount that Turkey intends to triple.[96] By 2022, there were 751 hydropower plants in the country, with a total capacity of 31,571 MW; more were under construction.[97] However, electricity demand was growing at an even faster rate, so that the share of hydroelectricity by 2022 had fallen to 20.4 per cent of total generation.[98] Turkey acknowledges that dam-building costs are prohibitive, preventing the country from reaching its 'theoretical hydropower potential' of 180,000 MW.[99] It seems that the current plan for Turkey is for hydropower to grow at a rate of one large dam per year through to 2035.[100] The country does not provide any further information on these matters in its NDC or in related instruments.

3.2.3 Lesser promoters

Several other countries have announced plans in their NDCs to expand hydropower by a significant amount relative to their size and stage of development, without offering too many specifics. In Argentina, in 2020,

92 Inclusive Development International (IDI) and International Rivers (IR), *Reckless Endangerment: Assessing Responsibility for the Xe Pian-Xe Namnoy Dam Collapse* (Asheville, NC, and Oakland, CA: IDI and IR, 2019).

93 Ansar et al (n 2) 51–2, 55.

94 Turkey, *Intended Nationally Determined Contribution* (Ankara, 2021), 3; see also IEA (n 1) 9–10 ('Turkey … is expected to drive the largest expansion in capacity in Europe').

95 Turkey, *Seventh National Communication under the UNFCCC* (Ankara: Ministry of Environment and Urbanization, 2018), 18.

96 Ibid, 35.

97 Turkey, *Eighth National Communication and Fifth Biennial Report under the UNFCCC* (Ankara: Directorate of Climate Change, 2023), 42.

98 Ibid.

99 Turkey (n 95) 35.

100 Turkey (n 97) 114.

large-scale hydropower's installed capacity was 10,792 MW.[101] The country plans for the 'energy matrix by 2030 [to] be more … sovereign … based on the significant potential of … hydroelectric' power.[102] By 2030, installed capacity is slated to have grown to 13,242 MW.[103] Guinea (population 13.9 million), in its NDC, estimates its unrealized hydroelectric potential at more than 5,280 MW, all of which it intends to realize. Only about 720 MW has been developed there to date.[104] These installations generate about 60 per cent of the country's electricity. With international support, Guinea's plan is to raise hydroelectricity's share to 80 per cent by 2030.[105] Mozambique's NDC notes that its largest hydropower plant (2,075 MW installed capacity) is responsible for 71 per cent of national electricity generation. It estimates potential capacity for the technology at 19,000 MW, of which 5,645 MW will be 'prioritized' (no further detail is offered).[106] Myanmar's pre-coup government produced an NDC that identified 19,567 MW of additional potential hydropower for development.[107] For growth of this magnitude, the government presumably had plans to impound the Irrawaddy River, which until now has remained free-flowing.[108] The government had allocated US$2.5 billion to the building programme through to 2030.[109] (One must assume, in light of the current chaos in the country, that Myanmar's NDC is just words on paper.) Nigeria's NDC is largely conditional on international financial support. For its mitigation 'contribution' it requests international climate finance for hydropower development, which is to attain a total installed capacity of 15,500 MW.[110] Current installed capacity in Nigeria is just 1,938 MW.[111] Tanzania includes a one-liner in its NDC on its intention to grow hydropower to more than eight times its current size (reaching 4,700 MW),

---

[101] Argentina, *Cuarto Informe Bienal de Actualización de Argentina a La Convención Marco de Las Naciones Unidas Para El Cambio Climático* (Buenos Aires: Ministro de Ambiente y Desarrollo Sostenible, 2021), 264 (original in Spanish).

[102] Argentina, *Segunda Contribución Determinada a Nivel Nacional* (Buenos Aires: Ministerio de Ambiente y Desarrollo Sostenible, 2020), 20 (original in Spanish).

[103] Argentina (n 101) 263.

[104] Guinea, *Contribution Déterminée au niveau National de la République de Guinée* (Conakry, 2021), 13–14 (original in French).

[105] Ibid, 27.

[106] Mozambique, *Update of the First Nationally Determined Contribution to the United Nations Framework Convention on Climate Change* (Maputo, Ministry of Land and Environment, 2021), 59.

[107] Myanmar (n 82) 11.

[108] Grill et al (n 8) 218.

[109] Myanmar (n 82) 11.

[110] Nigeria, *First Nationally Determined Contribution (2021 Update)* (Abuja: Federal Ministry of Environment, 2021), 22.

[111] Nigeria, *Second Biennial Update Report to the United Nations Framework Convention on Climate Change* (Abuja: Federal Ministry of Environment, 2021), 12.

without saying how it will do so or by when.[112] Venezuela also plans to grow its hydropower strongly, but from an already significant base (more than 15,000 MW).[113]

3.2.4 Brazil

I have left Brazil till last because it is an unusual case. It is a predominantly lowland country with dams of high emission intensity.[114] Although Brazil already has a vast infrastructure of dams (installed capacity in 2018 was 104,139 MW, inclusive of 281 large dams and 427 small ones),[115] it plans to build more. The country has been silent in its national reports to the UNFCCC about the greenhouse gas emissions from its network of dams.[116] This is to be expected, in light of Brazil's outstanding success in using funding from the Clean Development Mechanism (CDM) to subsidize its dam-building programme.[117] (To have admitted high emissions from dams would have risked termination of those subsidies.) Brazil's reports are also silent on the severe reduction in river connectivity that its river-damming programme has caused.[118] The country's NDC praises the technology without offering an indication of how its expansion plans will further develop.[119] It would seem from Brazil's 2020 BUR, however, that its future hydropower development will be limited to small-reservoir dams and run-of-river plants.[120] It is likely that the last of the large dams in Brazil's remarkably ambitious history of damming its rivers were those being completed at the time of writing, which are due to come on line around 2030.[121]

In suggesting, as I have throughout this chapter, that countries should be more open about their reliance on hydroelectricity generation for mitigation purposes, because of the downsides of the technology, the alternatives

[112] Tanzania, *Nationally Determined Contribution* (Dodoma: Ministry of Environment, 2021), 13–14.

[113] Venezuela, *Segunda Comunicación Nacional Ante La Convención Marco de Las Naciones Unidas Sobre Cambio Climático* (Caracas: Ministerio del Poder Popular para Ecosocialismo y Aguas, 2018), 122 (original in Spanish).

[114] Almeida et al (n 16) 2–3.

[115] Brazil (n 82) 222.

[116] Fearnside (n 19) 231.

[117] Fearnside (n 18).

[118] Alexander S. Flecker, Qinru Shi et al, 'Reducing Adverse Impacts of Amazon Hydropower Expansion', 375 *Science* 753 (2022), 755.

[119] Brazil, *Nationally Determined Contribution* (Brasilia, 2022), 9–10.

[120] Brazil, *Fourth Biennial Update Report to the United Nations Framework Convention on Climate Change* (Brasilia: Ministry of Foreign Affairs and Ministry of Science, Technology and Innovation, 2020), 64–5.

[121] See Alexander C. Lees et al, 'Hydropower and the Future of Amazonian Biodiversity', 25 *Biodiversity Conservation* 451 (2016), 452, 462.

reasonably open to them must of course also be kept in mind. In Brazil's case, the country could turn to its untapped wind and solar potential for additional power, but it could just as easily turn to its vast deposits of onshore and offshore oil and gas, discovered in the 2010s.[122] The advantage of the latter as an energy source is that, like hydropower, they do not suffer from intermittency. Another advantage of oil and gas is that power plants fuelled by them occupy little land compared with wind and solar, so that the biodiversity opportunity costs for Brazil (supremely well-endowed as it still is in this regard) would be close to nil if the fossil-fuel option were chosen. Brazil's energy pathway of least resistance when hydropower is struck off the list of mitigation options is not necessarily much better for climate change mitigation than it would be if the option were kept in play.

3.3 Will power dams deliver the promised additional mitigation?

We have seen in the previous section that several countries have based their mitigation plans on the assumption that hydropower growth will continue and can indeed be accelerated. This group of countries is significant in population size, and the amount of mitigation that hydropower is supposed to deliver for this group is also significant on a global scale.

In general, any rapid roll-out of a mitigation technology will face implementation obstacles, for it cannot avoid running up against established interests. In a world that no longer offers any unmanaged spaces, mitigation technology must vie for space that is already occupied. This is emphatically true of the newer renewables (wind, solar, biofuel), which are space-hungry and are often installed on land proximate to human settlements. Working land in these instances must be repurposed to accommodate the hardware, or, in the case of biofuel, the changed agricultural practices. Hydropower does not face quite the same problem, because its large installations can efficiently be placed at a distance from settlements. But there, they will likely intrude on protected nature or on relatively unspoiled areas, where communal interests are invested not in particular land uses but in leaving the land unused and the rivers free-flowing for environmental reasons. These interests have law on their side, at least in developed countries. Remotely installed hydropower has opportunity costs for biodiversity that are difficult to calculate, including because many of the affected ecosystems are still only partially described.[123] In accordance with the precautionary principle, the potential biodiversity penalty of hydropower must not be discounted for this reason (the gaps in our knowledge).

[122] Ibid, 454.

[123] Ibid, 459–60; K. O. Winemiller et al, 'Balancing Hydropower and Biodiversity in the Amazon, Congo, and Mekong', 351(6269) *Science* 128 (2016); and Grill et al (n 8) 220.

Nevertheless, the issues for power dams go well beyond biodiversity loss. Soon after this century-old technology was repackaged as a mitigation measure in the late 1990s and embraced by the CDM,[124] scholars began investigating and questioning its mitigation credentials. A debate ensued, which continues to this day.[125] Any technology that causes methane emissions (as hydropower does) is assured a slice of controversy, as it raises the question of which global warming potential to use for the gas. Opponents of the technology will naturally choose the highest potential that the Intergovernmental Panel on Climate Change provides: 86 times the warming potential of carbon dioxide over a 20-year time horizon.[126] This would weigh heavily against the mitigation benefits of power dams. Consequently, the NDCs' promise of a renaissance in dam-building must contend not only with established interests of the environmental kind (which in some places enjoy legal backing), but also with scepticism concerning whether what is being promised will deliver significant net mitigation benefits before the mid-century milestone. Additional issues – financial viability, the possibility of economic dependency, structural safety under the punishing impacts of climate change, and so on – have been highlighted in this chapter.

It is indeed remarkable that a technically challenging, expensive and intensely nature-altering technology such as hydropower is accommodated with such ease in the climate change treaties, especially in the Paris Agreement through its NDC system. It should lead us to question the amount of real substance underlying the fine ambition expressed in NDCs through their embrace of the technology of hydropower.

---

[124] See n 15.

[125] For example, Philip M. Fearnside, 'Do Hydroelectric Dams Mitigate Global Warming? The Case of Brazil's Curuá-Una Dam', 10(4) *Mitigation and Adaptation Strategies for Global Change* 675 (2005); Edenhofer et al (n 2) 442, 462, 471–2; Philip M. Fearnside, 'Carbon Credit for Hydroelectric Dams as a Source of Greenhouse-Gas Emissions: The Example of Brazil's Teles Pires Dam', 18(5) *Mitigation and Adaptation Strategies for Global Change* 691 (2013) (critiquing the CDM's support for tropical dams); Philip M. Fearnside, 'Impacts of Brazil's Madeira River Dams: Unlearned Lessons for Hydroelectric Development in Amazonia', 38 *Environmental Science and Policy* 164 (2014) (on the role of Amazonian dams in stimulating deforestation); Fearnside (n 19) 229 (critiquing the IPCC for not providing default values for states to report fully on hydropower emissions in its 2006 guidelines on emission accounting); Berga (n 4) (the author, a member of the International Commission on Large Dams, and therefore necessarily partial, defends hydropower as comparatively 'clean'); Almeida et al (n 16) 2; Abid Hussain et al, 'Hydropower Development in the Hindu Kush Himalayan Region: Issues, Policies and Opportunities', 107 *Renewable and Sustainable Energy Reviews* 446 (2019), 448; Mike Muller, 'Dams Have the Power to Slow Climate Change', 566 *Nature* 315 (2019), 316; IEA (n 1) 8; and Flecker et al (n 118).

[126] Intergovernmental Panel on Climate Change, *Climate Change 2013: The Physical Science Basis: Working Group I Contribution to the Fifth Assessment Report of the Intergovernmental Panel on Climate Change* (New York, NY: Cambridge University Press, 2013), 714.

4

An Essential Zero-Carbon Solution or a Perilous Energy Technology? Exploring the Nuclear Debate in EU Law

Kaisa Huhta

4.1 Introduction

The global effort to combat global warming requires extensive measures in the energy sector, which is responsible for roughly a third of global greenhouse gas emissions.[1] A monumental scale-up of low-carbon energy sources must play a central role in reducing this contribution.[2] Renewable energy sources are typically cited as an obvious route to address the issue,[3] yet they are not enough in themselves, as their intermittency requires other supporting energy solutions.[4] What is more, electricity demand will increase substantially over the coming decades due to electrification and overall

[1] Jim Skea and others, 'Summary for Policymakers', in P.R. Shukla and others (eds), *Climate Change 2022: Mitigation of Climate Change. Contribution of Working Group III to the Sixth Assessment Report of the Intergovernmental Panel on Climate Change* (Cambridge University Press 2022), 12.

[2] Kaisa Huhta and Seita Romppanen, 'Why is energy law resistant to changes required by climate policies?' (2023) 4 *Energy and Climate Change* 100096; Sam Fankhauser and others, 'The meaning of net zero and how to get it right' (2022) 12 *Nature Climate Change* 15, 17.

[3] Volker Quaschning and Herbert Eppel, *Renewable Energy and Climate Change* (2nd edn, Wiley 2020).

[4] Kaisa Huhta, 'Energy Security in the Energy Transition: A Legal Perspective', in Geoffrey Wood and others (eds), *The Palgrave Handbook of Zero Carbon Energy Systems and Energy Transitions* (Springer International 2022).

growth in energy demand despite efforts to reverse this trend.[5] As a low-carbon means of producing electricity, nuclear energy has been suggested as a potentially influential solution.[6] The Intergovernmental Panel on Climate Change (IPCC) has held that nuclear energy 'could make an increasing contribution'[7] to low-carbon electricity and heat, and continues to include nuclear energy in its climate change mitigation scenarios.[8]

However, no form of climate change mitigation technology is without its trade-offs, as demonstrated throughout the chapters of this book. Nuclear energy, in particular, faces significant opposition from environmental groups and civil society, which voice loud concerns over its safety and feasibility.[9] Acknowledging this opposition, the IPCC has highlighted that the major obstacles to nuclear energy include 'long-term fuel resource constraints without recycling; economics; safety; waste management; security; proliferation, and adverse public opinion'.[10]

This chapter focuses on the interplay between the arguments in favour of nuclear energy and those against, as two sides of an enduring debate. It does not seek to discredit or to support one or the other of the opposing views on nuclear energy, nor does it take a legal, social or

5 International Energy Agency, 'Outlook for energy demand' (International Energy Agency 2022) <www.iea.org/reports/world-energy-outlook-2022/outlook-for-energy-demand> accessed 1 June 2023.

6 William C. Sailor and others, 'A nuclear solution to climate change?' (2000) 288 *Science* 1177; Vladimir Knapp, Dubravko Pevec and Mario Matijević, 'The potential of fission nuclear power in resolving global climate change under the constraints of nuclear fuel resources and once-through fuel cycles' (2010) 38 *Energy Policy* 6793; Michael Buchdahl Roth and Paulina Jaramillo, 'Going nuclear for climate mitigation: an analysis of the cost effectiveness of preserving existing U.S. nuclear power plants as a carbon avoidance strategy' (2017) 131 *Energy* 67; Vladimir Knapp and Dubravko Pevec, 'Promises and limitations of nuclear fission energy in combating climate change' (2018) 120 *Energy Policy* 94; John Parsons and others, 'A fresh look at nuclear energy' (2019) 363 *Science* 105; Jacopo Buongiorno and others, 'Nuclear energy in a carbon-constrained world: Big challenges and big opportunities' (2019) 17 *IEEE Power & Energy Magazine* 69.

7 R.E.H. Sims and others, 'Energy supply', in B. Metz and others (eds), *Climate Change 2007: Mitigation. Contribution of Working Group III to the Fourth Assessment Report of the Intergovernmental Panel on Climate Change* (Cambridge University Press 2007), 253; T. Bruckner and others, 'Energy Systems', in O. Edenhofer and others (eds), *Climate Change 2014: Mitigation of Climate Change. Contribution of Working Group III to the Fifth Assessment Report of the Intergovernmental Panel on Climate Change* (Cambridge University Press 2014), 517.

8 Jim Skea and others (n 1).

9 John F. Ahearne, 'Prospects for nuclear energy' (2011) 33 *Energy Economics* 572, 574; Jaesun Wang and Seoyong Kim, 'Comparative analysis of public attitudes toward nuclear power energy across 27 European countries by applying the multilevel model' (2018) 10 *Sustainability* 1518.

10 IPCC, n 7 above, 253.

economic (let alone personal) stand on them. Instead, this chapter reviews the key facts and arguments typically invoked by nuclear opponents and supporters and demonstrates, using the European Union (EU) as an illustrative case study, how this sensitive and politically charged debate is reflected in law and in legal governance of the low-carbon energy transition.

The chapter acknowledges, as its starting point, that nuclear energy is one of the most controversial means by which to produce energy. Not just public opinion but also research on nuclear energy is extraordinarily divided. Depending on the viewpoint taken, research has come out both in support of nuclear energy and warning against it.[11] A recent literature review conducted in this area shows that 'articles in scientific journals disagree on the role that nuclear energy should play in the low carbon future'.[12] As a result, both opponents and supporters of nuclear energy can find facts to support their arguments. This chapter shows that even when there is no agreement over the merits or perils of nuclear energy, transnational legal frameworks can be designed and adopted for nuclear energy. In fact, despite the fundamental controversies surrounding nuclear energy, the legal governance of nuclear activities, referred to as 'nuclear law', has been in existence globally since the 1960s.[13]

The focus on the EU is justified by several reasons. The founding member states signed the Treaty establishing the European Atomic Energy Community in 1957 (hereafter the 'Euratom Treaty').[14] Overall, governing the exploitation of steel, coal and nuclear resources has been at the heart of European economic and political integration from the outset as two of the three founding treaties focused on these resources.[15] Nuclear energy

[11] Critical views in Lutz Mez, 'Nuclear energy: Any solution for sustainability and climate protection?' (2012) 48 *Energy Policy*, 56–63; and favourable views in Knapp et al (n 6), as well as Knapp and Pevec (n 6).

[12] Nikolaus Muellner and others, 'Nuclear energy: The solution to climate change?' (2021) 155 *Energy Policy* 112363 <https://doi.org/10.1016/j.enpol.2021.112363> accessed 1 June 2023.

[13] Jakub Handrlica, 'Nuclear law revisited as an academic discipline' (2019) 12 *Journal of World Energy Law & Business* 52, 55; Jakub Handrlica, 'The unbearable lightness of nuclear jurisprudence' (2021) 39 *Journal of Energy & Natural Resources Law* 375.

[14] Consolidated version of the Treaty establishing the European Atomic Energy Community [2008] OJ C327/1.

[15] The other two founding treaties were the Treaty establishing the European Coal and Steel Community (no longer in force) and the Treaty establishing the European Economic Community. See Kaisa Huhta, 'The scope of state sovereignty under Article 194(2) TFEU and the evolution of competences in the energy sector' (2021) 70 *International and Comparative Law Quarterly* 4.

has been politically controversial in Europe especially since the 1970s,[16] with significant increases in nuclear opposition after the nuclear accidents in Ukraine in 1986 and Japan in 2011.[17] However, there have also been periodical increases in support for nuclear power, with some commentators even describing or predicting a 'nuclear renaissance'.[18] Most importantly from the point of view of the themes of this book, a focus on the EU is justified in the context of its ambitious climate goals, which rely on the success of low- and zero-carbon energy solutions.

The political and legal approaches to nuclear energy within the EU's member states have undergone continual evolution over a period of more than 60 years to the present day. Some member states rely on nuclear power extensively and consider it to be an inseparable part of the low-carbon transition, while others regard it as an unacceptably risky means of producing electricity.[19] In 2021, nuclear energy constituted around a fifth of the total electricity produced in the EU[20] and 13 EU member states had operational nuclear reactors.[21] Member states such as France and Finland rely heavily on

16 Patrick Hartmann and others, 'Nuclear power threats, public opposition and green electricity adoption: Effects of threat belief appraisal and fear arousal' (2013) *Energy Policy* 1366; Astrid Mignon Kirchhof and Jan-Henrik Meyer, 'Global protest against nuclear power: Transfer and transnational exchange in the 1970s and 1980s' (2014) 39 *Historical Social Research* 165.

17 On the history of nuclear opposition, see Horace Herring, 'Opposition to Nuclear Power: A Brief History', in David Elliott (ed), *Nuclear or Not?* (Palgrave Macmillan 2007); and Stephen C. Whitfield and others, 'The Future of Nuclear Power: Value Orientations and Risk Perception' (2009) 29 *Risk Analysis* 425.

18 Benjamin Sovacool, 'Critically weighing the costs and benefits of a nuclear renaissance' (2010) 7 *Journal of Integrative Environmental Sciences* 105; Tuula Teräväinen, Markku Lehtonen and Mari Martiskainen, 'Climate change, energy security, and risk: Debating nuclear new build in Finland, France and the UK' (2011) 39 *Energy Policy* 3434, 3436; Geordan Graetz, 'Uranium mining and First Peoples: The nuclear renaissance confronts historical legacies' (2014) 84 *Journal of Cleaner Production* 339.

19 Mez (n 11) 56; Fabienne Gralla and others, 'The role of sustainability in nuclear energy plans: What do national energy strategies tell us?' (2016) 22 *Energy Research & Social Science* 94; Wang and Kim (n 9); Teräväinen, Lehtonen and Martiskainen (n 18); Ilina Cenevska, *The European Atomic Energy Community in the European Union Context: The 'Outsider' Within* (Brill 2016), 2; Anna Södersten, *Euratom at the Crossroads* (Edward Elgar 2018), 3.

20 Cenevska (n 19) 1–2; Eurostat, 'Nuclear energy statistics, December 2022' (Eurostat 2022) <https://ec.europa.eu/eurostat/statistics-explained/index.php?title=Nuclear_energy_statistics> accessed 1 June 2023.

21 Belgium, Bulgaria, Czechia, Germany, Spain, France, Hungary, the Netherlands, Romania, Slovenia, Slovakia, Finland and Sweden. See Södersten (n 19) 2.

nuclear energy.[22] Finland has just opened a new nuclear reactor,[23] and France and the Czech Republic are planning massive new nuclear investments.[24] The United Kingdom, while no longer a member of the EU, is constructing new nuclear facilities.[25] In contrast, Belgium has decided to entirely phase out nuclear, and Germany completed its phase-out in spring 2023.[26] Other states, such as Austria, have no nuclear power plants because of widespread opposition to nuclear energy.[27] Overall, nuclear energy production in the EU is decreasing.[28]

Section 4.2 examines the key arguments for and against nuclear energy. While some of these arguments also apply in the context of small modular nuclear reactors, the analysis here focuses on large-scale, centralized nuclear energy production.[29] Section 4.3 then analyses the constitutional framework for nuclear energy in the EU and demonstrates how, by providing member states the right to choose their own energy mix, it creates a compromise between their conflicting opinions in the nuclear debate. To demonstrate how this compromise operates in a legal context, Section 4.4 analyses the nuclear approaches in two EU legislative initiatives. Finally, Section 4.5 concludes the chapter with insights on the potential and likely roles of nuclear in the EU decarbonization efforts from a legal point of view.

---

[22] Eurostat, Nuclear energy statistics, 2021, <https://ec.europa.eu/eurostat/statistics-explained/index.php?title=Nuclear_energy_statistics#Nuclear_heat_and_gross_electricity_production> accessed 1 June 2023.

[23] TVO, 'OL3, <www.tvo.fi/en/index/production/plantunits/ol3.html> accessed 1 June 2023.

[24] *The Guardian*, 'France to build up to 14 new nuclear reactors by 2050, says Macron', 10 February 2022, <www.theguardian.com/world/2022/feb/10/france-to-build-up-to-14-new-nuclear-reactors-by-2050-says-macron>; Reuters, 'Three companies vie to build new Czech nuclear plant', 1 December 2022, <www.reuters.com/business/energy/cez-gets-3-initial-bids-build-new-unit-dukovany-nuclear-plant-2022-11-30/> accessed 1 June 2023.

[25] In fact, a hefty state aid package was approved for a new UK power plant in Hinkley Point in 2014. See Commission Decision (EU) 2015/658 of 8 October 2014 on the aid measure SA.34947, OJ L 109, 28.4.2015, pp 44–116.

[26] Eurostat, Nuclear energy statistics, 2021, <https://ec.europa.eu/eurostat/statistics-explained/index.php?title=Nuclear_energy_statistics#Nuclear_heat_and_gross_electricity_production> accessed 1 June 2023.

[27] However, see Euronews, 'Nuclear energy: Belgium postpones phase-out by 10 years due to Ukraine war' (online, 19 March 2022) <www.euronews.com/2022/03/19/nuclear-energy-belgium-postpones-phase-out-by-10-years-due-to-ukraine-war> accessed 1 June 2023; Deutsche Welle, 'Belgium looks to extend lives of oldest nuclear reactors', (online, 2 March 2023, <www.dw.com/en/belgium-looks-to-extend-lives-of-oldest-nuclear-reactors/a-64606438> accessed 1 June 2023.

[28] Eurostat (n 26).

[29] Mark Cooper, 'Small modular reactors and the future of nuclear power in the United States' (2014) 3 *Energy Research & Social Science* 161.

4.2 Reviewing the facts and arguments of the nuclear debate

4.2.1 Core reasons against nuclear energy: an unsafe and risky energy form that should not be used in the modern world

The anti-nuclear movement has a long history, dating back to the 1940s and 1950s. Its roots lie in the activism that arose at the end of the Second World War after nuclear bombs were dropped on Hiroshima and Nagasaki, alerting the world to the dangers of nuclear weapons and radiation. The movement gained significant momentum in the 1960s and 1970s, with large anti-nuclear protests taking place in the United States, Europe and elsewhere and concerns arising over a wide spectrum of issues throughout the lifecycle of nuclear activities. Leaving aside the production of nuclear weapons, issues include prospecting for nuclear fuels – such as uranium – and their mining, refining and enrichment, as well as the production of energy through nuclear fission, end-of-life measures for decommissioning nuclear facilities and the management of nuclear waste.[30]

The core arguments and facts relied upon by the anti-nuclear movement revolve around issues of risk and safety. The risk and safety concerns about different activities in the lifecycle of nuclear energy production can be further dissected, but the umbrella argument remains the same: that prospecting for, mining and enriching nuclear fuels, producing nuclear energy and managing nuclear waste are activities that carry a whole host of environmental and health safety risks that, from both a financial and an ethical point of view, are not worth taking.[31]

The first and perhaps most oft-cited concern over the safety risks of nuclear activities relates to nuclear energy production itself and, in particular, to the potential for accidents or disasters.[32] The potential human and environmental consequences of radioactive pollution and contamination can be illustrated by the 1986 Chernobyl accident.[33] The explosion of one of the reactors

---

30 For an overview of these arguments, see Benjamin Sovacool, *Contesting the Future of Nuclear Power: A Critical Global Assessment of Atomic Energy* (World Scientific 2011), 1–12.

31 Ioannis N. Kessides, 'Nuclear power: Understanding the economic risks and uncertainties' (2010) 38 *Energy Policy* 3849; Behnam Taebi and Sabine Roeser (eds), *The Ethics of Nuclear Energy: Risk, Justice, and Democracy in the Post-Fukushima Era* (Cambridge University Press 2015); Alexander Gilbert and others, 'Cost overruns and financial risk in the construction of nuclear power reactors: A critical appraisal' (2017) 102 *Energy Policy* 644.

32 It is estimated that around 20 nuclear accidents have occurred since 1950. See Remus Prăvălie and Georgeta Bandoc, 'Nuclear energy: Between global electricity demand, worldwide decarbonisation imperativeness, and planetary environmental implications' (2018) 209 *Journal of Environmental Management* 81.

33 Benjamin K Sovacool, 'Questioning the safety and reliability of nuclear power: An assessment of nuclear incidents and accidents' (2011) 20 *Gaia* 95.

caused the uncontrolled release of radioactive isotopes for more than a week, contaminating vast areas in Eastern and Northern Europe.[34]

Second, the anti-nuclear movement expresses fundamental concerns over nuclear fuel activities. Uranium mining can have significant impacts on the environment, including the contamination of air, water and soil, as well as the disruption of ecosystems and wildlife.[35] The opponents of nuclear energy also point out research that shows that the management of waste generated during uranium mining and processing poses environmental and health risks.[36] The risks presented by uranium mining and processing also stretch beyond its environmental impacts to the health and safety of workers and local communities, as well as their land rights.[37]

Third, there are also arguments against nuclear activities that relate to the management of nuclear waste.[38] After nuclear fuel is spent, it remains highly radioactive for centuries and can cause significant environmental contamination and pose health risks.[39] Finally, nuclear opposition has, from the very beginning of the anti-nuclear movement, highlighted the linkages between nuclear activities and the prospect of nuclear weapons proliferation. While nuclear reactors themselves are not the primary proliferation risk, the concern is that states lacking nuclear weapons and interested in nuclear armament will utilize enrichment and uranium processing facilities for energy use as a front for building nuclear weapons.[40]

When assessing the potential of nuclear activities for climate change mitigation, these facts and arguments against nuclear energy are often placed in the context of the economic (un)profitability of nuclear energy. The construction of a nuclear power plant is a highly capital-intensive project, with high upfront costs required for construction, licensing and regulatory compliance.[41] It is generally accepted that the perceived risk of nuclear activities increases the cost of capital and the investment uncertainties of nuclear energy are generally higher than for other energy sources.[42]

---

34 Prăvălie and Bandoc (n 32).

35 Doug Brugge and Virginia Buchner, 'Health effects of uranium: New research findings' (2011) 26 *Reviews on Environmental Health* 231.

36 Sovacool (n 17).

37 Ibid; Fernando P. Carvalho, 'Mining industry and sustainable development: Time for change' (2017) 6 *Food and Energy Security* 61.

38 Prăvălie and Bandoc (n 32).

39 Ahearne (n 9) 574.

40 Ibid.

41 The high up-front cost of nuclear compared to other energy sources has been highlighted by the IPCC (n 8), 25.

42 IPCC 2007 (n 7), 260.

In the context of these economic factors, the opponents of nuclear energy argue that because of the increasing profitability and fast uptake of renewable energy sources, nuclear energy is not a competitive means of producing energy compared to renewable energy. For example, this argument was frequently cited as a reason for Germany's nuclear phase-out.[43] Similarly, Greenpeace, which is well known for its opposition to nuclear, states that nuclear energy 'isn't just bad for the environment, it's bad for our economy'.[44] It argues that 'nuclear plants are more expensive and take longer to build than renewable energy sources' and that, in the context of the urgency of climate measures, nuclear energy should not be relied upon because it is neither rapidly deployable nor affordable.[45] In stark contrast to these arguments, the supporters of nuclear energy are in favour of not only keeping existing nuclear facilities operational but also of constructing new ones. The arguments and facts relied upon by this movement are explored in the next subsection.

4.2.2 Core reasons in favour of nuclear energy: a necessary energy technology to support the low-carbon transition and supply growing demand

At the end of 2021, there were 439 nuclear power reactors in operation in 32 countries.[46] The total electricity production from these facilities accounts for roughly a tenth of global electricity generation, with countries such as France meeting most of their electricity consumption with nuclear power.[47]

The core arguments and facts relied upon by the supporters of nuclear energy revolve around climate change, energy security and reliability. When assessing the potential of nuclear activities to further climate change mitigation, supporters of nuclear energy emphasize its potential as a low-carbon form of energy production.[48] While the lifecycle of nuclear activities from the mining of uranium to the final disposal of nuclear waste

---

43 See for instance Politico, 'Why Germany won't give up on giving up nuclear', 28 April 2022 <www.politico.eu/article/politics-behind-germany-refusal-reconsider-nuclear-phaseout/> accessed 1 June 2023.

44 Greenpeace, 'Nuclear energy', <www.greenpeace.org/usa/fighting-climate-chaos/issues/nuclear/> accessed 1 June 2023.

45 Ibid.

46 International Energy Agency, 'Nuclear Power and Secure Energy Transitions', 2022, <www.iea.org/reports/nuclear-power-and-secure-energy-transitions> accessed 1 June 2023. The number of plants and the contribution of nuclear energy production to total energy consumption has remained roughly the same for some time now. See Paul L. Joskow and John E. Parsons, 'The economic future of nuclear power' (2009) 138 *Daedalus* 45.

47 International Energy Agency (n 46) 8.

48 Teräväinen, Lehtonen and Martiskainen (n 18) 3436; Prăvălie and Bandoc (n 32); International Energy Agency (n 46) 7.

does produce some greenhouse gas emissions, the production of electricity through nuclear power itself does not. With at least a 40-year operating lifespan, nuclear energy could provide a long-lasting low-carbon means to produce electricity.[49] It has been estimated that nuclear energy has avoided significant amounts of greenhouse gas emissions compared to the hypothetical scenario in which all of the nuclear energy had been produced using fossil fuels.[50] In Germany, for instance, the increased emissions caused by the shutdown of nuclear facilities have been relied on to argue in favour of keeping the power plants operational,[51] with some politicians arguing that nuclear should be used as a bridge technology to assist in the transition toward renewable energy sources.[52] In this context of the contribution nuclear energy could make to climate change mitigation, it has even been asserted that states have 'an ethical obligation to invest in nuclear energy to mitigate climate change'.[53]

It is also often pointed out that the benefits of nuclear energy include the reliability of nuclear electricity production, a quality that contributes to energy security.[54] Once up and running, nuclear power is steady and stable and can run uninterrupted at near full capacity for months and sometimes for years and can therefore complement an energy system based on more intermittent renewable energy sources. As a reliable source of fairly continuous energy supply, nuclear energy is a much less carbon-intensive solution to mitigate the intermittency of renewable energy sources than fossil-fuel-based energy generation.[55] In terms of energy security, advocates of nuclear energy highlight that the role of nuclear power is important in diversifying energy supplies, decreasing dependence on fossil fuels and keeping energy prices reasonable.[56] These facts and arguments have been

49 The period of 40 years is a rough estimate of the planned lifespan of a nuclear reactor. However, many nuclear facilities have operated safely for much longer. For instance, Finland's first nuclear reactor started operations in 1977 and is still fully operational with a licence in force until the end of the 2020s. The company operating the reactor has applied to continue operations until 2050.

50 International Energy Agency (n 46) 15.

51 World Nuclear News, 'Germans asked to keep reactors in operation, 15 October 2021', <www.world-nuclear-news.org/Articles/Germans-asked-to-keep-reactors-in-operation> accessed 1 June 2023.

52 Deutsche Welle, 'Germany's Scholz seeks to end coalition's nuclear squabble, 16 October 2022', <www.dw.com/en/germanys-scholz-seeks-to-end-coalitions-nuclear-squabble/a-63456925> accessed 1 June 2023.

53 Simon Friederich and Maarten Boudry, 'Ethics of nuclear energy in times of climate change: Escaping the collective action problem' (2022) 35(2) *Philosophy & Technology* 30.

54 Teräväinen, Lehtonen and Martiskainen (n 18) 3436.

55 Ibid, 10.

56 International Energy Agency (n 46) 8 and 13.

more frequently invoked in light of Russia's invasion of Ukraine and the related energy crisis in the EU.[57]

Finally, nuclear energy's supporters point out that nuclear energy production is highly energy efficient, in that a relatively small amount of nuclear fuel can produce a large amount of energy. This characteristic has a number of benefits. Among other things, a nuclear energy installation produces much more electricity per square metre than any other energy production method.[58]

Governments that support nuclear power often employ what is called a 'necessitation strategy'.[59] At the core of this strategy is the argument that there are no other alternatives to nuclear power given the urgency of climate change, the intermittency of renewables, the increasing demand for electricity and the limited range of existing energy technologies.[60] Because of the lack of other real alternatives to nuclear energy in light of the urgency of the need for climate measures, nuclear energy production merits taking the risks associated with the mining of uranium, the safety concerns associated with nuclear energy production and the challenges related to the management of nuclear waste.[61]

4.3 The legal side of the nuclear debate: 60+ years of nuclear law in the EU

4.3.1 Constitutional foundations: Euratom and the EU member states' right to choose their energy mix

The Euratom Treaty seeks to 'contribute to raising the standard of living in the Member States' by 'creating the conditions necessary for the speedy establishment and growth of nuclear industries'.[62] While the other founding

---

57 Ibid.

58 It has therefore also been argued to be potentially beneficial, indirectly, for improving biodiversity. See Barry W. Brook and Corey J. A. Bradshaw, 'Key role for nuclear energy in global biodiversity conservation' (2015) 29 *Conservation Biology* 702. On the land-use intensity of different energy production methods, see United Nations Economic Commission for Europe, 'Lifecycle Assessment of Electricity Generation Options', 2021, <https://unece.org/sed/documents/2021/10/reports/life-cycle-assessment-electricity-generation-options> accessed 1 June 2023, 75.

59 Heike Hermanns, 'South Korean nuclear energy policies and the public agenda in the 21st century' (2015) 7 *Asian Politics & Policy* 265; Teräväinen, Lehtonen and Martiskainen (n 18) 3437.

60 Sailor and others (n 6); Teräväinen, Lehtonen and Martiskainen (n 18) 3437.

61 Teräväinen, Lehtonen and Martiskainen (n 18); Friederich and Boudry (n 53).

62 Euratom Treaty Art 1. For the purposes of this chapter, nuclear power is the most important component of 'nuclear industries', but it can also include nuclear research and development and using nuclear technologies in medicine, for instance.

treaties established a wide range of rights and obligations that could be successfully invoked in courts, the spirit of the Euratom Treaty is more promotional and facilitative, creating a framework for advocating the safe use of nuclear technologies in Europe.[63]

The Euratom Treaty established the Atomic Energy Community as an autonomous legal entity[64] and stipulates numerous tasks for the Community.[65] It is expected to contribute to the promotion of research and the dissemination of information. Its responsibilities include establishing safety standards to protect the health of workers and the general public, facilitating investment and ensuring the establishment of the basic installations necessary for the development of nuclear energy. It has tasks relating to the regular and equitable supply of nuclear fuels, as well as fostering progress in the peaceful uses of nuclear energy. While the other founding treaties have been amended and revised numerous times over the years, the substance of the Euratom Treaty has never been revisited or changed.[66]

When first established, there were high expectations for Euratom among the founding member states. They hoped that it would set in motion a rapid growth of the nuclear industry and was seen by some as the 'main instrument for integration'[67] between EU member states. Over the years, it became clear that Euratom would not be able to achieve this kind of nuclear revolution. Some argued that the Euratom Treaty was neglected compared to other legal bases that were used to pursue the internal energy market[68] and others even called for the repeal of the Euratom Treaty.[69] With the latest revision of the EU's founding treaties through the Treaty of Lisbon in 2009, several member states supported the idea of bringing the Euratom Treaty 'up to date'.[70] Even though the Euratom Treaty has been unable to facilitate significant growth in nuclear activities, it nevertheless became the

[63] Thomas F. Cusack, 'A tale of two treaties: An assessment of the Euratom Treaty in relation to the EC Treaty' (2003) 40 *Common Market Law Review* 117, 125. The other two founding treaties are the Treaty establishing the European Coal and Steel Community (no longer in force) and the Treaty establishing the European Economic Community.

[64] Södersten (n 19) 1–2.

[65] Euratom Treaty Art 2.

[66] Rasa Engstedt, *Handbook on European Nuclear Law: Competences of the Euratom Community Under the Euratom Treaty* (Kluwer Law International 2020) 11; Cenevska (n 19) 3.

[67] Södersten (n 19) 1; Jean Monnet, *Mémoires* (Fayard 1976).

[68] Leigh Hancher, 'Energy and the Environment: Striking a Balance?' (1989) 26 *Common Market Law Review* 475; Cusack (n 63).

[69] Södersten (n 19) 1–2.

[70] See, for example, 54th declaration annexed to the Treaty on the Functioning of the European Union. The signatories to the declaration include both opponents and supporters of nuclear energy, so it remains unclear what changes these member states advocate in practice.

constitutional foundation for a body of secondary law governing the safe operation of nuclear activities.

It was the Chernobyl accident that provided the justification for the adoption of a substantive body of secondary nuclear law in the EU.[71] That nuclear disaster demonstrated that the transboundary nature of radioactive contamination required transnational governance and cross-border cooperation. As a result, several secondary law instruments were adopted to address the transboundary safety issues stemming from nuclear activities.[72]

Euratom and the related secondary EU law in force impose onerous safety requirements along the entire lifecycle of nuclear activities. The secondary EU legal framework for nuclear energy is extensive and, consequently, its examination is beyond the scope of this chapter.[73] Rather than obligating or promoting the uptake of nuclear energy, the secondary legal framework focuses on safety, radiation protection, the management of nuclear waste, decommissioning of nuclear facilities and safeguards to avoid misuse.[74]

The detailed legal framework for ensuring the safety of nuclear activities under Euratom and the secondary legal framework should be interpreted in the context of the Treaty of the Functioning of the European Union (TFEU),[75] which takes no explicit stand on whether its member states should rely on nuclear energy in their energy mix. According to Article 194 of the TFEU, the EU's energy policy aims to ensure the functioning of the energy market and the security of energy supply, to promote energy efficiency and energy savings and the development of new and renewable forms of energy, and to promote the interconnection of energy networks.[76] However, the EU's pursuit of these objectives may not 'affect a Member State's right to determine the conditions for exploiting its energy resources, its choice between different energy sources and the general structure of its energy supply'.[77] This constitutional delimitation of the EU's powers seems inconsistent with Euratom's function as the facilitator and promoter of nuclear activities in the EU. The relationship between the two constitutional

[71] European Commission, 'Outline communication from the Commission to the Council on the consequences of the Chernobyl accident' COM(86) 327 final; Södersten (n 19) 2; Hancher (n 68) 486.

[72] Rasa Engstedt (n 66) 12; Södersten (n 19) 2.

[73] In addition to the Euratom Treaty itself, the legislative framework consists of nearly 20 secondary legislative instruments.

[74] For an overview, see European Parliament, 'Nuclear energy', 2022, <www.europarl.europa.eu/factsheets/en/sheet/62/nuclear-energy> accessed 1 June 2023.

[75] Consolidated version of the Treaty on the Functioning of the European Union, OJ [2012] C326/47.

[76] TFEU, Art 194(1).

[77] TFEU, Art 194(2). See also Huhta (n 15).

treaties is a delicate balance, which demonstrates the sensitivity of the nuclear debate and its legal dimensions in the EU.

4.3.2 The nuclear debate playing out in case-law: opposing nuclear energy is not an argument that can be won in the EU courts

The controversial nature of nuclear activities has been acknowledged and explored in EU law and policy and in legal scholarship from very early on. The nexus and potential trade-offs between nuclear activities and environmental protection were already recognized in the European Commission's Fourth Environmental Action Programme, in 1986, following the Chernobyl accident.[78] In the Action Programme, the Commission focused on the safe use of nuclear power and emphasized the importance of EU-level action to ensure common rules to address the potential transboundary harm caused by nuclear activities.[79] It held that the management of nuclear waste 'remains a central issue of environmental concern'[80] and that the EU must adopt measures to manage those issues.

In the Action Programme, the Commission emphasized its commitment to the safe use of nuclear power within the framework of the Euratom Treaty. On that basis, it announced its intention to issue a series of legislative proposals to address the safety issues associated with nuclear energy as illustrated by the Chernobyl disaster and to keep a close surveillance on the questions of environmental protection in the context of nuclear power.[81]

The Chernobyl disaster and the Environmental Action Programme that followed were the nudge and the justification for further legislation on nuclear activities in the EU. Against this developmental background, it is clear that the risk and safety concerns embedded in the nuclear debate are already innately part of the EU legislative framework for nuclear activities.

The debate is also visible in legal scholarship analysing this legislative framework, with some scholars criticizing the EU for its legislative action in the nuclear sphere and others taking the opposite view and criticizing the EU for its inaction under the Euratom Treaty. For instance, the Commission was criticized in the 1970s for interpreting the facilitative and promotional character of the Euratom Treaty to justify an allocation of the EU's energy

---

[78] European Commission, 'EEC Fourth Environmental Action Programme (1987–1992)' COM(86) 485 final. See also European Commission, 'Outline communication from the Commission to the Council on the consequences of the Chernobyl accident' COM(86) 327 final.

[79] European Commission, 'EEC Fourth Environmental Action Programme (1987–1992)' COM(86) 485 final.

[80] Ibid.

[81] Ibid.

research budget to nuclear research 'at the expense of renewable or other alternatives and energy conservation', which were argued to be much 'more favourable to the environment'.[82]

The EU's environmental law and policy has also been criticized for major omissions relating to nuclear activities, including the EU's 'relative reluctance to regulate or even arbitrate in trans-frontier pollution disputes' and its 'almost total failure to introduce measures on nuclear power'. The 'relative neglect of nuclear energy' has been considered remarkable in the context of Euratom, which explicitly gives the Community the competence to control 'health- and safety-related aspects of nuclear production'.[83]

The sensitivity of nuclear activities has brought about friction between member states to this day, which is reflected in nuclear-themed legal disputes. Nuclear energy has been the topic in two court cases concerning state aid for nuclear energy. Because the Euratom Treaty does not provide any mechanisms by which to block nuclear projects in other member states, EU state aid law enshrined in the TFEU offers an interesting instrument through which to challenge the nuclear choices of other member states.[84] Austria, which is an opponent of nuclear power, plays a key role in both cases, which comprise the General Court case of *Austria v. Commission*, its appeal from the General Court to the European Court of Justice (decided in 2020), and the most recent case of *Austria v. Commission* (decided in late 2022).[85] The first of these, which arose from the Commission's approval of the UK's state aid scheme in respect of a new nuclear power plant at Hinkley Point, was the first time a member state challenged a Commission state aid decision in favour of another member state.[86] The unprecedented nature of this line of case-law alone shows the political sensitivity of nuclear power.

Austria challenged the Commission's favourable decision inter alia on the ground that if public funds were allocated to nuclear power, they would not be available for the development of renewable energy sources.[87] In its review of the legality of the Commission's decision, the General Court found that 'the Commission did not err in finding that

[82] N. J. D. Lucas, *Energy and the European Communities* (Europa 1977); Hancher (n 68).

[83] Hancher (n 68) 491.

[84] Leigh Hancher, 'Euratom, state aid and environmental protection: Hinkley Point' (2021) 58 *Common Market Law Review* 1491, 1491–1492.

[85] Cases T-356/15, *Austria v Commission* [2018] ECLI:EU:T:2018:439; C-594/18 P, *Austria v Commission*, [2020] ECLI:EU:C:2020:742; and T-101/18, *Austria v Commission* [2022] ECLI:EU:T:2022:728. The Commission's decision was also appealed by other parties, but Austria's claim was the only one that was ultimately considered admissible. See, for instance, Case T-382/15, *Greenpeace and others v Commission* [2016] ECLI:EU:T:2016:589.

[86] Hancher (n 84) 1493.

[87] Case T-356/15, *Austria v Commission* (n 81) paras 377 and 383.

the United Kingdom was entitled to decide upon the promotion [of] nuclear energy and, more specifically, incentives for the creation of new nuclear energy generating capacity'.[88] It based this argument on Article 194(2) of the TFEU as the constitutional foundation of EU energy law by highlighting that 'it is apparent from the second subparagraph of Article 194(2) TFEU that each Member State has the right to choose between the different energy sources those they prefer'.[89] Furthermore, the General Court went on to confirm that '[i]n the light of the United Kingdom's right to determine its own energy mix and to maintain nuclear energy as a source in that mix … the decision to maintain nuclear energy in the supply structure cannot be considered to be manifestly disproportionate'[90] given the objective of security of supply enshrined in Article 194(1) of the TFEU and confirmed by the CJEU.[91] The UK was thus entitled to choose its emphasis as between the various energy sources available.[92] The UK's choice could not be called into question even though the state funds used to support Hinkley Point were, as a result, unavailable for other types of energy projects.[93] Furthermore, the General Court clarified the relationship between the TFEU and the Euratom Treaty by highlighting that 'the provisions of the Euratom Treaty constitute special rules in relation to the provisions of the FEU Treaty and therefore derogate from the latter provisions in the event of any conflict'.[94]

Displeased with the General Court's findings, Austria appealed the decision to the European Court of Justice, which not only confirmed the General Court's ruling but also strengthened the argument that member states were indeed free to choose their energy mix under Article 194(2) of the TFEU.[95] It held that because of this specific competence limitation, the choice of nuclear energy is a matter for member states to decide.[96]

A close reading of Advocate General Hogan's opinion in C-594/18 P *Austria v. Commission* exposes the politically loaded nature of nuclear energy and how the political nature of the nuclear debate can be echoed in legal disputes. Hogan described the core of the case as 'the legal side of a dispute

88 Ibid para 97.
89 Ibid.
90 Ibid para 507.
91 Cases C-121/15, *ANODE* [2016] EU:C:2016:637, para 48; and C-411/17, *Inter-Environnement Wallonie and Bond Beter Leefmilieu Vlaanderen* [2019] EU:C:2019:622, para 156.
92 Case T-356/15, *Austria v Commission* (n 81) para 105.
93 Ibid para 526.
94 Ibid para 73.
95 Case C-594/18 P *Austria v Commission* (n 81).
96 Ibid paras 48–9.

between Member States that are in favour of nuclear power and those that are not' and pointed out that '[b]oth sides claim that they pursue their course with a view to protecting the environment'.[97] In his opinion, Austria's line of thinking was based on the assumption that because it 'is resolutely opposed to the construction of nuclear power stations, the granting of aid for such projects by other Member States who support nuclear power is either expressly or implicitly precluded by the various Treaties governing the EU (including the Euratom Treaty)'.[98]

Like the General Court before him and the European Court of Justice after him, Advocate General Hogan rejected Austria's claims. While he acknowledged that there is no political agreement over the merits or perils of nuclear power, he took the view that 'this Court has neither the competence nor the democratic legitimacy to rule on such issues'.[99] He maintained that because the development of nuclear power is a clearly defined objective of EU law provided for in the Euratom Treaty, that objective cannot be subordinated to other objectives of EU law, such as the protection of the environment, even if they are conflicting.[100] He emphasized that the case should be interpreted in the context of Article 194(2) of the TFEU, which extends 'to the right of each Member State to develop nuclear power as part of its energy supply sources'.[101] As noted, the European Court of Justice decided the case in line with the Advocate General's suggestion.

The political debate surrounding state aid for nuclear energy was also reflected in the second case, in late 2022, in a ruling by the General Court in *Austria v. Commission*.[102] Again, Austria brought an action for annulment of the Commission's decision approving state aid for nuclear energy, this time in Hungary. Austria argued inter alia that the approved aid would lead to disproportionate distortions of competition and unequal treatment which would result in the exclusion of producers of renewable energy from the liberalized internal electricity market. Again, the General Court rejected Austria's position, observing that, in line with Article 194(2) of the TFEU, member states are free to determine the composition of their energy mix and the Commission cannot require state financing to be allocated to alternative energy sources.

[97] Opinion of Advocate General Hogan in case C-594/18 P *Austria v Commission* [2020] ECLI:EU:C:2020:352, para 1.
[98] Ibid para 3.
[99] Ibid para 151.
[100] Ibid.
[101] Ibid.
[102] Case T-101/18 *Austria v Commission* (n 81).

4.4 Legislative developments in the EU nuclear debate

The previous section showed that, because of competence limitations in the EU energy policy area, the EU cannot oblige EU member states to increase the use of nuclear energy in their national energy mixes. In light of the member states' conflicting policy approaches to nuclear energy, it seems highly unlikely that this constitutional compromise could be radically changed in the future, because any change would require approval by all member states. It seems therefore likely that the approaches to nuclear energy production within EU member states will remain diverse. However, there are two highly publicized legislative initiatives where the EU has taken a facilitative stand on nuclear energy.

The first was the EU's Taxonomy Regulation of 2020, which establishes the criteria for sustainable finance in the EU.[103] It is not a financial instrument in itself, but a classification system to inform companies, investors and policy makers of economic activities that can be considered environmentally sustainable.[104] When it was first proposed by the European Commission in 2018, the approach and wording of the Regulation were heavily debated and criticized for including nuclear energy in the list of energy sources that are considered sustainable.

In particular, non-governmental organizations publicly pushed back on the approach adopted in the Taxonomy Regulation. The World Wide Fund for Nature (WWF) argued that the inclusion of nuclear energy in the list of environmentally sustainable activities meant that 'billions of euros are at risk of being diverted from wind, solar, and other green technologies to … nuclear energy, worsening the climate crisis, harming our environment, and risking increased energy bills'.[105] Greenpeace decided to challenge the inclusion of nuclear energy before EU courts, arguing that the Taxonomy Regulation 'wrongly designates nuclear energy as a climate mitigation activity' and leads 'to the curtailment of renewables'.[106] Defending the

---

[103] Regulation (EU) 2020/852 of the European Parliament and of the Council on the establishment of a framework to facilitate sustainable investment, and amending Regulation (EU) 2019/2088 [2020] OJ L198/13.

[104] Christos Gortsos, 'The Taxonomy Regulation: More Important Than Just an Element of the Capital Markets Union', in Danny Busch, Guido Ferrarini and Seraina Grünewald (eds), *Sustainable Finance in Europe: EBI Studies in Banking and Capital Markets Law* (Palgrave Macmillan 2021).

[105] WWF, '#StopFakeGreen', <www.wwf.eu/campaigns/stopfakegreen/> accessed 1 June 2023.

[106] Greenpeace, 'Greenpeace to take European Commission to court over controversial gas and nuclear greenwashing' (press release, Greenpeace European Unit, 9 February 2023) <www.greenpeace.org/eu-unit/issues/climate-energy/46570/greenpeace-to-take-european-commission-to-court-over-controversial-gas-and-nuclear-greenwashing> accessed 1 June 2023.

Taxonomy Regulation, the Commission argued that its approach follows scientific advice and that, on this basis, nuclear energy 'can play a role in the transition towards climate neutrality in line with the European Green Deal',[107] a green growth strategy published by the Commission in 2019.[108] The Taxonomy Regulation entered into force in 2020 and, despite the heavy criticism, continued to include nuclear energy in the list of sustainable activities at the time of writing.

The second initiative where the EU took a stand on nuclear energy was the legislative proposal to reform the EU electricity market. Published by the Commission on 14 March 2023, the proposal aims to address the shortcomings of the existing electricity market design that were exposed by the radical energy supply shock resulting from the Russian invasion of Ukraine. With particular relevance to nuclear energy, the proposal suggests new rules regarding support schemes for new investments in electricity generation. It allows direct price-support schemes for new investment in not just renewable energy but also nuclear energy.[109] The proposal is likely to face similar criticism for its nuclear approach as the Taxonomy Regulation. However, the final wording of the proposed legal instrument will be a matter of negotiation between the EU institutions and member states.

Finally, the EU legal framework for nuclear energy must also be placed in the context of the ongoing Russia-Ukraine war and the related energy crisis. Some member states that had decided to phase out nuclear energy cancelled or postponed their plans. For instance, Belgium announced delays in their planned phase-out of nuclear energy, and Germany, for a while, had similar plans.[110] In countries such as Finland, the construction of additional nuclear power has gained the support of the majority of members of parliament and it is possible that the nuclear reactor that was finalized in spring 2023 will not

[107] European Commission, 'Questions and Answers on the EU Taxonomy Complementary Climate Delegated Act covering certain nuclear and gas activities' (press release supplement, 2 February 2022) <https://ec.europa.eu/commission/presscorner/detail/en/QANDA_22_712> accessed 1 June 2023.

[108] European Commission, 'The European Green Deal' COM(2019) 640 final.

[109] Art 19b of the Proposal for a Regulation of the European Parliament and of the Council amending Regulations (EU) 2019/943 and (EU) 2019/942 as well as Directives (EU) 2018/2001 and (EU) 2019/944 to improve the Union's electricity market design, COM(2023) 148 final.

[110] Euronews, 'Nuclear energy: Belgium postpones phase-out by 10 years due to Ukraine war' (online, 19 March 2022) <www.euronews.com/2022/03/19/nuclear-energy-belgium-postpones-phase-out-by-10-years-due-to-ukraine-war> accessed 1 June 2023; Kate Connolly, 'Germany to delay phase-out of nuclear plants to shore up energy security' *The Guardian* (Berlin, 5 September 2022) <www.theguardian.com/world/2022/sep/05/germany-to-delay-phase-out-of-nuclear-plants-to-shore-up-energy-security> accessed 1 June 2023.

be the last one.[111] These changes suggest that the energy supply shock caused by the Russian invasion and the resulting need to urgently and dramatically decrease dependence on Russian energy has influenced public opinion on nuclear energy, which can affect the role of nuclear energy in the future.

4.5 Concluding thoughts

This chapter has explored the interplay between the arguments in favour of nuclear energy and those against it in a legal context. Instead of taking a stand on either side of the debate, this chapter has examined the opposing views on nuclear power and shown, using EU law as a case study, how this politically sensitive and controversial energy technology is addressed in law and in legal governance of the low-carbon energy transition.

A review of the arguments for and against nuclear energy seems to suggest that the debate eventually boils down to how the various pros (effective climate change mitigation, energy efficiency, energy security, and so on) and cons (risk of environmental contamination, availability of uranium, nuclear weapons proliferation, etc.) of nuclear energy are evaluated by those making political decisions on nuclear power. This is to say that the debate revolves around risk perception and the evaluation of not just the arguments and counterarguments but also the valuation of objective facts.[112]

In a legal context, the EU provides a powerful example of this dynamic. The governance of nuclear energy in the EU can be traced back to the very beginning of European integration in the 1950s, with the political and environmental sensitivities of nuclear energy emerging in the 1970s. While the secondary legal framework governing the safety of nuclear power is robust in the EU, the role of nuclear energy as a climate change mitigation technology in EU law triggers heated debate. Since 2010, this has been demonstrated by the Taxonomy Regulation, the proposed electricity market reform, and the cases brought before the EU courts to challenge member states' choices in favour of nuclear energy.

The constitutional legal framework for nuclear energy involves two key features that reflect the significant diversity within member states on nuclear

---

[111] Elli Piirainen, 'Lisäydinvoima saa vahvan tuen eduskunnasta – Ylen kyselyn mukaan yli puolet kansanedustajista kannattaa' (Translated title: 'Additional nuclear power gains strong parliamentary support – based on a survey a majority of the Members of Parliament approve') *Yle* (online, 31 November 2022) <https://yle.fi/a/3-12669270> accessed 1 June 2023.

[112] Annukka Vainio, Riikka Paloniemi and Vilja Varho, 'Weighing the risks of nuclear energy and climate change: Trust in different information sources, perceived risks, and willingness to pay for alternatives to nuclear power' (2017) 37 *Risk Analysis* 557; Stephen C. Whitfield and others (n 17).

energy and establish a compromise on which both supporters and opponents of nuclear energy have been able to agree. First, EU member states are free to choose their energy mix and decide whether to include nuclear energy. Second and relatedly, arguments opposing the development of nuclear energy in other member states have been and are likely to continue to be unsuccessful in EU courts, reflecting states' sovereignty over their natural resources.[113] Through these two features, the EU legal framework for nuclear energy cannot be used to oblige member states to increase the uptake of nuclear energy. Instead, as demonstrated by the legislative initiatives and the case-law discussed in this chapter,[114] the EU legal framework can be used to defend the right of member states to choose nuclear energy and to facilitate opportunities to use nuclear energy as a climate change mitigation technology. Overall, the analysis suggests that the EU legal framework does not place constraints on the expansion of nuclear energy as a climate mitigation technology. Instead, it seems that the obstacles for pursuing nuclear energy in Europe are political rather than legal in nature.

Acknowledgement

This work was supported by the Academy of Finland (340998).

[113] Huhta (n 15).

[114] Cases T-356/15 *Austria v Commission* [2018] ECLI:EU:T:2018:439; C-594/18 P *Austria v Commission*, [2020] ECLI:EU:C:2020:742; T-101/18 *Austria v Commission* [2022] ECLI:EU:T:2022:728.

5

Critical Raw Materials in the Anthropocene: Regulatory Perspectives on their Promise and Pitfalls

Seita Vesa

5.1 Introduction

Effecting the low-carbon transition which is urgently needed to limit the most drastic impacts of climate change demands a swift exit from our heavily fossil-fuel-dependent economy. This in turn relies on the comprehensive use of clean energy technologies, which enable a shift away from high-carbon mineral fuels, such as coal and petroleum, as well as from natural gas. With clean energy technologies, such as electric vehicles, wind turbines and solar panels, being developed and deployed apace, the demand for the required raw materials – critical raw materials (CRMs) among them – is surging in tandem.[1] In fact, the manufacture of many of the technologies is heavily reliant on specific CRMs.

[1] Davide Castelvecchi, 'Electric cars *and batteries: how will the world produce enough?' (2021) 596* Nature 336; International Energy Agency, *World Energy Outlook Special Report: The Role of Critical Minerals in Clean Energy Transitions* (IEA, 2021), available at: www.iea.org/reports/the-role-of-critical-minerals-in-clean-energy-transitions/mineral-requirements-for-clean-energy-transitions (accessed 22 April 2022); Economic and Social Council, Economic Commission for Europe, Committee on Sustainable Energy, 'Sustainable Management of Critical Raw Materials Required for the Low-Carbon Energy Transition – Draft for discussion', prepared by the United Nations Resource Management System Sub-group of the Expert Group on Resource Management, ECE/ENERGY/2022/6, 1.

Overall, CRMs, such as lithium, cobalt and rare-earth metals, are indispensable for strategic sectors, from renewable energy to the digital industry and the space and defence sectors. For instance, the International Energy Agency (IEA) forecasts that within the next 20 years, the clean energy sector's need for specific CRMs could surge from 20-fold to over 40-fold. Moreover, based on the projected rate of the clean energy transition, the overall demand for minerals, encompassing both the energy sector and others, is anticipated to increase approximately four to six times from 2020 to 2030.[2]

CRMs – a collective term utilized in criticality assessments[3] – encompasses various groups of materials such as rare earths, precious metals, iron and ferro-alloy metals, and non-ferrous metals.[4] CRMs are extracted from mineral[5] ores,[6] and most are metals.[7] The life-cycle of a CRM usually starts in primary sources, that is, mining and mineral processing,[8] but CRMs can also be harvested from seawater.[9] They can also be recovered from secondary sources, such as recycled materials (for example, waste streams and electronic waste).[10] The supply of CRMs faces heightened geopolitical, environmental

---

2 IEA, *The Role of Critical World Energy Outlook Special Report Minerals in Clean Energy Transitions* (IEA, 2021), available at: https://iea.blob.core.windows.net/assets/ffd2a83b-8c30-4e9d-980a-52b6d9a86fdc/TheRoleofCriticalMineralsinCleanEnergyTransitions.pdf (accessed 15 December 2023), 5.

3 The term criticality assessment refers to the methodology used, for example, by the EU, to assess the economic importance and supply risk in the context of critical raw materials. See European Commission, Critical Raw Materials, available at: https://single-market-economy.ec.europa.eu/sectors/raw-materials/areas-specific-interest/critical-raw-materials_en (accessed 15 December 2023).

4 Milan Grohol and Constanze Veeh, *Study on the Critical Raw Materials for the EU 2023. Final Report* (European Commission 2023), 2.

5 A mineral is a solid, naturally occurring inorganic substance which can contain a metal or metal compound of economic importance. European Commission, Commission staff working document, impact assessment report accompanying the document 'Proposal for a Regulation of the European Parliament' and of the Council establishing a framework for ensuring a secure and sustainable supply of critical raw materials and amending Regulations (EU) 168/2013, (EU) 2018/858, 2018/1724 and (EU) 2019/1020, SWD (2023) 161 final, 3–4.

6 An ore is a naturally occurring solid material from which a metal or valuable mineral can be extracted profitably. European Commission, n 5, 3–4.

7 Metals are often extracted from mineral ores through a process called metallurgy or metallurgical processing. Metals are elementary substances, such as gold, silver, copper, iron, aluminium, zinc and lithium. European Commission, n 5, 3–4.

8 Richard Herrington, 'Mining our green future' (2021) *Nature Reviews Materials* 6, 457.

9 See, for example, V. Vallés et al, 'Polymeric and inorganic sorbents as a green option to recover critical raw materials at trace levels from sea saltwork bitterns' (2023) *Green Chemistry* 25(2), 700–19.

10 See, for example, Gian Andrea Blengini et al, 'Recovery of critical and other raw materials from mining waste and landfills' (2019) *Science for Policy report by the Joint Research Centre (JRC)*, 6.

and social challenges and risks. CRMs are described as 'critical' for two reasons: they have considerable economic importance; and they entail a high supply risk, as their unique properties mean that (viable) substitutes are lacking.[11] The Organisation for Economic Co-operation and Development (OECD) defines CRMs as those raw materials which have been 'found to be used intensely in the green transition technologies'.[12] There are over 40 materials that have been classified as critical by Canada, the European Union (EU), Japan and the United States, for example – although the list of CRMs differs from country to country.[13]

CRM supply chains are global, as well as complex due to their extensive networks, and fragile because they are easily disrupted. Consequently, they are laden with risks arising from geopolitical tensions, political instabilities and trade restrictions in the regions where the materials are obtained.[14] For example, over the past decade, export restrictions on CRMs have surged fivefold worldwide.[15] The CRM life-cycle, spanning as it does acquisition, refining, finished products, waste management and potential recycling, generates supply chains with a complex mix of stakeholders and technologies.[16] Although sourced from various locations globally, CRMs and their procurement and the trade in CRMs are concentrated in certain countries and regions. Geographical dependencies can lead to over-reliance, and any disruptions (for example, political disturbances) may undermine the entire supply chain.[17] In another concern, increasing demand for CRMs

[11] European Commission, Proposal for a Regulation of the European Parliament and of the Council establishing a framework for ensuring a secure and sustainable supply of critical raw materials and amending Regulations (EU) 168/2013, (EU) 2018/858, 2018/1724 and (EU) 2019/1020, COM(2023) 160 final, 1.

[12] Przemyslaw Kowalski and Clarisse Legendre, *Raw Materials Critical for the Green Transition: Production, International Trade and Export Restrictions*. OECD Trade Policy Paper 269 (OECD Publishing), 11.

[13] Yina Su and Dewen Hu, 'Global Dynamics and Reflections on Critical Minerals' (2022) *E3S Web of Conferences* 352, 1–3.

[14] Grohol and Veeh, n 4, 7; European Parliament, Securing Europe's Supply of Critical Raw Materials: The Material Nature of the EU's Strategic Goals, briefing, available at: www.europarl.europa.eu/RegData/etudes/BRIE/2023/739394/EPRS_BRI(2023)739394_EN.pdf (accessed 10 May 2023); Herrington, n 8, 457.

[15] Kowalski and Legendre, n 12, 11.

[16] See, for example, Melanie Müller et al, From Competition to a Sustainable Raw Materials Diplomacy: Pointers for European Policymakers. SWP Research Paper 1, available at: www.swp-berlin.org/publications/products/research_papers/2023RP01_RawMaterialsDiplomacy.pdf (accessed 12 May 2023), 13.

[17] European Commission, Communication from the Commission to the European Parliament, the Council, the European Economic and Social Committee and the Committee of the Regions, A Secure and Sustainable Supply of Critical Raw Materials in Support of the Twin Transition, COM(2023) 165 final, 1.

exacerbates human-rights issues, such as the use of child labour, which prevail in many regions from which the materials are sourced.[18]

What is more, depending on the process used, the extraction and processing of CRMs in countries with weak environmental laws may impact water, fauna or flora and aggravate climate change.[19] Negative environmental impacts can be caused by geological factors (for example, heavy metals in ores), mining techniques that disrupt or destroy ecosystems (such as open-pit mining[20]), or a site's natural conditions (for example, water scarcity).[21]

Underlying the idea of replacing fossil-fuel-powered technologies with electric counterparts is the ambition of attaining a generally closed-loop and sustainable circular economy, one in which raw-material demand is fully satisfied by reuse and recycling.[22] Although there is a consensus on the ideal of an economy where scarce raw materials, such as CRMs, circulate endlessly, functioning as essential 'industrial vitamins',[23] such a goal is still far from being realized.[24] In other words, most CRMs will continue to be sourced from mining and, moreover, more mines will be needed to meet the growing demand. CRM supplies in the future will rely on a combination of primary (mining) and secondary (recycling) supply.[25]

The discussion thus far shows that the transition to net zero is distinctly raw materials intensive.[26] Against this background, this chapter raises the question: To what extent should the low-carbon transition allow – indeed, encourage – business as usual practices that maintain consumption of, and

---

18 Ibid; Kristi Disney Bruckner, 'Improving environmental and social practices in the mining sector is essential in the transition to renewable energy' (2022) *Journal of Energy & Natural Resources Law*, 2; BBC News, 'Top tech firms sued over DR Congo cobalt mining deaths', 16 December 2019, available at: www.bbc.com/news/world-africa-50812616 (accessed 11 August 2023).

19 European Commission, n 11, 14.

20 Liang Tang and Tim Werner, 'Global mining footprint mapped from high-resolution satellite imagery' (2023) *Communications Earth & Environment* 4, 2.

21 John Bachér et al, *Environmental aspects related to the use of critical raw materials in priority sectors and value chains* (European Environmental Agency, 2020), 2.

22 European Commission, n 11, 14.

23 Ajay Patil, Rudolf Struis and Christian Ludwig, 'Opportunities in critical rare earth metal recycling value chains for economic growth with sustainable technological innovations' (2022) *Circular Economy and Sustainability*, 1.

24 Herrington, n 8, 456.

25 Christian Hagelüken and Daniel Goldmann, 'Recycling and circular economy: towards a closed loop for metals in emerging clean technologies' (2022) *Mineral Economics* 35, 540.

26 OECD, Global Material Resources Outlook to 2060: Economic Drivers and Environmental Consequences, highlights, available at: www.oecd.org/environment/waste/highlights-global-material-resources-outlook-to-2060.pdf (accessed 4 April 2023), 3; Yanan Liang et al, 'Material requirements for low-carbon energy technologies: a quantitative review' (2022) *Renewable and Sustainable Energy Reviews* 161, 1.

dependence on, CRMs and intensify exploitation of multiple-use metals and minerals resources?[27]

The global race for these resources will be fierce in the next ten to 20 years as governments take action to reach net-zero emissions and reassess what raw materials they need to meet the pressing climate targets. The laws and regulatory frameworks for CRMs are still in their early stages, with regions such as the EU and the United States only starting to draft laws and regulations[28] in response to the growing need and interest. While the chapter touches on multiple jurisdictions, it centres on the EU's recent legal advancements in CRMs, given its pioneering role in this nascent field. For example, in 2024, the EU adopted the Critical Raw Materials Act, marking one of the first pieces of comprehensive CRM-specific regulation.[29] At the same time, however, CRMs are a prime example of the 'speed at scale' conundrum, in which the need for rapid production and deployment is interlinked with the paramount challenge of procuring and managing scarce resources sustainably. The paradigm of Anthropocene law[30] – a legal framework aligned with robust climate change mitigation – should be fundamentally different from traditional law. In addressing the multifaceted challenges of climate change, comprehensive solutions inevitably give rise to additional issues or concerns that jurisprudence has traditionally sought to avoid or ignore. The onus then lies on legal scholarship to flesh out the nuances and implications of what will be a transformative legal transition.

---

[27] Laura Cole, Raw Materials: Upcoming Act an Opportunity for 'Systemic Change', ENDS Europe, 24 November 2022, available at: www.endseurope.com/article/1806280/raw-materials-upcoming-act-opportunity-systemic-change (accessed 17 April 2023).

[28] See, for example, the Executive Order 13817 of 20 December 2017, 'A Federal Strategy to Ensure Secure and Reliable Supplies of Critical Minerals'.

[29] Regulation (EU) 2024/1252 of the European Parliament and of the Council of 11 April 2024 establishing a framework for ensuring a secure and sustainable supply of critical raw materials and amending Regulations (EU) No 168/2013, (EU) 2018/858, (EU) 2018/1724 and (EU) 2019/1020, OJ[2024] L 2024/1252. See also European Commission, n 11, 1–2. An interinstitutional agreement on the Critical Raw Materials Act was reached in November 2023 between the European Parliament and the Council. The Parliament adopted the agreement in December 2023. The Act will have to be formally endorsed by the Council before it is adopted. See also Alessandra Hool, Christoph Helbig and Gijsbert Wierink, 'Challenges and opportunities of the European Critical Raw Materials Act' (2023) *Mineral Economics* 1, 2.

[30] The Anthropocene requires an evolution in legal discourse and governance challenging the methods by which traditional legal frameworks address complex issues such as climate change. This necessitates a shift toward structures more attuned to our changing socio-ecological realities and demands a critical re-engagement of legal thinking. Louis Kotzé, 'Preface: Discomforting Conversations in the Anthropocene', in Louis Kotzé (ed), *Environmental Law and Governance for the Anthropocene* (Bloomsbury Publishing 2017), viii.

The regulatory context described above suggests two salient legal perspectives. The first, narrower, substantive and crucial in the shorter term, hinges on the answers to questions such as how should the immediate concerns of rapidly increasing CRM deployment be governed (that is, geographical dependencies, security of supply, human-rights problems, negative environmental aspects and challenges related to secondary supply) and how can regulatory incentives and frameworks be quickly and comprehensively established for the full deployment of CRMs? For example, ensuring the sustainability of the supply chain for CRMs poses a significant challenge. To avoid sustainability dead-ends,[31] policy makers must carefully consider how to proceed in a coherent yet transparent and sustainable manner.[32] The second perspective, a more principled, longer-term one, centres on the question of what is the role of law and regulatory approaches in steering resource efficiency, that is, reducing consumption while enhancing circularity?

The chapter proceeds to explore these two regulatory perspectives. First, Section 5.2 briefly describes the CRM landscape to pinpoint the relevant policy settings and explain the singular regulatory complexities involved. This then provides the necessary backdrop for the analysis proper of the perspectives in Section 5.3. Section 5.4 concludes the chapter with a synthesis of the arguments elaborated in the preceding sections, presenting the issues surrounding CRMs as a paradox calling for radical regulatory action and as a field of legal research that merits urgent scholarly attention.

5.2 CRMs in the low-carbon transition: balancing the promise and pitfalls of the regulatory complexities

5.2.1 Unpacking the dynamic CRM landscape: key facts

CRMs as the 'transition minerals' are essential for clean energy technologies and, as explained earlier, they can be defined in various ways.[33] The term

---

[31] The policy on sustainable biofuels illustrates an alternative fuel policy that has badly faltered. Initially intended to provide robust regulatory incentives for the adoption of biofuels, it instead transformed into a burdensome policy focused on rectifying previous errors stemming from various sustainability issues associated with conventional biofuels. See, for example, Peter Fairley, 'The biofuel course correction' (2022) *Nature* 611, 15.

[32] Benjamin Sovacool et al, 'Sustainable minerals and metals for a low-carbon future: policy coordination is needed for global supply chains' (2020) *Science* 367 (6473), 30; Nihan Karali and Nihar Shah, 'Bolstering supplies of critical raw materials for low-carbon technologies through circular economy strategies' (2022) *Energy Research & Social Science* 88, 2; Datu Buyung Agusdinata, Hallie Eakin and Wenjuan Liu, 'Critical minerals for electric vehicles: a telecoupling review' (2022) 17 *Environmental Research Letters* 2.

[33] IEA, Critical Minerals Market Review 2023, available at: https://iea.blob.core.windows.net/assets/afc35261-41b2-47d4-86d6-d5d77fc259be/CriticalMineralsMarketReview2023.pdf (accessed 21 July 2023), 12.

'critical raw material' hence references a dynamic concept, which is continually revised with advancements in research and science, as uptake of the materials depends on evolving technical solutions.[34] The criticality of a specific CRM is also contingent upon its applicability in various industrial sectors, technological demands, regional differences and its particular life-cycle. The fluctuating determinations of criticality make it challenging to put forward cogent future scenarios for the deployment of CRMs.[35] While by no means exhaustive, the following discussion aims to demonstrate the intricacies of the present policy domain.

Since 2008, the EU has undertaken criticality assessments of raw materials (that is, assessments that identify materials with high economic importance for the EU that also entail a supply risk) as a principal action under the 2008 EU Raw Materials Initiative (RMI),[36] whose aims include establishing an EU-wide list of CRMs. The 2023 assessment, the fifth to be carried out, evaluated 67 individual materials as well as 20 others in three categories: ten heavy rare-earth elements, five light rare-earth elements and five platinum-group metals. Following the assessment, 34 of the 87 materials were placed on the EU list of CRMs.[37] The criticality assessment has been complemented by a foresight study[38] focusing on selected strategic technologies and industries. The study explores the supply-chain structure, potential vulnerabilities, and dependencies of 15 key technologies[39] in five sectors of strategic significance for the EU economy: renewables, electric mobility (e-mobility), industry, information and communications technology (ICT), aerospace and defence. As the EU's assessment represents the most recent and comprehensive overview of CRMs, the discussion that follows employs the EU's sector and key technology framework.

In the *renewable-energy sector*, technologies such as solar photovoltaic systems (PV), wind turbines, electrolysers, batteries, fuel cells and residential heat pumps all rely on CRMs. For example, PV technology requires silicon metal for traditional systems, and gallium and germanium for thin-film

---

34 Herrington, n 8, 456.

35 See. for example, OECD, n 27, 4; Walter Leal Filho et al, 'Understanding rare earth elements as critical raw materials' (2023) *Sustainability* 15(1919), 2.

36 European Commission, 'Raw Materials Initiative: Meeting our Critical Needs for Growth and Jobs in Europe', COM(2008) 699 final.

37 Grohol and Veeh, n 4, 3.

38 Samuel Carrara et al, *Supply chain analysis and material demand forecast in strategic technologies and sectors in the EU: A foresight study* (JRC Science for Policy Report 2023), 14.

39 The technologies assessed include the following: Li-ion batteries; fuel cells (FCs); electrolysers; wind turbines; traction motors; solar PV; heat pumps; hydrogen direct reduced iron and electric arc furnaces (H2-DRI and EAF); data transmission networks; data storage and servers; smartphones, tablets and laptops; additive manufacturing; robotics; drones; and space launchers and satellites.

applications. Other critical materials for solar PV include antimony, arsenic, fluorspar and phosphorus.[40]

In 2020, the global solar PV capacity stood at 710 GW. Even the more cautious predictions for global solar power capacity see it expanding from 2,950 GW in 2030 and 7,500 GW in 2050. The most optimistic scenario anticipates capacity increasing ten-fold relative to 2020 levels by 2030, and 16-fold by 2050.[41] In recent developments, China ramped up its solar PV capacity by nearly 100 GW in 2022, marking an increase of around 70 per cent from 2021. The country was projected to add another 150 GW in 2023.[42] The EU also registered a significant rise in solar PV installations, recording growth of approximately 50 per cent in 2022, compared to 2021.[43] These scenarios put pressure on several materials, in particular germanium, tellurium, indium, selenium and silicon. Solar PV in the EU relies on a high proportion of CRMs that are imported from third countries. For instance, China processes almost 100 per cent of most of the world's rare-earth elements.[44] Such dominance considerably amplifies the supply-chain constraints impinging on the manufacturing of solar PV.

Electrification is a central strategy for reducing carbon emissions in the *transport sector*.[45] Essential elements for electric mobility include batteries, electric motors, and fuel cells.[46] Rechargeable batteries, commonly used in portable electronic devices, are becoming prevalent in powering electric vehicles (EVs) and are also crucial for energy storage. The continuously improving performance of lithium-ion batteries is a significant factor in rising demand for these batteries. The primary driver for this demand, especially when compared to other types of batteries, such as lead-acid batteries, will be the automotive industry.[47] Its demand for lithium-ion batteries surged to 550 GWh in 2022, roughly 65 per cent up from around 330 GWh in 2021, mainly due to the growth in passenger EV sales. Over three years, the proportion of EV sales has more than tripled, from about 4 per cent in 2020 to 14 per cent in 2022, representing over 10 million electric vehicles globally.[48] Yet, to meet worldwide climate objectives, this proportion needs

---

[40] Carrara et al, n 44, 143.

[41] Ibid, 142.

[42] IEA, n 34, 13.

[43] Ibid, 5.

[44] Grohol and Veeh, n 4, 6.

[45] Meng Yuan et al, 'The electrification of transportation in energy transition' (2021) 236 *Energy*, 1; EIA, Global EV Outlook 2023 (EIA, 2023), available at: www.iea.org/reports/global-ev-outlook-2023/trends-in-batteries (accessed 22 July 2023).

[46] Carrara et al, n 44, 154.

[47] EIA, n 34.

[48] EIA, n 34.

to rise more rapidly, to about 40 per cent by 2030, and the growth should be matched by the electrification of light commercial vehicles, buses and freight trucks.[49]

Of the materials commonly used in lithium-ion batteries, cobalt, lithium and natural graphite are classified as CRMs, but the specific needs for different CRMs differ greatly based on the particular battery's electrode chemistry.[50] Between 2017 and 2022, the demands of the energy sector led to a three-fold increase in the demand for lithium, and a 70 per cent rise in that for cobalt.[51] China holds a commanding position in the supply chain of raw and refined materials (especially cobalt and lithium) essential for batteries.[52]

The *energy-intensive industries sector* stands to play a pivotal role in decarbonization. Industrial heat pumps, hydrogen direct reduced iron technologies (used to produce carbon-free steel), and carbon capture for cement are among the technologies used in these decarbonization efforts. For example, the cement and concrete industries are highly CO_2-intensive. In this context, carbon capture involves capturing the CO_2 emissions from a cement kiln and storing or using them. However, these technologies are still emerging, which makes the prediction of future demand for CRMs for their manufacturing challenging.[53] Heat pumps are more efficient than boilers, promoting renewable-energy use and overall system flexibility. One constraint is that they utilize CRMs such as neodymium, a rare-earth element primarily mined in China, as is the case with many other rare-earth elements, as noted.[54]

Information and communication technologies (ICT) permeate all sectors of the economy and society. Ubiquitous devices, such as smartphones, tablets and laptops, have seen both technical advances and widespread adoption. Their batteries are made using nickel, cobalt, aluminium, manganese and lithium, while their audio components rely heavily on rare-earth elements. Smartphones, tablets and laptops consist of hundreds of components, supported by an intricate worldwide supply chain. For instance, a smartphone might be assembled in China, incorporate semiconductors designed in the United States but manufactured in Taiwan, and utilize raw materials and chemicals from Japan along with European-made parts. Disruptions at any point in this supply chain could have ripple effects, affecting not just the EU but the entire global market.[55]

---

49 Ibid.

50 See, for example, Grohol and Veel, n 4, 2.

51 Ibid.

52 Carrara et al, n 44, 25.

53 Ibid, 161.

54 Grohol and Veel, n 4, 1; Carrara et al, n 44, 71; EIA, n 34, 11.

55 Carrara et al, n 44, 101.

The reliance on and demand for such devices is growing, with the number of various mobile devices rising to nearly 15 billion in 2021 and projected to reach 18 billion by 2025.[56]

Finally, the *aerospace and defence sector* remains strategically vital, spearheading technological innovation and driving economic growth. Many of the technologies deployed in aeronautics and space exploration, such as advanced batteries, robotics, drones, satellites and 3D printing, have applications in defence as well. Drones and robotic systems in particular figure prominently in that sector, which accounted for 54 per cent of the global drone market in 2020. Autonomous and robotic systems, especially, are anticipated to significantly transform the nature of military operations at both a global and national level within the 2021–40 timeframe.[57]

5.2.2 Emerging challenges associated with CRMs

The rising technological and economic significance of CRMs, coupled with concerns about their future availability due to geopolitical, geological and sustainability issues, have put CRMs in the spotlight.[58] Today's fast-growing CRM sourcing and deployment will exacerbate both supply-chain vulnerabilities, referred to earlier, and sustainability concerns.[59]

China dominates the green transition market, from refining minerals such as cobalt and lithium to manufacturing solar panels and EV batteries.[60] Many Chinese-funded projects in resource-rich countries such as Indonesia, Peru and the Democratic Republic of the Congo are conducted in contexts lacking human-rights or environmental due diligence legislation. This leaves victims of abuses, such as child labour, poor working conditions, work-related death and injury, unequal and discriminatory treatment of workers, negative impacts on livelihoods and environmental harm, without any recourse. Since

---

[56] Statista, 'Forecast number of mobile devices worldwide from 2020 to 2025 (in billions)', available at: www.statista.com/statistics/245501/multiple-mobile-device-ownership-worldwide/ (accessed 26 July 2023).

[57] Carrara et al, n 44, 175.

[58] Aude Pommeret, Francesco Ricci and Katheline Schubert, 'Critical raw materials for the energy transition' (2022) *European Economic Review* 141, 1; Magdalena Skipper, 'The global fight for critical minerals is costly and damaging' (2023) *Nature* (editorial 19 July 2023), 619, 436.

[59] OECD, *Global Material Resources Outlook to 2060. Economic Drivers and Environmental Consequences* (OECD, 2019), available at: https://read.oecd-ilibrary.org/environment/global-material-resources-outlook-to-2060_9789264307452-en#page4 (accessed 13 May 2023), 16, 184.

[60] International Energy Agency, *Energy Technology Perspectives 2023* (IEA, 2023), available at: www.iea.org/reports/energy-technology-perspectives-2023 (accessed 21 August 2023), 81–2.

2021, Chinese companies have faced over 100 global allegations of abuse related to human-rights and environmental violations.[61]

The entire life-cycle of CRMs – mining, extraction, refining and production – presents a spectrum of potential environmental hazards. The mineralogical and geological features of a given extraction site may give rise to risks of pollution locally. For example, certain CRMs are commonly accompanied by heavy metals, which may be toxic to both humans and the natural environment if released during the extraction process. A related concern is that the extraction and refining of CRMs found in low concentrations in multi-metal deposits require harsh chemical processes.[62]

Notwithstanding these issues, if managed responsibly, the mining of CRMs can be contained and carried out sustainably.[63] However, despite efforts to enhance governance and corporate management, the sourcing of numerous mineral and metal resources still takes place in regions widely recognized for mismanagement.[64] The environmental repercussions of such procurement are often unpredictable, and in certain instances resource extraction sites become sources of conflict.[65] In an additional caveat, raw-material extraction and processing produce a significant proportion of global greenhouse gas emissions.[66] Thus, although CRMs represent only part of the raw materials extracted globally, increased procurement may boost emissions, intensifying the problem that use of the materials aims to mitigate.[67]

The more the reliance placed on CRMs, the more problems related to their availability and management may be encountered, as well as increased concerns about their environmental and other negative impacts.[68] The

---

61 Business & Human Rights Resource Centre, *Unpacking Clean Energy: Human Rights Impacts of Chinese Overseas Investment in Transition Minerals* (Business & Human Rights Resource Centre, 2023), available at: www.business-humanrights.org/en/from-us/briefi ngs/ (accessed 21 August 2023), 8–12.

62 Ibid, 79–80.

63 Kristi Bruckner, 'Improving environmental and social practices in the mining sector is essential in the transition to renewable energy' (2022) *Journal of Energy & Natural Resources Law*, 2; Sovacool et al, n 33, 33.

64 Business & Human Rights Resource Centre, n 69, 12.

65 Sovacool et al, n 33, 30; Daniel Franks et al, 'Conflict translates environmental and social risk into business costs' (2014) *Proceedings of the National Academy of Sciences (PNAS)* 111(21), 7576.

66 See for example, Dirk Nelen et al, *Improving the climate impact of raw material sourcing.* EEA Report No 08/2021 (European Environmental Agency, 2021), 7.

67 Mehdi Azadi et al, 'Transparency on greenhouse gas emissions from mining to enable climate change mitigation' (2020) *Nature Geoscience* 13, 102; Bruno Oberle et al, *Global Resources Outlook. 2019: Natural Resources for the Future We Want* (International Resources Panel, 2019), available at: www.resourcepanel.org/reports/global-resources-outlook (accessed 21 July 2023), 8.

68 Sovacool et al, n 33, 30.

'enormous appetite for resources' is putting extreme pressure on the planet.[69] The environmental impacts and risks associated with CRMs present a dual challenge. On the one hand, certain products and technologies that rely on these materials, such as wind turbines and solar PV, were initially developed to mitigate environmental harm; on the other, the mining and refining processes involved in sourcing CRMs can have substantial negative environmental impacts. Offsetting the latter to some extent are potential positive impacts, such as the jobs created locally and other economic development.[70]

Reuse, recycling and circularity emerge as the 'silver bullet'[71] in the policies for a sustainable CRM supply. The EU, for example, plans to protect the environment by improving circularity, and measures to increase circularity and efficient use of CRMs will contribute crucially to reducing its import dependencies.[72] However, meeting even the current net-zero pledges will require more raw-material extraction (mining), at least in the near term, if industry is to have enough material to reuse and recycle later (the key assessments referred to earlier project demand through 2030 and 2050). As Section 5.2.1 demonstrates, secondary sources of CRMs are paramount for the low-carbon transition, but the stocks of secondary supplies and recycling rates will fall markedly short of demand in the short and medium term.[73] The shortfall stems from factors such as supply-chain complexities, insufficient production capacity, export restrictions and import dependencies.[74] Even for materials for which the overall recycling rates are relatively high, for example nickel and zinc, the contribution of recycling to meeting demand is relatively low because demand is high.[75] The recycling rates for CRMs vary considerably. However, overall, these rates are much lower than what the high demand for these materials would justify, often due to the challenges posed by underdeveloped recycling technologies, leading to the discarding of these valuable resources.[76]

---

[69] European Commission, Critical Raw Materials Resilience: Charting a Path towards Greater Security and Sustainability, COM(2020) 474 final, 1.

[70] John Bachér et al, n 21, 3.

[71] Frédéric Simon, 'Critical minerals: recycling "not a silver bullet", industry says', EURACTIV 8 March 2023, available at: www.euractiv.com/section/circular-economy/news/critical-minerals-recycling-not-a-silver-bullet-industry-says/ (accessed 17 May 2023).

[72] European Commission, n 11, 2–3.

[73] Grohol and Veel, n 4; Carrara et al, n 44; IEA, n 34; Bruckner, n 71, 2.

[74] IEA, n 34; Kowalski and Legendre, n 12, 52.

[75] Magnus Gislev et al, *Report on Critical Raw Materials in the Circular Economy* (European Commission, 2018), 24.

[76] See, for example, T.E Graedel et al, *Recycling Rates of Metals – A Status Report* (UNEP, International Resource Panel 2011), 9–22; Karali and Shah, n 33, 1; European Commission, n 5, 164; Hagelüken and Goldmann, n 26, 540.

As the UN Economic and Social Council (ECOSOC) notes, 'Most of the global non-fossil fuel industrial ecosystem has yet to be constructed' and '[w]hat has yet to be built cannot be recycled.'[77] The EU is setting a legally binding target for the proportion of recycled material in the consumption of strategic raw materials, but such plans have been criticized as 'unrealistic for the foreseeable future' – even if the EU's circular economy action plan and other relevant policies were to be rapidly implemented.[78] In short, such plans would require 'enormous technological progress and a massive expansion of the circular economy'.[79] Research shows that enhanced collection and recycling could yield secondary materials meeting 37–91 per cent of the demand for CRMs by 2050.[80] As things stand, mining, despite its negative environmental and social impacts, will not only continue to play a key role in CRM sourcing but is also expected to supply the majority of raw materials for the green society, especially as demand for CRMs increases.[81] In any event, the strong increase in demand for CRMs implies that the use of both primary and secondary materials will increase at roughly the same rate.[82]

The key technologies rapidly developing across all industrial sectors rely on the unique physical properties of certain raw materials. CRMs are 'inextricably linked to all industries across all supply chain stages'.[83] Many CRMs (such as rare-earth elements) are found mostly in low concentration in multi-metal deposits, making their extraction difficult and expensive.[84] Furthermore, as demand has increased only recently, their production remains low.[85] China, a major supplier, is an economy with low costs, lax environmental standards and a focus on profits from foreign exchange.[86] The CRM market is dominated by only a few countries, one being China, and the materials often have unique and dependable properties, as seen in certain rare earths, which have superior magnetic properties that other elements cannot easily replicate.[87] Countries and regions, such as the EU, face a complex dilemma in relying on CRMs from China, even as they

[77] Economic Commission for Europe (ECOSOC), 'Sustainable management of critical raw materials required for the low-carbon energy transition', ECE/ENERGY/2022/6, 3.

[78] Müller et al, n 12, 38.

[79] Ibid.

[80] Karali and Shah, n 33, 2.

[81] ECOSOC, n 87, 3; Bruckner, n 71, 2.

[82] Herrington, n 8, 456; Karali and Shah, n 33, 2.

[83] ECOSOC, n 87, 1.

[84] ECOSOC, n 87, 3; Filho et al, n 41, 1.

[85] IEA, n 34, 5–6.

[86] Ibid.

[87] See, for example, Josefine Marx et al, 'Comparative life cycle assessment of NdFeB permanent magnet production from different rare earth deposits' (2018) *ACS Sustainable Chemistry & Engineering* 6(5), 5858.

increasingly scrutinize these regimes, for example, for controversial policies and human-rights violations. This creates a tension between immediate economic needs in the low-carbon transition and long-term ethical and geopolitical considerations. It serves as another reminder that the 'critical' in 'critical raw materials' does not refer primarily to their scarcity, but rather emphasizes their economic significance coupled with supply challenges and the absence of viable substitutes.

5.2.3 The legal and regulatory context of CRMs

Major global institutions, such as the OECD, the International Resources Panel, the World Bank, the International Renewable Energy Agency (IRENA), the United Nations (UN), the United Nations Economic Commission for Europe (UNECE) and the IEA have contributed research on CRMs underlining their escalating economic importance but also highlighting their negative environmental and social consequences.[88] Despite the increasing scientific interest and enquiry surrounding CRMs, sparked particularly by their economic significance, comprehensive legislative and regulatory measures in response have remained scant; the few exceptions are the measures developed in EU.[89] In a further shortcoming, the legal and regulatory considerations pertaining to CRMs in the domains of mining, extraction, refining and production have yet to be addressed on a comprehensive scale.[90]

Extraction of CRMs is generally part of the broad regime of natural resources law.[91] Given that mining is the predominant source of CRMs, the sourcing of most CRMs falls under mining legislation. The domestic

---

88 OECD, n 27; IEA, n 34; IRENA, 'Critical Materials for the Energy Transition', Technical Paper 5/2021 (IRENA, 2021), available at: www.irena.org/Technical-Papers/Critical-Materials-For-The-Energy-Transition (accessed 21 December 2023); UNECE Framework for Critical Raw Materials, available at: https://unece.org/unece-and-sdgs/critical-raw-materials (accessed 21 December 2023); The UN Secretary-General's Panel on Critical Energy Transition Minerals, Resourcing the Energy Transition: Principles to Guide Critical Energy Transition Minerals Towards Equity and Justice (United Nations, 2024), available at: www.un.org/en/climatechange/critical-minerals (accessed 12 September 2024).

89 The EU Critical Raw Materials Act and the EU European Commission, n 11.

90 Ana Bastida, *The Law and Governance of Mining and Minerals: A Global Perspective* (Hart Publishing, 2020), 5.

91 Legal frameworks in the context of natural resources can be understood as referring to legal norms and rules that have a range of aims in regulating natural resource exploration, exploitation and conservation. Elisa Morgera and Kati Kulovesi, 'Preface', in Elisa Morgera and Kati Kulovesi (eds), *Research Handbook on International Law and Natural Resources* (Edward Elgar, 2016), x.

legal frameworks regulating mining and minerals extraction are long-established, with deeply rooted laws and customs going back hundreds of years. For example, norms providing miners with access to or permission to extract minerals have existed for a long time.[92] The regulatory focus in legal frameworks concerning mining is somewhat different from that governing other natural resources: the former emphasize the environmental impacts of mining and processing more so than the minerals per se, whereas the latter consider the sustainable utilization of the natural resource in question.[93]

There is no international treaty on mining or mineral resources, nor is there an international legal instrument that specifically addresses the environmental impacts of mining.[94] Mining law has yet to systematically broaden to 'encompass the international, transnational, regional, national and local levels of normative orderings of social relations concerning mining and minerals'.[95]

There are several relevant international customary rules and principles that merit mention. Perhaps the most important principle is that international environmental law applies to transboundary impacts of mining activities (as it does to other activities) as well as to mining in areas beyond national jurisdiction and the mining of specific minerals. For example, the principle of state sovereignty, coupled with the responsibility of not causing harm to other states, applies to negative transboundary environmental impacts caused by mining activities.[96] In addition, the obligation to carry out comprehensive environmental impact assessments (EIA) for activities that may have significant environmental impacts extends to mining activities.[97]

Furthermore, several multilateral environmental agreements (MEAs) have been concluded that regulate activities that may have a direct or indirect impact on mining.[98] These include the Basel Convention on the Control of Transboundary Movement of Hazardous Wastes and their Disposal,[99] which

---

[92] Ibid, 5; Thomas Wälde, 'Mineral development legislation: result and instrument of mineral development planning' (1988) *Natural Resources Forum* 12(2), 175.

[93] Michael Kidd, 'Minerals', in Elisa Morgera and Kati Kulovesi (eds), *Research Handbook on International Law and Natural Resources* (Edward Elgar, 2016), 327.

[94] Kidd, n 103, 327; Bastida, n 100, 11.

[95] Bastida, n 100, 5.

[96] Trail Smelter (*United States v Canada*) (1941) 3 RIAA 1905.

[97] Case Concerning Pulp Mills on the River Uruguay (*Argentina v Uruguay*) (Judgment) [2010] ICJ Reports 14. EIA is the subject of specific international conventions as well as regional and domestic legislation. See, for example, Convention on Environmental Impact Assessment in a Transboundary Context (Espoo) (1971) 1989 UNTS 309.

[98] Kidd, n 103, 335.

[99] Basel Convention on the Control of Transboundary Movement of Hazardous Wastes and their Disposal (adopted 22 March 1989, entered into force 5 May 1992) 1673 UNTS 57.

applies to mining waste.[100] Minerals in the international deep seabed fall under the specific resource regime set out in the United Nations Convention on the Law of the Sea (UNCLOS),[101] which defines such minerals as belonging to the common heritage of mankind. The Espoo Convention on Environmental Impact Assessment in a Transboundary Context aims to prevent environmental harm caused by extractive industry operations that span across national borders.[102] Some MEAs can have an indirect regulatory impact on mining operations, such as the Convention on Biological Diversity,[103] which requires parties to take various measures to conserve biodiversity and its components; in many cases these conservation measures would conflict with mining activities.[104] Another important instrument in this connection is the General Agreement on Tariffs and Trade (GATT). Its free-trade provisions regulate the international trade in minerals,[105] the principal rule being that minerals are not treated any differently from other tradable products or substances.[106] Finally, there is also a growing number of standards related to the use of CRMs[107] and pledges calling for scaling up international cooperation and regulation on the sustainable use of CRMs.[108]

---

100 For a more detailed overview on the Basel Convention and its provisions applicable to mining waste, see Kidd, n 103, 336.

101 Art 136, UNCLOS (adopted 10 December 1982, entered into force 16 November 1994) 1833 UNTS 3.

102 Convention on Environmental Impact Assessment in a Transboundary Context (adopted 25 February 1991, entered into force 10 September 1997) 1989 UNTS 309.

103 Convention on Biological Diversity (adopted 5 June 1992, entered into force 29 December 1993) 1760 UNTS 382.

104 Kidd, n 103, 337.

105 General Agreement on Tariffs and Trade 1994 (adopted 15 April 1994, entered into force 1 January 1995).

106 It merits mentioning that there are some noteworthy exceptions that can be attributed to specific GATT Articles, such as Art XX on general exceptions, which allows member countries to adopt certain measures that would otherwise violate GATT rules. Specifically, Art XX allows for exceptions relating to the conservation of exhaustible natural resources as long as the measures taken are implemented in conjunction with restrictions on domestic production or consumption. For further discussion in the context of minerals, see Kidd, n 103, 331–4.

107 See, for example, 'General Method to Declare the Use of Critical Raw Materials in Energy-Related Products', EN 45558:2019, available at: https://standards.cencenelec.eu/dyn/www/f?p=CENELEC:110:::::FSP_PROJECT,FSP_ORG_ID:65687,2240017&cs=1D6C1C93A4A4B6F6D1A19F72399DC406C (accessed 22 August 2023).

108 See, for example, UN, Secretary-General's remarks to G77+China COP28 Leaders' Summit [as delivered 2 December 2023], available at: www.un.org/sg/en/content/sg/statement/2023-12-02/secretary-generals-remarks-g77china-cop28-leaders-summit-delivered?_gl=1%2Azn56n7%2A_ga%2AMTY5NjUwNjQyMy4xNzAzMTY0ODk3%2A_ga_TK9BQL5X7Z%2AMTcwMzE2NTc0MS4xLjAuMTcwMzE2NTc0MS4wLjAuMA..%2A_ga_S5EKZKSB78%2AMTcwMzE2NTc0MS4xLjAuMTcwMzE2NTc0My41OC4wLjA; UNECE, COP28: UN urges coordinated action to align

Mining and minerals fall under Agenda 2030[109] and several of the Sustainable Development Goals.[110]

5.2.4 *The development of CRM-specific regulatory frameworks*

Advances in legal regulation on CRMs are starting to emerge, especially in the context of the low-carbon transition and rapid technological development. The growing focus on CRMs and concerns over their availability, security of supply and environmental implications have brought them into the purview of policy makers and regulators.[111] The EU stands out in this domain, having taken regulatory action that shows increasing recognition of, and response to, the multifaceted challenges.

The EU launched its CRM policy prior to publishing its first list of CRMs, in 2011.[112] The list has been updated regularly. In 2024, the EU introduced the EU Critical Raw Materials Act,[113] designed to safeguard a secure and sustainable supply of the materials for the EU. Accompanied by publication of the fifth list of CRMs, containing 34 minerals, the Act reinforces all stages of the value chain and, in particular, diversifies imports to minimize strategic dependencies (the EU's reliance on imports of CRMs is extremely high, sometimes reaching 100 per cent for certain rare-earth elements).[114] The Act also bolsters the EU's ability to monitor and address supply disruptions and advance circularity and sustainability practices. As the EU Critical Raw Materials Act represents the first comprehensive attempt to regulate

soaring Critical Raw Materials extraction and use with sustainable development, available at: https://unece.org/climate-change/press/cop28-un-urges-coordinated-action-align-soaring-critical-raw-materials (accessed 21 December 2023).

109 United Nations General Assembly, Transforming Our World: the 2030 Agenda for Sustainable Development, A/RES/70/1.

110 World Economic Forum, Mapping Mining to the Sustainable Development Goals: An Atlas. White Paper (World Economic Forum, 2016), available at: www.undp.org/sites/g/files/zskgke326/files/publications/Mapping_Mining_SDGs_An_Atlas_Executive_Summary_FINAL.pdf (accessed 20 May 2023), 3.

111 Barry Barton, 'Building Resilience from the Ground Up. Local Supply and Demand Management with Renewables, Prosumers, Energy Efficiency, Critical Minerals, and the Circular Economy', in Catherine Banet et al (eds), *Resilience in Energy, Infrastructure, and Natural Resources Law* (Oxford University Press, 2022), 334.

112 European Commission, n 42.

113 See also European Commission, n 11. The Act is a regulation. Regulations are the strongest of EU legal instruments as they apply directly, automatically and uniformly to all EU countries as soon as they enter into force; they do not have to be transposed into national law. Art 288, Consolidated version of the Treaty on the Functioning of the European Union (TFEU), OJ [2012] C326/47.

114 European Parliament Briefing, n 14.

CRM management and guide related industry practices, it may be seen as establishing a regulatory benchmark for sustainable CRM governance.[115]

The EU Critical Raw Materials Act focuses on five key areas. First, it defines priorities and establishes clear objectives for strengthening the EU's policy on CRMs. This includes codifying the list of CRMs and introducing a new strategic raw materials list. The Act sets out to enhance domestic extraction, processing and recycling capacities for strategic raw materials in the EU and to guide diversification efforts. These benchmarks set minimum percentages for domestic EU extraction, processing and recycling of strategic raw materials, as well as limits on reliance on a single third country for these raw materials. Second, the Act introduces measures to bolster CRM capacities across the entire value chain. Member states are required to provide permits for all CRM projects with tight time limits for permitting procedures, facilitating the process with a one-stop-shop approach. Additionally, the Act promotes circularity and efficient use of CRMs by creating value chains for recycled materials as well as mandating that operators and EU member states improve recovery from products and waste in the EU market containing CRMs. Third, the Act outlines actions to enhance EU preparedness and mitigate supply risks related to CRMs. Fourth, the legislation establishes a common governing structure to oversee the implementation and enforcement of its provisions.[116] Last, the Act makes links to environmental legislation (especially legislation relating to permitting) and puts forward plans to develop a system for calculating the environmental footprint of activities undertaken.[117]

Member states must observe a number of EU-level legal instruments already in force that set out permitting conditions and procedures for projects along the raw-material value chain. These include the Industrial Emissions Directive,[118] Habitats and Birds Directives,[119] Water Framework Directive,[120] Environmental Liability

[115] Hool, Helbig and Wierink, n 30, 2.

[116] European Commission, n 11 (proposal), 1–17.

[117] Ibid, 13.

[118] Directive 2010/75/EU of the European Parliament and of the Council on industrial emissions (integrated pollution prevention and control). OJ [2010] L334, 17. The Industrial Emissions Directive is currently undergoing a revision, see Proposal for a Directive of the European Parliament and of the Council amending Directive 2010/75/EU of the European Parliament and of the Council of 24 November 2010 on industrial emissions (integrated pollution prevention and control) and Council Directive 1999/31/EC of 26 April 1999 on the landfill of waste, COM/2022/156 final. The proposal also contains new provisions that are inclusive of raw materials sourcing.

[119] Council Directive 92/43/EEC on the conservation of natural habitats and of wild fauna and flora, OJ [1992] L206/7.

[120] Directive 2000/60/EC of the European Parliament and of the Council establishing a framework for Community action in the field of water policy, OJ [2000] L327/1.

Directive[121] and the EU Conflict Minerals Regulation.[122] As CRMs are also sourced from secondary sources, such as waste streams containing critical minerals, waste legislation[123] is also of growing importance in regional CRM governance. Legislation on waste management includes rules governing the operation of mines with regard to handling of extractive waste and influencing the potential to recover and recycle CRMs from relevant waste streams. Furthermore, CRM sourcing in the European context must also consider relevant circular economy strategies and action plans.[124]

In the United States, concerns about the availability of critical minerals have been construed with a view to national security and safeguarding against potential disruptions caused by foreign adversaries. A 2017 executive order[125] introduced a policy aimed at mitigating this vulnerability, and a 2020 order called for action to address risks associated with critical mineral supply chains.[126] Furthermore, a 2021 executive order directed the US government to undertake a comprehensive review of critical US supply chains to identify risks, address vulnerabilities and develop a strategy to promote resilience.[127] Canada has a newly revised Critical Minerals Strategy,[128] as well as a list of critical minerals.[129] The country is home to almost half

[121] Directive 2004/35/CE of the European Parliament and of the Council on environmental liability with regard to the prevention and remedying of environmental damage, OJ [2004] L143/56.

[122] Regulation (EU) 2017/821 of the European Parliament and of the Council laying down supply chain due diligence obligations for Union importers of tin, tantalum and tungsten, their ores, and gold originating from conflict-affected and high-risk areas, OJ [2017] L 130/1; European Commission, n 5, 78.

[123] See, for example, Directive 2008/98/EC of the European Parliament and of the Council on waste and repealing certain Directives, OJ [2008] L312/3 and Directive 2006/21/EC of the European Parliament and of the Council on the management of waste from extractive industries and amending Directive 2004/35/EC – Statement by the European Parliament, the Council and the Commission, OJ [2006] L102/15.

[124] European Commission, Communication from the Commission to the European Parliament, the Council, the European Economic and Social Committee and the Committee of the Regions. A new Circular Economy Action Plan for a cleaner and more competitive Europe, COM(2020) 98 final.

[125] Executive Order 13817 of 20 December 2017, A Federal Strategy to Ensure Secure and Reliable Supplies of Critical Minerals.

[126] Executive Order 13953 of 30 September 2020, Addressing the Threat to the Domestic Supply Chain from Reliance on Critical Minerals from Foreign Adversaries and Supporting the Domestic Mining and Processing Industries.

[127] Executive order of 24 February 2021 on America's Supply Chains.

[128] The Canadian Critical Minerals Strategy – From Exploration to Recycling: Powering the Green and Digital Economy for Canada and the World, available at: www.canada.ca/content/dam/nrcan-rncan/site/critical-minerals/Critical-minerals-strategyDec09.pdf (accessed 22 May 2023).

[129] Ibid, 9.

of the world's publicly listed mining and mineral exploration companies, and the strategy recognizes critical minerals as a 'generational opportunity' for Canada's workers, economy and net-zero future.[130] Other countries, among them Japan,[131] China[132] and Australia,[133] have implemented strategic measures concerning CRMs, and have developed lists of CRMs as part of their approach. These initiatives reflect recognition of the economic and strategic importance of CRMs in various quarters, although there is varying acknowledgement – if at all – of the related sustainability concerns. By identifying and categorizing CRMs, countries intensify their efforts toward understanding their importance and the risks associated with their supply. This approach facilitates the development of specific strategies aimed at ensuring that industries have access to stable and sustainable supplies of these crucial materials. However, this process necessitates active movement toward establishing comprehensive regulatory frameworks and governance systems that can safeguard the stability and sustainable availability of these resources.

The regulatory approaches outlined above illustrate, albeit not exhaustively, the diverse measures adopted by different countries and regions. Building upon this preliminary examination, the next section goes on to discuss the two regulatory perspectives put forward in the introduction – governing quickly increasing CRM deployment and developing regulatory approaches that steer resource efficiency. In the light of the existing landscape of CRM governance, it becomes imperative to chart the path forward, asking what steps should be taken to address the challenges and opportunities associated with CRMs?

5.3 Contrasting perspectives on CRMs in the Anthropocene

5.3.1 Overcoming the imminent challenge: sustainable regulation of CRMs

CRMs present a distinct regulatory challenge. In principle, the sourcing of CRMs is covered by existing mining laws, but many were passed long

---

[130] Ibid, 5.

[131] See further Ministry of Economy, Trade and Industry, Agency for Natural Resources and Energy, Japan's new international resource strategy to secure rare metals (31 July 2020), available at: www.enecho.meti.go.jp/en/category/special/article/detail_158.html (accessed 20 May 2023).

[132] See further Su and Hu, n 13, 2–3.

[133] Australian Government, Department of Industry, Science and Resources, Critical Minerals Strategy 2022, March 2022, available at: www.industry.gov.au/sites/default/files/2022-09/2022-critical-minerals-strategy_0.pdf (accessed 20 May 2023).

ago and may not accommodate the concerns prompted by the increasing demand for these materials today. Once production capacities match demand, CRM reserves could satisfy global needs for decades, but the scale and speed required to accommodate the growing demand pose a challenge in their own right. The lengthy timeline for opening a new mine and substantial capital investment involved also suggest that gaps will persist in mining capacity.[134] Moreover, technical and economic limitations may render some known reserves non-extractable.[135] Contingencies such as these should be meticulously considered when assessing the reliability of existing mining regimes. What is more, the emerging debate surrounding CRMs is undoubtedly intensifying the pressure to scrutinize the applicability of mining law from a new, contemporary perspective.

In legal scholarship, the study of mineral (or mining) law has been 'circumscribed to domestic legal systems and generally limited to national boundaries'.[136] This observation signals an acute need for broader approaches, given the expansion of exploration and extraction, the complexity of global value chains, and uneven distribution of mineral deposits globally.[137] Any new perspective must capture the range of legal and normative phenomena – from local to global – relating to sourcing of CRMs. It will require a grasp of 'all geographical levels of relations and legal ordering concerning rules and procedures in order to access, use and manage minerals (and the rights and obligations derived therefrom).'[138]

As explained above, while mining remains the main source of CRMs currently, future strategies focus on minimizing this reliance by promoting a circular economy for reusing and recycling these scarce resources. Achieving the ambitious and urgent goal of sourcing CRMs predominantly through reuse and recycling will necessitate advancements in technology coupled with incentives and innovation.[139] It is imperative to develop regulatory approaches that acknowledge and address these needs. Although it is technically feasible to recover all CRMs, current recovery efforts are hindered by the lack of robust economic drivers and insufficient policies and regulations. Also lacking are clear economic incentives, support for a competitive secondary-materials market, and encouraging use of recycled materials.[140] For instance, existing

[134] For example, the time required for a mining project to set off in the EU is usually 10 to 15 years. See E. Righetti and V. Rizos, The EU's quest for strategic raw materials: what role for mining and recycling? (2023) *Intereconomics* 58, 69.

[135] Karali and Shah, n 33, 2.

[136] Bastida, n 100, 5–6.

[137] Ibid, 7–8.

[138] Bastida, n 100, 9.

[139] European Commission, n 11, 12.

[140] For discussion, see for example, Antoine Oger and Emma Watkins, 'We need a Critical Materials Regulation, not a raw materials' boosting act', Institute for European

waste-management legislation in the EU fails to provide adequate incentives for improving the circularity of CRMs or fostering the development of secondary raw-material markets.[141] While a diverse set of national measures on waste collection and waste recovery is in place, the present arrangement has been described as 'an incoherent regulatory framework for economic operators and producers'.[142] A salient challenge is the EU's aim of raising its circular-materials use rate, the proportion of recycled waste in the total of materials used in the economy. Plans call for raising the rate to 23.4 per cent by 2030,[143] yet between 2010 and 2021 it stood at less than 1 per cent, a gap indicating the substantial effort required to achieve this ambitious target.[144] The EU Critical Raw Materials Act, introduced in the previous section, will play a crucial role in attaining this objective if it enters into force in its proposed form. The Act introduces only minimum benchmarks for extraction, processing and recycling capacities bearing on the EU's annual consumption of CRMs by 2030. These benchmarks call for covering 10 per cent of EU consumption through domestic extraction, 40 per cent through domestic processing (including recycled content), and 15 per cent through recycling.[145]

An additional shortcoming is that the current regulatory approaches to CRMs focus on supply security, addressing vulnerabilities and managing bottlenecks. However, if regulation is to shift the sourcing of CRMs from primary (mining) to more sustainable secondary sources (reuse and recycling), governance should prioritize improvements in material efficiency, which will minimize the overall environmental impact. This can be achieved by scaling up domestic circularity capacities through product-design requirements geared to enhance material efficiency, implementing policies that encourage repair and reuse, and creating value from waste and safely recovering CRMs from end-of-life products for recycling.[146]

The emerging regulatory frameworks should also recognize the numerous liabilities stemming from adverse environmental effects and potential

Environmental Policy news (22 March 2023), available at: https://ieep.eu/news/we-need-a-critical-materials-regulation-not-a-raw-materials-boosting-act/ (accessed 16 May 2023).

141 European Commission, n 11, 3.

142 Sirja-Leena Penttinen, 'Batteries and the low-carbon energy transition: circularity and secondary supply approach highlighted in the EU's Policy Discourse' (2021) *European Energy and Environmental Law Review*, 236.

143 The EU Circular Economy Action plan envisions 'doubl[ing] its circular material use rate in the coming decade'. Doubling the CMUR would mean an increase from 11.7 per cent in 2021 to 23.4 per cent by 2030. European Commission, n 135, 1; Oger and Watkins, n 151.

144 Oger and Watkins, n 151.

145 European Commission, n 11, 17–18.

146 See, for example, Hagelüken and Goldmann, n 26, 539.

violations of human and labour rights linked to CRMs. To address these concerns, the EU references existing environmental legislation (and other instruments, such as the EU Conflict Minerals Regulation) in its Critical Raw Materials Act. The Act also mandates full adherence to relevant regulations when selecting and implementing strategic raw-materials projects geared to diversifying and securing a sustainable domestic supply of CRMs.[147] The EU Critical Raw Materials Act serves to complement the Directive on Corporate Sustainability Due Diligence (CSDD). While the CSDD might eventually apply to companies using CRMs, mandating that they address human-rights and environmental concerns in their operations and value chains, it will not require them to report on the environmental footprint of CRMs.[148] The effectiveness of the approaches and instruments described here in protecting the environment from the adverse effects of increased CRM sourcing is yet to be determined.

The multiple complexities associated with mining, extraction, refining and production of CRMs, combined with the evolving multilevel global governance framework, pose a significant challenge in establishing a coherent and sustainable governance framework for CRMs in a tight timeframe. As noted, regulation must address not only the rapid escalation in demand but also a wide spectrum of considerations relating to sustainable governance, such as the social impacts of increased mining activities and the greenhouse gas emissions associated with the energy-intensive production technologies described in previous sections. If regulation is to address these complex and interconnected challenges successfully, it will have to emanate from a comprehensive approach that balances economic, environmental and social dimensions of CRM governance.[149] The first step in developing regulatory approaches to CRMs would be to evaluate the applicability of existing laws, especially mining and waste legislation.

5.3.2 Striking a balance: a safe operating space for CRMs?

The current policy and regulatory frameworks show limited consideration of the targets for achieving an absolute reduction in resource consumption or addressing the environmental impacts associated with scaling up domestic (EU) capacities, particularly in the case of extraction.[150] In a future aligned

[147] European Commission, n 11, 23.

[148] Directive (EU) 2024/1760 of the European Parliament and of the Council of 13 June 2024 on corporate sustainability due diligence and amending Directive (EU) 2019/1937 and Regulation (EU) 2023/2859, OJ [2024] L2024/1760. See also Proposal for a Directive of the European Parliament and of the Council on corporate sustainability due diligence and amending Directive (EU) 2019/1937, COM(2022) 71 final.

[149] Sovacool et al, n 33, 33; Bruckner, n 71, 3; Herrington, n 8, 458.

[150] Oger and Watkins, n 151.

with the climate targets and guided by the Sustainable Development Goals, should the regulatory frameworks governing CRMs not emphasize circularity rather than supply vulnerabilities, the focus of the EU Critical Raw Materials Act? This would entail integrating circular-design principles, enhancing resource efficiency and encouraging reuse, remanufacturing and recycling.

Appropriate laws and regulations could play a crucial role in mitigating the criticality of CRMs by effectively reducing the demand for primary sourcing of these materials. This prompts the broader question: How should legal instruments be used to promote resource efficiency, specifically to decrease consumption and boost the circularity of raw materials? Law and regulatory frameworks could also be used to make CRMs more sustainable simply by reducing the need for primary sourcing. Whilst current low-carbon policies encourage subscribing to 'deeply transformative' policies that put sustainability 'at the centre of economic policy',[151] at the same time they incentivize rapid upscaling of CRM-intensive technologies.[152] Such a paradox raises a question of regulatory coherence.[153] In any event, as long as the supply of traceable materials is limited, encouraging the demand for CRMs 'only leads to shifting the problem to other value chains'.[154] Accordingly, regulations and incentives are needed to forestall such a trend and to increase responsible supply.

There is 'no benchmark or ultimate global limit of sustainable resource use',[155] but (sustainable) governance can be effected through law and regulatory instruments. Law is not merely an instrument in sustainability transition but a complex system that steers, slows and accelerates the forces affecting transition.[156] Exploring the potential of legal mechanisms enables

---

151 European Commission, Communication from the Commission to the European Parliament, the Council, the European Economic and Social Committee and the Committee of the Regions. The European Green Deal, COM(2019) 640 final, 3–4.

152 For example, the EU has a legal framework in place to ensure that all cars registered in the EU after 2035 will run exclusively on CO_2 neutral fuels. In practice, this means electric vehicles. Proposal for a Regulation of the European Parliament and of the Council amending Regulation (EU) 2019/631 as regards strengthening the CO_2 emission performance standards for new passenger cars and new light commercial vehicles in line with the Union's increased climate ambition, 10906/21 – COM(2021) 556 final.

153 Audrey Danthinne and Michael Picard, 'Assessing the compatibility of vehicle electrification with the EU's circular economy objective' (2022) *European Energy and Environmental Law Review*, 395.

154 ECOSOC, n 87, 18.

155 Chris Backes and Marlon Boeve, 'Envisioning the future of the circular economy: a legal perspective' (2022) *Environmental Policy and Law* 52, 259.

156 Niko Soininen et al, 'A brake or an accelerator? The role of law in sustainability transitions' (2021) *Environmental Innovation and Societal Transitions* 41, 71–3.

the development of strategies and policies that facilitate sustainable resource management and contribute to a shift toward a circular economy. For instance, setting a goal to reduce material use not only decreases demand but also bolsters the resilience of the low-carbon transition while shrinking its material footprint.[157] Without explicit objectives, the ambitious goals of the green transition run the risk of becoming empty promises, incurring significant costs without spurring substantial progress. In essence, the question that needs to be decided is: continue consumption-as-usual, transitioning from fossil-fuel-based to CRM-based technologies, or initiate a radical reform of regulatory approaches that lead from a low-carbon to a net-zero and green future? Answering the second question in the affirmative would entail moving beyond the current approach of incentivizing unsustainable practices to acknowledge the need for comprehensive legal reforms that address the challenges posed by CRMs. It would also require embracing a more sustainable and resilient framework that promotes responsible sourcing, efficient resource utilization and the development of alternative materials and technologies.

5.4 Conclusion

CRMs will be as important to us as oil and gas are in today's fossil-fuel-powered world.[158] However, it is well worth asking whether CRM-dependent technologies are any better for the development of low-carbon economies than the alternatives they are slated to replace. The chapter has demonstrated that while the unique role that CRMs play in accelerating low-carbon transitions is undeniable, ensuring sustainable and safe sourcing of CRMs as well as responsible CRM governance presents a crucial challenge in the upcoming years.

The Anthropocene[159] exerts new and partly unexplored pressures on Earth's mineral resources, which are fundamental raw materials for contemporary societies. Humans have been using minerals for thousands of years and human activities have permanently changed our geologic environment.[160] The urgency of mitigating the impacts of climate change requires a rapid shift away from an economy heavily dependent on fossil fuels. This transition is contingent

---

[157] Ibid.

[158] Skipper, n 66.

[159] The global environmental changes suggest that Earth has transitioned into a new geological epoch called the Anthropocene, characterized by human dominance. Simon Lewis and Mark Maslin, 'Defining the Anthropocene' (2015) *Nature* 519(7542), 171.

[160] Robert M. Hazen et al, 'On the mineralogy of the "Anthropocene Epoch"' (2017) *American Mineralogist* 102(3), 595–6.

upon the widespread adoption of clean energy technologies, a development requiring the extensive deployment of CRMs in various sectors.

The chapter's discussion of the legal and policy contexts of CRMs forms a basis for weighing the two regulatory perspectives outlined in the Introduction and reflected upon in Section 3. The current regulatory framework governing CRMs is complex, fragmented and falls short of fully addressing the complexities of sustainable CRM governance. At the end of the day, particularly in global multilevel governance, it remains challenging to obtain a comprehensive view of the applicable legal rules and norms on the emergent issue of CRMs. The first perspective centres on effectively governing the immediate repercussions of the rapid deployment of CRMs, including concerns related to geographical dependencies, supply security, human-rights issues, negative environmental implications, and challenges associated with secondary supply. The second perspective queries the role of law and regulatory approaches in guiding resource-efficiency efforts, including strategies to reduce resource consumption while enhancing the circularity of raw materials. The overview of the current regulatory framework reveals the fragmented nature of CRM governance and the need for its evolution.

On balance, the analysis of these perspectives demonstrates the scope and complexity of CRMs: as objects of regulation, critical raw materials pose wide-ranging and multilayered challenges, and the related governance requires a commensurably comprehensive and multilevel approach. Under current policies, the increasing use of CRMs entails a paradoxical, no-win situation, entangling economic necessity with ethical and environmental dilemmas. Approaches addressing the issue will have to accommodate perspectives spanning local as well as global levels. The need for debate is urgent, as regulatory objectives are swiftly shifting to facilitate and support the low-carbon transition.[161] Policy makers and regulators must step back and take a holistic approach to ensure that the transition to a low-carbon economy is sustainable and equitable, and does not cause adverse effects on the environment or society. The law-reform agenda is also somewhat different for energy and minerals, given that they are already subject to legal and regulatory measures in numerous countries.

Finally, the chapter has marked out CRMs as an emerging issue-area in legal research. Particularly salient is the substantive perspective, that is, how the various aspects of CRM sourcing and deployment should be regulated and governed. Also relevant are the foundational considerations of how the law should be employed to guide and incentivize the transition from a fossil-fuel-dependent economy to a CRM-dependent one. Furthermore,

[161] Barton, n 121, 340.

while there is copious scientific literature on CRMs, legal scholarship on the topic is scanty. The legal aspects and regulatory challenges relating to CRMs present fertile ground indeed for future research and academic inquiry.

Acknowledgement

This work was supported by the Strategic Research Council of Finland (358393 RELIEF and 335563 2035 Legitimacy).

6

Hydrogen: What Does its Deployment at Speed and Scale Mean for Legal Systems?

Ruven Fleming and Kelsey Pailman

6.1 Introduction

Hydrogen is a central 'battle ground' for the decarbonization of the gas sector. Hydrogen is increasingly viewed as an energy carrier that could help states with meeting their commitments to mitigate climate change, which were made in support of the 2015 Paris Agreement and at subsequent United Nations Climate Change Conferences.[1] This is primarily due to the potential use of low-carbon, renewable-energy-derived 'green' hydrogen as a replacement for less climate-friendly natural gas.[2] Hydrogen's potential to support the Just Transition is also recognized, as it may offer sustainable growth opportunities to regions with economies that have previously been reliant on the fossil fuel industry.[3] Consequently, many states are developing hydrogen strategies and polices.[4] However, these developments are, for

---

1 Mona Wappler, Dilek Unguder, Xing Lu, Hendrik Ohlmeyer, Hannah Teschke and Wiebke Lueke, 'Building the green hydrogen market: Current state and outlook on green hydrogen demand and electrolyzer manufacturing' (2022) 47 (79) *International Journal of Hydrogen Energy* 33551–70, at 3352 (hereafter: Wappler et al).

2 Ibid.

3 For more on that see Jennifer Campion 'Giving hydrogen the green light and putting it on the fast-track? Consenting hydrogen developments in Aotearoa New Zealand' in Ruven Fleming (ed) *Cambridge Handbook of Hydrogen and the Law* (pp 201–14) (Cambridge University Press, 2024).

4 Pasquale Marcello Falcone, Michael Hiete and Alessandro Sapio, 'Hydrogen economy and sustainable development goals: Review and policy insights' (2021) 31 *Current Opinion in Green and Sustainable Chemistry* 100506.

many states, still at a fairly early stage. It is also recognized that hydrogen presents challenges as well as opportunities, especially around pricing and safety concerns.[5] While some states are prioritizing green hydrogen in their strategies, this commitment is by no means universal. Nevertheless, the growing enthusiasm for hydrogen is apparent, with more than 26 countries having published hydrogen roadmaps.[6] With potential uses including heavy industry, transport, grid electricity production and heating homes and water,[7] hydrogen is seen as having a key role in the energy transition, in particular by offering a pathway to decarbonize 'hard to abate' sectors.

Hydrogen does not equal hydrogen. Most people by now have heard of the 'colourbook' of hydrogen, ranging from green hydrogen (produced via electricity from renewable sources) to grey hydrogen (produced in the traditional way from fossil fuels). The European Union (EU) has used broad terminology in describing different types of hydrogen and has introduced terms like 'clean' or 'renewable' hydrogen in its legislation, as opposed to definitions conforming with the 'colourbook'.[8]

There seems to be a general consensus that hydrogen is relevant for the decarbonization of heavy industries, including steel and cement factories, as this option is included in most hydrogen strategies of countries that have them.[9] But the integration of hydrogen in other applications is highly controversial and the subject of significant debate. In Germany, for example, an intense debate has been waged about inclusion of hydrogen for the decarbonization of gas in the heating sector, because of questions about heating as the most efficient way to use a precious resource like hydrogen.[10] What these examples highlight is the fact that the deployment of hydrogen is a highly controversial question in various respects. Resolving the tension between the rise of new technological innovations concerning hydrogen, on

---

5 Ibid.

6 Wappler et al, 3353.

7 Hydrogen Council, 'Hydrogen Scaling Up: A sustainable pathway for the global energy transition', November 2017, available online at https://hydrogencouncil.com/wpcontent/uploads/2017/11/Hydrogen-scaling-up-Hydrogen-Council.pdf [accessed 18 January 2024] at 8.

8 Ruven Fleming 'Clean or renewable: Hydrogen and power-to-gas in EU energy law', (2021) *Journal of Energy & Natural Resources Law* 39:1, 43–63, DOI: 10.1080/02646811.2020.1795382 [accessed 18 January 2024].

9 Wasserstoff-Kompass, 'International Wasserstoffstrategien im Vergleich – Länderanalyse' 17 & 19, available at: www.wasserstoff-kompass.de/fileadmin/user_upload/img/news-und-media/dokumente/2022_H2_Laenderanalyse.pdf [accessed 28 June 2023].

10 German Federal Government 'Climate-friendly heating: New Building Energy Act to be implemented', available at:www.bundesregierung.de/breg-en/news/new-building-energy-act-2185010 [accessed 28 June 2023].

the one hand, and the most optimal use of these innovations, on the other, becomes central for the future uptake of hydrogen in various industries.

The general move toward deployment of hydrogen in various sectors and areas at a much higher scale and speed than has been done so far means that legal systems need to adapt or be changed. This naturally creates friction, and the friction needs to be channelled and tackled. There are great and noteworthy examples from legal theory on how this can and should be done, including important contributions in this book.

But the theoretical consideration seems to be one thing, practical implementation a different matter. Unfortunately, legislators rarely consult academic textbooks before making decisions that pre-determine future legal frameworks on energy vectors, such as hydrogen, and often do not follow a well-thought-through process on how to come up with best possible laws. In fact, in almost all cases, laws are the result of political compromises and are shaped and formed in particular ways due to political considerations. This is too often overlooked in the legal literature, but it shapes the interplay between forward-thinking regulation and necessary adjustments to existing legal frameworks. As a result, we have seen recommendations and suggestions for improvement that are unrealistic and non-reflective of the actual needs and realities 'on the ground'.

To be clear, this chapter is certainly not going to solve this general tension between legal theory and practical realities in the law-making process. The ultimate aim is, rather, to provide thoughts from the hydrogen context on how to reconcile both worlds by inventorying and highlighting the frictions. In line with this, the chapter proceeds in two ways. In the following section an inventory concerning the main legal literature on hydrogen regulation is provided. Different aspects and viewpoints will be heard and an inventory of the 'state of the art' legal debate on hydrogen, and the core frictions, will be created. This part will primarily focus on the EU as a selected region, given that its regulatory debate is well underway. Moreover, the EU debate clearly highlights challenges that apply at any given jurisdiction that aims to regulate hydrogen.

Based on this inventory, a second step will be taken to catalyse three central 'battlegrounds' for future hydrogen regulation in any given jurisdiction. The core frictions highlighted by the EU debate will be contrasted and analysed, in order to gain a holistic view of the regulatory landscape governing hydrogen today and in the future. These frictions raise questions such as who is allowed to own hydrogen grids, which role hydrogen plays in the future heating markets, and the issue of what actually is sustainable in green, clean and renewable hydrogen. The eager hydrogen expert will, of course, be able to add many more to this list and the chapter does not claim to provide a comprehensive overview. Rather, it provides a first idea from which some general conclusions will be drawn on the underlying tension between rapid technology expansion and the law's general tendency to make normative compromises.

6.2 Current status: literature review on hydrogen regulation

Both hydrogen uptake and the body of legal literature on it, are at a nascent stage. Nevertheless, over the past few years, gradual increases in the adoption of hydrogen in various sectors across industry, transport and buildings have resulted in increased academic discussion and debate regarding the legal implications of a hydrogen economy. The following section highlights some of these discussions and debates across key legal implications of a hydrogen economy. To highlight the interplay between technology expansion, on the one hand, and the law's behaviour, gaps and issues, on the other, the following explanations are structured along the hydrogen value chain, zooming in on the three most important stages of production, transportation and supply to end-users.

6.2.1 Hydrogen production: defining the transition

Hydrogen production is currently undergoing its own (energy) transition. The 'colourbook' of hydrogen, ranging from grey to green, is illustrative of the moving spectrum from fossil-fuel-based hydrogen production to more sustainable alternatives.[11] The classification of hydrogen is particularly important, as it determines the energy production methods that would catalyse its uptake within the EU.[12] Riemer identifies two key challenges faced in evaluating key aspects of an (energy) transition, such as sustainability of the transition.[13] First, it is noted that clarity in legislative frameworks is required when ascertaining which hydrogen production (for example green and/or blue hydrogen) would be compatible with a climate-neutral system from a legal and policy perspective.[14] Second, once the forms of hydrogen have been selected, the actual content of the sustainability criteria will need

11 Ruven Fleming 'The hydrogen revolution and natural gas: A new dawn in the European Union?' p 125 in D. Olawuyi and E. Pereira (eds) *The Palgrave Handbook of Natural Gas and Global Energy Transitions* (pp 123–40) (Palgrave Macmillan, 2022) (hereafter: Fleming).

12 Catherine Banet 'Building Europe's Hydrogen and Renewable Gas Markets' (2023) Centre on Regulation for Europe, 24 (hereafter: Banet).

13 Matia Riemer, 'Lessons learnt from certifying biofuels for a future hydrogen certification scheme' (2022) 18th International Conference on the European Energy Market (EEM), Ljubljana, Slovenia, 1–7. DOI: 10.1109/EEM54602.2022.9921171 [accessed on 19 January 2024] (hereafter: Riemer).

14 Ibid.

to be evaluated, particularly in light of decarbonization targets and lessons learned from certifying biofuels.[15]

With a view to the first challenge, to accelerate the rapid uptake of hydrogen within the EU, imports of renewable and low-carbon hydrogen are anticipated to complement domestic production.[16] The trade-off here is found in allowing flexibility for diverse hydrogen pathways, while considering that certified fuels need to be climate neutral by 2050.[17] As discussed by Fleming, the second Renewable Energy Directive (RED II),[18] now contained in the revised third Renewable Energy Directive (RED III),[19] does not expressly classify 'renewable hydrogen' as renewable energy for the purposes of certification.[20] As noted by Banet, there is a need to ensure consistency across legislative acts, as there remains a risk of divergence in definitions.[21] Such divergence may impact both the commodity market, and the scale and speed of infrastructure development.[22]

With a view to certification, Tanase and Herrera Anchustegui showed that the newly revised gas Directive (rGD)[23] provides for a certification scheme for renewable and low-carbon gases in line with RED II.[24] However, here too the devil is in the details: with respect to the definition of 'low-carbon gases', the rGD defines 'low-carbon hydrogen' as hydrogen derived from

15 Ibid.

16 Directive (EU) 2023/2413 of the European Parliament and of the Council amending Directive (EU) 2018/1999 and Directive 98/70/EC as regards the promotion of energy from renewable sources and repealing Council Directive (EU) 2015/652 [2023] OJ L 2023/2413, Recital 9(B).

17 Banet (n13), 38.

18 Directive (EU) 2018/2001 of the European Parliament and of the Council of 11 December 2018 on the promotion of the use of energy from renewable sources (recast) [2018] OJ L 328.

19 Directive (EU) 2023/2413 of the European Parliament and of the Council of 18 October 2023 amending Directive (EU) 2018/2001, Regulation (EU) 2018/1999 and Directive 98/70/EC as regards the promotion of energy from renewable sources, and repealing Council Directive (EU) 2015/652 [2023] OJ L 2023/2413.

20 Fleming (n 15), 132.

21 Banet (n 13), 25.

22 Banet (n 13), 25.

23 Directive (EU) 2024/1788 of the European Parliament and of the Council of 13 June 2024 on common rules for the internal markets of renewable gas, natural gas and hydrogen, amending Directive (EU) 2023/1791 and repealing Directive 2009/73/EC [2024] OJ L 2024/1788 (hereafter: rGD).

24 Lavinia Tanase and Ignacio Herrera Anchustegui, 'EU Hydrogen and Decarbonised Gas Market Package: Unbundling, third-party access, tariffs and discounts rules at the core of transport of hydrogen', in Íñigo del Guayo Castiella and Lorenzo Mellado Ruiz (eds) *Retos Regulatorios de los Gases Renovables en la Economía Circular* (Marcial Pons, 2023) (hereafter: Tanase and Herrera Anchustegui).

non-renewable sources which meets a greenhouse gas (GHG) emission threshold of 70 per cent compared to the fossil fuel comparator for renewable fuels of non-biological origin (RFNBOs) set out in the Directive.[25] However, Riemer argues that such a relative reduction is not always transparent and it would be preferable to include more concrete emission footprints to catalyse further emission reductions.[26]

With regard to the second challenge, namely the content of the sustainability criteria for hydrogen, lessons may be drawn from the sustainability of biofuels which preceded that of renewable gases in the first Renewable Energy Directive[27] (RED I). In this regard, biofuel certification has had more than a decade of practical experience which can serve as a guideline for certification of hydrogen. For example, in the context of biofuels, the sustainability criteria in RED I did not adequately consider indirect environmental effects, such as that a large percentage of biofuels from crops in use were not satisfactorily sustainable.[28] As a result, the updated 2018 RED II took into account indirect emissions as a result of indirect land-use change. In the context of hydrogen, indirect emission effects can take place in instances where carbon leakage occurs during the production, storage or transportation of, for example, low-carbon hydrogen.[29] This concern has partly been addressed in Art 9 of the rGD providing for the certification of low-carbon gases. According to Art 9(5) of the rGD, the EU must adopt delegated acts setting out GHG emissions savings for low-carbon fuels and must take emission leakages due to hydrogen into consideration.

Taking these issues into consideration, it is evident that legal certainty is a key factor with respect to defining renewable and low-carbon hydrogen as well as the sustainability criteria in respect thereof. However, the law also needs to be adaptable in instances where revision is required to meet the purposes set out therein. One could frame this, indeed, as the underlying tension between rapid technology expansion – here the need to import a lot of hydrogen into the EU, and the law's tendency of making normative compromises – namely by stretching and tweaking the definitions on hydrogen with a view to sustainability. This tension, albeit in a different form, will also become clear from the following section, tackling the second part of the hydrogen value chain, the transportation of hydrogen through pipelines.

---

25 RgD (n 24), Art 2.

26 Riemer (n 14), 2.

27 Directive 2009/28/EC of the European Parliament and of the Council of 23 April 2009 on the promotion of the use of energy from renewable sources and amending and subsequently repealing Directives 2001/77/EC and 2003/30/EC [2009] OJ L 140 (hereafter: RED I), Arts 17 and 18.

28 Riemer (n 14), 2.

29 Ibid.

6.2.2 Hydrogen transmission and distribution: transporting the transition

Once hydrogen is produced, it needs to be transported before ultimately arriving at the end customer,[30] a third step that will be discussed further below in Section 6.2.3. As noted by Tanase and Herrera Anchustegui, the rGD, which forms part of the Hydrogen and Decarbonized Gas Markets Package, plays an important role in providing regulatory principles applicable to hydrogen in the context of, among other things, the transmission through pipelines.[31] The principles applicable to hydrogen have been adapted from existing frameworks within the natural gas and electricity sectors.[32] Both the repurposing of existing gas pipelines containing either exclusively hydrogen or a mingled stream, as well as the establishment of dedicated hydrogen pipelines, are envisioned in the rGD.[33]

Scheibe and Poudineh opine that the system of hydrogen transmission through pipelines possesses characteristics akin to natural monopolies, particularly due to the high costs of developing and repurposing infrastructure.[34] As a result, experience from the electricity and natural gas sectors has emphasized the importance of ensuring a competitive market through unbundling measures as well as ensuring non-discriminatory third party access to pipeline networks.[35] At the same time, exceptions to these rules are required to stimulate more private investments in the sector.[36] Barnes highlights a chicken and egg dilemma given that the framework developed by the rGD will only become practically effective upon the development of a hydrogen market and associated infrastructure.[37] Yet, it is essential for the regulatory framework to be in place to facilitate this development.[38] There

---

30 Fleming (n 12).

31 Tanase and Herrera Anchustegui (n 25), 8. The Hydrogen and Decarbonized Gas Markets Package consists of the rGD and Regulation (EU) 2024/1789 of the European Parliament and of the Council of 13 June 2024 on the internal markets for renewable gas, natural gas and hydrogen, amending Regulations (EU) No 1227/2011, (EU) 2017/1938, (EU) 2019/942 and (EU) 2022/869 and Decision (EU) 2017/684 and repealing Regulation (EC) No 715/2009 (recast) [2024] OJ L 2024.

32 Tanase and Herrera Anchustegui (n 25), 8.

33 Fleming (n 12), 128.

34 Alexander Scheibe and Rahmatallah Poudineh, 'Regulating the future European hydrogen supply industry: A balancing act between liberalization, sustainability, and security of supply?' (2023), OIES Paper: ET, No. 26, ISBN 978-1-78467-218-8, Oxford Institute for Energy Studies, Oxford, 1.

35 Tanase and Herrera Anchustegui (n 25), 7.

36 Ibid.

37 Alex Barnes 'The EU Hydrogen and Gas Package: Help or hindrance for the development of a European hydrogen market?' (2023), OIES Paper: ET, No. 22, Oxford Institute for Energy Studies <www.oxfordenergy.org/publications/regulating-the-future-european-hydrogen-supply-industry-a-balancing-act-between-liberalization-sustainability-and-security-of-supply/> [accessed on 23 January 2024], 21.

38 Ibid.

is a certain tension arising from this chicken and egg dilemma, particularly with respect to the law needing to be both clear, on the one hand, and flexible, on the other.

In the development of a hydrogen market, questions arise as to who will own and operate hydrogen distribution and transmission networks. There are strikingly different opinions on this. The EU wants to use the so-called strict ownership unbundling model (OU)[39] as the default model for the regulation of hydrogen grids.[40] According to Barnes, the OU model can dissuade further private investment.[41] Barnes provides an example in this regard that the strict OU model prevents risk sharing that was common in the gas pipeline and LNG markets.[42] In line with that, even the Agency for the Cooperation of European Energy Regulators (ACER), according to Tanase and Herrera Anchustegui, advocated for usage of the so-called independent transmission operator model[43] (ITO) beyond 2030.[44] This would allow for flexibility in the law in accordance with the pace of the development of the hydrogen market.[45]

In addition to this debate about so-called vertical unbundling, the rGD also proposes a new unbundling model, horizontal unbundling. Baumgart and Lavrijssen explain horizontal unbundling in the following terms: where a hydrogen transmission network operator is part of an undertaking active in transmission or distribution of natural gas or electricity, it shall undergo at least legal unbundling, as well as accounting unbundling, in effect making it impossible for gas and electricity companies to cross-subsidize their future hydrogen grid business.[46] Horizontal unbundling, however, has been subject to the criticism that it may hamper synergies between gas and hydrogen networks, and may also result in an inefficient duplication of structures.[47] Tanase and Herrera Anchustegui counter that a lighter version of unbundling has been employed for horizontal unbundling, given that

---

39 Hydrogen producers and gas suppliers cannot own hydrogen transmission network companies and vice versa. See Art 60(1) of the rGD (n 24).

40 Alexander Scheibe and Rahmatallah Poudineh (n 41), 27.

41 Alex Barnes (n 37), 16.

42 Ibid.

43 The hydrogen transmission network company remains within the vertically integrated company, subject to certain stringent rules applying to prevent production and supply branches from interfering with running the network. See Art 61(4) of the rGD.

44 Tanase and Herrera Anchustegui (n 25), 50.

45 Ibid.

46 Max Baumgart and Saskia Lavrijssen, 'Exploring regulatory strategies for accelerating the development of sustainable hydrogen markets in the European Union' (2023) *Journal of Energy and Natural Resources Law*, available online at: https://doi.org/10.1080/02646811.2023.2257528 [accessed on 20 January 2024].

47 Ibid.

the risks of conflicts of interest stemming from combined operatorship is less likely than in vertical integration.[48] In this respect it has been noted that any remaining risks can be mitigated through monitoring and approval by regulatory authorities.[49]

This debate on the correct type of unbundling highlights the underlying tension between rapid technology expansion, here hydrogen grids, and the law's general tendency of making normative decisions – here the idea of creating liberalized and competitive energy markets. This tension is also clear from the third and final part of the regulation of the hydrogen value chain concerning hydrogen end-use.

6.2.3 Hydrogen customer end-use: (em)powering the transition

While the current uptake of hydrogen by end-users is in its infancy, the supply of hydrogen to customers has been the topic of some discussion, particularly regarding key themes of customer empowerment, on the one hand, and hydrogen end-use in sectors such as heating, on the other.

With a view to customer empowerment, Heidecke and others note that the rGD places a particular emphasis on consumer protection rules concerning supplier switching, price comparisons and obtaining accurate data on consumption.[50] With regard to supplier switching in particular, the rGD provides customers with a right to switch suppliers for natural gas and hydrogen, and sets protective measures on pricing.[51]

With a view to hydrogen end-use, the EU Hydrogen Strategy has already identified several early-adopter sectors in which a 'switch' to renewable or low-carbon hydrogen may take place.[52] These sectors include the industrial sector using hydrogen in the production of ammonia, methanol and steel making.[53] However, dedicated hydrogen infrastructure can also be utilized in providing heat for residential and commercial buildings. As noted by Scheibe, the use of hydrogen in the heating of buildings has been the subject

48 Tanase and Herrera Anchustegui (n 25), 11.

49 Ibid, 11.

50 Laura Heidecke, Irina Kustova, Michael Flickenschild, Yoeri Dijkhof, Iris Larmi, Harry Van Til et al, 'The Revision of the Third Energy Package for Gas' (2022) Publication for the Committee on Industry, Research and Energy (ITRE) www.europarl.europa.eu/RegData/etudes/STUD/2022/734009/IPOL_STU(2022)734009_EN.pdf [accessed on 17 January 2024].

51 rGD (n 24), Art 12.

52 European Commission, 'EU Strategy on Hydrogen' COM (2020) 301 final, 17 (hereafter: EU Hydrogen Strategy).

53 Ibid, 10.

of uncertainty and debate.[54] Many debates, as of late, concern the extent to which the Energy Performance of Buildings Directive[55] should support the use of hydrogen as a heating solution in buildings. Scheibe noted that hydrogen strategies across member states differ in respect of their treatment of heating in buildings. For example, the German Hydrogen Strategy anticipates a limited role for hydrogen in district heating, whereas the Dutch Hydrogen Strategy foresees a more prominent role for using hydrogen as a means of decarbonizing the heating sector.[56]

Hydrogen Europe, however, has argued that the advantage of using renewable hydrogen, for example in the heating sector, is that renewable energy may be stored and used at a later stage for the heating of homes in an efficient and cost-effective manner.[57] The Association of the European Heating Industry proposed a 'multi-vector' approach, meaning that heating solutions are provided on a case-by-case basis depending on the placement and particular construction of a building.[58] This may include hybrid solutions, consisting of both electricity and hydrogen as heating solutions.[59] Ultimately, the revised Energy Performance and Buildings Directive does not mention hydrogen explicitly in relation to the promotion of zero-emission buildings, but does refer to energy derived from the combustion of renewable fuels.[60]

These two items illustrate the tension between the protection of end-use customers and enabling free choice, on the one hand, and the steering of these for achieving the most efficient use of hydrogen in the energy sector, on the other. This, once again, shows how the rapid expansion of a certain type of technology (here large-scale production of hydrogen) can sit squarely with normative ideas like customer protection and the willingness of the state to keep a firm grip on the wheel to steer developments.

Taking the aforementioned literature review as well as the identification of three central themes into consideration, the following section summarizes

---

54 Alexander Scheibe, 'Regulating the future European hydrogen supply industry: A balancing act between liberalisation, sustainability, and security of supply' (2023) OIES Paper: ET26 <www.oxfordenergy.org/publications/regulating-the-future-european-hydrogen-supply-industry-a-balancing-act-between-liberalization-sustainability-and-security-of-supply/> [accessed on 20 January 2024] (hereafter: Scheibe).

55 Directive (EU) 2024/1275 of the European Parliament and of the Council of 24 April 2024 on the energy performance of buildings (recast) [2024] OJ L, 2024/1275, recital 22 (hereafter: Energy Performance Directive).

56 Scheibe (n 56), 17.

57 Hydrogen Europe Position Papers, 'Hydrogen technologies can boost the energy performance of buildings', May (2022) 6.

58 European Heating Industry 'EHI position on the review of the Energy Performance of Buildings Directive' (2021).

59 Ibid.

60 Energy Performance Directive (n 57), recital 20.

these three key 'battlegrounds' of hydrogen regulation and explains why they require further consideration in any given jurisdiction that is aiming to regulate its emerging hydrogen economy.

6.3 Three central battlegrounds for the creation and implementation of hydrogen-specific legislation

Having provided an overview of the legal academic debate and the frictions, we will now catalyse these contributions into three central points currently being debated. All three will be examined and discussed in turn. Their relevance is likely to extend beyond the EU context and these will be the key debates about hydrogen regulation in any given jurisdiction (of course with some local variations).

6.3.1 Who owns the hydrogen grids? State-owned grid company vs private companies (and who pays for them?)

The chicken and egg dilemma exists in hydrogen use, just as it exists in electric vehicle regulation, for example. The solution to this dilemma is the creation of the infrastructure first, to then be able to create demand. But this immediately opens up an interesting Pandora's box – for, if we need to build new hydrogen pipelines and receiving terminals and need to repurpose existing natural gas infrastructure, who is going to do it? The state or private companies? A state-owned infrastructure company might be quick to create hydrogen pipelines, as there would be taxpayers' money involved. This could be done in the form of Transmission (TSO) and Distribution System Operators (DSO)[61] who already own natural gas infrastructure.

When making a comparison to the currently existing natural gas grid, we find that in Europe the state is directly or indirectly in charge of the TSOs (and most DSOs). Usually this is done, as in the Netherlands and Germany, by the creation of a public limited company, where the state is either the only shareholder[62] or, as in France, an indirect shareholder in this public limited company.[63]

The EU, with its new Hydrogen and Decarbonized Gas Markets Package, essentially decided to continue putting state-owned, or at least

---

[61] For general explanations on what TSOs and DSOs are and their effective roles, please consider Hans Vedder et al, 'EU Energy Law' in Martha Roggenkamp et al (eds) *Energy Law in Europe*, 3rd edition (Oxford University Press, 2016) para 4.224 and following.

[62] Gasunie, 'Organisation', available at: www.gasunie.nl/en/organisation [accessed 18 January 2024].

[63] GRT Gaz 'Governance', available at: www.grtgaz.com/en/our-group/governance [accessed 18 January 2024].

state-controlled, companies in charge of the hydrogen infrastructure. For existing and newly built infrastructure this will be either new-to-create TSOs or, if the member states individually allow for it, the existing gas TSOs, if the provided exemptions of the rGD are being used.[64]

It will be interesting to see how the EU rules, which are about to come into existence, will be transposed to national law. Germany is one of the countries where various options for the operation of hydrogen grids have been contemplated, but ultimately one option has been selected.[65] The applicable legislation is, at the time of writing, being drafted, but the underlying idea will be that the existing gas TSOs build and repurpose the pipelines, either directly or perhaps through subordinate companies.

Another question could be asked when contemplating the friction between legal 'catching up' and developments on the ground: Is there a certain area where more room could be given to private companies/players? Imagine a country deciding to build a separate hydrogen grid for the supply of gas to industry. If there is an area where gas pipelines exist, but instead of converting those, new hydrogen pipelines are being built, then the state might be more willing to give more room to private companies in the hydrogen context. But this seems limited to very specific (and niche) use-cases, as the trend is to consider conversion of existing natural gas pipelines to hydrogen pipelines rather than building new ones.[66]

One way or the other, the interesting question for the future is: If we know who owns the grids, who is paying for them? In the beginning they will not have many customers that want to move hydrogen via these pipelines, as the hydrogen economy needs to grow. To incentivize the switch of end-customers from natural gas to hydrogen and the usage of these newly built and repurposed pipelines, the TSOs could charge discounted tariffs to 'first mover customers'.[67] In essence, the TSOs will, therefore, have to bear a large share of the costs. In the future when more customers use the pipelines, the 'late coming' customers will then be charged tariffs above the amount

---

64 For more details see the debate in Section 6.2.

65 Art 28r Energy Industry Act (Energiewirtschaftsgesetz).

66 The Netherlands even wants to repurpose about 85 per cent of its existing natural gas grid for hydrogen, see Dutch State Secretary of Economic Affairs and Climate Policy – Climate and Energy Policy, 'Development transport network for hydrogen' ('*Ontwikkeling transportnet voor waterstof*') (letter to the Dutch Parliament, 30 June 2021), available at: https://open.overheid.nl/documenten/ronl-66d67edc-8d97-42e5-9f61-c4bc4bf5a1c6/pdf [accessed on 22 October 2023] and Germany about 60 per cent, according to the Economic Ministry of Germany, see: www.bmwk.de/Redaktion/DE/Pressemitteilungen/2023/11/20231115-gesetz-zur-wasserstoff-netzplanung-und-kernnetz-finanzierung-beschlossen.html [accessed 18 January 2024].

67 This is currently being discussed in the German parliament. It shall be included in Art 28s of the EnWG, see: https://dserver.bundestag.de/btd/20/100/2010014.pdf

needed to maintain the networks.[68] The idea being that the late comers will pay for the first movers.

6.3.2 Hydrogen in the heating sector

The decarbonization of the housing sector is a key element of climate change policies in many countries.[69] There are two possible routes for decarbonization. The first route would be a more or less all-electric heating system with little to no gas molecules involved anymore. Such a system would mainly run on heat pumps or geothermal heating for individual residential homes, as well as on, for example, district heating for bigger housing units. The key argument, in terms of hydrogen, against this system is that hydrogen is too expensive and precious (meaning it could be better employed elsewhere) to be used for heating purposes.[70]

The competing model to the all-electric heating world is one where solar panels, for example, play a significant role in heating, but where hydrogen is also part of the picture. The arguments for that model are essentially two-fold: first, the individual applications in houses that are currently running on natural gas heating would require minor adjustments to be able to burn hydrogen.[71] As a result, costs of conversion would be much lower than in the 'all-electric' scenario.[72] Second, the use of different energy carriers for heating purposes (electricity and hydrogen) is preferable from an energy security perspective.[73] As price hikes in Europe for natural gas in the context of Russia's war on Ukraine aptly demonstrated, it is never a good idea to rely heavily on one energy carrier. Diversification of energy carriers but also of

---

68 Ibid.

69 For Europe see European Commission, 'Heating and Cooling', available at: https://energy.ec.europa.eu/topics/energy-efficiency/heating-and-cooling_en#:~:text=Decarbonising%20the%20heating%20and%20cooling,cannot%20be%20achieved%20cost%2Deffectively [accessed 18 January 2024].

70 See debate earlier in this chapter and, for example, Hydrogen Insight, 'Hydrogen is too scarce, too expensive and too inefficient for heating, warns German state body', available at: www.hydrogeninsight.com/policy/hydrogen-is-too-scarce-too-expensive-and-too-inefficient-to-use-for-heating-warns-german-state-body/2-1-1485409 [accessed 18 January 2024].

71 See DVGW, 'Gasgeräte bereit für 20% Wasserstoff', available at: www.dvgw.de/medien/dvgw/leistungen/publikationen/gasgeraete-bereit-fuer-20_-h2-factsheet-dvgw-avacon.pdf [accessed 18 January 2024].

72 DVGW, 'Heizen mit Wasserstoff', available at: www.dvgw.de/medien/dvgw/leistungen/publikationen/heizen-mit-h2.pdf [accessed 18 January 2024].

73 IRENA, 'Geopolitics of the Energy Transformation: The Hydrogen Factor', available at: www.irena.org/Digital-Report/Geopolitics-of-the-Energy-Transformation [accessed 18 January 2024].

supplies is, thus, key. Laws on heating and on energy security, however, are based mainly on the usage of natural gas, so there will need to be adjustments to cater for this new option.

6.3.3 What actually is sustainable in green, clean and renewable hydrogen?

When can hydrogen be considered 'sustainable'? This is an important issue, as it raises the question whether there are particular criteria that can be applied to categorize hydrogen as green or clean and which role does sustainability play in hydrogen? If we want to deploy hydrogen at speed and scale, what are the legal measures that we need to take to make sure we do not use just any hydrogen, but a sustainable one?

Sustainability criteria for fuels and energy carriers are well known in EU law, particularly in the context of the import of bioenergy into the EU. Whether or not this can be used as an example, or an analogy, to solve the issue of sustainability criteria for hydrogen, is the subject of this section.

Sustainability criteria for the production and import of bioenergy are region-specific in the sense that the importing regions will all set their own, individual criteria. To give an indication, however, we will briefly discuss the legal situation in Europe concerning sustainability criteria to enable a comparison to other regions to be carried out by others.

Sustainability criteria were included in the law for the first time in the 2009 Renewable Energy Directive (hereinafter RED I).[74] The scope of this Directive was limited to biofuels and bioliquids. According to the Directive, both are produced from biomass, but biofuels are liquid or gaseous fuels for transport, while bioliquids are liquid fuels for energy purposes other than for transport, including electricity, heating and cooling.[75]

The mandatory sustainability criteria of the EU have two components. First, biofuels and bioliquids must achieve a certain threshold of GHG emissions savings in comparison to the use of fossil fuels.[76] Second, the raw materials cultivated for the production of biofuels or bioliquids must not come from land with high biodiversity value, land with high carbon stock or from peatlands.[77] If these criteria are not met, the biofuels or bioliquids can still enter and be sold in the EU market, but they cannot receive financial support or count toward the renewable energy targets of EU member states.[78]

---

[74] RED I (n 29). The same criteria were also included in parallel in a revision to the Directive 98/70/EC of 13 October 1998 relating to the quality of petrol and diesel fuels [1998] OJ L 350.

[75] Ibid, Art 2(h) and (i).

[76] Ibid, Art 17(2).

[77] Ibid, Art 17(3)–(5).

[78] Ibid, Art 17(1).

With a view to sustainability criteria in hydrogen, the starting point is REDII/RED III as well as the rGD.[79] The rGD features sustainability criteria for renewable hydrogen, at least indirectly. Indeed, Art 9(1) of the rGD requires renewable gases, hence also renewable hydrogen, to be certified in accordance with Arts 29, 29a and 30 of the Recast 2023 Renewable Energy Directive (RED III).[80] Renewable gases encompass biogas and the renewable gaseous fuels part of RFNBOs. RFNBOs are themselves defined as liquid and gaseous fuels, the energy content of which is derived from renewable sources other than biomass, and to guarantee that RFNBOs are indeed of renewable origin, they have to comply with a 2023 Delegated Act from the Commission.[81] This Delegated Act was adopted as required by RED III.[82] For hydrogen to be considered renewable, the electrolyser must consume electricity either through a direct connection to a generation plant using renewable sources, or through a connection to the grid on the condition that it runs overwhelmingly on renewable sources, or otherwise through power-purchase agreements with generation from renewable sources and, additionally, temporal correlation and geographical correlation rules. These rules all apply to hydrogen whether produced inside or outside the EU.[83]

All this essentially means that EU law adapted RED's bioenergy sustainability criteria, that have been devised for bioenergy, and then enlarged them and applied them to renewable hydrogen. Whether or not such an approach is appropriate for other regions to enable the development of hydrogen at speed and scale needs to be determined individually.

6.4 Conclusion

Acknowledgement of the importance of hydrogen in decarbonization efforts across sectors in the EU has gained significant traction over the last few years. This acknowledgement has been marked by the development of EU and national policies facilitating the transition from 'grey' to 'green' hydrogen. However, the key challenge lies in the practical implementation, particularly given that renewable and low-carbon hydrogen uptake in various industries

79 rGD (n 24).

80 While Art 9(1) provides that renewable hydrogen will be certified in accordance with RED II, Arts 29, 29a and 30 are contained in the consolidated version of RED III.

81 Commission Delegated Regulation (EU) 2023/1184 of 10 February 2023 supplementing Directive (EU) 2018/2001 of the European Parliament and of the Council by establishing a Union methodology setting out detailed rules for the production of renewable liquid and gaseous transport fuels of non-biological origin (hereafter: Delegated Act).

82 Ibid. As the Delegated Act was adopted before RED III, it still refers to Art 27(3) RED II, but in RED III this provision is now numbered 27(6).

83 Ibid, Delegated Act, Art 1.

is still at a nascent stage. Only careful and reasonable implementation can strike a good balance between rapid technology expansion and the (energy) law's general tendency of being driven by a normative impetus.

The 'chicken and egg' dilemma arising from the need for regulatory frameworks to facilitate uptake (and vice versa) presents an interesting opportunity for hydrogen regulation. The three battlegrounds for hydrogen highlighted key tensions and considerations for successful implementation. Evidently, the development of hydrogen infrastructure, and tailored legal frameworks governing such development, are key to overcoming the chicken and egg dilemma. Sufficient financial incentives need to be in place, particularly for increased (private?) investment in hydrogen infrastructure.

Another piece to the hydrogen puzzle is that both domestically produced and imported hydrogen need to meet policy commitments toward net-zero. Periodic revisions to the methodology for assessing sustainability criteria for hydrogen will need to take into consideration whether countries in and outside of the EU can meet the requirements that have been set out.

With a view to hydrogen supply to end-customers, the use of hydrogen largely depends on customers' ability to switch, both from a cost- and an energy-efficiency perspective. For heating, the choice between hydrogen, electrification or both largely depends on political interests and steering at a national level. Despite differences in the pace and scale at which hydrogen as an energy carrier is being adopted in various industries, the legal framework governing this adoption must keep up with developments, and will set the pace. Legal frameworks are therefore required to be both sufficiently certain and agile for the deployment of hydrogen to be facilitated both in scale and in speed around the globe.

Acknowledgement

This chapter is part of the THERESA Project (GA no. 101073195), which is funded by the EU Horizon programme under the Marie Skłodowska-Curie Actions (MSCA).

7

Genetically Modified Organisms: Is the Proof in the Pudding or in the Process?

Mirta Alessandrini and Josephine van Zeben

7.1 Introduction

Since its earliest applications, the use of genetic engineering in agriculture, food and energy-crop production has sparked significant debate among scientists, policy makers and the general public. An important driver of this debate was the uncertainty surrounding the potential effects of DNA mutations on human, animal and plant health, and on the environment more generally. However, since the early 2010s,[1] the increased use of gene-editing techniques as an alternative to genetic engineering, in crop development, environmental management, climate change mitigation and the overall governance of a more sustainable agriculture, has led to a resurgence of this debate.

While genetic engineering involves the manipulation of an organism's genetic material by introducing foreign DNA into its genome, gene editing enables precise modifications of an organism's genome by means of a small, controlled edit of its DNA. Through gene editing, the gene is modified in a predictable manner by way of regulatory RNA molecules and/or site-specific DNA-modifying enzymes, without the insertion of foreign genetic material.[2] This means that it is nearly impossible to detect whether an

---

1 Ahmad M. Khalil, 'The Genome Editing Revolution: Review' [2020] 18(68) *Journal of Genetic Engineering and Biotechnology* 1.

2 Thorben Sprink et al, 'Regulatory Hurdles for Genome Editing: Process- vs. Product-based Approaches in Different Regulatory Contexts' [2016] 35 *Plant Cell Reports* 1493, 1494.

organism's DNA has been altered, as the changes may be indistinguishable from naturally occurring mutations.

The regulation and use of Genetically Modified Organisms (GMOs) is therefore an especially important part of this debate. With population growth and climate change, many consider GMOs as a, or *the*, solution to feeding the world. Despite having been consistently proven safe by authoritative regulatory and scientific authorities such as the American Food and Drug Administration (FDA)[3] and the European Food Safety Authority (EFSA),[4] the public debate surrounding GMOs extends far beyond the realm of science.[5]

Importantly, GMO regulation differs strongly between jurisdictions, in part because the public perception of GMOs and their risk and desirability varies strongly between countries. France stands out as a large economy with the strongest resistance and most deep-seated fears regarding the hazards associated with GMOs.[6] Japan, the United States and the United Kingdom display varying levels of consumer apprehension, with Americans generally expressing the lowest level of risk toward GMOs in terms of human health and environmental concerns.[7] Traditionally, consumers in countries such as Belgium, France, the Netherlands, Spain and the UK, are willing to pay a premium to avoid purchasing GM-labelled food.[8] Italians have been reticent in embracing GMOs and, due to the perceived potential risks that GM foods may pose to human health and the environment, have advocated for stringent labelling requirements for GM food.[9] Interestingly, Eurobarometer findings from 2022 indicate that Europeans, as a whole, have become less concerned about GMOs; across the European Union (EU), fewer than one in ten individuals expressed specific concerns about GMOs when asked about them.[10] Nevertheless, EU regulation regarding GMO use, development and

3 FDA, 'How GMO Crops Impact Our World' (19 September 2023) <www.fda.gov/food/agricultural-biotechnology/how-gmo-crops-impact-our-world> accessed 16 June 2023.

4 See, for instance, the latest renewed authorization for oilseed rape MS8, RF3 and MS8 × RF3, EFSA 'Assessment of genetically modified oilseed rape MS8, RF3and MS83RF3 for renewal authorisation under Regulation (EC) No 1829/2003 (application EFSA-GMO-RX-024)' [2023] 21(4) *EFSA Journal* 1.

5 Peter Rembischevski and Eloisa Dutra Caldas 'Risk Perception Related to Food' [2020] 40(4) *Food Science and Technology* 779.

6 Keiko Komoto et al, 'Japanese Consumer Perceptions of Genetically Modified Food: Findings from an International Comparative Study' [2016] 5(3) *Interact Journal of Medical Research* 23.

7 Ibid.

8 Anne-Cécile Delwaide et al, 'Revisiting GMOs: Are There Differences in European Consumers' Acceptance and Valuation for Cisgenically vs Transgenically Bred Rice?' [2015] 10(5) *Plos One* 1.

9 Shahla Wunderlich and Kelsey A. Gatto, 'Consumer Perception of Genetically Modified Organisms and Sources of Information' [2015] 6 *Advances in Nutrition* 842.

10 EFSA, 'Special Eurobarometer Wave EB97.2 Food safety in the EU' [2022], Report.

labelling continues to be highly risk averse.[11] Unsurprisingly, American regulation shows a greater openness to scientific innovation and technology, even in the context of the agri-food sector.[12]

In this chapter, we focus on debates surrounding the EU GMO legislation,[13] which has come under great scrutiny with respect to its ability to regulate the biotechnology progress in DNA editing.[14] Notwithstanding the socio-cultural and regulatory specificity of this example, and the many important developments in other jurisdictions, the European debate brings together key scientific, ethical, environmental, regulatory and socio-economic considerations and highlights the need for balancing potential benefits with risk assessment, responsible innovation and inclusive decision-making processes.[15] As such, we find this debate to be illustrative of the debate regarding GMOs writ large.

The primary aim of the EU's GMO legislation is to ensure a high level of protection of human, animal, plant and environmental health, irrespective of the cost of such protection.[16] The EU approach is based on a complex risk regulatory architecture setting different levels of control mechanisms. It consists of an *ex-ante* authorization procedure for the cultivation of GMOs[17] or their placement on the market, based on a meticulous safety assessment;[18]

---

[11] Zaki Laïdi, 'Europe as a Risk Averse Power' in Daniel Innerarity and Javier Solana (eds), *Humanity at Risk: The Need for Global Governance* (Bloomsbury 2013).

[12] Parul Goyal and Stuti Gurtoo, 'Factors Influencing Public Perception: Genetically Modified Organisms' [2011] 2(1) *GMO Biosafety Research* 1.

[13] Directive 2001/18/EC of the European Parliament and of the Council of 12 March 2001 on the deliberate release into the environment of genetically modified organisms and repealing Council Directive 90/220/EEC – Commission Declaration [2001] OJ L 106 [hereinafter: GMO Release Directive]; Regulation (EC) No 1829/2003 of the European Parliament and of the Council of 22 September 2003 on genetically modified food and feed [2003] OJ L 268; Regulation (EC) No 1830/2003 of the European Parliament and of the Council of 22 September 2003 concerning the traceability and labelling of genetically modified organisms and the traceability of food and feed products produced from genetically modified organisms and amending Directive 2001/18/EC [2003] OJ L 268; and Directive 2009/41/EC of the European Parliament and of the Council of 6 May 2009 on the contained use of genetically modified micro-organisms [2009] OJ L 125; Directive (EU) 2015/412 of the European Parliament and of the Council of 11 March 2015 amending Directive 2001/18/EC as regards the possibility for the Member States to restrict or prohibit the cultivation of genetically modified organisms (GMOs) in their territory [2015] OJ L 68.

[14] Thorben Sprink et al (n 2) 1504.

[15] Dennis Eriksson et al, 'Options to Reform the European Union Legislation on GMOs: Scope and Definitions' [2020] 38(3) *Trends in Biotechnology* 231.

[16] Art 1, Regulation 1829/2003.

[17] Art 6, GMO Release Directive.

[18] Art 7, Regulation 1829/2003.

an *ex-post* monitoring of the potential effects on human and environmental health;[19] compliance with rigorous traceability and labelling requirements;[20] and the recognition of ample flexibility to member states to opt out of the authorization process for the cultivation of specific GMOs within the national territory on the basis of compelling grounds set by law.[21]

Despite its clear risk-regulation-based rationale, the EU's approach has become increasingly controversial since the early 2000s, as it has failed to address the relentless technological progress in this area and the urgent social needs presented by climate change and the need for food security. As a result, the political debate around GMOs has become increasingly polarized. The suitability and relevance of the EU's regulatory approach have increasingly been questioned, including the legitimacy of the EU's risk governance,[22] while continuing public hostility to GMOs has led to increasing divergence between the public's risk perception regarding GMOs and the EU's increasing (scientific) acceptance of GMOs, as demonstrated for example by EFSA's positive assessment of GMO safety.[23]

Following the introduction of the opt-out clause in 2015, a large proportion of EU member states chose to implement bans or restrictions on GMO cultivation based on various factors, including public opinion, concerns about potential environmental impacts, and the desire to protect organic or GM-free agricultural sectors. A total of 20 member states and the United Kingdom[24] banned the cultivation of approved GMOs in their territory, curbing the de facto geographical scope of GMO authorization and application in the EU.[25] On the other hand, Spain, Portugal, Czechia, Romania, Slovakia, Estonia and Finland have decided to allow for the

[19] Arts 4, 13 and Annex II, GMO Release Directive.

[20] Art 21, GMO Release Directive.

[21] Art 26b, Directive (EU) 2015/412.

[22] Susana Borrás, 'Legitimate Governance of Risk at the EU Level? The Case of Genetically Modified Organisms' [2006] 73 *Technological Forecasting and Social Change* 61.

[23] Yann Devos et al, 'EFSA's Scientific Activities and Achievements on the Risk Assessment of Genetically Modified Organisms (GMOs) During its First Decade of Existence: Looking Back and Ahead' [2014] 23 *Transgenic Research* 1.

[24] Austria, Region of Wallonia (Belgium), Bulgaria, Croatia, Cyprus, Denmark, France, Germany, Greece, Hungary, Italy, Latvia, Lithuania, Luxembourg, Malta, the Netherlands, Poland, Slovenia, Sweden, United Kingdom (Northern Ireland).

[25] <https://food.ec.europa.eu/plants/genetically-modified-organisms/gmo-authorisation/gmo-authorisations-cultivation/restrictions-geographical-scope-gmo-applicationsauthorisations-eu-countries-demands-and-outcomes_en#at> accessed 29 June 2023 and <https://stuff.coffeecode.net/www.loc.gov/law/help/restrictions-on-gmos/sweden.php#:~:text=Sweden%20is%20generally%20seen%20as,is%20currently%20not%20being%20cultivated> accessed 18 August 2023.

cultivation of authorized GMOs within their national borders.[26] Sweden has thus far approved only the GM Amflora potato for commercial production, though it is not currently being cultivated.

The key tension in the GMO debate on which this chapter focuses is the possible – and typically assumed – juxtaposition between safety and sustainability with regard to GMOs in our agri-food supply. In many ways, the growing importance of GMOs in our agri-food supply is the epitome of humanity's shift into the Anthropocene, with humans impacting ecosystems and species at a cellular level. This chapter presents these two sides of the debate on genetic engineering and gene-editing techniques from a legal perspective by focusing on three sub-topics. First, in Section 7.2, we discuss the legal definition of GMOs. Diverging views on GMOs and their risks and/or desirability, as well as the incredibly fast-moving development of new technologies, has also meant that a consistent legal definition of GMOs has yet to develop.[27] This section will summarize the main issues regarding GMO terminology. Second, in Section 7.3, we will analyse the regulation of GMO risk and the role of precaution, specifically given the debate regarding the 'optimal' level of precaution in light of perceived and actual risks of GMOs. Third, Section 7.4 will consider the potential roles of GMOs in food security and food sustainability, and more generally the shift toward a sustainability discourse. Section 7.5 concludes with an outline of a research agenda to steer the debate a step further.

7.2 Defining GMOs legally

7.2.1 Defining the definition

Depending on the type of DNA alteration techniques used, GM and non-GM can be indistinguishable from a biological perspective.[28] Legally, Art 2(2) of Directive 2001/18/CE on the deliberate release of GMOs into the environment ('GMO Release Directive') defines a GMO as: 'an

[26] <www.europarl.europa.eu/news/en/headlines/society/20151013STO97392/eight-things-you-should-know-about-gmos#:~:text=Only%20one%20GMO%20has%20been,Czech%20Republic%2C%20Romania%20and%20Slovakia> accessed 29 June 2023.

[27] Piet van Der Meer et al, 'The Status under EU Law of Organisms Developed through Novel Genomic Techniques' [2023] 14(1) *European Journal of Risk Regulation* 93.

[28] GM and non-GM organisms can be highly similar or even identical in terms of their basic biological characteristics. This is because genetic modifications made through genetic engineering techniques are often designed to target specific genes or traits while leaving the overall biological functioning of the organism largely intact. These biological similarities can pose challenges in distinguishing between GM and non-GM organisms solely based on their biological characteristics. It highlights the need for specialized testing methods, such as DNA analysis or genetic markers, to accurately determine the presence of genetic modifications and identify whether an organism is genetically modified or not.

organism, with the exception of human beings, in which the genetic material has been altered in a way that does not occur naturally by mating and/or natural recombination'.[29] Annex I A of the Directive provides an open-ended list of techniques through which genetic modification may be considered to occur or not.[30] In addition, Annex I B lists a number of techniques that are expressly excluded from the scope of the Directive, in particular mutagenesis.[31]

The EU's legal definition of GMOs refers to both process-based (for example, recombinant nucleic acid techniques) and product-based characteristics (for example, drought-tolerant crops, disease-resistant plants, or crops with increased vitamin content). Despite this, GMO legislation is frequently interpreted as strictly process-based.[32] This means that GMOs are regulated solely with reference to the technology through which they are produced, regardless of whether traces of modified DNA are present in the final crop, food or feed. A product-based approach focuses instead on the end product's safety, nutritional value and overall impact, rather than on the technological process, to determine its suitability for consumption or use. This approach is intended to ensure that the GM product meets the same safety and quality standards as non-modified products, regardless of the methods used to create it.

The process of DNA alteration of an organism can, in fact, be designed in a way that the final crop, food or feed product does not contain modified DNA. This can be achieved through a process called 'gene silencing' or 'gene suppression'.[33] Gene silencing involves targeting the specific gene that was modified or introduced into the organism and suppressing its activity. This is often done using naturally occurring mechanisms within the organism's cells. One common method is to introduce a double-stranded RNA molecule

[29] GMO Release Directive.

[30] Annex I A, parts 1 and 2, respectively.

[31] Annex I B also includes 'cell fusion of plant cells of organisms that can exchange genetic material through traditional breeding methods'. Mutagenesis is a process by which an organism's DNA is changed by the production of a mutation. This change may occur spontaneously in nature, as a result of exposure to mutagens (chemical or physical mutation-causing agents), or through laboratory procedures. There are a variety of mutations that can occur in DNA, such as changes in the DNA sequence or rearrangement of the chromosomes. Mutagenesis has been extensively used for the improvement of microorganisms of agricultural and industrial importance and has gained popularity in plant genetics research in the last century as a powerful tool for inducing genetic variation.

[32] Thorben Sprink et al (n 2) 1496.

[33] Huw D. Jones, 'Gene Silencing or Gene Editing: the Pros and Cons' in Bruno Mezzetti, Jeremy Sweet and Lorenzo Burgos (eds), *RNAi for Plant Improvement and Protection* (CABI 2021).

that is complementary to the gene's sequence.[34] This molecule triggers a cellular response that leads to the degradation or inhibition of the gene's messenger RNA (mRNA), which is required for protein synthesis. Without functional mRNA, the modified gene's protein product is not produced. As a result, even though the organism's genetic material was initially altered, the gene modification is effectively suppressed in the final product. Therefore, the final crop, food or feed derived from this modified organism does not contain detectable levels of modified DNA or the associated proteins.

Gene-editing techniques are increasingly championed as the solution for creating more efficient and more sustainable agricultural systems.[35] This push has highlighted the rigidity, and possible obsolescence, of the EU's legal definition. Specifically, genetic engineering and gene (or genome) editing are both used to modify the genetic material of organisms, but, as highlighted above, the way and extent to which they result in genetic modification differ in important ways. The rigidity of the GMO legal definition, therefore, raises implementation and enforcement challenges, as it could be difficult or even impossible to differentiate, for instance, plants obtained by targeted mutagenesis, cisgenesis or intragenesis from plants from conventional breeding.

Targeted mutagenesis is an umbrella term that refers to the intentional and specific modification of an organism's genetic material at a predetermined location within its genome. This allows scientists to introduce precise changes to the DNA sequence of an organism with the aim of creating desired genetic modifications, such as introducing or knocking out specific genes, without introducing foreign DNA from unrelated organisms. One of the most prominent techniques used for targeted mutagenesis is CRISPR-Cas9 (Clustered Regularly Interspaced Short Palindromic Repeats and CRISPR-associated protein 9).[36] Cisgenesis is a specific type of genetic modification

---

[34] Rima Kumari and Devendra Pratap Singh, 'Nano-Biotechnological Approach of Plant Genetic Engineering' in Pardeep Singh et al (eds), *Policy Issues in Genetically Modified Crops: A Global Perspective* (Academic Press 2021).

[35] Felicity Keiper and Ana Atanassova, 'Enabling Genome Editing for Enhanced Agricultural Sustainability' [2022] 4 *Frontiers in Genome Editing, Journal of Risk Regulation* 1.

[36] CRISPR-Cas9 is a specific variant of the CRISPR-Cas (Clustered Regularly Interspaced Short Palindromic Repeats-CRISPR associated) gene-editing system. It is a precise and efficient tool for editing DNA and consists of two main components: the Cas9 enzyme and the guide RNA (gRNA). The Cas9 enzyme is a protein that acts as a pair of molecular scissors capable of cutting DNA strands at specific locations. The gRNA serves as a molecular guide that directs the Cas9 enzyme to the desired target sequence in the genome. To use CRISPR-Cas9, scientists design a specific gRNA that is complementary to the target DNA sequence they want to modify. The gRNA is then combined with the Cas9 enzyme and introduced into the cells of an organism. Inside the cell, the gRNA binds to the target DNA sequence, allowing the Cas9 enzyme to make a precise cut at that location.

that involves transferring genes from the same species or one related to the target organism. The genetic material introduced into the target organism is derived from within its own gene pool or from sexually compatible species. Both targeted mutagenesis and cisgenesis are techniques used in genetic modification with the aim of improving crops, enhancing traits and addressing agricultural challenges.[37] Intragenesis, like cisgenesis, involves transferring genes between closely related species or within the same species. However, while in cisgenesis the new sequence contains an exact copy of the sequence already present in the breeders' gene pool, intragenesis produces a new sequence containing a rearranged copy of sequences already present in the breeders' gene pool.[38] As will be discussed below, in such cases, applicants might not be able to provide a specific detection method required for the pre-market authorization as required by law.

If a GMO falls within the scope of the Directive, it must undergo a pre-market authorization procedure before the organism can be released into the environment or cultivated.[39] If the GMO is created with an aim to be used in or as food or feed,[40] it will fall within the scope of Regulation 1829/2003 on genetically modified food and feed and it will be required to undergo an authorization procedure similar to that under the GMO Release Directive.[41] If the GMO is to be used in food or feed without cultivation, applying for food and feed purposes is enough. If the GMO is not to be used in food or feed, applying for authorization for cultivation suffices. However, if the GMO is to be used in food or feed with cultivation in the EU, companies need to apply for both cultivation and food/feed purposes.

The application for GMO authorization is submitted by the applicant or developer to the competent national authority of a member state first and then forwarded to the EU Commission. The application includes detailed information about the GMO, its characteristics, intended use and potential risks.[42] The application is then submitted to EFSA, which is responsible

Once the DNA is cut, the cell's natural DNA repair mechanisms are activated to repair the break. Depending on the repair process, different modifications can occur. See Jian-Hua Zhang et al, 'Optimization of Genome Editing through CRISPR-Cas9 Engineering' [2016] 7(3) *Bioengineered* 166.

37 EFSA, 'FAQ: Criteria for risk assessment of plants produced by targeted mutagenesis, cisgenesis and intragenesis' (EFSA Official website, 23 May 2023) <www.efsa.europa.eu/en/news/faq-criteria-risk-assessment-plants-produced-targeted-mutagenesis-cisgenesis-and-intragenesis> accessed 18 August 2023.

38 Ibid.

39 Arts 6–19, GMO Release Directive.

40 Art 3(1), Regulation 1829/2003.

41 Arts 15–23, Regulation 1829/2003.

42 Art 6, GMO Release Directive and Art 5, Regulation 1829/2003.

for performing the risk assessment.[43] In the case of GMOs for cultivation purposes, the assessment process begins with a comprehensive risk assessment conducted by the applicant and evaluated by the national competent authority of the member state where the application originates first and then by all the other member states, before being submitted to EFSA.[44] EFSA evaluates the data provided by the applicant, the national authority and the member states, assessing potential risks to human and animal health, as well as environmental impacts, potential allergenicity and antibiotic resistance. EFSA issues a scientific opinion on the GMO's overall safety, which includes recommendations for risk management and monitoring.[45] This opinion is made publicly available.[46]

Based on EFSA's opinion and other relevant factors, the EU Commission drafts a proposal for the authorization or refusal of the GMO. The proposal is then put forward to the EU's Standing Committee on Plants, Animals, Food, and Feed.[47] If the Standing Committee approves the proposal, the EU Commission issues a final authorization decision. In the case of GMOs for release into the environment or cultivation purposes, this decision is valid for a period of ten years, after which a new risk assessment must be performed and the authorization renewed.[48] Moreover, the GMO Release Directive respects each member state's right to restrict or prohibit the cultivation of GMOs on their territory, even if those GMOs have already received EU-wide authorization from the EU Commission.[49] This measure ensures a degree of flexibility and autonomy to national governments in deciding whether to allow the cultivation of GMOs within their borders, taking into consideration their specific agricultural and environmental conditions, as well as public concerns. In the case of GM food, however, there is no opt-out option nor the requirement to renew the authorization every ten years.

7.2.2 Applying definitions

While the GMO Release Directive provides the contours of a legal definition for GMOs, this definition refers to the unnatural altering of genetic material without exhaustively defining those processes. The exceptions of Annex I B are presently too vague; problematically, Annex I B excludes 'mutagenesis' from the scope of the Directive without detailing which mutagenesis

43 Art 28, GMO Release Directive.

44 Art 11, GMO Release Directive.

45 Art 6(6), GMO Release Directive and Art 18(6), Regulation 1829/2003.

46 Art 9, GMO Release Directive and Art 5(b), Regulation 1829/2003.

47 Art 15, GMO Release Directive and Arts 7 and 35, Regulation 1829/2003.

48 Art 15(4), GMO Release Directive.

49 Art 26(b), GMO Release Directive.

techniques this refers to. The scoping of the Directive has been further complicated with the arrival of new gene-editing techniques.[50]

This legal uncertainty has led to the creation of a significant body of case law in the European courts. In the *Confédération paysanne* case (Case C-528/16), one of the first landmark cases on this topic, the Court of Justice of the EU (CJEU) ruled that any organisms altered by gene-editing techniques should be subject to the same (stringent) regulation as GMOs under the GMO Release Directive.[51] The CJEU based its judgment on Recital (17) of the GMO Release Directive,[52] stating that only 'traditional' mutagenesis techniques, namely those with a *substantial application* and a *long history of safe use* may be exempted from the GMO legislation, while the 'newest' techniques – which mostly refers to targeted mutagenesis and cisgenesis techniques – should fall within the GMO legal definition and therefore be subject to the GMO regulation's stringent requirements.[53] In light of the Court's restrictive interpretation, new gene-editing techniques such as CRISPR-Cas9[54] and TALENs[55] are not covered by the 'mutagenesis exemption' and should therefore be considered techniques that create GMOs, in a legal sense.[56]

---

[50] Emerging plant breeding techniques, such as genome editing, expedite the creation of crop varieties with enhanced precision and efficiency compared to traditional methods. These advancements offer potential solutions to climate change, sustainability and global food security challenges. Nevertheless, the adoption of these techniques faces obstacles due to regulatory uncertainties surrounding genome edited crops. See Penny A. C. Hundleby and Wendy A. Harwood, 'Impacts of the EU GMO Regulatory Framework for Plant Genome Editing' [2019] 8(2) *Food and Energy Security* 1.

[51] Case C-528/16 Confédération paysanne and others vs Premier ministre and Ministre de l'agriculture, de l'agroalimentaire et de la forêt [2018] ECLI:EU:C:2018:583.

[52] 'This Directive should not apply to organisms obtained through certain techniques of genetic modification which have conventionally been used in a number of applications and have a long safety record'.

[53] Hanna Schebesta, 'Confédération paysanne case (C-528/16): Legal Perspective on the GMO Judgment of the European Court of Justice' [2020] Revue européenne de droit de la consommation (R.E.D.C.) 369, 372.

[54] Jian-Hua Zhang et al (n 36).

[55] TALENs (Transcription Activator-Like Effector Nucleases) are a type of genome-editing tool that enable targeted modifications of DNA sequences. Once the TALEN construct is introduced into cells, the TALE DNA-binding domain guides the nuclease domain to the desired target DNA sequence. See J. Keith Joung and Jeffry D. Sander, 'TALENs: A Widely Applicable Technology for Targeted Genome Editing' [2013] 14 *Nature Reviews Molecular Cell Biology* 49.

[56] The CJEU's position on gene-editing techniques was confirmed in Confédération paysanne II (C-688/21), which centred on *in vitro* random mutagenesis, where the mutagenic agents are applied to plant cells, and the whole plant is then artificially reconstituted. In this 2023 ruling, the Court ruled that organisms obtained by the *in vitro* application of a technique/method of mutagenesis which has conventionally been

In 2019, in view of this case law and other developments, the Council tasked the EU Commission with conducting a study on the current status of 'new genomic techniques' (NGTs) and submitting a legislative proposal, if appropriate.[57] In the EU Commission's study, NGTs are referred to as techniques capable of modifying an organism's genetic material which have been developed since 2001.[58] The study mainly focuses on targeted mutagenesis and cisgenesis in plants, clarifying that they are currently still both subject to GMO legislation.

The study acknowledges the potential benefits of NGTs – such as contributing to sustainable agri-food systems – but also underlines concerns regarding safety, biodiversity conservation, coexistence with organic/conventional agriculture, and consumers' right to information.[59] The study concludes that the current authorization procedure and risk assessment requirements are not adaptable to the variety of potential plant products that can be obtained by targeted mutagenesis, cisgenesis and intragenesis, as the current risk assessment protocols may struggle to distinguish between naturally occurring variations and modifications introduced through these techniques. The result is a disproportionate or inadequate regulatory framework that cries out to be adjusted to better reflect scientific and technological progress, as this approach may treat gene-edited organisms with the same level of scrutiny as transgenic organisms, even though the risks and characteristics can differ significantly.[60]

The EU Commission's conclusions were endorsed by EFSA, which received a mandate to develop specific risk assessment criteria for plants produced by targeted mutagenesis, cisgenesis and intragenesis.[61] EFSA stated that similar risk profiles can be achieved in plant products using conventional breeding techniques, suggesting that, from a scientific point of view, it may not be justifiable to subject NGT products to the same regulatory

used in a number of *in vivo* applications and has a long safety record with regard to those applications are excluded from the scope of the Release Directive.

57 Council Decision (EU) 2019/1904 of 8 November 2019 requesting the Commission to submit a study in light of the Court of Justice's judgment in Case C-528/16 regarding the status of novel genomic techniques under Union law, and a proposal, if appropriate in view of the outcomes of the study [2021] OJ L 293.

58 European Commission, Commission Staff Working Document, Study on the status of new genomic techniques under Union law and in light of the Court of Justice ruling in Case C-528/16 (2021) 116.

59 Ibid.

60 Ibid.

61 European Commission, 'Request in accordance with Article 31 of Regulation (EC) No 178/2002 for an EFSA statement on criteria for risk assessment for plants produced by targeted mutagenesis and cisgenesis' [2022] (Ref. Ares(2022)3302985).

requirements as conventional GMOs.[62] Based on these findings, the EU Commission proposed the development of a legal framework for plants obtained by targeted mutagenesis and cisgenesis and for their food and feed products in July 2023. The EU Commission's explicit aim for the new legal framework is to enable agri-food system-related innovation so as to achieve the sustainability targets set by the EU Green Deal (EUGD)[63] and the Farm to Fork Strategy (F2F Strategy),[64] while upholding a high level of protection for human and animal health and the environment.[65]

In 2021, the EU Commission presented the Inception Impact Assessment on the 'Legislation for plants produced by certain new genomic techniques'.[66] It presents five policy options, of which the EU Commission deems Option 4 the most groundbreaking. Under this option, the legislative initiative would address key issues, such as the lack of clear legal definitions for terms such as 'mutagenesis', 'techniques/methods of mutagenesis', 'conventional use in a number of applications', or 'long safety record'. Moreover, plants obtained by targeted mutagenesis or cisgenesis that could also occur naturally or be produced by conventional breeding would only require a notification procedure.

The preferred option is a combination of Option 4 for products that could also occur naturally or be produced by conventional breeding and Option 2 for all other products. Option 2 entails that an authorization would still be required and the organism would remain subject to the same traceability and labelling conditions as 'traditional' GMOs; the risk assessment would, however, be adapted to diverse risk profiles and detection challenges.[67] This combination would maximize the safety of NGT plants and their food/feed products, making them comparable in safety to conventional products, without imposing unnecessary regulatory burdens.[68] Regardless of the options chosen, the proposal is expected to enhance legal certainty by providing important clarifications,

[62] EFSA, 'Criteria for Risk Assessment of Plants Produced by Targeted Mutagenesis, Cisgenesis and Intragenesis' [2022] 20(10) *EFSA Journal* 12.

[63] Communication from the Commission to the European Parliament, the European Council, the Council, the European Economic and Social Committee and the Committee of the Regions – The European Green Deal [2019] COM/2019/640 final.

[64] Communication from the Commission to the European Parliament, the European Council, the Council, the European Economic and Social Committee and the Committee of the Regions – A Farm to Fork Strategy for a fair, healthy and environmentally-friendly food system [2020] COM/2020/381 final.

[65] Art 191, TFEU.

[66] European Commission, Inception Impact Assessment, Legislation for plants produced by certain new genomic techniques, COM(2021) 5835503 5.

[67] FDA (n 3).

[68] Sprink et al (n 2).

albeit no specific reference has been made to a possible change in the legal definition of GMOs.[69]

At the time of writing, the Proposal for a Regulation on plants obtained by certain NGTs and their food and feed[70] had been published. The key regulatory innovation included in the Proposal centres on the simplification of the administrative procedure around authorization, notably through an adapted risk assessment and a new verification procedure for products that fulfil the criteria for equivalence to conventional breeding. This simplification of the authorization process for NGTs should facilitate access to Small and Medium Enterprises (SMEs), reducing costs and expediting the development of new products.

The draft Proposal combines Option 4 and Option 2 as presented in the Inception Impact Assessment. The deliberate release and placing on the market of NGT plants will be subjected to one of two procedures: the notification procedure to establish equivalence with conventional products for plants and their derivatives obtained by targeted mutagenesis or cisgenesis that could also occur naturally or be produced by conventional breeding (that is, 'category 1 NGT plants');[71] or the authorization procedure in accordance with Directive 2001/18 or Regulation 1829/2003 for all the other products ('category 2 NGT plants').[72] From the proposal, it is clear that the GMO legal definition will not be amended but rather juxtaposed with a legal definition for NGT plants. Though the NGT terminology was already commonplace in the GMO debate and regulatory landscape, the proposal provides the first legal definition. The definition provided defines an NGT plant as a 'genetically modified plant obtained by targeted mutagenesis, cisgenesis, intragenesis, or a combination thereof, on the condition that the NGT plant does not contain any genetic material originating from outside the breeders' gene pool that temporarily may have been inserted during the development of the NGT plant'.[73] This means that, contrary to the traditional genetic engineering, in the case of NGTs the changes in the DNA are made using only the plant's own genes or those of its close relatives. No genes from outside sources are added, even temporarily, during the process.

Plants falling within the 'category 1 NGT plants' will be exempted from the GMO legal requirements and will be subjected to those applicable to

---

69 Sprink et al (n 2).

70 Proposal for a Regulation of the European Parliament and of the Council on plants obtained by certain new genomic techniques and their food and feed, and amending Directives 68/193/EEC, 1999/105/EC, 2002/53/EC, 2002/55/EC, and Regulation (EU) 2017/625 [2023] Draft.

71 Title II, Proposal.

72 Title III, Proposal.

73 Art 3(2), Proposal.

conventional products.[74] Nonetheless, 'category 1 NGT plants' will be treated as GMOs for the purposes of organic production,[75] as the Organic Regulation prohibits the use of GMOs in the production or cultivation of organic food and crops.[76] This category is determined having regard to the fulfilment of the criteria of equivalence to conventional plants set out in Annex I and to the intended characteristic(s) or property(ies) conveyed by the genetic modification.[77] The notification procedure will also apply in the case of the placing on the market of products containing or consisting of category 1 NGT plants or food or feed produced from such plants.[78] For all the other NGT plants (that is, 'category 2 NGT plants'), and food and feed derived from them, the pre-market authorization will still be required.[79] Interestingly, the Proposal dedicates specific incentives for 'category 2 NGT plants' and their NGT food- and feed-containing traits relevant for sustainability.[80] Incentives will consist in an accelerated procedure for risk assessment as regards applications handled by a fully centralized procedure (food and feed products) and enhanced pre-submission advice to help developers prepare the dossier for the purpose of the environmental and food and feed safety assessments.[81] In general, applicants will be supported by pre-admission advice provided by EFSA. For example, SME applicants will be exempted from the payment of specific financial contributions and will receive suggestions on the structure of the scientific study required to support the application for the pre-market authorization procedure when necessary.[82]

7.2.3 State of the debate

Much of the definitional debate has revolved around the understanding of the second part of the GMO legal definition. Particularly, the question whether the indication of 'altered in a way that does not occur naturally by mating and/or natural recombination'[83] pertains to the technique employed, the resulting novel genetic combination, or both, has proven challenging.[84]

[74] Art 5, Proposal.

[75] Art 5(2), Proposal.

[76] Art 11, Regulation (EU) 2018/848 of the European Parliament and of the Council of 30 May 2018 on organic production and labelling of organic products and repealing Council Regulation (EC) No 834/2007 [2018], OJ L 150.

[77] Art 5(1)(a), Proposal.

[78] Art 7, Proposal.

[79] Arts 13–24, Proposal.

[80] Art 25, Proposal.

[81] Recital (34), Proposal.

[82] Art 25(2)(b) and (3)(b), Proposal.

[83] Art 2(2), GMO Release Directive.

[84] Piet Van Der Meer et al (n 27) 100.

Art 2(2) of the GMO Release Directive, read in conjunction with Annex I A, suggests that the legal definition should cover both, as the legal text refers to the term 'techniques' in combination with the requirement of the identification of a resulting organism whose genetic material has been 'altered'. Therefore, based on the letter of the law, one could conclude that, in order to be classified as a GMO, two cumulative conditions must be fulfilled: the use of a technique that modifies the organism's DNA as well as the creation of an organic product with altered DNA.

However, the use of a genetic technique does not always result in an organism with altered genetic material that is different from an organism whose genetic alteration has occurred naturally. This similarity arises because genetic techniques, including targeted mutagenesis, cisgenesis and intragenesis, aim to mimic and harness natural genetic variations that could occur through regular processes such as mutation or recombination. As explained, NGTs often work by introducing small-scale and controlled changes that are similar to what could happen naturally, avoid introducing foreign genes, and work within the boundaries of what is genetically possible for a given organism. Currently, and in light also of the CJEU's ruling on gene editing, the two situations are regulated equally, and this proves that the EU legislature gives much more relevance to the techniques used than to the final characteristics of the resulting product.

In the gene-editing case law, the CJEU performed a more traditional task of testing and validating regulatory systems and acting as an impetus for regulatory revisions.[85] While in *Confédération paysanne* the CJEU provided a clear (strict) indication of what should fall within the scope of the mutagenesis exemption, '[t]he challenge thus remains to determine the precise scope of the GMO definition itself, [under which] both process (genomic techniques employed) and product (genetic alteration realised) need to be cumulatively met'.[86] This can, therefore, only be evaluated on a case-by-case basis.

The Inception Impact Assessment[87] criticizes the existing state of regulatory uncertainty due to a lack of clear legal definitions. Ironically, it still fails to clarify the legal definition of GMOs or its 'mutagenesis exemption'. The Impact Assessment solely refers to '[m]echanisms to be able to rapidly adjust elements of the legislation and its implementation over time, as warranted by scientific and technological progress, for a future-proof regulatory framework'[88] without addressing the elephant in the room, namely, the

85 Eleonora Sirsi, 'Le regole degli OGM nello spazio globale: un'agenda per i governanti del futuro' [2010] 3 *Rivista di Diritto Agrario* 466, 467.

86 Tarja Laaninen, 'New Genomic Techniques: European Commission Study and First Reactions' [2021], European Parliamentary Research Service, PE 698.760, p 10.

87 European Commission (n 66).

88 FDA (n 3).

regulation of genome-editing techniques that do not necessarily lead to mutagenesis as defined by the GMO Release Directive. The NGT study, therefore, should be used as a basis to clarify further which level of genetic alteration would be considered as 'not occur[ring] naturally by mating and/or natural recombination'.[89] However, the proposal fails to address this issue.

The 2022 public consultation incorporates a range of views: academics, breeders, farmers (excluding organic and GM-free practitioners), other participants in the agri-food chain, and public authorities largely advocate for revising the current legislation to establish a more flexible framework. Conversely, the majority of environmental organizations, NGOs and retail and consumer associations favour maintaining the existing status quo.[90] Moreover, the fast-paced developments in this field have underlined the much slower pace of legislative processes, which has led to legal disruption.[91]

Some have suggested the introduction of a definition of GMOs that emphasizes the characteristics of the final product rather than the type of technique used, shifting from a process-based to a more product-based approach.[92] This legal definition could take inspiration from the definition of a Living Modified Organism (LMO) as provided in the Cartagena Protocol on Biosafety (CPB) to the Convention on Biological Diversity,[93] which states that an LMO is 'any living organism that possesses a novel combination of genetic material obtained through the use of modern biotechnology'.[94] This would also aid the enforcement of the GMO legislation.[95]

At this stage, the EU Commission has clarified the objective of the Proposal. The new Regulation on NGT will aim at ensuring that 'only plants with a risk profile comparable to that of naturally occurring or conventionally bred plants are not subject to the requirements of the current GMO legislation',[96] while, at the same time, 'plants featuring complex sets of modifications [will] continue to be under the regulatory oversight of the GMO legislation'.[97]

---

89 Art 2(2), GMO Release Directive.

90 European Commission, 'Legislation for plants produced by certain new genomic techniques Public Consultation Factual Summary Report' [2022] Ref. Ares(2022)6392169.

91 Kai P. Purnhagen, 'You Want It Extra CRISPERY? Legal Disruption through New Plant Breeding Technologies in the EU' [2021] 40(1) *Yearbook of European Law* 374.

92 Dennis Eriksson et al (n 15).

93 Cartagena Protocol on Biosafety (CPB) to the Convention on Biological Diversity (CBD), opened for signature 29 January 2000, 2226 UNTS 208, entered into force 11 September 2003 [hereinafter: CPB].

94 Art 3(g), CPB.

95 Piet Van Der Meer et al (n 27) 100.

96 Recital (14), Proposal.

97 Ibid.

7.3 Managing uncertainty: risk and the precautionary principle

7.3.1 The risk–precaution dichotomy

When it comes to scrutinizing the risk–precaution dichotomy in the genetically engineered organisms setting, the debate around the regulation and use of GMOs quickly becomes polarized. The precautionary principle plays a pivotal role in shaping the risk regulatory architecture of GMOs, and, in general, in defining the controversial relationship between law and science. By advocating a cautious and proactive stance in managing risks, the precautionary approach has resulted in a risk-averse attitude from the EU Commission. As shown by our analysis of the EU's legal definition, this precautionary approach has also been applied to GMOs based on concerns about their potential long-term effects on human and animal health and the environment.[98]

The EU's risk regulatory architecture is built on several pillars that aim at minimizing or mitigating potential harm to public health, safety and the environment. The first pillar is the setting up of a pre-market authorization procedure before a GM product can be released into the environment or marketed as food and feed.[99] Second, as described above, there is a system for an *ex-ante* assessment of the environmental risks associated with releasing the GMO performed case-by-case (that is, 'environmental risk assessment').[100] Third, also mentioned earlier, there is the *ex-post* monitoring of the GM product after its release into the environment and placement on the market as food and feed.[101] Fourth, there is a mechanism that allows the member state to modify, suspend or terminate deliberately released GMOs once information regarding the risks of releasing becomes available (the safeguard clause),[102] or that revokes the authorizations for placing it on the market for food and feed if the product no longer meets the conditions set by the law.[103] Last, as previously described, there is the possibility of an opt-out clause, under which member states have the right to adopt measures to restrict or ban the cultivation of an authorized GMO within their territory, based on compelling grounds related to national environmental or agricultural policy objectives, socio-economic impacts or public policy concerns.[104]

---

98 Alessandra Guida, 'The Precautionary Principle and Genetically Modified Organisms: A Bone of Contention between European Institutions and Member States' [2021] 8(1) *Journal of Law and the Biosciences* 1.

99 Arts 6 *et seq.*, GMO Release Directive and Arts 3 *et seq.*, Regulation 1829/2003.

100 Art 4 in conjunction with Annex II, GMO Release Directive.

101 Art 20 in conjunction with Annex VII, GMO Release Directive.

102 Art 23, GMO Release Directive.

103 Arts 10 and 22, Regulation 1829/2003.

104 Art 26b, Directive 2015/412.

This *safety-conscious* approach is complemented by the principle of coexistence under Art 26a(2) of the GMOs Release Directive. The principle requires member states in which GMOs are cultivated to take appropriate measures in border areas to avoid the possible cross-border contamination into neighbouring countries in which the cultivation of those GMOs is prohibited, and especially to avoid the unintended presence of GMOs in conventional and organic crops. The principle of coexistence also allows member states to set their own stricter technical standards and leaves considerable freedom to individual growers and companies to manage the minimizing of unintended presence of GMOs in non-GM crops. On the basis of the EU Commission's guidelines for the development of national coexistence measures issued in 2010,[105] member states adopted different protectionist measures.

On the one hand, the principle of coexistence allows the setting up of stringent rules to ensure that genetically modified, conventional and organic crops can be kept segregated from one another during cultivation, harvest, transport, storage and processing. On the other, it aims to preserve the freedom of choice for farmers, consumers and other stakeholders involved in agricultural production, to protect the integrity of conventional and organic production systems, and to mitigate potential economic impacts on farmers and economic losses that may occur due to cross-contamination or market rejection of crops that contain an unintended GMO presence.

The principle of coexistence brings up the issue of how socio-economic assessment and environmental risk assessment are connected.[106] Specifically, it raises the question of the extent to which the socio-economic implications of introducing GM crops into agricultural land should be taken into account in the risk assessment procedure – *science-based* by definition – and what is their weight and relevance in the final decision. Following the risk analysis definition under Art 6 of Regulation 178/2002,[107] socio-economic consequences are an aspect considered only in the risk management stage, performed by the EU Commission; the latter follows the risk assessment – performed by EFSA – and precedes the risk communication – performed by both – in the risk analysis framework. Nevertheless, with the introduction of the opt-out clause, they now play a major role in the decision of a member

[105] Commission Recommendation of 13 July 2010 on guidelines for the development of national co-existence measures to avoid the unintended presence of GMOs in conventional and organic crops [2010] OJ C 200/1.

[106] Eleonora Sirsi (n 85).

[107] Regulation (EC) No 178/2002 of the European Parliament and of the Council of 28 January 2002 laying down the general principles and requirements of food law, establishing the European Food Safety Authority and laying down procedures in matters of food safety [2002] OJ L 31.

state to restrict or ban the cultivation of an authorized GMO within the national territory.

7.3.2 *State of the debate*

Over the years, the traditional risk-averse position on GMOs adopted by the EU Commission, and supported by a vast majority of the member states, has been bolstered by an anti-GMO campaign conducted by NGOs.[108] Drawing heavily from an anti-corporate narrative, this campaign has widened the gap between public opinion – mostly based on the perception of the risk and the intuitive perception of GMOs' potential harmfulness by the public – and scientific evidence regarding the potential harms.

The precautionary approach adopted by the EU Commission, based on its reading of the precautionary principle, has had several consequences. First, while the introduction of an opt-out clause managed to create a bypass to the de facto moratorium on GMO authorization, this has also exacerbated the already fragmented stance of the member states. Some scholars have therefore suggested shifting from an 'opt-out' to an 'opt-in' legal mechanism.[109] This would allow individual member states to authorize the cultivation of a GM crop in the national territory after an EFSA risk assessment, rather than banning or limiting it after the GM crop has been authorized at the EU level.

Second, the current interpretation of the precautionary approach appears to pre-empt the safeguard clause under the GMO Release Directive. The safeguard mechanism allows member states to adopt emergency measures – based on their own assessments of potential risks – to restrict or prohibit the cultivation or sale of a specific GMO within their territory.[110] This provision can be invoked when a member state has new scientific evidence or information that suggests a potential risk to human health or the environment that was not previously identified during the EU's risk assessment process.

Once informed, the EU Commission then assesses the situation and may either request additional information or take measures to ensure the uniformity of the internal market. The application of the safeguard clause has been restricted to prevent the unilateral application of the precautionary principle by single member states.[111] The safeguard clause has been invoked by many member states in various circumstances. In 2011 alone, there were 22 active bans in place across six member states: Austria, France, Germany,

[108] Robert Paarlberg, 'A Dubious Success: The NGO Campaign against GMOs' [2014] 5(3) *GM Crops & Food* 223.

[109] Dennis Eriksson et al, 'Why the European Union Needs a National GMO Opt-in Mechanism' [2018] 36(1) *Nature Biotechnology* 18.

[110] Art 23, Directive.

[111] Eleonora Sirsi (n 85) 468.

Luxembourg, Greece and Hungary.[112] This has led scholars to claim that the opt-out option 'will have exactly the opposite effect to that intended, allowing the creation of arbitrary GM-free zones in Europe that will cause untold damage to the EU economy and its global scientific standing'.[113] Nevertheless, on review, many of these bans have been considered scientifically unjustified by EFSA.[114] The precautionary stance has often been used to put the conflict on GMO authorization on hold for too long, rather than to circumscribe risks without a consequent hampering of opportunities.

Third, the precautionary principle may obscure the relevance of the final product characteristics. To overcome this hurdle, regulatory requirements could also be adjusted based on the type of genetic change, and the risk assessment could take into account the specific characteristics and potential risks associated with different GMOs, as suggested by the Proposal for a Regulation on plants obtained by certain NGTs and their food and feed. This approach can differentiate between GMOs with varying levels of risk and focus regulatory scrutiny on higher-risk GMOs.

Fourth, the authorization is granted on the basis of a qualified majority voting system that ensures great flexibility to the member states, as the procedure can easily be halted by a few votes against any GM crop approval, regardless of the recommendation of the EFSA GMO Committee.[115] Some scholars have regarded this process as dysfunctional,[116] due to the decentralization of competences – from the EU to the national level – with respect to the implementation of the precautionary principle.[117]

Last, the strict EU regulatory framework and its precautionary stance have created obstacles to research and development activities, including with regard to GM energy crops, as addressed later in the chapter. Obtaining authorization for field trials and conducting research on GM energy crops is a complex and resource-intensive process. In particular, the current GMO legislation may affect the availability and accessibility of GM energy crops with desired traits, such as enhanced biomass production, improved tolerance to environmental stress, or increased efficiency in biofuel production. Despite the fact that GM energy crops currently fall within the scope of

---

[112] Maite Sabalza et al, 'EU Legitimizes GM Crop Exclusion Zones' [2011] 29 *Nature* 315.

[113] Ibid, 315.

[114] <www.europarl.europa.eu/EPRS/EPRS-AaG-545708-Member-State-bans-on-GMOs-FINAL.pdf> accessed 29 June 2023.

[115] Nigel G. Halford, 'Legislation Governing Genetically Modified and Genome-Edited Crops in Europe: The Need for Change' [2019] 99 *Journal of the Science of Food and Agriculture* 8.

[116] Ibid, 11.

[117] Jale Tosun and Ulrich Hartung, 'Decentralising Competences in Multi-level Systems: Insights from the Regulation of Genetically Modified Organisms' [2018] 41(3) *West European Politics* 803.

the GMO legislation, discussions over tailor-made criteria and procedures for risk assessment and management are already ongoing.[118] Therefore, the characteristics of newly developed GM plants intended for non-food purposes may require the consideration of revisions of risk assessment and risk management criteria, confinement and/or containment measures, as well as the regulation of coexistence, due to the potential risks posed by outgrowing or gene flow from non-food crops.

The principle of coexistence was developed with a view to counterbalancing some of these effects. In the scholarship, the term 'coexistence' has been given two different meanings: coexistence based on the mutual recognition of agriculture's methods (organic, conventional, GM), on the one hand, and coexistence that calls for non-existence of GM crops in the national territory, on the other.[119] This second meaning has been identified in some member states' legislation, in which one of the most frequent types of coexistence measures consists in the isolation distances. For instance, Hungary employs the use of minimum compulsory buffer zones, which are established in border regions between member states that allow GMO cultivation and those that do not. In other countries, including Czechia, Germany and Austria, crops in buffer zones must be treated as GM crops after harvesting for the purpose of labelling requirements.[120]

7.4 GMOs in food security and sustainable agriculture

7.4.1 Feeding tomorrow's world in a more sustainable way?

Developments in biotechnology, specifically through agricultural applications, have shifted the focus to genetic engineering as the answer to food security, and gene editing as an important ally for sustainable agriculture.[121]

Food security has been part of the GMO debate since the late 1990s. In 1999, the Nuffield Council on Bioethics[122] released a report that emphasized the potentially beneficial connection between food security and GM

[118] Danielle Bütschi et al, 'Genetically Modified Plants and Foods: Challenges and Future Issues in Europe' [2009] European Parliamentary Technology Assessment (EPTA) 1.

[119] Eleonora Sirsi, 'Coexistence: A New Perspective, a New Field' [2016] 8 *Agriculture and Agricultural Science Procedia* 449.

[120] Pauline Verrière, 'Preventing GMO contamination: an overview of national "coexistence" measures in the EU' [2014] IFOAM EU Group 1.

[121] Sigfrid Kjeldaas et al, 'With Great Power Comes Great Responsibility: Why "Safe Enough" Is Not Good Enough in Debates on New Gene Technologies' [2023] 44 *Agriculture and Human Values* 533.

[122] The Nuffield Council on Bioethics is a UK-based independent charitable body established in 1991 that examines and reports on ethical issues in biology and medicine <www.nuffieldbioethics.org/> accessed 29 June 2023.

technology.[123] By introducing genes from other organisms, scientists could enhance crop traits such as yield, resistance to pests and diseases, and tolerance to environmental stressors.[124] GM crops have long been viewed as a great opportunity to stabilize and ensure food supply for poor subsistence farmers, as well as to reduce the use of chemical pesticides and fertilizers.[125] The Nuffield report stated that it was a moral imperative to carefully investigate these potential benefits in order to help alleviate poverty, enhance food security and promote profitable agriculture in developing nations.[126]

Since the report's publication, climate change and continuing environmental pollution have exacerbated the need to improve existing varieties and increase agricultural production. However, the push for intensive agriculture without considering ecological requirements has led to further environmental problems, including soil degradation, erosion, pollution, salinization, desertification and the depletion of freshwater resources and agrobiodiversity.[127] Since the early 2010s,[128] scientific and academic research has redirected its attention toward exploring gene-editing techniques as a viable source of sustainability.

The alignment between current EU GMO legislation and the sustainability objectives of key strategies such as the EUGD, F2F and Biodiversity[129] has sparked a heated debate, especially regarding the possibility of GM technology to help achieve these aims.

While the EU Commission acknowledges the importance of new technologies,[130] sustainable solutions and disruptive innovation[131] in achieving

---

123 Nuffield Council on Bioethics, 'Genetically modified crops: the ethical and social issues' [1999] Report.

124 See, for instance, Hossein Azadi and Peter Ho, 'Genetically Modified and Organic Crops in Developing Countries: A Review of Options for Food Security' [2010] 28(1) *Biotechnology Advances* 160; Henrik Egelying, 'Managing Agricultural Biotechnology for Sustainable Development: The Case of Semi-arid India' [2000] 2(4) *International Journal of Biotechnology* 342; Robert Paarlberg, 'The Real Threat to GM Crops in Poor Countries: Consumer and Policy Resistance to GM Foods in Rich Countries' [2002] 27 *Food Policy* 247.

125 Henrik Egelying (n 124).

126 Nuffield Council on Bioethics (n 123), 67–8.

127 Agata Tyczewska et al, 'Towards Food Security: Current State and Future Prospects of Agrobiotechnology' [2018] 36(12) *Trends in Biotechnology* 1219.

128 Stuart J. Smyth, 'Contributions of Genome Editing Technologies Towards Improved Nutrition, Environmental Sustainability and Poverty Reduction' [2022] 4 *Frontiers in Genome Editing* 1.

129 Communication from the Commission to the European Parliament, the Council, the European Economic and Social Committee and the Committee of the Regions – EU Biodiversity Strategy for 2030 Bringing nature back into our lives [2020] COM/2020/380 final.

130 Art 7, Regulation 1829/2003.

131 EUGD Communication (n 63), 18.

the EUGD and the Biodiversity Strategy objectives, it falls short of providing concrete measures for their development and application. Similarly, with respect to the F2F Strategy, the EU Commission committed to exploring the sustainability benefits of new genomic techniques through the NGTs study,[132] but no specific legal measures have been included in the F2F Action Plan.

While some scholars emphasize the importance of regulatory reform that considers socio-economic factors, public engagement and ethical implications of biotechnology, there is a pressing need to also incorporate sustainability aspects.[133]

The urgency to strike a balance between sustainability, socio-economic considerations and ethics is heightened by the realization that relying solely on scientific information for risk management decisions is insufficient.[134]

7.4.2 State of the debate

The scientific and academic discussion on the perceived food security–sustainability dichotomy has shifted: gene-editing techniques are increasingly considered a promising tool for achieving sustainable agriculture rather than one solely appropriate for ensuring food security. However, issues such as Intellectual Property Rights, costly research and development activity, and technology accessibility need careful attention.

Championing local food production and embracing a diverse range of enhanced crop plants have been considered as ways to promote – rather than inhibit – biodiversity.[135] Advocates of existing GMO legislation argue that strict regulations safeguard human and environmental health, promote biodiversity and support organic and conventional farming practices in alignment with the objectives of the F2F and Biodiversity Strategies.[136] Similarly, there is concern about corporate control and high costs associated with the technology, limiting access for small-scale farmers who are already disadvantaged.[137] 'The worldwide Intellectual Property Rights regime

132 F2F Communication (n 64), 8.

133 Rosa Binimelis and Anne Ingeborg Myhr, 'Inclusion and Implementation of Socio-Economic Considerations in GMO Regulations: Needs and Recommendations' [2015] 8(62) *Sustainability* 1.

134 Artem Anyshchenko, 'The Precautionary Principle in EU Regulation of GMOs: Socio-Economic Considerations and Ethical Implications' [2019] 32 *Journal of Agricultural and Environmental Ethics of Biotechnology* 855.

135 Carsten Hjort et al, 'European Genome Editing Regulations: Threats to the European Bioeconomy and Unfit for Purpose' [2021] 1 *EFB Bioeconomy Journal* 1.

136 <www.eurovia.org/wp-content/uploads/2021/03/2020-09-17-Public-GMO-letter-to-Kyriakides.pdf> accessed 19 June 2023.

137 Henrik Egelying (n 124).

has already been extended to cover plants and animals, including their genes, making the technology expensive and thus, inaccessible for the subsistence farmers who already have limited opportunities for using the GM technology.'[138] To ensure progress and local innovation, it is crucial to address these challenges, considering equitable access, adequate education and training for farmers. If not addressed properly, the aforementioned aspects may hinder rather than stimulate both global and local innovation processes.[139]

7.5 Conclusions and outlook

Most individuals hold predetermined opinions about GMOs, which are heavily influenced by societal and cultural norms. Pro-technology narratives are deeply rooted in capitalist ideology, emphasizing security and efficiency to maximize economic gains. These approaches are particularly entrenched in the US, where there is a greater tendency to embrace risks associated with technological advances. Conversely, the EU's resistance to technological development in the GMO sector is grounded in ethical considerations and a precautionary approach.[140] Although, as shown above, Europeans, especially younger generations, are not inherently more sceptical of biotechnology than Americans, the EU has experienced significant mobilization against GMOs since the 1990s, driven by concerns over bio-safety and bio-property threats.[141] 'The idea of genes escaping into the environment—transgenic plants are "released", while other seeds are simply planted—resonated with a discourse of corporate irresponsibility and environmental risk.'[142]

Due to the scientific complexity of the issue, the political framing of GMOs in mainstream media has significantly influenced public opinion and legislation. A recent study indicates that both traditional and social media have tended toward increasing favourability and less polarized discussions on GMOs.[143] However, despite these improvements and the shift toward more sustainable considerations, the scientific community continues to face

[138] Hossein Azadi and Peter Ho (n 124), 164.

[139] Henrik Egelying (n 124).

[140] Jacob Hunter, ' GMO Framing in the United States and European Union' [2022] 4(1) *North Carolina Journal of European Studies* 1.

[141] George Gaskell et al, 'Europeans and Biotechnology in 2005: Patterns and Trends. A report to the European Commission's Directorate-General for Research' [2006] Eurobarometer 64.3 EU Commission.

[142] Ronald J. Herring, 'Opposition to Transgenic Technologies: Ideology, Interests and Collective Action Frames' [2008] 9 *Nature Reviews Genetics* 458.

[143] Sarah Evanega et al, 'The State of the "GMO" Debate: Toward an Increasingly Favorable and Less Polarized Media Conversation on Ag-biotech?' [2022] 13(1*) GM CROPS & FOOD* 38.

significant challenges in bridging the gap between media debates and the scientific consensus on the safety of GMOs.

Therefore, the European GMO debate continues to be divided into two camps, one that supports the current, restrictively precautionary, approach with a one-size-fits-all view of potential risks, and another that advocates for a more tailored regulatory approach that wants to consider the nature and size of individual risks.

The GMO debate takes place against a backdrop of continued scientific uncertainty – within which the known unknowns are continuously shifting – and a growing need to ensure greater sustainability in a safe food system. Adding to the resulting puzzle is the significant divergence between scientific and legal progress, with the law often lagging behind the fast-paced technological developments in this domain in troubling ways. If the aim is to create more holistic, adaptive, and integrative regulatory approaches to the GMO-NGT dispute, several avenues of future research are particularly important.

First, decision-making processes must more meaningfully incorporate environmental, social and economic considerations in order to holistically assess the risks and benefits of emerging technologies. The Inception Impact Assessment on the legislation for plants produced by certain NGTs shows the EU Commission's first attempt to steer the conversation in this direction, by highlighting the failure of the current GMO legislative framework to take into account whether GM plants, food or feed have the potential to contribute to societal challenges, notably sustainability. The draft Proposal seeks to respond to the objectives of climate neutrality as set by the European Climate Law,[144] recognizing the importance of utilizing genetic diversity and non-harmful plant genetic resources, supported by the latest scientific advancements, as crucial solutions to address climate risks faced by farmers and land managers. Moreover, we envisage the debate to be also affected by the legislative framework for sustainable food systems.[145]

Second, alternative governance models, such as adaptive regulatory and management approaches, should be considered in order to find other ways to incorporate the precautionary principle. An adaptive regulatory approach[146] would more easily keep pace with ongoing research, incorporating adjustments

[144] Regulation (EU) 2021/1119 of the European Parliament and of the Council of 30 June 2021 establishing the framework for achieving climate neutrality and amending Regulations (EC) No 401/2009 and (EU) 2018/1999 [2021] OJ L 243.

[145] The proposal was meant to be presented in the fourth quarter of 2023 but has been put on hold. European Commission, Inception Impact Assessment – Sustainable food system framework initiative [2021] COM/2020/381 final.

[146] Pablo Rozas et al, 'Genetically Modified Organisms: Adapting Regulatory Frameworks for Evolving Genome Editing Technologies' [2022] 55(31) *Biological Research* 1.

in the legal texts based on emerging scientific evidence. Flexible regulatory frameworks will prioritize safety and sustainability while accommodating advancements. Moreover, an adaptive management approach[147] would empower agile decision-making, striking a balance between sustainability, socio-economic considerations and ethics. The urgency is heightened by the realization that relying mostly (or even solely) on scientific information for risk management decisions is insufficient. Under both approaches, the precautionary principle is used as a backstop, aiming at minimizing harmful effects on human health and the environment, while balancing potential benefits to agriculture and trade.[148]

In closing, we suggest that future research focuses on participatory and integrative approaches,[149] which could help bridge gaps between society, law and science through knowledge exchange and public engagement. Addressing both food security and the human impact on climate change requires a multifaceted approach, in which genetic engineering techniques could be one of the viable avenues to explore.[150] Empowering collective problem-solving and boosting scientific literacy are therefore key to unlocking these complex challenges.

[147] Jamie Benidickson et al, 'Practicing Precaution and Adaptive Management: Legal, Institutional and Procedural Dimensions of Scientific Uncertainty' [2005] Report 1.

[148] Alessandra Guida, 'The Precautionary Principle and Genetically Modified Organisms: A Bone of Contention between European Institutions and Member States' [2021] 8(1) *Journal of Law and the Biosciences* 1.

[149] Rosa Binimelis and Anne Ingeborg Myhr (n 133).

[150] Joanna K. Sax, 'Genetically Engineered Food, Food Security, and Climate Change' [2022] 6(1) *The Business, Entrepreneurship & Tax Law Review* 1.

8

Cultivated Meat and Dairy as a Game-Changing Technology in the Agricultural and Food Transition in the EU: What Role for Law?

Jonathan Verschuuren

8.1 Introduction

Conventional meat and dairy production and consumption have a large impact on the health of the planet and its inhabitants. Climate change, land system change, biodiversity loss, fresh water depletion, disruption of nitrogen and phosphate cycles, ocean acidification as well as a range of direct impacts on human health are all to some extent attributed to livestock-keeping and meat and dairy consumption. In addition, there are serious animal welfare concerns about large-scale livestock-keeping for human consumption. A great food transformation is inevitable, as advocated by the EAT-Lancet Commission on healthy diets from sustainable food systems,[1] as well as by the Intergovernmental Panel on Climate Change (IPCC),[2] the

[1] W Willet et al, 'Food in the Anthropocene: the EAT–Lancet Commission on Healthy Diets from Sustainable Food Systems' (2019) 393 *The Lancet* 447–92.

[2] GJ Nabuurs et al, 'Agriculture, Forestry and Other Land Uses (AFOLU)' in PR Shukla et al (eds), Climate Change 2022: Mitigation of Climate Change. Contribution of Working Group III to the Sixth Assessment Report of the Intergovernmental Panel on Climate Change (CUP 2022).

Intergovernmental Science-Policy Platform on Biodiversity and Ecosystem Services (IPBES)[3] and the Food and Agriculture Organization (FAO).[4]

To date, the EU's regulatory approach to reduce the negative environmental, health and animal welfare impacts of livestock-keeping has been diverse and haphazard. Nitrogen deposition in protected areas is regulated through the Habitats Directive,[5] methane emissions are regulated through the Effort Sharing Regulation[6] and through the Industrial Emissions Directive[7] for the larger livestock farms (as are nitrous oxide emissions), nitrate and phosphate discharges into surface and ground waters are regulated through the Nitrates Directive[8] and the Water Framework Directive,[9] there are some rules on animal welfare[10] and organic livestock farming,[11] and the Common Agricultural Policy (CAP)[12] tries to persuade farmers to reduce environmental and climate impacts through conditional income support, and so on. Under the EU's Farm to Fork Strategy for a fair, healthy and environmentally friendly food system and the EU's climate policy,[13] many of

3 S Diaz et al, Summary for Policymakers of the Global Assessment Report on Biodiversity and Ecosystem Services of the Intergovernmental Science-Policy Platform on Biodiversity and Ecosystem Services (IPBES 2019) 18.

4 FAO, Transforming Food and Agriculture to Achieve the SDGs. 20 Interconnected Actions to Guide Decision-makers (FAO 2018).

5 Directive 92/43/EEC of 21 May 1992 on the conservation of natural habitats and of wild fauna and flora [1992] OJ L 206/7.

6 Regulation (EU) 2018/842 of the European Parliament and of the Council of 30 May 2018 on binding annual greenhouse gas emission reductions by Member States from 2021 to 2030 contributing to climate action to meet commitments under the Paris Agreement and amending Regulation (EU) No 525/2013 [2018] OJ L 156/26.

7 Directive 2010/75/EU on industrial emissions (integrated pollution prevention and control) [2010] OJ L 334, 17.

8 Council Directive 91/676/EEC of 12 December 1991 concerning the protection of waters against pollution caused by nitrates from agricultural sources [1991] OJ L 375/1.

9 Directive 2000/60/EC of the European Parliament and of the Council of 23 October 2000 establishing a framework for Community action in the field of water policy [2000] OJ L 327/1.

10 Council Directive 98/58/EC of 20 July 1998 concerning the protection of animals kept for farming purposes [1998] OJ L 221/23.

11 Regulation (EC) No 834/2007 of 28 June 2007 on organic production and labelling of organic products and repealing Regulation (EEC) No 2092/91, [2007] OJ L 189/1.

12 Including through Regulation (EU) 2021/2115 of the European Parliament and of the Council of 2 December 2021 establishing rules on support for strategic plans to be drawn up by Member States under the common agricultural policy (CAP Strategic Plans) and financed by the European Agricultural Guarantee Fund (EAGF) and by the European Agricultural Fund for Rural Development (EAFRD) and repealing Regulations (EU) No 1305/2013 and (EU) No 1307/2013 [2021] OJ L 435/1.

13 Communication from the Commission to the European Parliament, the Council, the European Economic and Social Committee and the Committee of the Regions, 'A Farm

these rules are being tightened without really addressing the core issue: the production and consumption of animal-based food.

Although the agricultural and food transition is about many different things, it generally revolves around moving from animal-based foods toward plant-based foods. Achieving such a transition requires a fundamentally different regulatory approach. This is not just about limiting some side effects of livestock-keeping, this is about reinventing the livestock sector altogether as well as about changing the dietary choices of European consumers. To make that happen, a much more comprehensive and robust regulatory approach will be needed.

Recently, technology seems to have come to the rescue of the policy makers who are struggling to develop such a comprehensive and robust regulatory approach: cultivated meat and dairy. These have the advantage that not all European consumers have to turn to fully plant-based diets. Cultivated meat and dairy do not have the negative environmental and climate change impacts of conventional meat and dairy. This new technology might very well be a game-changer that may accelerate the agricultural and food transition. The emergence of this technology, however, does not mean that the transition will happen unassisted. For the introduction of a wide-scale adoption of cultivated meat and dairy production and consumption, laws and policies will be needed.

This chapter answers the question of how law can facilitate a smooth agricultural and food transition, with a focus on the emergence of cultivated meat and dairy. It will do so through a literature survey and an assessment of the current, proposed and potential laws and regulations at the level of the EU in terms of their current and potential roles to achieve a transition to a food system without the large-scale involvement of animals. The literature on cultivated meat and dairy is mostly from such fields as food science, agricultural and environmental sciences, and resource economics. Legal scholarship on the topic is still lacking. The chapter will first briefly sketch the impact of meat and dairy production on the planet so as to show the urgency of the agricultural and food transition (Section 8.2). It will then briefly map the current and proposed EU laws that seem to aim for a more sustainable livestock sector (Section 8.3). Section 8.4 introduces cultivated meat and dairy and discusses the pros and cons of this new type of food. In Section 8.5, I discuss the laws and regulations needed to facilitate a large-scale introduction of cultivated meat and dairy: laws aimed at limiting the production of and demand for conventional meat and dairy, laws aimed at stimulating the production of and demand for cultivated meat and dairy, and

to Fork Strategy for a fair, healthy and environmentally-friendly food system' COM(2020) 381 final.

laws aimed at ensuring that cultivated-meat production is safe, nutritious, fair and sustainable.

The focus of this chapter is on cultivated meat and dairy as these are considered the most promising meat and dairy alternatives.[14] However, most of what is discussed in this chapter and certainly the role of the legal instruments discussed here, also applies to other unconventional meat and dairy alternatives that are currently being developed, such as microbial proteins and edible fungus as well as to more conventional plant-based meat alternatives.[15] I do not discuss cultivated seafood as this would require me to discuss the fisheries sector and fisheries laws. The transition from live seafood to cultivated seafood is an important and complex topic in itself that warrants a separate discussion.

8.2 Impact of meat and dairy and the need for a food transition

Over the past 11,000 years or so, humans have kept animals to produce meat and dairy for consumption.[16] Over the millennia, livestock-keeping for food production has grown tremendously, with an acceleration taking place since around 1950. With higher incomes and urbanization, people have been moving away from relatively monotonous diets of varying nutritional quality (based on indigenous staple grains or starchy roots, locally grown vegetables, other vegetables and fruits, and limited foods of animal origin) toward more varied diets that include more pre-processed foods, more foods of animal origin, more added sugar and fat, and often more alcohol.[17] This has caused a rapid increase in body weight and obesity and in diet-related chronic diseases,[18] and to public health effects that are directly associated with livestock-keeping, such as those caused by air pollution (especially ammonia and particulate matter such as $PM_{2.5}$), antibiotic resistance, added hormones and zoonotic diseases.[19]

14 Changtai Zhang et al, 'Production of Meat Alternatives Using Live Cells, Cultures and Plant Proteins' (2022) *Current Opinion in Food Science* 43, 44.

15 See Zhang et al (n 14) for an overview of all meat alternatives.

16 Albano Beja-Pereira et al, 'The Origin of European Cattle: Evidence from Modern and Ancient DNA' (2006) 103:21 PNAS 8113–18.

17 LEAD/FAO, Livestock's long shadow: Environmental issues and options (LEAD 2006), 10.

18 Ibid.

19 Randall S Abate, 'Anthropocene Accountability Litigation: Confronting Common Enemies to Promote a Just Transition' (2021) 46 *Columbia Journal of Environmental Law* 225, 237–8. On the air-pollution health issues caused by livestock, see Emmanuelle Lavaine et al, 'Health, Air Pollution, and Animal Agriculture' (2020) 101 *Review of Agricultural, Food and Environmental Studies* 517–28. On infectious diseases, see Romain Espinosa, Damian Tago and Nicolas Treich, 'Infectious Diseases and Meat Production' (2020) 76 *Environmental and Resource Economics* 1019–44.

Agriculture is the major driver of the transgression of four of the five planetary boundaries that already are transgressed or are at a high risk of being so:[20]

- land system change (almost 50 per cent of ice-free land surface is used for agriculture, of which 77 per cent is in use for livestock, mostly for growing concentrated feeds for livestock);
- biosphere integrity (80 per cent of changes in biosphere integrity occurred through agriculture; of the 24,845 species that the IUCN has identified as threatened with extinction, 19,421 are threatened through our food system[21]);[22]
- fresh water use (agriculture is the largest user of green water in soils; for one kilogram of beef, a stunning 15,415 litres of water are needed; this equals 112 litres per gram of protein, which is more than four times that needed for vegetables: 26 litres per gram of protein);[23]
- biochemical flows (85 per cent of nitrogen flows and 90 per cent of phosphate use are in agriculture, both mostly in connection with livestock).

Agriculture also has a large impact on two other planetary boundaries:[24]

- climate change (25 per cent of greenhouse gas emissions are from agriculture if land-use change is included; 12 per cent without the inclusion of land-use change, the emissions being mostly methane and nitrous oxide from livestock);
- ocean acidification (the contribution from agriculture is 25 per cent).

---

[20] Bruce M Campbell et al, 'Agriculture Production as a Major Driver of the Earth System Exceeding Planetary Boundaries' (2017) 22:4 *Ecology and Society*; Nicholas Bowles, Samuel Alexander and Michalis Hadjikakou, 'The Livestock Sector and Planetary Boundaries: A "Limits to Growth" Perspective with Dietary Implications' (2019) 160 *Ecological Economics* 128–36.

[21] David Tilman and David R Williams, Preserving Global Biodiversity Requires Rapid Agricultural Improvements (The Royal Society 2022), available online, https://royalsociety.org/topics-policy/projects/biodiversity/preserving-global-biodiversity-agricultural-improvements/.

[22] For more detail on the biodiversity impacts of agriculture, see Nigel Dudley and Sasha Alexander, 'Agriculture and Biodiversity: A Review' (2017) 18:2–3 *Biodiversity*.

[23] Data from the Water Foodprint Network, available through https://waterfootprint.org/en/water-footprint/product-water-footprint/water-footprint-crop-and-animal-products/. For more detailed underlying data, see PW Gerbens-Leenes, MM Mekonnen and AY Hoekstra, 'The Water Footprint of Poultry, Pork and Beef: A Comparative Study in Different Countries and Production Systems' (2013) 1–2 *Water Resources and Industry* 25–36.

[24] Campbell et al (n 20).

Besides human health impacts and environmental impacts, the current focus on animal-based foods also has large animal welfare implications. In 2018, an estimated 69 billion chickens, 1.5 billion pigs, 656 million turkeys, 574 million sheep, 479 million goats and 302 million cattle were killed for meat production.[25] Many of these were, prior to being slaughtered, kept under unnatural and cruel conditions and in stressful situations.

It is clear that the current emphasis on animal-based foods in our diets is not sustainable. A transition to plant-based diets would solve most, if not all, of the above mentioned negative impacts of livestock-keeping.[26] A simple illustration of this is the fact that the 77 per cent of the agricultural land that is used for livestock-keeping gives us only 18 per cent of our calories and 37 per cent of our protein supply. Eighty-two per cent of our calories and 63 per cent of our proteins are obtained from the remaining 23 per cent of agricultural land.[27] A plant-based diet, or a diet in which cultivated meat plays a major role, needs much less space.

The need for such a transition is increased by projected population trends. A growing world population that is also becoming more wealthy, requires more food. Between 2020 and 2070, a 70–100 per cent increase in global food demand is projected.[28] This could require around 750 million hectares of additional cropland, which will mostly come at the expense of biodiverse areas in the tropics and subtropics,[29] most of these being forests.[30] Future dietary choices will determine the impact of such increased food demand. Under business-as-usual scenarios, total meat consumption is projected to nearly double by 2050,[31] with all the associated impacts discussed above. To achieve a sustainable food system, with room to increase food production so as to protect the right to food while remaining within (or getting back to) the planetary boundaries and transitioning to healthier and more nutritious food for all, regulatory interventions are inevitable.[32]

An agricultural and food transition has been advocated by many international organizations:

---

25 Hannah Ritchie, Pablo Rosado and Max Roser, Meat and Dairy Production, Our World in Data (2017–2019), available online, https://ourworldindata.org/meat-production.

26 Bowles et al (n 20) 133.

27 Hannah Ritchie and Max Roser, Land Use, Our World in Data (2019), available online, https://ourworldindata.org/land-use.

28 Tilman and Williams (2022). See also Bowles (n 20).

29 Tilman and Williams (n 28).

30 Bowles et al (n 20) 131.

31 Tara Garnett, Peter Scarborough and Jessica Finch, Focus: the difficult livestock issue (Table 2016), available online, www.tabledebates.org/chapter/focus-difficult-livestock-issue.

32 Bowles et al (n 20) 133.

- A major policy focus on the livestock sector is needed when dealing with land degradation, climate change, air pollution, water shortage, water pollution and loss of biodiversity (LEAD/FAO).[33]
- A global transition to a healthy and sustainable food system is needed (FAO).[34]
- The agriculture and food system is key to global climate change responses;[35] GHG emission reductions can be achieved through sustainable crop and livestock management, carbon sequestration in agriculture, demand-side and material substitution measures, such as shifting to balanced, sustainable healthy diets, reducing food loss and waste, and using bio-materials, which can also free up land for reforestation and restoration, and the production of renewable energy (IPCC).[36]
- Feeding humanity and enhancing the conservation and sustainable use of nature are complementary and closely interdependent goals that can be advanced through sustainable agriculture (IPBES).[37]
- A Great Food Transformation to sustainable food production and healthy dietary patterns for everyone is needed to achieve the SDGs and the Paris Agreement (EAT-Lancet 2019).[38]

8.3 Current and proposed regulatory approaches to reducing emissions from livestock in the EU: pollution control, climate change mitigation, agricultural income support

Around the world, including in the EU, regulatory approaches to limiting the adverse environmental impacts of livestock-keeping have primarily focused on technological fixes and changes to farm management. Examples of the first are the use of air washers in concentrated animal feeding operations, the injection of manure into the soil and changes to livestock feeds, while examples of the latter are rotational grazing and reducing the animal-to-land ratio. The legal instruments used to achieve such changes are focused on farmers and are diverse. They range from command-and-control type instruments such as environmental permits prescribing the use of a certain

[33] LEAD/FAO (n 17).

[34] FAO (n 4).

[35] C. Mbow et al, 'Food Security' in PR Shukla et al (eds), Climate Change and Land: an IPCC special report on climate change, desertification, land degradation, sustainable land management, food security, and greenhouse gas fluxes in terrestrial ecosystems (CUP 2019), 440.

[36] Nabuurs et al (n 2).

[37] Diaz et al (n 3).

[38] Willet et al (n 1).

technology or achieving certain environmental standards, to subsidies granted to farmers who adopt a certain new or altered practice.

At the EU level, we find these instruments in the Industrial Emissions Directive (hereafter IED) and the CAP, respectively. GHG emissions from large poultry and pig farms currently fall under the scope of the IED and hence should be regulated through the permits,[39] as is also indicated by the Best Available Techniques (BAT) Reference Document (BREF) for the poultry and pig sector which pays ample attention to technologies and practices aimed at reducing methane and nitrous oxide emissions.[40] Under the revised CAP, which entered into force in 2023, 40 per cent of the budget is allocated to finance climate-change-related actions. Member states need to achieve climate change mitigation goals in their National CAP Strategic Plans.[41] There are, however, not many specific requirements for the livestock sector, except that member states 'may choose' this sector to achieve objectives that contribute to climate change mitigation and adaptation.[42]

The IED and the CAP are the two most obvious examples. Many other legal instruments may to some extent limit the environmental impact of livestock-keeping as well, such as legal instruments regulating water and air quality (Water Framework Directive,[43] Nitrates Directive[44], Air Quality Directive[45]), instruments on organic farming (Organic Agriculture Regulation[46]), nature conservation law (Habitats Directive[47]) and methane emissions (Effort Sharing Regulation[48]).

The European Commission planned to gradually tighten most of these instruments with the aim to limit the negative impact of livestock-keeping. This is clearly visible when looking at the legislative proposals that were published in 2022 and that target the livestock sector. The proposal for a

39 See n 7. Annex I, category 6.6 mentions intensive rearing of poultry or pigs (a) with more than 40,000 places for poultry; (b) with more than 2,000 places for production pigs (over 30 kg), or (c) with more than 750 places for sows.

40 Germán Giner Santonja et al, Best Available Techniques (BAT) Reference Document for the Intensive Rearing of Poultry or Pigs, EUR 28674 EN (European IPPC Bureau 2017), passim (for example, 34–5).

41 See n 12, see especially Arts 5, 6, 12, 13.

42 Art 66. Note that this provision has weaker language than the proposal, see Art 59 of the proposal, COM/2018/392 final.

43 See n 9.

44 See n 8.

45 Directive 2008/50/EC of the European Parliament and of the Council of 21 May 2008 on ambient air quality and cleaner air for Europe [2008] OJ L 152, 1.

46 See n 11, Art 14(1)(b)(iv).

47 See n 5.

48 See n 6.

revision of the IED not only expanded the permit requirement to cover farms for rearing cattle (as well as smaller pig or poultry farms and farms with mixed animal categories),[49] it also inserted several provisions specifically aimed at this sector.[50] These provisions included an aggregation requirement (to end the widespread practice of dividing one large livestock farm into separate facilities run by different people, usually family members, in order to remain below the IED threshold) and a requirement to include emission-limit values in the permit.[51] In 2023, however, the proposal to expand the permit requirement was rejected. The new Nature Restoration Law, adopted in June 2024, requires member states to put in place the restoration measures necessary to enhance biodiversity in agricultural ecosystems, with specific targets – for example, on the presence of butterflies and birds[52] – which will be hard to achieve in areas with large-scale livestock-keeping. The revised LULUCF Regulation expands the scope of the regulation to cover GHG emissions from enteric fermentation, manure management and urea application, be it only as a 'regulatory option', to be decided later.[53] The proposal for a revised Air Quality Directive aims at a 50 per cent reduction of $PM_{2.5}$ levels by 2030, which means that emissions from livestock need to be reduced as well.[54]

There are, however, also signs that more radical changes will be required. The Farm to Fork Strategy acknowledges that the food system remains one of the key drivers of climate change and environmental degradation.[55] The strategy points to the urgent need to reduce dependency on pesticides and antimicrobials, reduce excess fertilization, increase organic farming, improve animal welfare and reverse biodiversity loss.[56] What is particularly interesting in the Farm to Fork Strategy is that it also addresses the consumer:

49 Proposal for a Directive of the European Parliament and of the Council amending Directive 2010/75/EU of the European Parliament and of the Council of 24 November 2010 on industrial emissions (integrated pollution prevention and control) and Council Directive 1999/31/EC of 26 April 1999 on the landfill of waste, COM(2022) 156 final, Annex Ia.

50 Art 70a/70i of the proposal.

51 Art 70b and 70i(1)(a), respectively.

52 Regulation (EU) 2024/1991 of the European Parliament and of the Council on nature restoration OJ L 2024/1991, Art 11.

53 In the version that was adopted in 2023, this proposal was converted to a 'regulatory option' to be decided later. See the revised Art 17(2)(g), Regulation (EU) 2023/839 [2023] OJ L 107, 1.

54 Proposal for a Directive of the European Parliament and of the Council on ambient air quality and cleaner air for Europe (recast), COM(2022) 542 final, Annex II (from 10 $\mu g/m^3$ currently to the WHO advisory level of 5 $\mu g/m^3$).

55 See n 13 at 5.

56 Ibid.

> It is clear that the transition [to sustainable food systems] will not happen without a shift in people's diets. … While about 20% of the food produced is wasted, obesity is also rising. Over half of the adult population are now overweight, contributing to a high prevalence of diet-related diseases (including various types of cancer) and related healthcare costs. … If European diets were in line with dietary recommendations, the environmental footprint of food systems would be significantly reduced.[57]

The European Commission is currently working on a proposal for a regulation on a framework for sustainable food systems (FSFS), which it calls one of the flagship initiatives of the Farm to Fork Strategy.[58] A proposal is not expected to be published before 2025.[59] It is as yet unclear what the legislative framework will look like. If it is really to add to existing instruments, such as the ones described earlier, there should be a much larger emphasis on stimulating consumers to shift their dietary choices. A tightening of the technological fixes through such changes as the revised IED, and a greater emphasis on changes to farm management practices such as those supported by the CAP, will only marginally change dietary choices. It is possible that meat and dairy prices will rise due to the additional requirements, which may decrease demand.

Additional price instruments can be adopted to further drive up the price of animal-based food products and to decrease the price of plant-based food products. Examples are the introduction of carbon taxes on nitrogen fertilizers or cattle populations,[60] or the expansion of the EU Emissions Trading System (ETS) to cover major meat and dairy processing installations.[61] A meat or dairy tax is under discussion in several European capitals, especially in Germany, the Netherlands, Sweden and Denmark, while Switzerland introduced a meat tax in 2016 and Spain slightly raised its VAT tariff on meat (from 8 to 10 per cent) in 2011.[62] The only example that has been well researched so far is the Danish tax on saturated fat, which

---

57 Ibid.

58 Labelled as such here, https://food.ec.europa.eu/horizontal-topics/farm-fork-strategy/legislative-framework_en.

59 The European Parliament's website states that the proposal 'is mentioned neither in the Commission work programme for 2024 nor in the tentative agenda for forthcoming Commission meetings. It remains uncertain when this proposal will be tabled', www.europarl.europa.eu/legislative-train/spotlight-JD%2023-24/file-sustainable-eu-food-system.

60 H Guyomard et al, 'Review: Why and How to Regulate Animal Production and Consumption: The Case of the European Union' (2021) 15 *Animal* 100283, 7.

61 Jonathan Verschuuren, 'Achieving Agricultural Greenhouse Gas Emission Reductions in the EU Post 2030: What Options Do We Have?' (2022) 31 *RECIEL* 246, 254.

62 According to a 2021 overview by the NGO True Animal Protein Price Coalition (TAPPC), Increasing number of countries start taxing meat and dairy, available online,

impacted a large number of meat products as well as butter.[63] The tax, which was introduced in 2011, led to a reduction of consumption of saturated fat of around 10 to 15 per cent. In 2013, the tax was abolished because of high administrative costs, controversies over its inflationary effects, cross-border purchases and negative impacts on the economic results of firms.[64] An increase in cross-border purchases has also been seen after the Swiss introduction of a meat tax.[65]

Other options to change dietary choices by consumers are the use of informational and behavioural instruments.[66] While these are important and can be very influential, I will not discuss them here because the focus of this chapter is on the use of legal instruments.

The remainder of this chapter will focus on the emergence of cultivated meat and animal-free dairy, which is competing with the current and discussed regulatory options analysed earlier and which has a much more far-reaching impact not only on the agricultural sector but also on the environment, since replacing animal-grown meat and animal-produced milk with lab-grown meat and dairy has the potential to reduce livestock emissions to zero, reduce the current nitrogen and phosphate overload, and transform landscapes through the allocation to other uses of space now used for grazing and for the production of animal feed, without any need for a transition to plant-based food. What does this new option mean for the regulatory package that is currently being developed across Europe to get the food transition going?

8.4 What is cultivated meat and dairy and what are the benefits and pitfalls?

Cultivated meat is meat produced in a controlled environment from a few cells taken from a live animal (in a biopsy) or from a cell bank. These cells grow and multiply in a production facility through added embryonic stem cells as well as a range of other substances, such as carbohydrates, amino acids, fats, vitamins, minerals and growth stimuli (hormones). Inside a cultivator (or bio-reactor), the cells mature into muscle and fat, and into the desired texture and form. After five to seven weeks, the muscle tissue,

---

www.tappcoalition.eu/nieuws/16831/increasing-number-of-countries-start-taxing-meat-and-dairy-.

63 Guyomard et al (n 59) 9; Celine Bonnet et al., 'Viewpoint: Regulating Meat Consumption to Improve Health, the Environment and Animal Welfare' (2020) 97 *Food Policy* 101847, 5.

64 Jorgen Dejgard Jensen and Sinne Smed, 'The Danish Tax on Saturated Fat: Short Run Effects on Consumption, Substitution Patterns and Consumer Prices of Fats' (2013) 42 *Food Policy* 18–31.

65 TAPPC (n 62).

66 Bonnet et al (n 63) 5–8.

which is identical to animal flesh, can be harvested.[67] This process has been shown to work for a large range of domestic and wild land and sea animals, from cows, sheep and kangaroos to salmon, tuna and lobster.[68] Cultivated meat is also called cultured meat, lab-grown meat, *in-vitro* meat, cellular (or cell-based) agriculture, alternative protein, or alt-meat.[69]

Cultivated dairy products are dairy products (milk, cheese, ice cream) produced through precision fermentation technology. Cultivated dairy is also called fermented milk or animal-free dairy. Microorganisms are fed plant-based nutrients and sugars. They then grow and produce proteins that are identical to those found in milk from cows. In this way, all types of dairy products can be produced without the involvement of animals.[70]

There are several positive environmental impacts associated with switching from animal meat and dairy products to cultivated meat and dairy, especially the avoidance of methane and nitrous oxide emissions from livestock. This is an obvious positive impact and probably the main reason why cultivated meat gets so much attention. However, this advantage is less straightforward than it looks at the outset. Cultivated meat production requires an enormous amount of energy. Lynch and Pierrehumbert calculated that almost 23 per cent of the total global energy production would be needed for cultivated-meat production, should it aim to replace all animal-grown meat by cultivated meat.[71] Since cultivated-meat production potentially will replace methane emissions from livestock with carbon dioxide (CO_2) emissions from energy production, and since CO_2 is a much longer-lived greenhouse gas, the climate change impact of cultivated meat in the long run is not much less than that of normal meat.[72] An important precondition for moving from meat to cultivated meat, therefore, is that this move goes hand-in-hand with the transition to electricity generated from renewable energy sources.

A similar precondition applies to the land issue. As stated earlier, 77 per cent of current agricultural land is used for livestock. Much of that land sequesters

---

67 For a full overview, the website of the knowledge platform 'what is cultivated meat' is useful, see www.whatiscultivatedmeat.com/.

68 Ibid. As explained in Section 8.1, cultivated seafood falls outside the scope of this chapter.

69 The latter is an umbrella term for both plant-based and cultivated meat.

70 Netsanet Shiferaw Terefe, 'Recent Developments in Fermentation Technology: Toward the Next Revolution in Food Production' in Pablo Juliano et al (eds), *Food Engineering Innovations Across the Food Supply Chain* (Elsevier 2022) 89–106.

71 John Lynch and Raymond Pierrehumbert, 'Climate Impacts of Cultured Meat and Beef Cattle' (2019) 3 *Frontiers in Sustainable Food Systems* 9. Earlier research, however, indicated that the production of cultivated meat has a much lower carbon footprint than regular meat, of 78–96 per cent less CO_2-eq., Hanna L Tuomisto and Joost Teixeira de Mattos, 'Environmental Impacts of Cultured Meat Production' (2011) 45(14) *Environmental Science and Technology* 6117–23.

72 Ibid.

carbon and provides for plant and animal biodiversity.[73] The environmental benefits associated with cultivated-meat production depend on what happens with the land that is freed up. When converted back to nature, for example under a rewilding policy, the impacts on carbon sequestration and biodiversity will be positive, whereas when paved over to make room for new cities or new industrial sites, the environmental services of the land will be lost. Jepson and Blythe advocate 'agricultural rewilding', which entails commercial breeding and ranching of mega-herbivores, such as aurochs and bison, with associated tourism opportunities, increased carbon sequestration, regenerative agriculture and pasture-fed livestock-rearing that restores soil fertility and microorganisms and that enhances biodiversity in addition to producing sustainable high-quality premium meat for which there will be a market next to cultivated meat.[74] This is an interesting proposal that seems to merge agriculture, nature conservation and climate change mitigation.

For other environmental factors, cultivated meat generally performs better than conventional meat. Air pollution from livestock, such as ammonia and particulate matter, will cease, assuming that production of cultivated meat takes place in controlled and regulated facilities, and soil and groundwater pollution through excess amounts of manure will be sharply reduced.[75] The downside of the latter is that an increase in synthetic fertilizers may be needed to grow crops in areas that are now fertilized through animal manure. The water use associated with conventional meat and cultivated meat seem comparable,[76] although widely different views are apparent in literature.[77] Cultivated meat also scores much better with a view to climate change adaptation since cultivated meat is produced in controlled environments that are less impacted by droughts and heat waves than livestock.[78] The risk of the emergence of zoonotic diseases becomes substantially less with much fewer animals living around human populations.[79]

It is generally concluded that cultivated meat is healthier than conventional meat because contamination by bacteria in the slaughtering

[73] Sghaier Chriki, Marie-Pierre Ellies-Oury and Jean-François Hocquette, 'Is "Cultured Meat" a Viable Alternative to Slaughtering Animals and a Good Comprise between Animal Welfare and Human Expectations?' (2022) 12(1) *Animal Frontiers* 38.

[74] Paul Jepson and Cain Blythe, *Rewilding: The Radical New Science of Ecological Recovery* (Icon 2020), 151 and 165.

[75] Farley Simon Nobre, 'Cultured Meat and the Sustainable Development Goals' (2022) 124 *Trends in Food Science & Technology* 145–7.

[76] Chriki (n 73).

[77] Nobre (n 75) 145, refers to an older publication (from 2011) that seems to argue that cultivated meat has the potential to save 82 to 96 per cent of the water used in livestock systems.

[78] Nobre (n 75) 145.

[79] Ibid 147.

and processing phases is considerably less likely, as is the use of antibiotics (although both of these depend on the production method used for cultivated meat). The design of the cultivated-meat product can be such that all important and healthy micronutrients are added, in case these do not succeed to grow in the cultivated meat, and the amount of saturated fat is controlled.[80]

A huge advantage of cultivated meat and dairy over conventional meat and dairy is the fact that many fewer animals will be needed, thus ending the poor living conditions of many animals in the intensive livestock industry. Depending on the production process, some animals may still be needed for biopsies to extract stem cells, unless genetic modification processes are used. The first method may still generate some animal welfare concerns.[81] Animals may also still be needed for the production of other products, such as hide, blood and viscera, and products used in feed, pharmaceuticals and cosmetics.[82] Animal welfare concerns may also arise as to the fate of the current farm animals.[83]

It can be assumed that small-scale subsistence animal-keeping especially in developing countries will not be touched by the large-scale adoption of cultivated meat and dairy products in developed countries.[84] Were that not the case, livelihoods may be negatively affected. The debate on the socio-economic impact of the transition away from keeping farm animals is still very much ongoing and without clear answers.[85] What is clear is that besides cultivated meat, plant-based food products will play a major role in the food transition as well, so not only will new jobs emerge in the cultivated meat and dairy sector, many farmers will remain occupied in the farming sector to grow all the plants that are needed for plant-based food products.[86] A first assessment of the costs involved with large-scale cultivated-meat production is that these are very high (US$63/kg).[87] Initially, therefore, cultivated meat will be a niche product for developed-country consumers and a product that may need government subsidies.

---

80 Chriki (n 73) 37.

81 Ibid 39.

82 Ibid.

83 Ibid.

84 Chriki seems to assume (n 73) 39.

85 Nobre (n 75) 147; Nicolas Treich, 'Cultured Meat: Promises and Challenges' (2021) 79 *Environmental and Resource Economics* 45–6.

86 Charlotte E. Blattner, 'Just Transition for Agriculture? A Critical Step in Tackling Climate Change' (2020) 9(3) *Journal of Agriculture, Food Systems, and Community Development* 53–8.

87 Greg L Garrison, Jon T Biermacher and B Wade Brorsen, 'How Much Will Large-Scale Production of Cell-Cultured Meat Cost?' (2022) 10 *Journal of Agriculture and Food Research* 100358.

Finally, it should be noted that there also are ethical debates about cultivated meat and dairy. Besides positive ethical considerations based on the environmental and animal welfare and animal dignity benefits, negative ethical considerations are voiced as well. Cultivated meat continues to fetishize meat, may result in a non-meat-eating elite and decontextualizes sustainability.[88] Cultivated meat can also be seen as a threat to food and culinary heritage and to the traditional agri-food industry. This argument has been used in Italy to draft legislation banning cultivated meat.[89]

8.5 What role for law?

The EU's Farm to Fork Strategy does not explicitly mention cultivated meat and dairy. It only generally advocates moving to a more plant-based diet with less red and processed meat,[90] and it advocates research on 'increasing the availability and source of alternative proteins such as plant, microbial, marine and insect-based proteins and meat substitutes'.[91] It is unlikely, therefore, that the various regulatory proposals to implement this strategy, such as the regulation on an FSFS, will focus on the actual production and marketing of cultivated meat and dairy. The market, however, seems to move much faster than the drafters of the Farm to Fork Strategy had in mind. As of 1 January 2023, 31 companies producing cultivated meat had been established and were operating around the world, delivering their products to supermarkets and restaurants.[92] It is, therefore, safe to assume that policy makers will have to catch up sooner rather than later and seize this opportunity to promote cultivated meat and dairy as one of the tools in the toolbox for the agriculture and food transformation, a transition which seems inevitable, as was discussed earlier.

What would be the role of law to achieve this policy goal? More specifically, what regulatory interventions are needed to achieve large-scale production and consumption of cultivated meat as one of the ways to reduce the production and consumption of conventional meat and dairy? There are

---

88 For a brief overview of the ethics literature, Neil Stephens et al, 'Bringing Cultured Meat to Market: Technical, Socio-Political, and Regulatory Challenges in Cellular Agriculture' (2018) 78 *Trends in Food Science & Technology* 160–1. Treich reviews the literature on moral issues raised by cultivated meat, including potential religious issues (is cultivated meat halal/kosher?), see n 85, 52–3.

89 Francesco Planchenstainer, '"Meat me in Italy": The Italian Ban on Meat Sounding Names and Cell Cultured Meat' (2024) 19(2) *European Food and Feed Law Review*, 66–71.

90 See n 13, 13.

91 Ibid 15.

92 Dale Cudmore, 25+ Lab Grown Meat Companies: The Complete List (2023), available online, https://vegfaqs.com/lab-grown-meat-companies/.

roughly three interconnected categories of regulatory interventions which are relevant: laws aimed at limiting the production of and demand for conventional meat and dairy, laws aimed at stimulating the production of and demand for cultivated meat and dairy, and laws aimed at ensuring that cultivated-meat production is safe, fair and sustainable. I will discuss these three categories in turn. I will assess to what extent such laws already exist or are in the making, with a focus on the EU.

8.5.1 Regulatory approaches to reduce meat and dairy production and consumption

It is important to develop a broader regulatory approach aimed at reducing conventional meat and dairy to avoid the outcome that cultivated meat and dairy does not replace conventional meat and dairy but instead becomes just a new product in the food market and does not reduce negative impacts of conventional meat and dairy.[93] Such a broader regulatory approach is now missing in the EU.[94] A recent paper found that public funding in the EU promoting animal products is 1,200 times higher than public funding for novel technologies producing alternatives to animal-sourced products.[95] On the other hand, the regulatory pressure on conventional livestock farming is mounting as was discussed in Section 8.3. It can be assumed that European climate laws, in particular, will increase the price of conventional meat and dairy considerably in the coming ten years or so, with the adoption of strict emission-reduction targets for methane under the LULUCF Regulation and a carbon pricing mechanism for agriculture as part of the EU ETS, for example. In regions with high livestock density, environmental standards for water quality, air quality and nature conservation are already forcing member state governments to reduce livestock numbers. Member states can be required to do so through various legal instruments, such as the requirement to reduce nitrates and phosphates to reach a good water quality under the Water Framework Directive,[96] the requirement to achieve $PM_{2.5}$ standards under the Air Quality Directive,[97] and the requirement to

---

[93] Stephens calls this the 'addition effect' (n 88) 162.

[94] Verschuuren (n 61).

[95] Simona Vallone and Eric F Lambin, 'Public Policies and Vested Interests Preserve the Animal Farming Status Quo at the Expense of Animal Product Analogs' (2023) 6 *One Earth* 1–14.

[96] Mark Wiering et al, 'The Wicked Problem the Water Framework Directive Cannot Solve: The Governance Approach in Dealing with Pollution of Nutrients in Surface Water in the Netherlands, Flanders, Lower Saxony, Denmark and Ireland' (2020) 12(5) *Water* 1240.

[97] Lavaine et al (n 3) 525 show that in the EU there is an under-regulation of air pollution caused by livestock farms despite serious public health impacts when compared to industry.

substantially reduce nitrogen deposition in Natura 2000 sites to achieve a favourable conservation status for species and habitats under the Habitats Directive.[98] Such mounting pressure may stimulate farmers to transition from conventional livestock farming to more sustainable modes of farming and, more generally, prepare other players in the food system, such as banks and supermarkets, for a stronger focus on non-animal food products. In addition, there are proposals to impose taxes or otherwise increase the price of meat and dairy in order to reduce their consumption, as was shown in Section 8.3, which may also have an impact on farmers.

A more orchestrated approach toward reducing meat and dairy production and consumption is needed.[99] The instruments just mentioned need to be aligned with the aim to achieve the ambition to drastically reduce livestock numbers and a dietary change within the population away from products involving the keeping of large numbers of animals. The agricultural and food transition is likely more complex than the energy transition, so it is likely that even more laws and regulations are needed to get that transition going. International trade, for example, will have to be carefully regulated. Some form of border tax adjustment, similar to that of the CBAM Regulation,[100] will be required to prevent the replacement of expensive European meat and dairy products with cheaper imported products. Laws are also needed to scale up the availability of affordable plant-based and cultivated alternatives and to ascertain that the transition is safe, fair and sustainable. These topics will be discussed in Sections 8.5.2 and 8.5.3, respectively.

8.5.2 Regulatory approaches to stimulate production and consumption of cultivated meat and dairy

Legislation aimed at reducing conventional meat and dairy production and consumption has to go hand-in-hand with legislation aimed at stimulating the production and consumption of plant-based and cultivated alternatives. In the transition phase, blended products (in which both conventional and cultivated meat are used) may be produced. This will help to avoid shocks throughout the food chain, not just with regard to the availability and affordability of meat and dairy products in supermarkets, but also with regard to the economic viability of livestock farms (on the latter, see further in Section 8.5.3).

98 David De Puea and Jeroen Buysse, 'Safeguarding Natura 2000 habitats from nitrogen deposition by tackling ammonia emissions from livestock facilities' (2020) 111 *Environmental Science and Policy*, 74–82.

99 Verschuuren (n 61); Vallone and Lambin (n 95).

100 Regulation 2023/956/EU of the European Parliament and of the Council of 10 May 2023 establishing a carbon border adjustment [2023] OJ L 130, 52.

Rising prices for conventional meat and dairy caused by the adoption of pricing instruments such as a meat tax or by tightened regulations on emissions associated with conventional meat and dairy production previously discussed will encourage consumers to look for cheaper alternatives. Regulatory instruments to make these alternatives even more attractive are, again, price mechanisms such as subsidizing cultivated meat and dairy or differentiation in taxation between cultivated and conventional meat and dairy products. Subsidizing cultivated meat and dairy may indeed be inevitable given the expected high initial price of these products. Tax differentiation will also contribute to mitigating the high price of cultivated meat compared to conventional meat.

Although the EU does not have competence on indirect taxation (such as through VAT differentiation), some cooperation and alignment would be useful, for example through applying the enhanced cooperation provisions of the Treaty on the Functioning of the EU, which allows a minimum of nine EU member states to adopt joined goals in a particular field without binding the others in case the latter cannot achieve the adopted goals.[101] For that, unanimity is needed in the European Council.[102] One important reason for the EU to be involved is the risk that unilateral taxes will lead to cross-border purchases, as happened following domestic imposition of a meat tax (see Section 8.3). For the transition to be a success, a level playing field within and outside the EU is needed. As already indicated, a CBAM-like mechanism will be needed to avoid the import of cheap meat and dairy products.

Research indicates that non-regulatory approaches are important as well, especially the provision of information to improve the willingness to accept cultivated meat.[103] Although animal welfare and ecological concerns are the most influential positive drivers of consumers' attitudes toward cultivated meat, there are also emotions and preconceptions about 'non-natural' novel food technology ('food neophobia').[104] The perceived naturalness or unnaturalness of cultivated meat, for example, plays a major role in consumers' acceptance of cultivated meat.[105] Governments, as well as the wider food industry will have to address these perceptions and possible concerns in information campaigns and in the public debate.

[101] Arts 326–34 TFEU.

[102] Art 329(2) TFEU.

[103] Ashkan Pakseresht, Sina Ahmadi Kaliji and Maurizio Canavari, 'Review of Factors Affecting Consumer Acceptance of Cultured Meat' (2022) 170 *Appetite* 105829, 4–6.

[104] Ibid 7–9.

[105] Ibid.

8.5.3 Regulatory approaches to guide safe, nutritious, fair and sustainable production of cultivated meat and dairy

The EU already has comprehensive rules in place that aim to guarantee that novel food on the market is *safe and nutritious* if the novel food is intended to replace the original food. Cultivated meat and dairy will have to be authorized by the European Commission under the Novel Foods Regulation,[106] and, in case genetic modification is used, under the EU's GMO legislation.[107] (See Chapter 7 of this book.) Authorization of novel foods largely depends on the advice given to the European Commission by the European Food Safety Authority (EFSA). In its assessment, EFSA focuses on two aspects: is the novel food safe and is it not nutritionally disadvantageous compared with the original food?[108] Unsafe, for example, would be a product that creates a biological hazard or that risks being contaminated in a bio-reactor. Whether cultivated meat is nutritionally disadvantageous will be tested in comparison with conventional meat. As at the time of writing, no cultivated meat or dairy products had yet gone through the authorization phase. Therefore, no cultivated meat or dairy products had been placed on the EU market.[109]

The accuracy of information provided to consumers on food products is already regulated in detail under EU consumer protection law.[110] Stephens has warned of the risk of food fraud: cultivated meat might be passed off as conventional meat or vice versa.[111] This is an enforcement issue that needs special attention from the competent authorities.[112] The debate on the question whether 'meaty' and 'milky' words can be used to market cultivated meat and dairy is relevant here as well, just like with vegetarian

---

[106] Regulation 2015/2283/EU of the European Parliament and of the Council of 25 November 2015 on novel foods [2015] OJ L 327/1.

[107] Including Regulation 1829/2003/EC of the European Parliament and of the Council of 22 September 2003 on genetically modified food and feed [2003] OJ L 268/1 and Directive 2009/41/EC of the European Parliament and of the Council of 6 May 2009 on the contained use of genetically modified micro-organisms [2009] OJ L 125/75.

[108] Art 11(2).

[109] Information provided on EFSA website in funding call (23 April 2023 deadline); www.efsa.europa.eu/en/funding-calls/cultured-meat-and-cultured-seafood-state-play-and-future-prospects-eu (page removed by February 2024). There are concerns that the authorization process is too demanding: Anu Lähteenmäki-Uutela et al, 'Alternative Proteins and EU Food Law' (2021) 130 *Food Control* 108336, 1–11 at 8–9. Similarly, Astrid Seehafer and Marvin Bartels, 'Meat 2.0: The Regulatory Environment of Plant-Based and Cultured Meat' (2019) 14(4) *European Food and Feed Law Review*, 323–31 at 331.

[110] Regulation 1169/2011/EU of the European Parliament and of the Council of 25 October 2011 on the provision of food information to consumers [2011] OJ L 304/18.

[111] Stephens (n 88) 163.

[112] Stephens suggests the development and use of a 'protein tracker' for cultivated meat that could provide inspectors with a tool for product verification: ibid.

burgers and sausages.[113] In 2020, the European Parliament rejected a proposal to ban such descriptions of plant-based alternatives for meat (for example, 'vegetarian burger' is allowed), but stricter national rules do exist in several member states.[114] The rules on descriptions of plant-based alternatives for milk, however, are much stricter (for example, 'soy milk' is not allowed), as was already decided by the Court of Justice of the EU in 2017.[115] To avoid similar issues with cultivated meat and dairy, it is important that EU regulation provides clarity on the terminology soon.[116]

Moreover, legislation is needed to make the agricultural and food transition *fair*, especially for those with the highest stakes, that is, farmers and low-income consumers. The position of consumers was already discussed in Section 8.5.2 when we looked at the role of price incentives to bring about affordable food products. Special attention to the position of developing-country consumers is needed here too.[117]

The role of farmers is important, for in the transition process sufficient economically viable options need to be present, including that farmers play a role in the production process of cultivated meat and dairy. For plant-based meat and dairy products, the opportunities for farmers are quite obvious, as the growing demand for plant-based alternatives for meat and dairy leads to an increased demand for crops such as soy, lentils, peas and other legumes. For cultivated meat, however, this is more uncertain at the moment as it is unclear what feedstock will be used by the cultivated meat and dairy industry. Research indicates that it may well be that traditional agricultural crops will be used as input (for example, barley, beets, corn, peas, soy, sugarcane, wheat), although less traditional crops may also play a role (such as algae, fungi, seaweed, yeast, and a range of bacteria for the cultivated-milk fermentation processes).[118] In addition, some farmers may keep a small herd of animals to provide the cells necessary

---

[113] See, extensively, Anu Lähteenmäki-Uutela (n 104) 6–7. The legal rules have been laid down in Regulation (EU) No 1308/2013 of the European Parliament and of the Council of 17 December 2013 establishing a common organisation of the markets in agricultural products and repealing Council Regulations (EEC) No 922/72, (EEC) No 234/79, (EC) No 1037/2001 and (EC) No 1234/2007, [2013] OJ L 347/671.

[114] Anu Lähteenmäki-Uutela (n 109) 7. A preliminary ruling in a French case is currently pending before the CJEU, case C-438/23.

[115] CJEU (Seventh Chamber), 14 June 2017, Case C-422/16, ECLI:EU:C:2017:458 (*Verband Sozialer Wettbewerb eV v TofuTown.com GmbH*).

[116] Anu Lähteenmäki-Uutela (n 109) 7.

[117] Stephens (n 88) 161.

[118] Peter Newton and Daniel Blaustein-Rejto, 'Social and Economic Opportunities and Challenges of Plant-Based and Cultured Meat for Rural Producers in the US' (2021) 5 *Frontiers in Sustainable Food Systems* 624270, 4.

for cultivated-meat production.[119] It may even be possible to place the cultivators on farms in the case of more localized production, aiming for consumers who demand locally produced products.[120] Finally, producers of cultivated milk may deliver their product to farmers who will then produce cheese and other dairy products on their farm.[121] All of these possibilities are primarily ruled by the market, although regulators may choose to intervene, for example by requiring cultivated meat producers to involve (local) farmers. Such interventions may limit the risk that a small number of large multinational corporations monopolize the production of cultivated meat and dairy. Ideally, such market intervention should happen at the EU level so as to avoid, or at least reduce, potential conflicts with competition law. In general, intellectual property law plays a dominant role in the development of cultivated meat and dairy, which leads Johnson to conclude that 'statutory warranties and other such responses would be needed beyond reform of patent law itself, if regulators wanted to prevent anti-competitive and anti-innovation dynamics'.[122]

It should be stressed here that for the current livestock farmers, their role in the production of cultivated meat and dairy (and plant-based products for that matter) is only one of many ways to transition to a more sustainable farming business. In fact, these farmers will have to completely reinvent their farms. Many, if not all of them, will need to be assisted in this transition toward a drastically different type of farming business. They will need financial and practical support and sufficient time to make the transition. As there are many different roads to a more sustainable farming business for each individual farm, based on local environmental circumstances, individual preferences and expertise, local and regional markets, and so on, an individualized approach will be necessary. The government could, for example, roll out a large programme that allows farmers to hire consultants to help them with the transition. Newton and Blaustein-Rejto also point out that the government has the option 'to create and/or support wider policies and programs that support just transitions for farmers and rural communities, including debt forgiveness, compensating for losses incurred, and funding (re)training initiatives'.[123]

Part of such a wider policy will also be the instruments aimed at rewarding farmers for those environmental, climate- and biodiversity-related measures,

[119] Ibid 4–5.

[120] Ibid 5.

[121] Ibid.

[122] Hope Johnson, Regulating Cell-Cultured Animal Material for Food Systems Transformation: Current Approaches and Future Directions (2021) 13(1) *Law, Innovation and Technology* 108–40, at 134.

[123] Newton and Blaustein-Rejto (n 118) 8.

especially in the CAP and the LIFE programme. Member states can use the National Strategic Plans under the CAP to implement their policy to help livestock farmers diversify their farms.[124] Other market opportunities for farmers are the selling of carbon credits obtained through carbon sequestration measures on their land under private and public carbon farming schemes.[125] Increasing attention across Europe to rewilding, especially its focus on large herbivores, also offers opportunities for livestock farmers to contribute to such plans through agricultural rewilding, such as breeding and ranching reintroduced species, possibly with associated eco-tourism activities.[126]

Finally, the production of cultivated meat and dairy has to happen *sustainably*. As indicated in Section 8.4, the main environmental concern is the energy use and associated carbon footprint of cultivated meat and dairy. Reliance on renewable energy for the production of cultivated meat and dairy is to be preferred. The other potentially unsustainable development in the wake of the transition to cultivated meat and dairy is unsustainable use of the agricultural lands that become available through decreased pastures and decreased animal feed production. Land-use (planning) laws can be used to guide this development. Linking the agricultural and food transition to biodiversity policies and rewilding will, as indicated in the previous paragraph, limit the emergence of this risk.

8.6 Conclusion

This chapter has addressed the question of how law can facilitate a smooth agricultural and food transition with a focus on the emergence of cultivated meat and dairy and related unconventional meat alternatives, such as microbial proteins, as a potentially disruptive technology in the food market with large benefits for the climate, the environment, animal welfare and human health. I have showed that there are three interconnected categories of regulatory interventions which may accommodate a large-scale introduction of cultivated meat and dairy. First, the production of and demand for conventional meat and dairy can be reduced through drastically reducing livestock numbers and raising retail prices through higher production costs. EU climate and environmental laws are already being tightened, with the aim to limit negative impacts on the climate and the environment, but

[124] See Section 8.3.

[125] In 2022, the European Commission published a proposal for a Regulation of the European Parliament and of the Council establishing a Union certification framework for carbon removals, COM/2022/672 final.

[126] See Section 8.4.

without a clear end-goal or overarching ambition in mind. The emergence of cultivated meat and dairy as a real alternative for conventional meat and dairy allows for the adoption of such an ambition. Without a clear policy on the phasing down of conventional meat and dairy, there is the risk that cultivated meat and dairy will remain a niche product on the market.

Second, and closely linked to the previous category, laws can be adopted that stimulate the production of and demand for cultivated meat and dairy and related alternatives. The price mechanism is the most obvious instrument here: raising the retail price of conventional meat and dairy and lowering that of cultivated meat and dairy, and, in the transition phase, allowing for blended products to avoid price shocks. An important element of a policy aimed at increasing the demand for cultivated meat and dairy is information and deliberation, as research shows that there is a real risk that a large group of consumers may not automatically accept this new technology.

Last, but certainly not least, the transition to cultivated meat and dairy has to be safe, nutritious, fair and sustainable. The safe and nutritious test is primarily aimed at the consumer and is part of the authorization process under the EU Novel Foods Regulation. Given the fact that cultivated meat and dairy is essentially identical to meat and dairy from live animals, it is not expected that there will be any issues in this process. To make the transition fair is more challenging, as it requires the authorities to help a large group of livestock farmers across the EU to reinvent their farming business. It is important to understand that farmers will need to play an important role in the production of cultivated meat and dairy. Agricultural crops will be used as input in the cultivation process, some farmers may keep a flock of animals to provide the cells necessary for cultivated-meat production, the cultivators can be placed on farms and be managed by farmers, and farms can be used for producing and marketing locally produced products. In addition, livestock farmers may have surplus agricultural lands which they no longer need for grazing animals or growing animal feed. These can be used, for example, for carbon sequestration and rewilding projects, provided there are legal rules in place that reward farmers for these projects. The current CAP and LIFE programmes and the new EU rules on carbon farming are a start, but it would seem that more robust programmes are needed. Finally, production needs to happen sustainably, which means, in particular, that the large amount of energy needed for the production of cultivated meat and dairy must be generated from renewables and not from fossil fuels.

Acknowledgement

This research was made possible through funding from the Netherlands Research Council NWO under grant number 406.18.RB.004.

9

Regulating Geoengineering

Gareth Davies

9.1 Introduction

This chapter considers the regulation of geoengineering, both as a distinctive form of climate technology and as a technological product of the Anthropocene. Its aim is to show some of the constraints on that regulation which arise from embedded attitudes and pre-existing epistemic communities. The question is whether these attitudinal constraints are a welcome introduction of values and complexity into an otherwise over-technocratic approach to regulation, or whether they confuse and obstruct an optimal approach.

Section 9.2 sketches an intuitive and simple approach to regulation, drawing on well-established ideas about the regulation of technology. Section 9.3 moves away from this, and explores how geoengineering is regarded in public and policy discourse, and why. The conclusion, in Section 9.4, brings these together, showing how the values and attitudes represented in the discourse would not easily fit in with the model of the first part. The embeddedness of technology in wider social politics complicates attempts to regulate it in a decontextualized way.

9.2 Regulating geoengineering: a naive approach

The Anthropocene is all about scale.[1] If humans are making a distinctive mark on the Earth it is not so much because they are suddenly doing new things, but because they are now doing certain things to such an enormous

[1] Paul J Crutzen, 'The "Anthropocene"' (2002) 12(10) *Journal de Physique* IV, 1.

extent that the impact begins to compete with, and in the short term even to exceed, natural processes.[2] The emission of greenhouse gases is a perfect example.

As that scale implies, the actors involved in creating these emissions causing climate change are not just a few central powers but are multiple and diffuse; in fact, most of humanity. Moreover, they have good reason for what they are doing: fossil fuels are uniquely flexible, compact, easy-to-use, transportable sources of energy which have contributed hugely to human welfare. They only become problematic when too much of them is used, for too long, as is now the case, but this over-use is nothing but a reflection of their deep embeddedness in societies and ways of life.

These considerations entail that addressing climate change is not realistically possible through a few simple decrees. First, there are no bodies or persons with the kind of central decreeing power that would be required, and, second, the interests involved are too complex and multifaceted.[3] A rule that 'the burning of oil is prohibited' is not going to happen, and would not be respected, except where it is the culmination of an intense prior process of regulation, and it is there that the real work is done. The prior process is one of management and manipulation, sometimes highly technical, sometimes less so, but always nudging, steering and constraining.[4]

The law involved in regulating the Anthropocene, therefore, does not just consist of the simple normative prohibitions of Old Testament law or criminal law. It is rather social engineering, using both incentives and punishments to turn around the behavioural supertanker that is humanity.[5] The question will be what should be incentivized and what should be excluded, bearing in mind that striking a balance is part of law's effectiveness.[6] An over-reliance on prohibition typically creates non-compliance. An over-reliance on incentives typically creates a risk of unexpected and unwanted effects.

---

2 Paul Crutzen and Will Steffen, 'How long have we been in the Anthropocene era?' (2003) 61(3) *Climatic Change* 251.

3 Charles F Sabel and David G Victor, 'Governing global problems under uncertainty: making bottom-up climate policy work' (2017) 144 *Climatic Change* 15.

4 Ibid.

5 Jessica C Lawrence, 'Managing the Environment: Neoliberal Governmentality in the Anthropocene' in Pasi Heikkurinen (ed), *Sustainability and Peaceful Coexistence for the Anthropocene* (Routledge, 2017) 68; Louis J Kotzé, 'Rethinking global environmental law and governance in the Anthropocene' (2014) 32(2) *Journal of Energy & Natural Resources Law* 121, 146.

6 Lawrence, ibid; Kotzé, ibid; Gareth Davies, 'Privatisation and de-globalisation of the climate' (2013) 7(3) *Carbon and Climate Law Review* 187.

Geoengineering is a typically Anthropocenic phenomenon,[7] for it aims at changing the natural environment on Earth, not just as a side-effect – as in the case of greenhouse gas emissions, ecosystem destruction, and pollution – but as a primary goal.[8] The Anthropocene is associated with global change as a side-effect, a product almost of humanity's carelessness, but it is logical that, faced with the consequences of its own power, humanity should seek to take compensating actions. Stopping or limiting harmful actions is one obvious response. Developing techniques to counter their harm is another.[9] Prima facie, both are rational responses. The optimal balance between them is a case-by-case matter which, in a state of developing knowledge and technology, is not likely to be evident *ab initio*.[10]

Technology lawyers have long thought about this problem of regulating the new, and in particular addressing situations where the possibilities and consequences of technology are not yet fully clear and are almost certain to change.[11] If the regulator waits to see how the technology develops and which risks emerge, it faces the risk of a social and economic fait accompli, a status quo which is hard to change.[12] If society had been fully aware of the risks to privacy, health and politics that would emerge from the internet and social media, a regulator might have intervened early, when it was still possible, rather than letting apparently harmless entertainment facilities develop into part of the infrastructure of life, such that the public is simultaneously aware of their dark side and unwilling to allow them to be constrained. The cat is out of the bag.

On the other hand, if the regulator intervenes too early and too harshly, branches of science and technology that might have done great good for humanity will be deprived of the chance to develop.[13] A choice of strong

---

7 Jeremy Baskin, *Geoengineering, the Anthropocene and the End of Nature* (Palgrave Macmillan, 2019); Ina Möller, *The Emergence of Geoengineering: How Knowledge Networks Form Governance Objects* (CUP, 2023) chapter 4.3 'Anthropocene Narratives'.

8 Ibid; Richard Owen, 'Solar radiation management and the governance of hubris' in RM Harrison and R Hester (eds) *Geoengineering of the Climate System* (Royal Society of Chemistry, 2014) 212.

9 See, for discussion, Möller (n 7).

10 Louis J Kotzé, 'Rethinking global environmental law and governance in the Anthropocene' (2014) 32(2) *Journal of Energy & Natural Resources Law* 121–56; Sabel and Victor (n 3).

11 David Collingridge, *The Social Control of Technology* (St Martin's Press, 1980); Audley Genus and Andy Stirling, 'Collingridge and the dilemma of control: Towards responsible and accountable innovation' (2018) 47(1) *Research Policy* 61; see also Jack Stilgoe, 'Geoengineering as collective experimentation' (2016) 22 *Science and Engineering Ethics* 851; Jesse L Reynolds, 'Earth system interventions as technologies of the Anthropocene' (2021) 40 *Environmental Innovation and Societal Transitions* 132.

12 Collingridge, ibid.

13 Ibid.

regulation is an act of steering, away from certain kinds of development and toward others. Given that it is almost an axiom in a market-based society, and using a classically scientific mode of thinking, that the law-maker does not know everything, and that the products and technologies that will ultimately make a valuable contribution are most effectively found through a process of experimentation, there is reason to be cautious about banning or excluding new technologies. In other words, there is reason to try to keep options open.[14]

A compromise is typically found in the use of generalizable normative constraints: general principles reflecting values that are (arguably) not context specific. Taking medicine as an example, a regulator does not know which new therapies will ultimately be most effective, and does not take a stance on this. But nor does it just abdicate the regulatory role. Rather, any new medicines must comply with generalized rules on safety and effectiveness. It is then up to the market-like structure of competitive research to explore the many different therapeutic paths, within the framework of safety and effectiveness. The regulator does not choose the therapeutic pathway, but it sets the limits within which all pathways must pass.

Will the artificial and technology-based manipulation of the climate, via solar-radiation management or absorption of atmospheric greenhouse gases, play any kind of significant or constructive role in dealing with climate change? From any rational perspective, it is too early to tell. It depends on how the techniques themselves develop (or do not) and on how the alternatives (emission-reduction techniques) progress.[15] Nobody currently knows for sure how either of these will evolve.

A naive regulator might therefore seek to produce a generalized framework that does not preclude potentially positive developments, but constrains them within normative limits. Following the medical example, it might require any use of geoengineering techniques to comply with safety and effectiveness constraints. These might be ratcheted up according to the stage of development with no limits on purely theoretical discussion, other than the peer-review process which de facto excludes public funding for essentially random and unsubstantiated claims. There would be a degree of regulatory severity for initial field experiments and a higher degree for approval of any at-scale usage. As with medicine, a hundred promising ideas may fail to meet the challenges of reality and be extinguished before they pass the last test,

[14] Alfred Nordmann, 'Responsible innovation, the art and craft of anticipation' (2014) 1(1) *Journal of Responsible Innovation* 87.

[15] See, for example, Patrick Moriarty and Damian Honnery, 'Renewable energy and energy reductions or solar geoengineering for climate change mitigation?' (2022) 15 *Energies* 7315.

while some might make it all the way and contribute to increased well-being. As with medicine, even for those approved as potentially beneficial, a cost-benefit analysis will be required before deciding whether the treatment is actually something we want to use. Expense, side-effects, effectiveness and alternatives are all part of the balance.

This is a rather banal frame, hard to argue with: normative constraints should encourage the positive while preventing the negative. In the context of geoengineering, the discussion is currently framed in terms of governance, with a widespread insistence that adequate governance is a necessary prerequisite for geoengineering experimentation and use.[16] However, the demands made by any governance framework determine whether it facilitates geoengineering, blocks it (in practice), or something in between.[17] The striking fact about discussions on geoengineering governance is the widespread desire for a level of governance intensity and quality that would, in practice, exclude geoengineering from any serious role in climate responses in the coming years.[18] That can only be understood as the expression of an a priori preference that geoengineering not be part of a climate change response package. Why?

9.3 Geoengineering as a political act

9.3.1 Attitudes to geoengineering

Geoengineering gets bad press. There are a few who are openly supportive of it as an idea,[19] and a larger group who concede that it might become a useful, if regrettable, tool at some point in time,[20] but the most dominant view in public and academic discussion is that intellectual and scientific

---

16 Aarti Gupta and others, 'Anticipatory governance of solar geoengineering: conflicting visions of the future and their links to governance proposals' (2020) 45 *Current Opinion in Environmental Sustainability* 10, 13–14; Jesse L Reynolds, 'Solar geoengineering to reduce climate change: a review of governance proposals' (2019) *Proceedings of the Royal Society A*, 475 2019.0255.

17 Gupta, ibid; Reynolds, ibid.

18 Ibid.

19 David W Keith, 'Geoengineering' (2001) 409 *Nature* 420; Edward A Parson, 'Climate policymakers and assessments must get serious about climate engineering' (2017) 114.35 *Proceedings of the National Academy of Sciences* 9227.

20 Phillip Williamson and Ralph Bodle, *Update on Climate Geoengineering in Relation to the Convention on Biological Diversity: Potential Impacts and Regulatory Framework* (Secretariat of the Convention on Biological Diversity, Technical Series No. 84, 2016); Douglas G MacMartin, Ken Caldeira and David W Keith 'Solar geoengineering to limit the rate of temperature change' (2014) *Philosophical Transactions of the Royal Society* A 372 2014.0134; Virgoe (n 20); Freiman (n 20); David G Victor et al, 'The geoengineering option: a last resort against global warming?' (2009) 88(2) *Foreign Affairs* 64.

investment in it are a dangerous distraction from the essential, primary task of reducing greenhouse gas emissions.[21]

This hostility is prima facie a surprising reaction to an idea whose only apparent ambition is to use and develop science and technology to try to reduce the greatest current threat to humanity. One might think that such an attempt should instead be welcomed. In a crisis, anything that might help should be explored. Or maybe not.

The aim of this section is to unpack the default hostility to geoengineering. The conclusion is that the differences between advocates and opponents of geoengineering are best explained in terms of different values and preferences, different estimations of human behaviour, and, I will suggest, different professional backgrounds.

To make this point, I again draw a parallel with the regulation of medical treatment. Health is a context where, as with climate change, human behaviour creates risks and causes harm. In those situations one can, as with climate change, focus on changing the harmful behaviour, or one can try to develop techniques to reduce the harm caused. The balance between these choices and the reasons for them involve the same kinds of consideration as climate change: If we treat the symptoms, do we enable the underlying harmful behaviour? If we do not, do we produce more suffering?

The point of the comparison is not to find who is right, but to suggest that our collective responses to moral hazard – the risk that treating symptoms encourages the underlying harmful behaviour – vary according to context. Looking at the differences between the contexts might help us

---

[21] Frank Biermann and others, 'Solar geoengineering: the case for an international non-use agreement' (2022) 13(3) *Wiley Interdisciplinary Reviews: Climate Change* e754; Jennie C Stephens et al, 'The dangers of mainstreaming solar geoengineering: a critique of the National Academies Report' (2023) 32(1) *Environmental Politics* 157; BK Sovacool, CM Baum and S Low, 'Determining our climate policy future: expert opinions about negative emissions and solar radiation management pathways' (2022) 27 *Mitigation and Adaption Strategies for Global Change* 58; Ryan Gunderson, Brian Petersen and Diana Stuart, 'A critical examination of geoengineering: economic and technical rationality in a social context' (2018) 10 *Sustainability* 269; Tina Sikka, 'Activism and neoliberalism: two sides of the geoengineering discourse' (2020) 31 *Capitalism Nature Socialism* 84; Kevin Surprise, 'Stratospheric imperialism: liberalism, (eco) modernization, and ideologies of solar geoengineering research' (2019) 3(1) *Environment and Planning E: Nature and Space* 141; Christopher H Trisos et al, 'Potentially dangerous consequences for biodiversity of solar geoengineering implementation and termination' (2018) (2) *Nature Ecology and Evolution* 475; James R Fleming, *Fixing the Sky: The Checkered History of Climate and Weather Control* (Columbia University Press, 2010); Clive Hamilton, *Earthmasters: Playing God with the Climate* (Allen & Unwin, 2013); Naomi Klein, *This Changes Everything: Capitalism versus the Climate* (Simon & Schuster, 2014). See for overview Reynolds (n 16) section 6. Cf Gareth Davies, 'Climate change and reversed intergenerational equity: the problem of costs now, for benefits later' (2020) 10(3–4) *Climate Law* 266.

understand why there is more hostility to geoengineering than there is to, for example, blood-pressure medication or skin-cancer treatments, both of which conditions have a strong, if incomplete, lifestyle connection. That, in turn, may reveal something about the dynamics of environmentalism and climate change policy making.

9.3.2 The emergence of scientific 'solutionism'

The idea of geoengineering goes back to the last century.[22] The greenhouse effect was becoming more widely accepted and discussed, and the gravity of its consequences was gradually becoming apparent. Faced with a warming Earth, it is hardly surprising that some scientists began thinking about whether there might be techniques available, or imaginable, that could counter this – just as they do when any threat to humanity or well-being emerges.[23] Whether it is a pandemic or an economic or refugee crisis, the range of experts and expertise in the modern world ensures that responses are wide-ranging and variable – from those who seek to address root causes, to those who have ideas that might mitigate some part of the harm. For example, making lifejackets available on dangerous migration routes; or having schools in refugee camps; or delivering food to the immunosuppressed during the COVID-19 pandemic.

The climate change threat has the important distinguishing feature that it is entirely a result of human behaviour, so that the most obvious response to it is to change that behaviour.[24] However, the distinction between natural crises and those created by humanity is more of a contestable and gradual continuum than a clear line. Whatever the nature of the problem, generally those who see a place for their skill try to make use of it, whether they are addressing symptoms or causes.

Among those who first brought the idea of solar-radiation management to prominence were not environmentalists but a physicist, Cesare Micheletti, an engineer, David Keith, and, above all, an atmospheric chemist, Paul Crutzen. A 2006 article by Crutzen ignited the contemporary debate.[25] Nevertheless, he was not happy with the idea: he did not put it forward as a wonder-drug, for he could see that it brought with it considerable risks and uncertainties and was far from a perfect solution. Instead, he merely suggested that it could be

22 Cesare Marchetti, 'On geoengineering and the CO_2 problem' (1977) 1(1) *Climatic Change* 59. See also Stephen H Schneider, 'Geoengineering: Could—or should—we do it?' (1996) 33(3) *Climatic Change* 291.

23 See Möller (n 7).

24 Owen (n 8).

25 Paul Crutzen, 'Albedo enhancement by stratospheric sulfur injections: a contribution to resolve a policy dilemma?' (2006) 77 *Climatic Change* 211.

done – that inserting sulfur aerosols into the upper atmosphere was possible, and could, on the basis of fairly well-understood atmospheric science, be expected to have a cooling effect – and that, given the dangers of global warming, humanity might be wise to have this technique in reserve in case it ever became necessary to avert catastrophic temperature rise. It was conceived as an insurance policy, a second best, not a best-practice policy choice.[26]

Since then, geoengineering ideas have expanded, with the pace of theoretical innovation picking up in recent years.[27] Some, like Crutzen, have focused on the idea of reflecting away more sunlight, through use of techniques from the highly speculative to the mundane – from mirrors in space to painting rooftops white – with somewhere in-between the suggestion of modifying clouds to make them more reflective. In parallel, more and more scientists have begun to think about how carbon dioxide (CO_2) could be actively absorbed from the atmosphere. This is a case of taking emissions out of the atmosphere after the fact, rather than stopping them. Once again, techniques vary from the pedestrian to the high-tech, with planting trees and restoring woodland at one end, industrial CO_2-capture installations at the other, and ocean fertilization and enhanced rock weathering in-between.

What is distinctive about Negative Emission Technologies (NETs), as they tend to be called, is that the potential scale of their impact is very variable. The major solar-radiation management (SRM) techniques, which are the use of aerosols *à la* Crutzen as well as cloud modification, are both potentially enormously powerful, capable of cooling by several degrees.[28] By contrast, many NETs, such as restoring forests, protecting sea grass and seaweed growth, preventing soil degradation, and even the use of artificial trees, have modest maximum effects.[29] At best, they would make a small, albeit non-trivial, contribution to reducing CO_2 levels. On the other hand, NET techniques have been proposed that could be as impactful as SRM. The lively discussion on ocean fertilization includes the suggestion that it could entirely absorb excess

---

[26] Virgoe (n 20); Freiman (n 20); Victor and others (n 20); Ryan Gunderson, Diana Stuart and Brian Petersen, 'The political economy of geoengineering as Plan B: technological rationality, moral hazard, and new technology' (2019) 24(5) *New Political Economy* 696. Cf Augustin Fragniere and Stephen M Gardiner, 'Why Geoengineering Is Not "Plan B"' in Christopher J Preston (ed) *Climate Justice and Geoengineering: Ethics and Policy in the Atmospheric Anthropocene* (Rowman and Littlefield, 2016) 15.

[27] For a moderately recent overview see Williamson and Bodle (n 20).

[28] Royal Society, *Geoengineering the Climate: Science, Governance and Uncertainty* RS Policy document 10/09 (The Royal Society 2009); Jesse L Reynolds, *The Governance of Solar Geoengineering: Managing Climate Change in the Anthropocene* (Cambridge University Press, 2019).

[29] Royal Society, ibid.

human CO_2 emissions; and others have made similarly ambitious claims about the use of alkaline rocks to enhance CO_2 absorption by the oceans, as well as about enhanced rock weathering (spreading rock powder around to react with atmospheric CO_2).[30] Some have even taken reforestation beyond the comfort zone of most and have proposed reforesting the Sahara – and there have been other similarly dramatic interventions, which perhaps, if achieved, might absorb CO_2 on a scale that looks more like a solution than mere mitigation.[31] Despite their relatively low-tech nature, these ideas are perhaps most deserving of the name geoengineering. They truly do reimagine what the planet might look like, and how humanity could intervene to realize those dreams.

At the same time, the name 'geoengineering' has certainly become tainted over the years. Advocates of NETs, and particularly advocates of low-tech 'naturalistic' NETs, tend to avoid it and to describe their ideas in less controversial terms. Better, from the perspective of professional acceptance, to emphasize how restoring soils or under-sea forests to health can have a positive climatic impact than to say that you have an idea to engineer the planet – better to be an ecologist than a would-be master of the climate.[32] Hubris does not go down well.[33] On the other hand, there are also those who continue to be open about their desire to use technology to cure the planet from humanity's lifestyle disease. This range of ideological perspectives has led to fragmentation and proliferation of terms, from geoengineering to climate engineering to NETs to climate restoration to, simply, mitigation. The label, or the absence of one, often indicates where the proponents are positioning themselves in the debate.

9.3.3 Hostility to geoengineering

Yet, something that links all these techniques is the suggestion that their impact on climate change could be very significant, a game-changing intervention, and that this is the primary purpose of the measure, rather than just being a welcome side-effect of the restoration of nature. As a result, the label *geo*, with its implied planetary scale, is deserved, as is the label *engineering*, with its implied deliberateness and non-naturalness. These are met with a very significant degree of hostility in public discourse.[34]

---

30 Ibid.

31 Leonard Ornstein, Igor Aleinov and David Rind, 'Irrigated afforestation of the Sahara and Australian Outback to end global warming' (2009) 97 *Climatic Change* 409.

32 See Joana Castro Pereira, 'Geoengineering, scientific community, and policymakers: a new proposal for the categorization of responses to anthropogenic climate change' (2016) 6(1) *SAGE Open*; Möller (n 7).

33 Owen (n 8).

34 Ramit Debnath et al, 'Conspiracy spillovers and geoengineering' (2023) 26(3) *Iscience* 106166; Gerd Winter, 'Climate engineering and international law: last resort or the end

The discourse is public in the sense that it takes place in publicly accessible arenas, not in the sense that the public participate. Discussion of geoengineering has not entered popular debate, nor are politicians touching the subject, even those involved in or committed to addressing climate change.[35] The discussion takes place in newspapers and blogs, as well as in academic journals, and is between environmentalists – meaning people involved in environmental issues who are not officials in public institutions – policy academics and scientists.[36] Of course, a range of views is expressed,[37] including enthusiasm[38] and reasoned openness to research and experiment,[39] but the most dominant one is that the proposed techniques for large-scale manipulation of the climate should not be pursued, and that to take them seriously is a dangerous and irresponsible choice.[40]

---

of humanity?' (2012) 20 *Review European Community International Environmental Law* 277; Stephen Gardiner, 'Is "Arming the Future" with Geoengineering Really the Lesser Evil? Some Doubts about the Ethics of Intentionally Manipulating the Climate System' in Stephen Gardiner et al (eds) *Climate Ethics: Essential Readings* (Oxford University Press, 2010) 284. See also note 21.

35 Gupta et al (n 16) 16.

36 Ibid.

37 For overviews see Judith Kreuter, 'Technofix, plan B or ultima ratio? A review of the social science literature on climate engineering technologies' (2015) *Institute for Science, Innovation and Society Occasional Paper Series no. 2*, summarized in Williamson and Bodle (n 20) at 83; Duncan McLaren and Olaf Corry, 'The politics and governance of research into solar geoengineering' (2021) 12 *WIREs Climate Change* e707; Gupta et al (n 16).

38 David Keith, *A Case for Climate Engineering* (Review Books, 2013).

39 Williamson and Bodle (n 20). See also the several publications on this of the American National Academy of Sciences, available at nationalacademies.org, most recently: Committee on Developing a Research Agenda and Research Governance Approaches for Climate Intervention Strategies that Reflect Sunlight to Cool Earth et al, *Reflecting Sunlight Recommendations for Solar Geoengineering Research and Research Governance* (National Academies Press, 2021); Stefan Schäfer and others, *The European Transdisciplinary Assessment of Climate Engineering (EuTRACE): Removing Greenhouse Gases from the Atmosphere and Reflecting Sunlight away from Earth* (EuTRACE, 2015); Ron Baiman, 'Our two climate crises challenge: short-run emergency direct climate cooling and long-run GHG removal and ecological regeneration' (2022) 54(4) *Review of Radical Political Economics* 435; Claudia E Wieners et al, 'Solar radiation modification is risky, but so is rejecting it: a call for balanced research' (2023) 3(1) *Oxford Open Climate Change* kgad002; Olúfẹ́mi O Táíwò and Shuchi Talati, 'Who are the engineers? Solar geoengineering research and justice' (2022) 22(1) *Global Environmental Politics* 12; Marie-Valentine Florin, *Combatting Climate Change through a Portfolio of Approaches* (Lausanne, International Risk Governance Centre, 2021); 'Give research into solar engineering a chance', editorial (2021) 593 *Nature* 167; Jessie L Reynolds, Andy Parker and Peter Irvine, 'Five solar geoengineering tropes that have outstayed their welcome' (2016) 5 *Earth's Future* 562.

40 Discussing this phenomenon: Williamson and Bodle (n 20) 86; Clare Heyward and Steve Rayner 'A curious asymmetry: social science expertise and geoengineering' (2013) Climate Geoengineering Governance Working Paper Series 007, available at http://geoengineeringgovernanceresearch.org. For works taking this view, see notes 21 and 34; Mike Hulme,

An apparently more moderate but in practice similar view is that these techniques should only be investigated with extreme caution and after the establishment of extensive safeguards and global governance procedures.[41] This is not prima facie unreasonable, but the proposed conditions are sufficiently demanding that, in practice, they amount to kicking the geoengineering research ball a long way down the road – a de facto moratorium.[42] A desire for caution and optimal governance would, in most policy contexts, be balanced against the urgency of the crisis and the need for solutions – the risk of haste being matched by the risk of slowness[43] – but this is something that the 'extreme caution' advocates tend not to do, adopting in substance a one-sided application of the precautionary principle.[44]

These views have had a chilling effect. There is almost no empirical research at scale into such techniques.[45] Thus, while it is, of course, impossible to stop scientists from thinking and speculating and producing theoretical papers, there have been no real experiments to assess the potential of SRM or ocean fertilization, the two techniques for which the most grandiose claims are sometimes made. The science remains at the stage of modelling – sketching how, on the basis of understandings of the chemistry and biology of the air and sea, it might work. Yet, as is conceded by all parties to the debate, these are complex systems. Therefore, in order to know how those theories play out in practice, it is necessary to experiment.

Were this a less politicized field of science, the logical next step would be to try to modify some clouds, distribute an amount of sulfur aerosols in a certain area and fertilize an area of ocean with iron to see what happens to algal growth and CO_2 uptake. This is how science is done. Yet, here, no

---

Can Science Fix Climate Change? A Case Against Climate Engineering (Polity Press, 2014); Jennie C Stephens and Kevin Surprise (2020) 'The hidden injustices of advancing solar geoengineering research' (2020) 3 *Global Sustainability* e2; Klein (n 21); CIEL (Center for International Environmental Law), *Fuel to the Fire: How Geoengineering Threatens to Entrench Fossil Fuels and Accelerate the Climate Crisis* (CIEL, 2019).

41 Sikina Jinnah et al, 'Governing climate engineering: a proposal for immediate governance of solar radiation management' (2019) 11(14) *Sustainability* 3954. Discussion in Gupta et al (n 16).

42 See Gupta et al, ibid; Reynolds (n 16) section 7.

43 Keith (n 38).

44 For example, European Commission, *A new outlook on the climate and security nexus: Addressing the impact of climate change and environmental degradation on peace, security and defence*, JOIN(2023) 19 final, 20. Cf the more nuanced position of the UN Convention on Biodiversity in Williamson and Bodle (n 20).

45 Anna-Maria Hubert, 'A code of conduct for responsible geoengineering research' (2021) 12 *Global Policy* 82.

government is known to have allowed such 'outdoor' experiments to take place. There is even agitation for an international agreement to be reached on a moratorium on geoengineering experiments generally.[46] The sense that geoengineering is a path that should not be taken has had an impact on politicians. Their lack of engagement with the issue may not be a lack of interest but a sense that it is far too sensitive and dirty a topic for an elected official to be associated with. The politically safe course of action is unequivocally to not even touch geoengineering with a bargepole, but to strictly stick to the emission-reduction message.

It is certainly striking that serious and well-intentioned ideas about how a global crisis might be alleviated, put forward by qualified scientists, working at respected institutions, with no history of madness and no obvious radical or offensive agenda, and published generally in peer-reviewed journals, are met with the reaction: 'This idea must not be considered, and should not even be researched!' It is hard – I have found it impossible – to think of any other context where something similar happens.[47]

Nevertheless, there are coherent and understandable reasons for it. These are broadly of three types:[48] (1) geoengineering techniques are potentially highly dangerous; (2) geoengineering techniques are wildly speculative and unproven, and as such offer false promise (they are snake oil); and, relatedly and most importantly, (3) allowing people to think that there is a technological solution to climate change will distract humanity from the essential task of emission reduction, and thereby lead to doom. We cannot afford wrong turns and dead ends where the future of the planet is at stake. It is worth considering each of these reasons in turn, but it is in the last that the crux of the matter lies.

---

[46] See the open letter calling for non-use at www.solargeoeng.org/non-use-agreement/open-letter/; Biermann et al (n 21); ETC Group, *Geopiracy: The Case Against Geoengineering*, Communique No. 103 (ETC Group, 2010); Shinichiro Asayama and Mike Hulme 'Engineering climate debt: temperature overshoot and peak-shaving as risky as subprime mortgage lending' (2019) 19 *Climate Policy* 937; Stephens and Surprise (n 40); Michael Zürn and Stefan Schäfer, 'The paradox of climate engineering' (2013) 4 *Global Policy* 266; Albert C Lin 'The missing pieces of geoengineering research governance' (2016) 100 *Minnesota Law Review* 2509. See the extensive discussion and citations in Reynolds (n 16) section 7.

[47] See Megan Herzog and Edward A Parson, 'Moratoria for global governance and contested technology: the case of climate engineering' *UCLA School of Law, Public Law Research Paper* 16–17 (2016).

[48] There are sometimes others. See Aaron Tang, 'The slippery slopes of climate engineering research' (2023) 80 *Global Environmental Change* 102674; Debnath et al (n 34); Stephens and Surprise (n 40).

9.3.3.1 Geoenginering is dangerous

This argument is particularly applied to SRM techniques, where it is noted that the atmosphere and weather are complex processes, and that reflecting back 1 per cent (say) of the sunlight heading toward Earth will not result simply in a tidy reduction in temperature spread uniformly over the globe, but will affect wind flows and precipitation, and may cause more floods and droughts and storms. The overall net temperature may decrease, but it will mask diverse and complex effects, and we can be certain that even if humanity were to be a net winner from the overall resulting cooling effect, there would also be many losers.[49]

Oceanic interventions are considered particularly dangerous when they operate at scale. Even if ocean fertilization would lead to atmospheric CO_2 reduction, what would be its full effect on ocean ecosystems? Apparently benign interventions might turn out in the long run to have disrupted ocean biology on a scale or in a way that might cause dramatic harm – ocean death, human starvation, and even negative climate effects if the CO_2-absorption capacity of oceans were ultimately reduced by these disruptions.[50] The complexity of oceans means that small-scale experimentation can never provide conclusive evidence that large-scale actions would not turn out to be disastrous, a gamble that cannot be taken with a system so fundamental to global ecology.

Alongside these arguments, it is noted that the governance of geoengineering requires a degree of cooperation between governments that is generally not achieved, and seems implausible.[51] What temperature should humanity choose for the global thermostat, and, most importantly, how should that decision be taken? If a few nations, or even private individuals, take measures that affect global temperature and ecology, the potential exists for international conflict – which aside from its human costs would also be unlikely to have a positive climate effect.[52]

[49] Royal Society (n 28); Williamson and Bodle (n 20).

[50] Ibid.

[51] David Victor, 'On the regulation of geoengineering' (2008) 24(2) *Oxford Review of Economic Policy* 322; Stephen H Schneider, 'Geoengineering: could we or should we make it work?', (2008) 366 *Philosophical Transactions of the Royal Society A*, 3843, 3847; Virgoe (n 20); Reynolds (n 16); Shinichiro and Hulme (n 46); Owen (n 8). But see Thomas C Schelling, 'The economic diplomacy of geoengineering' (1996) 33 *Climatic Change* 303.

[52] Victor, ibid; Daniel Bodansky, 'The who, what, and wherefore of geoengineering governance' (2013) 121 *Climatic Change* 539; Olaf Corry, 'The international politics of geoengineering: the feasibility of Plan B for tackling climate change' (2017) 48(4) *Security Dialogue* 297; Sam Adelman, 'Geoengineering: rights, risks and ethics' (2017) 8(1) *Journal of Human Rights and the Environment* 119. Cf Holly J Buck, 'Environmental peacebuilding and solar geoengineering' (2022) 4 *Frontiers in Climate* 869774.

Finally, the uneven distribution of geoengineering's potential harms, with the greater burden probably being on the poor, the Global South and the already vulnerable, means that its riskiness is also often perceived as a justice issue: geoengineering would be an act of exploitation or even colonialism.[53]

These arguments are powerful, but they also have two weaknesses. One is that they apply equally to the emission of CO_2, which in its effects is akin to a particularly destructive and unguided form of geoengineering.[54] It too does not create uniform warming effects. It disrupts the climate, it acidifies the oceans and slowly kills them, it is caused by richer and more industrial nations and its greatest impact is on the poor. Moreover, it creates a frightening potential for local and global conflict and unmanageable migration. Danger is, therefore, only a decisive argument against geoengineering if it can be assumed that without it humanity will engage in timely and effective emission reduction. This, however, is unknown. It is not a plausible premise. To use risk to dismiss geoengineering is, therefore, to take a strong position on an essentially speculative issue of human behaviour. It is an apocalyptic gamble.

The second weakness is that risk is a common, almost ubiquitous, feature of policy solutions. Many medical treatments have side-effects, even potentially deadly ones. This is not a reason to dismiss them a priori, but is a factor to balance against benefits. Relevant is the extent to which the side-effects themselves can be managed. Interventions can be complex and may require developed approaches to risk assessment and management, but this can be done for geoengineering as it is done in other fields.[55] Limiting policy to riskless approaches to addressing climate change would paralyse policy and be without precedent.

There is thus a kind of symmetry to geoengineering and not-geoengineering: each brings risks. It should be noted that it is only a question of time before this is reflected in climate change litigation. This has increased rapidly in recent years, calling governments and firms to account

[53] Benjamin K Sovacool, Chad Baum and Sean Low, 'Climate protection or privilege? A whole systems justice milieu of twenty negative emissions and solar geoengineering technologies' (2022) 97 *Political Geography* 102702; Frank Biermann and Ina Möller, 'Rich man's solution? Climate engineering discourses and the marginalization of the Global South' (2019) 19(2) *International Environmental Agreements: Politics, Law and Economics* 151; Stephens and Surprise (n 40); cf Táíwò and Talati (n 39); Gareth Davies, 'Climate change and reversed intergenerational equity: the problem of costs now, for benefits later' (2020) 10(3–4) *Climate Law* 266.

[54] Gareth Davies, 'Framing the social, political and environmental risks and benefits of geoengineering: balancing the hard-to-imagine against the hard-to-measure', (2010) 46(2) *Tulsa Law Review* 261.

[55] Benjamin K Sovacool, Chad Baum and Sean Low, 'Beyond climate stabilization: exploring the perceived sociotechnical co-impacts of carbon removal and solar geoengineering' (2023) 204 *Ecological Economics* 107648; Táíwò and Talati (n 39); Florin (n 39).

for inaction.[56] It has been suggested that coming trends in this litigation may involve more attention to ocean protection, biodiversity and carbon sinks.[57] At some point, it may be expected that litigants will demand not just that actors do less harm, but that they take more active steps to protect, for example by combatting ocean acidification, by reforesting, or by promoting and recognizing new carbon sinks – geoengineering, in other words. There will be more and more situations where fundamental environmental rights are at issue, and merely reducing emissions does not offer an adequate solution, either in theory or in practice.

9.3.3.2 Geoengineering is speculative

Another critique is that none of the geoengineering techniques has been shown to work. To hold them up as potential saviours is fundamentally dishonest since they are just sketches on paper.[58] It would be madness to make something so speculative and flimsy a serious plank of global policy.[59] It would be equivalent to engaging in dangerous individual behaviour – smoking, say – on the basis that 'they'll probably find a cure in time'. They might, but people who have taken this approach systematically to known harmful behaviour are more likely to die than to get lucky.

Nevertheless, despite it being often emphasized in critiques of geoengineering how unproven its various techniques are, this does not obviously lead to the conclusion often drawn; namely that we should just forget about it. Rather, it would seem to be an argument for research:[60] if someone has a theoretically non-ridiculous idea which might help, then the obvious reaction is to test it to see if it holds up in practice. I made the point earlier that this is how science works, and to dismiss theoretical proposals because they have not been proven is, on the face of it, a rejection of science, or at the very least a misunderstanding of it. A survey of any field of applied science will reveal just how few proposals make it through from initial conception to final application. Since there is no means to distinguish the effective from the ineffective other than by experiment, the speculative

[56] Joana Setzer and Catherine Higham, *Global Trends in Climate Change Litigation: 2023 Snapshot* (London: Grantham Research Institute on Climate Change and the Environment and Centre for Climate Change Economics and Policy, London School of Economics and Political Science, 2023).

[57] Ibid, at 44.

[58] Owen (n 8) 212; Biermann et al (n 21); Fragniere and Gardiner (n 26).

[59] Winter (n 34).

[60] Philip Boyd and Chris Vivian, 'Should we fertilize oceans or seed clouds? No one knows' (2019) 570 *Nature* 155.

nature of untested geoengineering proposals is an irrelevant argument against it – for it is also a reason to speed up research.

9.3.3.3 Taking geoengineering seriously undermines emission reduction

The fear, however, is that geoengineering research will encourage the public to think that a solution is imminent, whereas in practice many a wonderful idea goes through years of testing and ends up amounting to nothing. It is not the speculativeness, as such, of geoengineering that is the problem, but the risk that the public will overestimate the chances of success and underestimate the timeframe for implementation and conclude that the crisis is nearly over. By the time they realize they are wrong, the situation will be even worse. This essentially is the third critique of geoengineering.

It is by far the most important one and it usually goes by the name of 'moral hazard', a term borrowed from insurance, where it refers to the risk that if someone is insured against a risk – wine on the carpet, house fire, unemployment – then they will make less of an effort to avoid the harm.[61] Ultimately, this leads to a less responsible world in which welfare for all is reduced.

The idea that looking for technological solutions to global warming will distract from the safe, proven and optimal approach of reducing emissions, by making it seem less important and less urgent, is a powerful argument.[62] If geoengineering experiments were carried out, it would increase funding and attention, and there would soon be headlines about the potential of new SRM and ocean fertilization techniques – and perhaps others too – which might indeed, imaginably, lead to cries of relief that the climate crisis is on the edge of being solved. The moral-hazard risk is real.

On the other hand, moral hazard is ubiquitous in policy making.[63] Wherever law addresses a risk – and that is surely the heart of modern regulation – it creates the new risk that the protected will take less responsibility upon themselves. This can be found in welfare, in security, in almost any field. To take medicine, again, the analogy with climate engineering is a good one. Many diseases are overwhelmingly the product of lifestyle, and an optimal medical outcome would be that people change their behaviour. It is harder to persuade them to do so if treatment is available that blunts the risk their

61 Katelyn Tsipiras and Will J Grant, 'What do we mean when we talk about the moral hazard of geoengineering?' (2022) 24(1) *Environmental Law Review* 27.

62 Albert C Lin 'Does geoengineering present a moral hazard?' (2013) 40 *Ecology Law Quarterly* 673. Jesse L Reynolds 'A critical examination of the climate engineering moral hazard and risk compensation concern' (2014) 2(2) *The Anthropocene Review* 1.

63 Gernot Wagner and Daniel Zizzamia, 'Green moral hazards' (2022) 25(3) *Ethics, Policy & Environment* 264.

behaviour creates.[64] Treatment may be a suboptimal solution, for it is still generally better to be healthy. But it makes it harder to find the motivation to make the choices required. This is not seriously put forward as a reason to limit medical research, though. Where diseases exist, attempts are made to treat or cure them.

There are good reasons why moral hazard is not generally treated as a policy trump card. For one thing, it can be addressed by countermeasures – communication and policy coordination.[65] Medical research does not exclude preventative programmes, and geoengineering research does not exclude intensive communication combatting over-enthusiasm and emphasizing the importance of emission reduction. There are tensions, certainly, but policy making in a complex field is usually about balancing tensions, not taking sides. Moreover, moral hazard remains in many contexts a speculative risk,[66] for whether people actually respond in practice as moral-hazard theory suggests is an empirical question which is far from proven.[67] Do people drive faster with seatbelts, as was once feared, or do seatbelts remind them of the importance of safety and make them careful? Many people refuse to engage with climate change, or even deny it is happening, because admitting the truth even to themselves would inspire feelings of despair and guilt that they would struggle to process. This draws on the idea of cognitive dissonance, namely that faced with two conflicting ideas people will suppress one, in order to remove the unpleasant conflict.[68] It may be that the possibility of

---

[64] Jonathan Klick and Thomas Stratmann, 'Diabetes treatments and moral hazard' (2007) 50(3) *The Journal of Law and Economics* 519; Dhaval Dave and Robert Kaestner, 'Health insurance and ex ante moral hazard: evidence from Medicare' (2009) 9 *International Journal of Health Care Finance and Economics* 367. Cf Jay Bhattacharya and Mikko Packalen, 'The other ex ante moral hazard in health' (2012) 31(1) *Journal of Health Economics* 135.

[65] Wagner and Zizzamia (n 63).

[66] Williamson and Bodle, (n 20) 96.

[67] Royal Society (n 28); Wim Carton et al, 'Is carbon removal delaying emission reductions?' (2023) 14(4) *WIREs Climate Change* e826; Talbot M Andrews, Andrew W Delton and Reuben Kline, 'Anticipating moral hazard undermines climate mitigation in an experimental geoengineering game' (2022) 196 *Ecological Economics* 107421; Christine Merk and others, 'Exploring public perceptions of stratospheric sulfate injection' (2015) 130 *Climatic Change* 299; Benjamin Hale, 'The World that Would Have Been: Moral Hazard Arguments Against Geoengineering' in Christopher J Preston (ed) *Engineering the Climate: The Ethics of Solar Radiation Management* (Lexington Books/Rowman & Littlefield, 2012) 113; Christopher J Preston, 'Ethics and geoengineering: reviewing the moral issues raised by solar radiation management and carbon dioxide removal' (2013) 4 *WIREs Climate Change* 23; Adam Corner and Nick Pidgeon, 'Geoengineering, climate change scepticism and the "moral hazard" argument: an experimental study of UK public perceptions' (2014) 372 *Philosophical Transactions of the Royal Society A*, 2014.0063.

[68] Cynthia M Frantz and F Stephan Mayer, 'The emergency of climate change: why are we failing to take action?' (2009) 9 *Analyses of Social Issues and Public Policy* 205; Hank Jenkins-Smith et al, 'Geoengineering and climate change polarization: testing a two-channel

climate engineering would offer hope, helping to avoid 'doomism', and this hope would increase engagement and preparedness to reduce emissions. If this is a problem we can solve, let us get on with it![69] It is the opposite result of what the moral-hazard critique suggests.

Social science has not yet settled the question of which of these dynamics is likely to be stronger, or whether geoengineering could persuade groups currently unreached by conventional discourse[70] that climate change is real and urgent.

That leaves moral hazard as an interesting and important factor in discussions of geoengineering, but one whose weight and impact are as speculative as the consequences of geoengineering itself. This then raises the question of why, in this context, is moral hazard taken so seriously? Why are potential climate-rescue techniques dismissed a priori on the basis of an unproven psychological theory which in no other context carries nearly so much weight? What is there about geoengineering and its relationship to alternatives that is different from other moral-hazard contexts, which might explain the distinctive reactions it inspires? I suggest that there are two features here worthy of note.

9.3.4 Geoengineering tries to rescue capitalism

One is that its success – assuming it were developed to the point that there is an importantly useful technique available – could be experienced as undesirable, since, while it might alleviate global warming, it would also, by definition, enable the energy transition to be delayed, or even, in the event of a radically successful NET technique, avoided entirely. It is imaginable, if implausible, that geoengineering might undo the climate consequences of current industrial practices and thereby remove the immediate need for them to change.[71]

Some might celebrate this – imagine, harmless fossil fuels! What a liberation![72] However, the burning of such fuels is not only a climate problem,

model of science communication' (2015) 658(1) *The ANNALS of the American Academy of Political and Social Science* 192.

69 See Frantz and Mayer, ibid; Rebecca Solnit and Thelma Young-Lutunatabua (eds) *Not Too Late: Changing the Climate Story from Despair to Possibility* (Haymarket Books, 2023); Anna Novoselov and Karlie Hayes, 'Sparking emotions conducive to action is critical to solving the climate crisis' (2022) 15(6) *Sustainability and Climate Change* 380.

70 Astrid Dannenberg and Sonja Zitzelsberger, 'Climate experts' views on geoengineering depend on their beliefs about climate change impacts' (2019) 9(10) *Nature Climate Change* 769; Hale (n 67); Corner and Pidgeon (n 67).

71 Gunderson et al (n 26); Kevin Surprise and Jean Philippe Sapinski, 'Whose climate intervention? Solar geoengineering, fractions of capital, and hegemonic strategy' (2022) 0(0) *Capital & Class*; Möller (n 7).

72 See discussion (not approving) in Möller (n 7).

but associated with a model of the economy that is often considered to be destructive and over-exploitative. Aside from its other environmental consequences, resulting from pollution, bad land-use choices, and destruction of areas with natural resources, it has also traditionally been the engine of modern capitalism, which is often considered to have created structurally exploitative societies.[73] Many environmentalists are to some extent left-wing, and issues of social justice and environmental protection are bound up with the preference for a socio-economic model that is less based on profit, exploitation, growth and fossil fuels. Climate change is one reason to develop a new model, perhaps the most urgent reason, but not the only one. By neutralizing the climate impact of fossil fuels, attempts to make the world a fairer place and the economy less destructive would be set back.[74] From this perspective, geoengineering is a dangerous distraction if it does not work – and it is a dangerous palliative if it does.[75] The very attempt to make the fossil-fuel economy harmless, rather than to seek to change it, is morally objectionable, and in its own way dangerous.[76]

This is somewhat different from the medical situation. If some new breakthrough treatment meant that we could all enjoy as much cake or alcohol as we wanted to consume without health consequences, that would certainly create a complicated situation for those committed to healthy lifestyles, but it would be hard to see it as a complete disaster. The moral objection to cake is that it causes ill health. If that objection is removed, it is possible to recategorize it as a non-guilty pleasure, without significant peripheral externalities. A world of unlimited cake, unlike a world of unlimited fossil-fuel use, is not inherently harmful for other reasons.

Thus geoengineering, far less than other areas of risk regulation, is judged not just by its effectiveness or not, but by its consequences for wider ideological ambitions. For those committed to degrowth, or deep ecology, it represents an abhorrent techno-fix, instrumentalizing nature at the expense of a meaningful and enriching relationship with it, and this objection remains valid whether or not geoengineering positively affects climate in the short

---

73 Klein (n 21); Aaron Saad, *Worlds at Stake: Climate Politics, Ideology, and Justice* (Fernwood Publishing, 2022); David Camfield, *Future on Fire: Capitalism and the Politics of Climate Change* (Fernwood Publishing, 2022).

74 Surprise and Sapinski (n 71); Klein (n 21); CIEL (n 40); Gunderson et al (n 26); Jean Philippe Sapinski, Holly Jean Buck and Andreas Malm (eds) *Has It Come to This? The Promises and Perils of Geoengineering on the Brink* (Rutgers University Press, 2020). See also Baiman (n 39).

75 Andreas Malm, 'The future is the termination shock: on the antinomies and psychopathologies of geoengineering. Part one' (2022) 30(4) *Historical Materialism* 3.

76 Malm, ibid; Jennie C Stephens, 'Beyond climate isolationism: a necessary shift for climate justice' (2022) 8 *Current Climate Change Reports* 83.

term.[77] For those committed to technological progress, it is to be embraced whether or not it is strictly necessary, because attempts to avoid technology are seen as futile and ultimately restrictive of human development: scientific and human progress are intertwined. In addition to the capitalist/anti-capitalist dynamic, there are as well decolonialist and feminist perspectives on geoengineering, which ask not just how it may affect temperature but what kind of power relationships it entrenches or dismantles and what kind of social order it encourages.[78] The novelty of geoengineering, and the potential scale of its impact, mean that no complete political ideology can fail to give it attention, with the most fundamental dividing line perhaps being between revolutionaries, who would like to see the existing global socio-economic order crumble, and those who prefer to see it buttressed against the chaotic winds of change.

9.3.5 Geoengineering disrupts the professional status quo

The other, closely connected, distinctive aspect of geoengineering is the professional commentators conducting the discourse about it. For medicine and health, the most authoritative and dominant voices are those of the medical experts, and they are involved primarily in treatment and cure, not prevention. That is not to say that prevention is frowned upon, but there is an enormous professional investment in developing new treatments, and the structure of science means that there is a professional advantage to coming up with new, breakthrough ideas. The careers of those who are able to speak with expertise about disease are dependent on the pursuit of better treatments, and a strong role for moral-hazard arguments in the discourse would be directly contrary to their personal and institutional interests.

By contrast, the position of vested authority in climate change discussions is occupied by those with expertise in environmental sciences and environmental governance, as well as by those with expertise in emission reduction. For the latter group, geoengineering as an object of research

[77] Barbara Muraca and Frederike Neuber, 'Viable and convivial technologies: considerations on climate engineering from a degrowth perspective' (2018) 197 *Journal of Cleaner Production* 1810; Frank Adloff and Iris Hilbrich, 'Practices of sustainability and the enactment of their natures/cultures: ecosystem services, rights of nature, and geoengineering' (2021) 60(2) *Social Science Information* 168.

[78] Holly J Buck, AR Gammon and CJ Preston, 'Gender and geoengineering' (2014) 29(3) *Hypatia* 651; Tina Sikka, 'Technology, gender, and climate change: a feminist examination of climate technologies' (2018) 8(4) *Societies* 109; Karsten Schulz, 'Decolonising the Anthropocene: The Mytho-politics of Human Mastery' in Marc Woons and Sebastian Weier (eds) *Critical Epistemologies of Global Politics* (e-international relations, 2017).

funding and eventual use is directly oppositional to their career interests and authority as a group.[79] For the environmental science experts – biologists, atmospheric scientists, and so on – and for environmental governance specialists – policy scientists and lawyers – that is not necessarily the case.

Public and scholarly discourse suggests that those involved in climate change research and policy have nonetheless come to form a coherent and self-sustaining body – an epistemic community.[80] Such a community has roughly aligned interests and a broadly shared world-view – what philosophers of science sometimes call a 'mindset'. In this case the world-view is built around a shared view of how climate change can and should be tackled, a view in which geoengineering has no part.[81]

The creation of such a mindset is fuelled by institutional factors. Inevitably, climate specialists work together in research and advice teams, and they also train the next generation. In doing so they create shared ideologies and cohesive policy commitments, which in turn make their community more robust and influential.[82] Thus, if the climate change community adopts common positions, and attacks common enemies, rather than engaging in internal debate, its position in institutions, policy advice and public debate is strengthened. This strength then enables it to become self-sustaining and maintain its position, in a process akin to conceptual capitalism: who owns the conversation, profits from it.

Such epistemic communities are powerful, but path-dependent, a risk-factor for their ultimate decline. The community of climate experts has a world-view built around emission reduction, and so much professional investment in this that it would be hard to change. The geoengineering suggestion that there might be a fundamentally different way to address climate change – at least to some extent – is methodologically and ideologically challenging.

Were geoengineering to be taken seriously and begin to show promise it would then be threatening to the interests and authority of the climate

79 Daniel Maliniak, Eric Parajon and Ryan Powers, 'Epistemic communities and public support for the Paris Agreement on climate change' (2021) 74(4) *Political Research Quarterly* 866; Mai'a K Davis Cross, 'Rethinking epistemic communities twenty years later' (2013) 39(1) *Review of International Studies* 137.

80 Claire Gough and Simon Shackley, 'The respectable politics of climate change: the epistemic communities and NGOs' (2001) 77(2) *International Affairs* 329; Cross, ibid.

81 Cross (n 79); Sean Low, Chad Baum and Benjamin K Sovacool, 'Undone science in climate interventions: contrasting and contesting anticipatory assessments by expert networks' (2022) 137 *Environmental Science & Policy* 249, on how IPCC methods are not well adapted to geoengineering assessment (cf McLaren and Corry (n 37) at 2.3); Benjamin K Sovacool et al, 'Actors, legitimacy, and governance challenges facing negative emissions and solar geoengineering technologies' (2023) *Environmental Politics*, forthcoming.

82 See n 75.

change community.[83] It would raise the prospect not just of a new mindset, but of new experts. From this perspective, the very heat of the opposition to it could be an indication that such threat is felt, and so, perhaps, an indication that experts consider – are afraid – that some schemes might work. If it was truly, inevitably, hopeless, it would hardly need attacking, and the easiest way of dealing with it might be just to let a few experiments go ahead and fail.

9.4 Rethinking regulation of geoengineering

Faced with a crisis of the magnitude of climate change, in which technology is inevitably important – be it better batteries, new energy-generating techniques, or climate technology – a population of rational people would inevitably pursue all options that might be useful to the process of climate repair, and their regulator would set outer normative limits to experiments while leaving space to encourage innovation and out-of-the-box thinking.[84]

However, the real world of politics and power disrupts this. For one thing, people are not trusted to resist false hope, so that there is pressure to pursue only the most well-understood and conservative pathways toward climate stability. In a radically anti-scientific move, many environmentalists are pushing the argument that they know a priori which lines of research and policy will lead to the best results; and others say that geoengineering should be ignored. There is an apocalyptic aspect to this point of view. The evidence seems to be that humanity is not reducing its emissions fast enough to prevent very harmful results. The resistance to developing alternative climate-management techniques seems almost like machismo: live CO_2-free, or die.

Then there is the institutional and professional momentum that environmental science has created. Intellectual, financial and ideological investment in emission reduction has created an expert community at the heart of climate discourse whose professional interests are opposed to the development of geoengineering. As technology lawyers have long known, if the regulator allows technology to develop without steering, then a status quo will emerge which is resistant to regulatory adjustment. In this case, the exercise of public power by universities and research funders has prevented geoengineering research moving to the phase of experiment, and so building up a critical professional mass or momentum. At the same time, law and funders have actively stimulated the development of emission-reduction technology of more conventional kinds. There is now an unbalanced status quo, and since the occupiers of that status quo are the same expert

[83] Low et al (n 81).

[84] Florin (n 39).

community who will have to be drawn upon for advice on future regulatory decisions, it is entrenched.

In any case, aspiring to regulate technology purely for its climate effects may be misguided.[85] Technologies that mitigate climate change are about more than climate – they are also about preservation or manipulation or disruption of a certain socio-economic system. Geoengineering may have consequences for CO_2 and global warming, but it also may have consequences for the development of capitalism. It is contestable, but not irrational, to take that into consideration, and on a view of environmental harm as resulting from modes of economic organization there is reason to see geoengineering as merely a form of temporary symptom management, delaying more effective treatment in a way that may ultimately do harm. If combatting climate change is about politics, and not just climate science, then a broad view of climate technology is inevitable.

Regulating geoengineering will then never be easy. What can perhaps be hoped for at this stage is more unpacking of the values and ideas involved, which has been the aim of this chapter. Decisions about what to fund or allow, like all important collective decisions, will be the product of compromise between competing values and world views. The quality of that compromise is likely to be improved if some of the more simplistic arguments – it is dangerous, it is speculative, it creates moral hazard – are replaced by more nuanced and socially aware points of view.

85 Sovacool, Baum and Low (n 55); Williamson and Bodle (n 20), 81–2.

10

Why Researching Solar Radiation Management Technologies Is Essential and Governable

Manon Simon, Kerryn Brent, Jeffrey McGee and Jan McDonald

10.1 Introduction

This chapter argues for urgent and substantial research and development (R&D) of technologies aimed at managing solar radiation, as a short-term and temporary response to dangerous climate change. Anthropogenic climate change is already affecting most regions of the world, with more extreme events (for example, intense heatwaves, precipitation and drought) resulting in wildfire, flooding and famine. Climate impacts are already challenging the resilience of socio-ecological systems globally.[1] The mitigation commitments of the Paris Agreement parties are currently insufficient[2] to achieve the objective of 'Holding the increase in the global average temperature to well below 2°C above pre-industrial levels'.[3] Scientists predict that the

---

1 Valérie Masson-Delmotte et al, 'IPCC, 2021: Summary for Policymakers' in: Climate Change 2021: The Physical Science Basis. Contribution of Working Group I to the Sixth Assessment Report of the Intergovernmental Panel on Climate Change (Cambridge University Press, 2021) 10.

2 See, for example, Conference of the Parties serving as the meeting of the Parties to the Paris Agreement, 'Synthesis Report on Nationally Determined Contributions under the Paris Agreement' (third session, Glasgow, Scotland, 31 October to 12 November 2021) FCCC/PA/CMA/2021/8; Masson-Delmotte et al (n 1) A4.3; United Nations Environment Programme, *Emissions Gap Report 2022: The Closing Window. Climate Crisis Calls for Rapid Transformation of Societies* (UN 2022) XVI.

3 *Paris Agreement*, opened for signature 12 December 2015 (entered into force 4 November 2016) Art 2(1)(a).

remaining global budget for carbon emissions that would provide a 50 per cent chance of limiting global warming to 1.5°C will be exceeded before 2030.[4] A climate 'overshoot' above 2°C is virtually certain and will likely have significant consequences, including extreme weather events, sea-level rise, ecosystem degradation, Antarctic ice sheet destabilization and loss of Arctic sea ice.[5] Anthropogenic carbon emissions continue to rise and in the period 2010–2019 were 'higher than in any previous decade on record'.[6] In 2019, global atmospheric carbon dioxide concentration averaged 410 parts per million (ppm), compared with an average of 390 ppm in 2011.[7] Average global atmospheric carbon dioxide concentrations reached 418 ppm in 2022.[8] The carbon dioxide that has built up in the atmosphere over the past centuries and the feedback effects on the climate system will delay the reversal of climate change effects.[9] Even if states were to implement the most ambitious of policies, stopping all greenhouse gas emissions immediately, it would not be enough to halt further heating.

It is against this backdrop that scientists are researching solar radiation management (SRM), or solar geoengineering techniques, to reduce radiative forcing.[10] SRM proposals aim to directly cool regional or global temperatures by altering the Earth's radiative balance between incoming solar radiation energy (sunlight) and outgoing infrared (long-wave) radiation. SRM proposals aim to do this either by reflecting a percentage of sunlight away

---

4 Piers Forster, 'Greenhouse Gas Emissions Are at an All-Time High and Earth Is Warming Faster than Ever – Report' (*The Conversation*, 8 June 2023) <http://theconversation.com/greenhouse-gas-emissions-are-at-an-all-time-high-and-earth-is-warming-faster-than-ever-report-207234> accessed 7 July 2023; summarizing the findings of Piers Forster et al, 'Indicators of Global Climate Change 2022: Annual Update of Large-Scale Indicators of the State of the Climate System and Human Influence' (2023) 15 *Earth System Science Data* 2295.

5 Masson-Delmotte et al (n 1) B2.2. See also generally Valérie Masson-Delmotte et al, *Global Warming of 1.5°C: An IPCC Special Report on the Impacts of Global Warming of 1.5° C Above Pre-Industrial Levels and Related Global Greenhouse Gas Emission Pathways, in the Context of Strengthening the Global Response to the Threat of Climate Change, Sustainable Development, and Efforts to Eradicate Poverty* (World Meteorological Organization, Geneva, Switzerland, 2018).

6 Masson-Delmotte et al (n 1) A1.4.

7 Ibid A1.3.

8 'Climate Change: Atmospheric Carbon Dioxide' (NOAA Climate.gov, 9 April 2024) <www.climate.gov/news-features/understanding-climate/climate-change-atmospheric-carbon-dioxide> accessed 11 December 2023.

9 See generally Susan Solomon et al, 'Irreversible Climate Change Due to Carbon Dioxide Emissions' (2009) 106 *Proceedings of the National Academy of Sciences* 1704–09.

10 See generally Mark Lawrence et al, 'Evaluating Climate Geoengineering Proposals in the Context of the Paris Agreement Temperature Goals' (2018) 9 *Nature Communications* 1–19.

from the Earth's surface or increasing the amount of terrestrial radiation that escapes to space.

SRM proposals are synonymous with the concept of a new geological age of human influence on the planet, 'the Anthropocene', and raise significant questions about the extent to which humans should exert power over the natural environment.[11] On the one hand, SRM might be criticized as a 'dangerous technofix' for climate change, a 'Promethean pipedream' that (if successful) would allow humans to establish mastery over nature.[12] These criticisms centre on the view that humanity is good at creating environmental problems, poor at fixing them, and more likely to create new problems in trying to do so. These criticisms are underpinned by concerns that SRM is bound to fail, given 'humanity's history of creating environmental disasters and its inability to control and predict nature'.[13] However, other authors offer a more hopeful perspective that SRM could be an important part of a global technological transition to a 'good' Anthropocene.[14] For example, Asayama et al suggest that SRM could help humanity achieve 'effective planetary stewardship' in this epoch.[15] Either way, there are significant concerns about the environmental and political risks of SRM proposals. It is also well recognized that innovative legal and governance approaches are required for SRM.[16] However, a key question is the extent to which law should facilitate SRM. There are growing calls not only to prohibit deployment, but also to ban the *development* of SRM.[17] Underlying these calls are claims that SRM is not just too risky, but ungovernable.

---

11 Kerryn Brent, 'Solar Geoengineering and the Challenge of Governing Multiple Risks in the Anthropocene', in Peter D. Burdon and Hames Martel (eds) *The Routledge Handbook of Law and the Anthropocene* (Routledge 2023) 257–67.

12 Clive Hamilton, *Earthmasters: The Dawn of the Age of Climate Engineering* (Yale University Press 2013) 179; Clive Hamilton, *Defiant Earth: The Fate of Humans in the Anthropocene* (John Wiley & Sons 2017) 91–2.

13 Jonas Anshelm and Anders Hansson, 'Battling Promethean Dreams and Trojan Horses: Revealing the Critical Discourses of Geoengineering' (2014) 2 *Energy Research & Social Science* 135–44, 142.

14 For an analysis of what is a 'good' Anthropocene, see Elena M Bennett et al, 'Bright Spots: Seeds of a Good Anthropocene' (2016) 14 *Frontiers in Ecology and the Environment* 441–8.

15 Shinichiro Asayama et al, 'Beyond Solutionist Science for the Anthropocene: To Navigate the Contentious Atmosphere of Solar Geoengineering' (2019) 6 *The Anthropocene Review* 19–37, 20.

16 For a detailed review, see Jesse L Reynolds, 'Solar Geoengineering to Reduce Climate Change: A Review of Governance Proposals' (2019) 475 *Proceedings of the Royal Society A: Mathematical, Physical and Engineering Sciences* 1–33.

17 See generally Carrol Muffett, 'Fuel to the Fire: How Geoengineering Threatens to Entrench Fossil Fuels and Accelerate Climate Change' (2019) 30 *Indigenous Policy Journal*; Frank Biermann et al, 'Solar Geoengineering: The Case for an International Non-use

In this chapter, we argue that SRM should be fully researched and developed to a state of deployment readiness because it might be needed to prevent dangerous overshoot of global temperature and provide humanity with time for more orderly mitigation and adaptation. The role of law, therefore, should be to facilitate responsible R&D of SRM. To be clear from the outset, we are not arguing in favour of *deployment* of large-scale SRM at this stage, or even in the near future. We instead argue that rigorous R&D is needed to ensure that decision-makers are informed of the risks of SRM and its potential benefits in attenuating the increasingly severe impacts of the climate crisis. This knowledge is essential in order to design robust governance at international, national and subnational levels. We argue that the greatest risk is that SRM will be implemented in a time of perceived crisis without an adequate understanding of its consequences and in the absence of robust governance mechanisms.

This chapter proceeds as follows. Section 10.2 explains key SRM proposals and their potential risks and benefits. Section 10.3 articulates and then responds to the key arguments against SRM. It argues that while the governance of SRM research and associated risks is exceptionally challenging, it does not preclude the development of realistic and pragmatic governance approaches for SRM research. Section 10.4 draws on natural law theory to argue that, in the context of a rapidly changing climate, living in ignorance of SRM's potential is not consistent with law's purpose of pursuing the common good of humanity. Finally, Section 10.5 concludes that research into diverse SRM proposals is the best way to determine which options will be worth supporting, and what safeguards are needed.

10.2 The potential benefits and risks of SRM

Scientists and policy makers are considering the potential for SRM to prevent global temperature overshoot and gain time for the world to achieve net negative emissions.[18] SRM technologies are being evaluated for deployment in three different regions: space, atmosphere and the surface of

---

Agreement' (2022) 13 *Wiley Interdisciplinary Reviews: Climate Change* 1–18; Aarti Gupta et al, 'Towards a Non-Use Regime on Solar Geoengineering: Lessons from International Law and Governance' (2024) *Transnational Environmental Law* 1.

18 See generally Paul Crutzen, 'Albedo Enhancement by Stratospheric Sulfur Injections: A Contribution to Resolve a Policy Dilemma?' (2006) 77 *Climatic Change* 211–20; Stefan Schäfer et al, 'The European Transdisciplinary Assessment of Climate Engineering (EuTRACE): Removing Greenhouse Gases from the Atmosphere and Reflecting Sunlight Away from Earth'; Christopher Field et al, 'Reflecting Sunlight: Recommendations for Solar Geoengineering Research and Research Governance' (National Academies of Sciences, Engineering, and Medicine 2021).

the Earth.[19] Space-based technologies include reflective mirrors or other sunshields and face considerable feasibility constraints.[20]

As for atmospheric SRM techniques, there are three main types. The most widely discussed option is stratospheric aerosol injection (SAI), which involves injecting minute reflective particles into the stratosphere. Scientific observations of the climatic effects of large-scale volcanic eruptions indicate that placing aerosols in the stratosphere could have a significant and rapid cooling effect on global temperatures. For example, the 1991 eruption of Mount Pinatubo in the Philippines injected 20 teragrams of sulfur dioxide into the stratosphere, cooling global mean surface temperatures by approximately 0.5°C for 12 months.[21] SAI is an attractive option for policy makers because it would likely be relatively inexpensive to test (in the order of tens of billion dollars annually) compared to the much higher financial costs of climate change impacts and mitigation activities.[22] If deployed on a large scale, SAI would likely reduce global temperatures, and scientists have also proposed deploying it to limit climate change impacts (for example, temperature increases) in regions such as the Arctic.[23]

Another SRM reflective technology that is being developed for regional application is marine cloud brightening (MCB).[24] This involves spraying minute sea-salt particles into low-lying ocean clouds to increase the amount of sunlight they reflect away from the ocean surface. Cirrus cloud thinning (CCT) is the main focus for increasing the escape of terrestrial heat. Cirrus clouds trap more energy in the Earth's atmosphere than they reflect, so scientists hypothesize that thinning them using cloud-seeding technology could allow more heat energy to escape the Earth's atmosphere.[25]

SRM could play a vital role in protecting vulnerable regions and ecosystems from irreversible climate change impacts, such as sea-ice loss and melting

[19] Lawrence et al (n 11) 9.

[20] See generally Chad Baum, Sean Low and Benjamin Sovacool, 'Between the Sun and Us: Expert Perceptions on the Innovation, Policy, and Deep Uncertainties of Space-Based Solar Geoengineering' (2022) 158 *Renewable and Sustainable Energy Reviews* 1–22.

[21] Alan Robock et al, 'A Test for Geoengineering?' (2010) 327 *Science* 530, 531.

[22] Anthony Harding and Juan Moreno-Cruz, 'Solar Geoengineering Economics: From Incredible to Inevitable and Half-way Back' (2016) 4 *Earth's Future* 569, 570.

[23] See generally Walker Lee et al, 'High-latitude Stratospheric Aerosol Injection to Preserve the Arctic' (2023) *Earth's Future* 1–19.

[24] See generally John Latham et al, 'Marine Cloud Brightening: Regional Applications' (2014) 372 *Philosophical Transactions of the Royal Society A: Mathematical, Physical and Engineering Sciences* 1–19.

[25] See generally Helene Muri et al, 'The Climatic Effects of Modifying Cirrus Clouds in a Climate Engineering Framework' (2014) 119 *Journal of Geophysical Research: Atmospheres* 1–18.

permafrost.[26] Scientists are also investigating the potential for MCB to cool water temperatures in the Great Barrier Reef in order to buffer the corals from marine heatwaves and reduce the risk of catastrophic coral bleaching events.[27] This research has progressed to small-scale outdoor experiments for the purpose of testing the delivery equipment and particle formation, without any planned climate impacts. Modelling suggests that regional shading (such as MCB) combined with crown-of-thorns starfish control could be 'relatively effective' at protecting the Great Barrier Reef until 2040, from which time onward other adaptation strategies might be sufficiently advanced to limit reef decline.[28]

However, no large-scale SRM experiment – that is, at a scale large enough to affect regional or global temperatures – has been conducted to date, and all technologies are a long way from being fully deployable.

SRM is fraught with uncertainty and carries serious risks of social and environmental impacts. It is feared that SAI could create regional climate disparities, disrupt precipitation patterns such as monsoons, and deplete stratospheric ozone.[29] In addition, SAI deployment would require a long-term commitment to maintain the cooling effect, with potentially serious intergenerational implications.[30] In particular, a sudden cessation of a large-scale SRM programme could result in a sudden spike in warming, with effects worse than gradual climate change.[31] Large-scale MCB presents similar risks, although they are lower because MCB could potentially be phased in and out more easily than technologies such as SAI.[32]

There are also concerns about the geopolitical and international security risks that SRM deployment presents. If SRM were to be conducted by

26 See generally Alistair Duffey et al, 'Solar Geoengineering in the Polar Regions: A Review' (2023) 11 *Earth's Future* 1–24.

27 Jeff Tollefson, 'Can Artificially Altered Clouds Save the Great Barrier Reef?' (2021) 596 *Nature* 476–8.

28 Condie et al suggest that, from 2040, thermally tolerant coral communities could be sufficiently established to slow the decline of the reef. Scott Condie et al, 'Large-Scale Interventions May Delay Decline of the Great Barrier Reef' (2021) 8 *Royal Society Open Science* 1, 23.

29 See generally Alan Robock, '20 Reasons Why Geoengineering May Be a Bad Idea' (2008) 64 *Bulletin of the Atomic Scientists* 14–18.

30 See generally William Burns, 'Climate Geoengineering: Solar Radiation Management and Its Implications for Intergenerational Equity' (2011) 4 *Stanford Journal of Law, Science & Policy* 39–55.

31 See generally Andy Parker and Peter Irvine, 'The Risk of Termination Shock from Solar Geoengineering' (2018) 6 *Earth's Future* 456–67.

32 Michael Diamond et al, 'To Assess Marine Cloud Brightening's Technical Feasibility, We Need to Know What to Study—and When to Stop' (2022) 119 *Proceedings of the National Academy of Sciences* 1, 5.

one country or small group of countries, without the consent of others, it could cause geopolitical tensions.[33] This could, in turn, provoke retaliation in the form of 'counter-geoengineering'.[34] There is also the potential for rogue actors to conduct SRM without state authorization. This concern moved closer to reality in 2022, when the US-based company *Make Sunsets* conducted SRM experiments in Mexico without the knowledge or approval of Mexican authorities, with a view to selling privately generated 'cooling credits'.[35]

These risks are compounded by the fact that SRM is not adequately governed by either national or international law at present. This has led to calls for moratoria on SRM deployment.[36] Given the risks of irreversible impact, a precautionary approach dictates that large-scale activities should not be allowed to proceed without further risk assessment and appropriate risk reduction. A more nuanced question, however, is whether SRM R&D should be encouraged at all, and, if so, how it should be governed. It is important that SRM R&D be governed ethically to ensure that the benefits outweigh both the associated risks and the risks of unabated climate change.

10.3 Refuting the arguments opposing SRM R&D

Views vary significantly on whether, and to what extent, SRM research activities should be prohibited.[37] Fervent opponents of SRM call for a strict prohibition on R&D on the basis that even field tests carry significant risks.[38] Principled opposition of this sort is often based on ideological

---

33 Michael Birnbaum, 'A "Climate Solution" That Spies Worry Could Trigger War' (*Washington Post*, 27 February 2023) <www.washingtonpost.com/climate-environment/2023/02/27/geoengineering-security-war/> accessed 10 July 2023.

34 See generally Daniel Heyen, Joshua Horton and Juan Moreno-Cruz, 'Strategic Implications of Counter-Geoengineering: Clash or Cooperation?' (2019) 95 *Journal of Environmental Economics and Management* 154–77.

35 'A Startup Says It's Begun Releasing Particles in the Atmosphere, in an Effort to Tweak the Climate' (*MIT Technology Review*, 24 December 2022) <www.technologyreview.com/2022/12/24/1066041/a-startup-says-its-begun-releasing-particles-into-the-atmosphere-in-an-effort-to-tweak-the-climate/> accessed 10 July 2023. See also generally Michael S Diamond, Kelly Wanser and Olivier Boucher, '"Cooling Credits" Are Not a Viable Climate Solution' (2023) 176 *Climatic Change* 96–105.

36 For a review of the literature on SRM moratoria, see Reynolds 'Solar Geoengineering to Reduce Climate Change' (n 16) 14.

37 Megan Herzog and Edward Parson, 'Moratoria for Global Governance and Contested Technology: The Case of Climate Engineering' [2016] UCLA School of Law, Public Law Research Paper 1, 3–4.

38 See generally Leila Marshy, *Geopiracy: The Case against Geoengineering* (ETC group, 2010); Clive Hamilton, 'No, We Should Not Just "At Least Do the Research"' (2013) 496 *Nature* 139.

considerations and assumptions about SRM risks.[39] For example, Pamplany et al show that most researchers and academics discussing the desirability of research on geoengineering take a cautious position: while a few advocates of SRM research may consider geoengineering as 'the best option', most frame geoengineering as a 'last resort option', rather than a Plan A or Plan B (and see also Chapter 9 of this book).[40]

There is widespread agreement among SRM proponents and opponents alike that, given gaps in scientific understanding and governance regimes, large-scale deployment of SRM should currently be prohibited, especially techniques such as SAI that could have global environmental impacts.[41] Some have advocated for a ban or moratorium on SRM field testing generally,[42] on field testing of certain kinds of SRM technologies,[43] or until appropriate governance frameworks are put in place.[44] In a recent report on the human rights impact of climate-intervention technologies, the UN Human Rights Council (HRC) recommended that states consider a moratorium on SRM development and deployment 'for as long as scientific uncertainty and the risk of causing serious, extensive and irreversible environmental and human damage remain high'.[45] The HRC further recommended regulation of SRM research based on a presumption that it is incompatible with states' human rights obligations, unless proven otherwise through meaningful and comprehensive assessments of risk, human rights and environmental impact.[46] In this section, we argue that research activities intended to assess the social and environmental risks of SRM are thus essential to rebut such

39 Herzog and Parson (n 37) 65.

40 Augustine Pamplany, Bert Gordijn and Patrick Brereton, 'The Ethics of Geoengineering: A Literature Review' (2020) 26 *Science and Engineering Ethics* 3069, 3096–9.

41 See generally SRMGI, 'Solar Radiation Management: The Governance of Research' (2011) 11112 The Royal Society, London <https://royalsociety.org/-/media/policy/projects/solar-radiation-governance/des2391_srmgi-report_web.pdf> accessed 10 July 2023; Anna-Maria Hubert, 'A Code of Conduct for Responsible Geoengineering Research' [2017] Geoengineering Research Governance Project, Interim Report; Steve Rayner et al, 'The Oxford Principles' (2013) 121 *Climatic Change* 499–512.

42 Albert Lin, 'The Missing Pieces of Geoengineering Research Governance' (2016) 100(6) *Minnesota Law Review* 2509, 2570–1.

43 Michael Zürn and Stefan Schäfer, 'The Paradox of Climate Engineering' (2013) 4(3) *Global Policy* 266, 274.

44 Stefan Schäfer et al, 'Field Tests of Solar Climate Engineering' (2013) 3 *Nature Climate Change* 766; Climate Overshoot Commission, 'Reducing the Risks of Climate Overshoot' (September 2023) 1, 15.

45 Note that the HRC refers to these technologies as 'new technologies intended for climate protection' (NTCPs). Human Rights Council, 'Impact of new Technologies Intended for Climate Protection on the Enjoyment of Human Rights', A/HRC/54/47 (HRC 2023) 1, 17.

46 Ibid 19.

a presumption and inform future decision-making. We do so by responding to the arguments advanced by opponents as follows.

10.3.1 Claim 1: SRM research creates a moral hazard

A common argument against SRM research is that speculative optimism about what SRM could achieve may detract from efforts to reduce the primary cause of climate change, the excess level of human-induced greenhouse gas emissions.[47] This is often referred to as the 'moral hazard' argument, that is, 'the lack of incentive to guard against risk when one is protected from its consequences'.[48] Moral hazard arguments have long been identified in academic literature as 'the most widespread basis for resistance to solar geoengineering'.[49]

The concept of moral hazard was first developed in the context of insurance law to refer to circumstances in which those having insurance will have an incentive to take less care to guard against the event for which they are insured.[50] In our view, however, the term affords an inappropriate analogy, in the context of SRM,[51] and is ambiguous and vague.[52] This is because SRM research and climate policies are not mutually exclusive.[53] The term moral hazard has, therefore, been replaced by such terms as 'emissions abatement displacement concern',[54] 'mitigation deterrence'[55] and 'mitigation

47 Biermann et al (n 17) 4.

48 Gernot Wagner and Christine Merk, 'Moral Hazard and Solar Geoengineering' in Robert N. Stavins and Robert C. Stowe (eds) *Governance of the Deployment of Solar Geoengineering* (Cambridge, MA: Harvard Project on Climate Agreements 2019) 15, 135.

49 Jesse Reynolds, *The Governance of Solar Geoengineering: Managing Climate Change in the Anthropocene* (Cambridge University Press 2019) 32. See also David Keith, 'Geoengineering the Climate: History and Prospect' (2000) *Annual Review of Energy and the Environment* 245, 276; Albert Lin, 'Does Geoengineering Present a Moral Hazard?' 40 *Ecology Law Quarterly* 673–712.

50 Tom Baker, 'On the Genealogy of Moral Hazard' (1996) 75(2) *Texas Law Review* 237, 239.

51 Martin Bunzl, 'Researching Geoengineering: Should Not or Could Not?' (2009) 4 *Environmental Research Letters* 1, 2.

52 Benjamin Hale, 'The World That Would Have Been: Moral Hazard Arguments against Geoengineering', in Albert Borgmann et al (eds) *Engineering the Climate: The Ethics of Solar Radiation Management* (Rowman & Littlefield 2012) 1–28; Katelyn Tsipiras and Will Grant, 'What Do We Mean When We Talk about the Moral Hazard of Geoengineering?' (2022) 24 *Environmental Law Review* 27, 28.

53 Lin (n 49) 688–9; Joseph Jebari et al, 'From Moral Hazard to Risk-Response Feedback' (2021) 33 *Climate Risk Management* 1, 3.

54 Reynolds, *The Governance of Solar Geoengineering* (n 49) 32.

55 Duncan McLaren, 'Mitigation Deterrence and the "Moral Hazard" of Solar Radiation Management' (2016) 4 *Earth's Future* 596, 596.

obstruction'.[56] The term moral hazard has also been disaggregated, distinguishing between individual, social and political moral hazard,[57] or between different types of hazard (for example, technical-dependence hazard or regulatory-capture hazard).[58]

Whatever term is used, the assertion that SRM research will or could undermine efforts to reduce greenhouse gas emissions is disputed. Bodansky and Parker argue that SRM research will provide the knowledge that scientists and policy makers need to determine whether a particular technique is safe and effective against climate change, and therefore constitutes 'the best defense against moral hazard'.[59] Similarly, proponents of SRM have claimed that conducting serious SRM research constitutes a 'moral imperative',[60] and they consider it morally perverse to withhold measures to reduce global warming under the argument that it would encourage more risky behaviours.[61] Others make clear that conducting R&D should be distinguished from the conditions that might attach to SRM deployment in order to reduce risks of mitigation displacement.[62]

Empirical research on the moral hazard concern has produced mixed results.[63] One study showed that how SRM is framed is an important factor: for instance, SRM could reduce mitigation efforts when presented as 'miraculous'.[64] However, most studies show that pursuing SRM research does not make the public[65] or

---

56 David Morrow, 'Ethical Aspects of the Mitigation Obstruction Argument against Climate Engineering Research' (2014) 372 *Philosophical Transactions of the Royal Society A: Mathematical, Physical and Engineering Sciences* 1, 1.

57 Adam Corner and Nick Pidgeon, 'Geoengineering, Climate Change Scepticism and the "Moral Hazard" Argument: An Experimental Study of UK Public Perceptions' (2014) 372 *Philosophical Transactions of the Royal Society A: Mathematical, Physical and Engineering Sciences* 1, 2–3.

58 Hale (n 52) 16–22.

59 See generally Daniel Bodansky and Andy Parker, 'Research on Solar Climate Intervention Is the Best Defense Against Moral Hazard' (2021) 37 *Issues in Science and Technology* 19–21.

60 Mark Lawrence and Paul Crutzen, 'Was Breaking the Taboo on Research on Climate Engineering via Albedo Modification a Moral Hazard, or a Moral Imperative?' (2017) 5 *Earth's Future* 136, 137.

61 David Keith, *A Case for Climate Engineering* (MIT Press 2013) 10.

62 Jesse Reynolds, 'A Critical Examination of the Climate Engineering Moral Hazard and Risk Compensation Concern' 2(2) *The Anthropocene Review* 174–91, 178.

63 Reynolds, 'Solar Geoengineering to Reduce Climate Change' (n 16) 32.

64 Kaitlin Raimi et al, 'Framing of Geoengineering Affects Support for Climate Change Mitigation' (2019) 13 *Environmental Communication* 300, 312.

65 See generally Talbot Andrews, Andrew Delton and Reuben Kline, 'Anticipating Moral Hazard Undermines Climate Mitigation in an Experimental Geoengineering Game' (2022) 196 *Ecological Economics* 1–8; Maura Austin and Benjamin Converse, 'In Search of Weakened Resolve: Does Climate-Engineering Awareness Decrease Individuals' Commitment to Mitigation?' (2021) 78 *Journal of Environmental Psychology* 1–11; Malcolm

experts[66] less willing to reduce greenhouse gas emissions, with slight variations based on ideological or political inclinations (for example, trust in climate science). Conversely, several studies show that SRM research could increase both individual and collective support for particular climate policies,[67] such as carbon-offset purchases[68] or a carbon tax.[69] This 'mitigation galvanization',[70] or 'reverse moral hazard',[71] suggests that research into climate intervention could actually encourage mitigation because it 'would persuade more people that the threat of global warming was serious and needed to be addressed'.[72] It is worth noting that most of these studies were conducted in the Global North. While this does not necessarily undermine the generality of these conclusions, further research is needed to assess whether similar results would be reached in Global South countries.

The risk that SRM might disincentivize mitigation should not be ignored, but nor should it distract from the need to conduct research into SRM options.[73] Jebari et al explain that 'the moral hazard framing of the debate over geoengineering is itself a moral hazard … by making it seem as though the use of carbon capture or solar geoengineering technologies can serve as potential substitutes for mitigation efforts'.[74] Andrews et al express similar concerns, arguing that 'moral hazard anticipation' may undermine mitigation efforts even when moral hazard does not. This is because, if SRM research has the potential to encourage mitigation efforts, hindering SRM research for fear of a speculative moral hazard might indeed delay further emission reductions.[75] The risk that vested interests may misuse SRM to maintain or

Fairbrother, 'Geoengineering, Moral Hazard, and Trust in Climate Science: Evidence from a Survey Experiment in Britain' (2016) 139 *Climatic Change* 477–89.

66 See generally Christine Merk, Gert Pönitzsch and Katrin Rehdanz, 'Do Climate Engineering Experts Display Moral-Hazard Behaviour?' (2019) 19 *Climate Policy* 231–43.

67 Todd Cherry et al, 'Climate Cooperation in the Shadow of Solar Geoengineering: An Experimental Investigation of the Moral Hazard Conjecture' (2023) 32 *Environmental Politics* 362, 363.

68 See generally Christine Merk, Gert Pönitzsch and Katrin Rehdanz, 'Knowledge about Aerosol Injection Does Not Reduce Individual Mitigation Efforts' (2016) 11 *Environmental Research Letters* 1–7.

69 See generally Todd Cherry et al, 'Does Solar Geoengineering Crowd out Climate Change Mitigation Efforts? Evidence from a Stated Preference Referendum on a Carbon Tax' (2021) 165 *Climatic Change* 1–8.

70 McLaren (n 55) 596.

71 Reynolds (n 49) 39.

72 UK House of Commons Science and Technology Committee, 'The Regulation of Geoengineering: Fifth Report of Session 2009–10' (House of Commons 2010) HC 221, 23, <https://publications.parliament.uk/pa/cm200910/cmselect/cmsctech/221/221.pdf> accessed 17 July 2023.

73 Wagner and Merk (n 48) 137.

74 Jebari et al (n 53) 3.

75 Andrews et al (n 65) 2.

increase greenhouse gas emissions simply points to the need for measures that facilitate accountability and oversight.[76] This requires shifting the discussion from its current concern about moral hazard to considering the types of measures needed to manage its effects.[77]

Numerous scholars have identified strategies needed to manage the possible effects of moral hazard. Morrow suggests that precautionary measures against mitigation obstruction include researching relevant alternatives, communicating results and engaging with relevant publics and policy makers.[78] Some also argue for linkages between SRM R&D and enhanced mitigation and adaptation measures.[79] More broadly, Reynolds argues that concerns over mitigation obstruction can be avoided if decisions are made with careful consideration of stakeholder interests and 'in ways that are expected to improve human well-being and foster sustainability'.[80] Whilst this will be important to consider once states start implementing official SRM policies, the concern that SRM may undermine mitigation commitments is relatively weak at the research stage. Quite to the contrary, incremental SRM research can stimulate discussions about the types of governance arrangements needed to reduce the possible effect of mitigation deterrence and to further encourage the reduction of greenhouse gas emissions.

10.3.2 Claim 2: The blurred line between SRM research and deployment

Some opponents of SRM oppose R&D because of a concern that meaningful and safe large-scale experiments of SRM are impossible.[81] One of the major issues with planetary SRM is that it would need to be tested at full scale – that is at a planetary scale with impact on global average temperatures – for the results to be meaningful and to distinguish its effects from natural climate variability.[82] While much can be learned from small-scale SRM experiments, there would come a time at which these are not sufficient to understand

76 Wagner and Merk (n 48) 137.

77 Jebari et al (n 53) 4.

78 Morrow (n 56) 10.

79 Edward Parson, 'Climate Engineering in Global Climate Governance: Implications for Participation and Linkage' (2014) 3 *Transnational Environmental Law* 89, 91; Jesse Reynolds, 'Linking Solar Geoengineering and Emissions Reductions: Strategically Resolving an International Climate Change Policy Dilemma' (2021) 22 *Climate Policy* 1, 3–4.

80 Reynolds (n 49) 46.

81 ETC Group, 'What Is Wrong with Solar Radiation Management?' (2017) <www.boell.de/sites/default/files/etc-group_briefing_why_srm_experiments_are_a_bad_idea_24.03.2017.pdf> accessed 17 July 2023.

82 Robock et al (n 21) 531.

the implications of potential deployment, and full-scale testing would be needed. Similar concerns may be attached to regional application (for example, Arctic or Antarctic SRM), that would need to be tested at scales large enough to produce temperature changes of significance. This means that experimental SRM implementation, beyond a certain scale, would carry the same risks as operational deployment.[83] While the line between SRM research and deployment is blurred, small-scale experiments are necessary in order to gain further scientific understanding.[84] This view was supported by the parties to the United Nations Convention on Biological Diversity, who agreed to a moratorium on climate-intervention techniques that could affect biodiversity, 'with the exception of small scale scientific research studies that would be conducted in a controlled setting'.[85] These studies should be aimed at assessing the risks of SRM and identifying risk-management options. This will require specifying thresholds beyond which experiments should be banned.[86]

The risks of SRM also need to be evaluated relative to the risks of climate change. The near-certainty of climate change impacts results in a 'risk-risk scenario', which requires balancing the risks associated with deploying SRM against the risks of non-deployment.[87] The relative weighting of these trade-offs largely depends on 'individual and societal differences in the tolerance and framing of risks'.[88] Arguments against SRM assume that it will be more dangerous than the consequences of climate change, or at least that combatting one problem with another potential problem is undesirable

83 Ralph Bodle, 'Geoengineering and International Law: The Search for Common Legal Ground' (2010) 46 *Tulsa Law Review* 305, 321.

84 Jane Long, Frank Loy and Granger Morgan, 'Policy: Start Research on Climate Engineering' (2015) 518 *Nature* 29, 30.

85 Decision adopted by the Conference of the Parties to the Convention on Biological Diversity at its Tenth Meeting x/33, Biodiversity and Climate Change, 10th mtg, Agenda Item 5.6, UNEP/CBD/COP/DEC/X/33 (29 October 2010).

86 Edward Parson and David Keith, 'End the Deadlock on Governance of Geoengineering Research' (2013) 339 *Science* 1278, 1279; Olivier Boucher et al, 'Rethinking Climate Engineering Categorization in the Context of Climate Change Mitigation and Adaptation' (2014) 5 *Wiley Interdisciplinary Reviews: Climate Change* 23, 32.

87 See generally Andy Parker, 'Governing Solar Geoengineering Research as It Leaves the Laboratory' (2014) 372 *Philosophical Transactions of the Royal Society A: Mathematical, Physical and Engineering Sciences* 1–17; Benjamin Sovacool, Chad Baum and Sean Low, 'Risk–Risk Governance in a Low-carbon Future: Exploring Institutional, Technological, and Behavioral Tradeoffs in Climate Geoengineering Pathways' [2022] *Risk Analysis* 1–22.

88 Nicholas Harrison, Janos Pasztor and Kai-Uwe Barani Schmidt, 'A Risk-Risk Assessment Framework for Solar Radiation Modification' (International Risk Governance Center 2021) 1, 3.

(that is, two wrongs do not make a right).[89] The HRC report, for instance, acknowledged the serious threats that climate change poses, but failed to put SRM risks in perspective against the rights infringement likely from climate change itself. It recognized that SRM could 'perpetuate and exacerbate' the human rights impacts of climate change,[90] but did not mention the potential benefits for the climate system and for the protection of human rights. The report's one-sided treatment of the risks of SRM is, in our view, regrettable, in that it does not promote high-quality debate.

Conversely, supporters of SRM research argue that SRM may be the lesser evil and could 'arm the future'[91] with appropriate policy options, if climate change impacts become too severe.[92] Human-rights-based values are also increasingly used in support of SRM R&D: from recognizing the freedom of scientific research and the fundamental right to enjoy the benefits of scientific knowledge,[93] to upholding the right to non-discrimination and procedural rights.[94] Reynolds and Fleurke also invoke the precautionary principle. They argue that, far from militating against SRM, a precautionary approach justifies conducting SRM research, including field experiments.[95] This is because research into SRM can reduce uncertainty, inform decisions about the risks and benefits of SRM, and determine which SRM options could be deployed safely.[96]

Information about risks and benefits is necessary to ensure that untested technologies are not deployed prematurely should the impacts of climate

---

89 Paula Curvelo and Ângela Guimarães Pereira, 'Geoengineering: Reflections on Current Debates', in Ana Delgado (ed) *Technoscience and Citizenship: Ethics and Governance in the Digital Society* (Springer 2016) 163, 171.

90 Human Rights Council (n 45) 19. See also Teea Kortetmäki and Markku Oksanen, 'Right to Food and Geoengineering' (2023) 36 *Journal of Agricultural and Environmental Ethics* 5.

91 See generally Stephen Gardiner, 'Is "Arming the Future" with Geoengineering Really the Lesser Evil? Some Doubts about the Ethics of Intentionally Manipulating the Climate System', Climate Ethics: Essential Readings (Oxford 2010) 1–19.

92 Dale Jamieson, 'Ethics and Intentional Climate Change' (1996) 33 *Climatic Change* 323, 333; Christopher Preston, 'Re-Thinking the Unthinkable: Environmental Ethics and the Presumptive Argument against Geoengineering' (2011) 20 *Environmental Values* 457, 466–70; Toby Svoboda, 'Geoengineering, Agent-Regret, and the Lesser of Two Evils Argument' (2015) 37 *Environmental Ethics* 207, 211.

93 Reynolds (n 49) 107–9.

94 See, for example, Toby Svoboda, Holly Jean Buck and Pablo Suarez, 'Climate Engineering and Human Rights' (2019) 28 *Environmental Politics* 397; Brian Citro and Patrick Taylor Smith, 'A Human Rights Framework for Climate Engineering: A Response to the Limits of Cost-Benefit Analysis', in Will Burns, David Dana and Simon James Nicholson (eds) *Climate Geoengineering: Science, Law and Governance* (Springer 2021) 111.

95 Jesse Reynolds and Floor Fleurke, 'Climate Engineering Research: A Precautionary Response to Climate Change' [2013] CCLR 101, 101.

96 Parker (n 87) 14; Reynolds and Fleurke (n 95) 101.

change escalate rapidly. Parker argues that a moratorium on field experiments would not only be impracticable and ineffective but could delay the learning necessary for responsible deployment. It could also hinder international cooperation, discourage research transparency, stigmatize safe research, or lead to premature rejection of SRM as a policy option.[97]

There are concerns that SRM R&D could cause technological lock-in and ultimately lead to inappropriate deployment.[98] However, we suggest that these arguments are weak. McKinnon contends that, far from justifying a moratorium, the challenges associated with SRM create an opportunity to develop new institutions to govern SRM in the interest of future generations.[99] She suggests that SRM research governance needs to include phase-out and exit mechanisms specifically to avoid undesirable technological lock-in.[100] Along these lines, Herzog and Parson consider that the conditions under which a moratorium on large-scale testing should be lifted would need to be adaptive to reflect new knowledge and socio-technological developments.[101]

10.3.3 Claim 3: SRM is ungovernable

Some commentators suggest that SRM development should be avoided because it is ungovernable. By 'governability', we mean the extent to which legal or non-legal mechanisms can mitigate not only physical risks, but also socio-economic, geopolitical and ethical concerns associated with SRM technologies.[102] A prominent example of these concerns is the call to develop a 'non-use agreement' for SRM. Biermann et al propose the negotiation of an international agreement to prohibit the development of global-scale SRM – including outdoor experimentation – and deployment, because it cannot be governed in a fair, just and democratic manner.[103] The proposal

[97] Parker (n 87) 10–11.

[98] Rob Bellamy and Peter Healey, '"Slippery Slope" or "Uphill Struggle"? Broadening out Expert Scenarios of Climate Engineering Research and Development' (2018) 83 *Environmental Science & Policy* 1, 9; Aaron Tang, 'The Slippery Slopes of Climate Engineering Research' (2023) 80 *Global Environmental Change* 1, 1.

[99] See generally Catriona McKinnon, 'Sleepwalking into Lock-in? Avoiding Wrongs to Future People in the Governance of Solar Radiation Management Research' (2019) 28 *Environmental Politics* 441–59.

[100] Ibid 452.

[101] Long, Loy and Morgan (n 84) 31; Herzog and Parson (n 37) 64–5.

[102] Governability is generally defined as 'The relative efficacy of government in meeting economic, social, and political challenges'. See Craig Calhoun, *Dictionary of the Social Sciences* (Oxford University Press 2002).

[103] Biermann et al (n 17) 3.

does not prohibit SRM research per se,[104] but the substantive principles could significantly stifle research activities because they include prohibiting outdoor experimentation, preventing public funds being used for SRM research, and banning private patents for SRM technology.[105]

According to Biermann et al, for SRM to be governed fairly, justly and democratically, vulnerable (that is, developing) states must have ultimate decision-making power over whether, and how, global-scale SRM is to be used. This would set a very high bar for any SRM research or development to proceed. Wealthier states with the capacity to develop SRM would be unlikely to grant this level of power to vulnerable states, which, they argue, makes fair, just and democratic governance impossible. They also doubt whether existing multilateral institutions, including the United Nations Environment Programme (UNEP), the UN General Assembly and the UN Framework Convention on Climate Change, could support achieving 'fair and effective political control' of SRM.[106] Even if appropriate governance mechanisms could be developed, the authors point out the lack of enforcement powers under international law. Therefore, they suggest that states and the UN should 'restrict the development of' SRM by implementing an international moratorium.

The call for a non-use agreement principally targets planetary-scale SRM and does not at this stage seek to prohibit regional SRM technologies, such as MCB. Biermann et al state that '[i]nterventions that are more regional or local in intent, such as marine cloud brightening to protect fragile ecosystems such as the Great Barrier Reef, are also conceivable, but they differ significantly in terms of governance, politics, and scale'.[107] However, a moratorium on SRM could create a false impression among civil society and other actors that *all* SRM R&D activities are similarly prohibited, regardless of scale, purpose, and so on, including regional-scale SRM and adaptation (such as MCB over the Great Barrier Reef).

The proposal for a non-use agreement has garnered some significant support, with close to 400 academics having signed an open letter.[108] At the same time, however, the proposal has been criticized for its stifling

---

[104] Duncan McLaren, 'Should There Be a "Non-Use" Agreement on Solar Geoengineering?' (*Legal Planet*, 24 February 2023) <https://legal-planet.org/2023/02/23/should-there-be-a-non-use-agreement-on-solar-geoengineering/> accessed 21 June 2023.

[105] Holly Buck, 'We Can't Afford to Stop Solar Geoengineering Research' (*MIT Technology Review*, 2022) <www.technologyreview.com/2022/01/26/1044226/we-cant-afford-to-stop-solar-geoengineering-research/> accessed 21 June 2023.

[106] Biermann et al (n 17) 3–4.

[107] Ibid, 2.

[108] 'Open Letter' (*Solar Geoengineering Non-Use Agreement*) <www.solargeoeng.org/non-use-agreement/open-letter/> accessed 9 May 2023.

effect on SRM research.[109] According to Buck, the non-use agreement proposal is a 'thinly veiled' effort to suppress research, by limiting research investment and creating intense social pressure against it.[110] Buck is not alone in advocating for SRM research to progress.[111] Over 130 experts have signed a call for balanced research and assessment of SRM, recommending a 'golden rule' for assessing SRM, where '[t]he risks of researching, developing and possibly implementing SRM must be balanced against the climate risks SRM would attenuate'.[112] More than 100 scientists have signed a separate open letter in support of SRM research.[113] The letter acknowledges the significant risks and uncertainties concerning SRM, but suggests that further research is needed to understand the role that SRM might play in climate policy, especially as pressure to consider SRM is likely to grow in coming years as climate change worsens.[114] According to Buck:

> Good science takes years to develop. If we put off research until the 2030s, we could find ourselves in a world that's made some uneven progress on the reduction of greenhouse-gas emissions but not enough, with temperatures still headed toward 3°C of warming. We can't then suddenly hope to produce rigorous science that would help us understand whether solar geoengineering is advisable.[115]

These perspectives focus on the scientific merit of pursuing research and do not expressly address questions of governability. Reynolds has refuted claims that SRM is 'ungovernable' as well as suggestions in reports of the Intergovernmental Panel on Climate Change (IPCC)[116] that it presents

[109] Buck (n 105); Janos Pasztor, 'Calls for an SRM "Non-Use Agreement" Underline the Need for Governance' (*C2G*, 28 January 2022) <www.c2g2.net/calls-for-an-srm-non-use-agreement/> accessed 21 June 2023.

[110] Buck (n 105).

[111] See generally Katharine Ricke, 'Solar Geoengineering Is Scary – That's Why We Should Research It' (2023) 614 *Nature* 391.

[112] 'Letter – call-for-balance.com' <https://call-for-balance.com/letter> accessed 21 June 2023.

[113] 'An Open Letter Regarding Research on Reflecting Sunlight to Reduce the Risks of Climate Change' (*climate intervention research letter*) <https://climate-intervention-resea rch-letter.org/> accessed 21 June 2023.

[114] Ibid.

[115] Buck (n 105).

[116] See, for example, Rajendra K Pachauri et al, *Climate Change 2014: Synthesis Report. Contribution of Working Groups I, II and III to the Fifth Assessment Report of the Intergovernmental Panel on Climate Change* (IPCC 2014) 24–5; Masson-Delmotte et al (n 5) 14–15.

'insurmountable' governance challenges.[117] The IPCC reports suggest that challenges include the risk that SRM might detract from mitigation efforts (moral hazard), that merely researching SRM could lead to 'inevitable and unwanted use' of the technologies (slippery slope), and that 'decisions to use it could be contrary to democratic norms'.[118] Reynolds suggests that these claims are largely speculative, and sometimes made contrary to existing peer-reviewed literature. He contends that the concern that SRM may be undemocratic is based on a selective reading of the literature, focusing on sources that support the claim rather than those that contest it. Horton et al have argued further that such 'incompatibility' claims rely on assumptions grounded in an ideal type of deliberative democracy.[119]

The governance of SRM research has also been considered by UNEP. In its 2023 expert review of SRM research and deployment, *One Atmosphere*,[120] UNEP recognizes the risk and governance challenges of SRM, including the fact that there is unlikely to be 'universal consensus in the broader community' on deployment.[121] It nevertheless suggests that the governance of small-scale outdoor experimentation could be pursued under the auspices of the UN, which could develop research governance alongside the evolution of scientific knowledge.[122] The report further suggests that 'appropriate governance mechanisms' could help address issues, such as environmental risks, asymmetry in technological capacity and power between developed and developing nations, and concerns about a slippery slope.[123]

We note that the HRC's position is unclear on this issue. While its report asserted that SRM is 'ungovernable', and the negotiation of a multilateral treaty unlikely, it also recommended that, in developing a 'proper governance framework', states should uphold human rights norms and standards as well as their duty to protect against human rights violations in SRM R&D.[124] The report recognized the importance of access to information, public participation, access to justice and effective remedies, particularly the

---

[117] Jesse L Reynolds, 'Is Solar Geoengineering Ungovernable? A Critical Assessment of Governance Challenges Identified by the Intergovernmental Panel on Climate Change' (2021) 12 *Wiley Interdisciplinary Reviews: Climate Change* 1–8, 2.

[118] Ibid 1.

[119] Joshua Horton et al, 'Solar Geoengineering and Democracy' (2018) 18 *Global Environmental Politics* 5, 9.

[120] United Nations Environment Programme (UNEP), 'One Atmosphere: An Independent Expert Review on Solar Radiation Modification Research and Deployment' <https://wedocs.unep.org/20.500.11822/41903> accessed 10 July 2023.

[121] Ibid 19.

[122] Ibid 22.

[123] Ibid 5.

[124] Human Rights Council (n 45) 12, 18.

inclusion of voices from the Global South, Indigenous peoples and other climate vulnerable communities.

Whether SRM is 'governable', therefore, appears to be a question of perspective. It depends not only on one's understanding of individual SRM governance issues but also on one's expectations of what governance can achieve. Legal and governance mechanisms rarely achieve perfect results, whether for SRM or for any other risky activity. They are usually the result of a negotiated compromise of the interests of various stakeholders, especially at the international level. Thus, in designing governance arrangements for SRM research, one must not let the perfect be the enemy of the good. According to Hathaway, 'the test of a legal rule – whether domestic or international – is not determined simply by whether it is violated. It is determined, too, by the *response* when it is violated'.[125] To that extent, SRM governance should prevent violation of agreed rules as far as practicable, but also ensure that any violation will receive an appropriate response.

10.3.4 Claim 4: SRM poses insurmountable risks of unilateral deployment

The risk of unilateral or minilateral deployment is advanced by those opposing *and* those proposing development of the technology.[126] Unlike mitigation options that require multilateral cooperation and buy-in, planetary-scale SRM could potentially be conducted by a single state or a small coalition of states.[127] SRM is relatively inexpensive – or at least within the affordability range of some nations – so it can be undertaken without international consensus.[128] These concerns raise serious legitimacy, security and stability issues,[129] including risks of counter-geoengineering and military

[125] Oona Hathaway, 'How Russia's Invasion of Ukraine Tested the International Legal Order' (Brookings, 3 April 2023) <www.brookings.edu/articles/how-russias-invasion-of-ukraine-tested-the-international-legal-order/> accessed 13 July 2023.

[126] John Virgoe, 'International Governance of a Possible Geoengineering Intervention to Combat Climate Change' (2009) 95 *Climatic Change* 103, 115–16; David Victor, 'On the Regulation of Geoengineering' (2008) 24 *Oxford Review of Economic Policy* 322, 327; Keith (n 58) 111.

[127] See generally Katharine Ricke, Juan Moreno-Cruz and Ken Caldeira, 'Strategic Incentives for Climate Geoengineering Coalitions to Exclude Broad Participation' (2013) 8 *Environmental Research Letters* 1–9; Zachary Dove, Joshua Horton and Katharine Ricke, 'The Middle Powers Roar: Exploring a Minilateral Solar Geoengineering Deployment Scenario' (2021) 132 *Futures* 1–8.

[128] Scott Barrett, 'The Incredible Economics of Geoengineering' (2008) 39 *Environmental and Resource Economics* 45, 46.

[129] Joshua Horton and Jesse Reynolds, 'The International Politics of Climate Engineering: A Review and Prospectus for International Relations' (2016) 18 *International Studies Review* 438, 443; Adam Lockyer and Jonathan Symons, 'The National Security Implications of

action.[130] These concerns have sparked model and scenario exercises analysing preferences, trade-offs and strategies.[131] One such scenario involves states deploying SRM as a matter of survival (for example, small island states affected by sea-level rise),[132] or because leaders want to appear as being responsive to climate emergencies to gain public support.[133] Another, more plausible, scenario involves SRM deployment by a great state power with financial, technical and geopolitical capacity and with greater incentives to deploy SRM than to mitigate (for example, a high greenhouse gas emitter).[134] In addition to the risks of unilateral state deployment is the possibility of unilateral deployment by non-state actors, such as a wealthy individual or corporation,[135] radical ecologic groups[136] or private citizens.[137]

Unilateral SRM deployment may have negative externalities on other countries, which raises procedural and distributive-justice issues.[138] This will require integrated assessment of ethical, legal and social implications and appropriate governance arrangements.[139] Biermann et al argue that international governance to address the risks of unilateral deployment is unlikely to attract support from developing countries; they consider a non-use agreement to be preferable to illegitimate or ineffective arrangements.[140]

Solar Geoengineering: An Australian Perspective' (2019) 73 *Australian Journal of International Affairs* 485, 495.

130 Andy Parker, Joshua Horton and David Keith, 'Stopping Solar Geoengineering through Technical Means: A Preliminary Assessment of Counter-geoengineering' (2018) 6 *Earth's Future* 1058, 1059; Muhammet Bas and Aseem Mahajan, 'Contesting the Climate: Security Implications of Geoengineering' (2020) 162 *Climatic Change* 1985, 1989.

131 Edward Parson and Jesse Reynolds, 'Solar Geoengineering: Scenarios of Future Governance Challenges' (2021) 133 *Futures* 1–12.

132 Adam Millard-Ball, 'The Tuvalu Syndrome: Can Geoengineering Solve Climate's Collective Action Problem?' (2012) 110 *Climatic Change* 1047, 1059–60; Felix Schenuit, Jonathan Gilligan and Anjali Viswamohanan, 'A Scenario of Solar Geoengineering Governance: Vulnerable States Demand, and Act' (2021) 132 *Futures* 1, 5.

133 David Victor, 'Governing the Deployment of Geoengineering: Institutions, Preparedness, and the Problem of Rogue Actors', in Robert N Stavins and Robert C Stowe (eds) *Governance of the Deployment of Solar Geoengineering* (Harvard Project on Climate Agreements 2019) 41, 43.

134 See generally Florian Rabitz, 'Going Rogue? Scenarios for Unilateral Geoengineering' (2016) 84 *Futures* 98–107.

135 Victor (n 126) 324; Daniel Bodansky, 'The Who, What, and Wherefore of Geoengineering Governance' (2013) 121(3) *Climatic Change* 539, 545.

136 Rabitz (n 134) 101.

137 See generally Anne Pasek et al, 'Reflections on a Hypothetical Decentralized Grassroots Deployment Solar Geoengineering Scenario' (2021) 132 *Futures* 1–8.

138 Johannes Urpelainen, 'Geoengineering and Global Warming: A Strategic Perspective' (2012) 12 *International Environmental Agreements: Politics, Law and Economics* 375, 376.

139 Joseph Aldy et al, 'Social Science Research to Inform Solar Geoengineering' (2021) 374 *Science* 815, 816.

140 Biermann et al (n 17) 3.

However, a global ban would not prevent sovereign states from acting unilaterally[141] and would only address part of the problem, given that non-state actors would not be bound by such a treaty.[142] It would also require monitoring to detect unlawful activities and attribute potential adverse impacts,[143] which would require understanding of SRM technologies. Honegger et al suggest that this is a further argument for research.[144] Concerns about unilateralism also arise from banning SRM because a moratorium would constrain responsible research and leave technological development to those actors that are the least likely to abide by international norms.[145]

The risk of unilateral action is real but demands consideration of how such a risk can be managed. Horton argues that 'the fear of unilateralism has become an *idée fixe* in conversations about geoengineering, in effect putting the brakes on more ambitious research efforts and deliberations about governance issues'.[146] He suggests that the governance arrangements, such as costs of retaliation (for example, trade sanctions, use of force and, eventually, counter-geoengineering) could create incentives to cooperate.[147] Developing strategies to foster international cooperation could reduce risks of unilateral action.[148] Weitzman goes further to suggest that the 'free-driver' theory could actually form 'the backbone of a governance architecture'.[149]

There has been limited discussion of what type of arrangement is needed to deter unilateral SRM deployment. Liability rules alone cannot adequately address risks of unilateral deployment, so '*ex ante* regulatory solutions like the nuclear non-proliferation regime need to be implemented in tandem with *ex post* liability schemes for there to be effective governance of SRM'.[150] Arrangements that facilitate international cooperation in SRM research,

141 Jesse L Reynolds and Gernot Wagner, 'Highly Decentralized Solar Geoengineering' (2020) 29 *Environmental Politics* 917, 927–8.

142 Antonia Eliason, 'Avoiding Moonraker: Averting Unilateral Geoengineering Efforts' (2021) 43 University of Pennsylvania Journal of International Law 429, 456.

143 Rabitz (n 134) 102.

144 Matthias Honegger, Axel Michaelowa and Sonja Butzengeiger-Geyer, 'Climate Engineering: Avoiding Pandora's Box through Research and Governance' 1, 6.

145 Victor (n 126) 325.

146 Joshua Horton, 'Geoengineering and the Myth of Unilateralism: Pressures and Prospects for International Cooperation' (2011) 4(1) *Stanford Journal of Law, Science & Policy* 56, 56.

147 Ibid 59. See also Daniel Heyen and Jere Lehtomaa, 'Solar Geoengineering Governance: A Dynamic Framework of Farsighted Coalition Formation' (2021) 1 *Oxford Open Climate Change* 1, 1–2.

148 Horton (n 146) 59–60; Horton and Reynolds (n 129) 444.

149 Martin Weitzman, 'A Voting Architecture for the Governance of Free-driver Externalities, with Application to Geoengineering' (2015) 117 *The Scandinavian Journal of Economics* 1049, 1066–7.

150 Eliason (n 142) 466.

combined with safeguards tailored to each technology, could address fears of unilateral deployment.[151]

10.4 The international community should avoid choosing to live in ignorance of SRM

Having rebutted the arguments in favour of limiting SRM R&D, this section advances a novel argument in support of SRM R&D, namely that natural law traditions support the advancement of knowledge.

Natural law provides a standpoint against which to assess how a system of positive law is responding to a human problem. In the classical version of natural law, St Thomas Aquinas defines law as 'an ordinance of reason for the common good, made by him who has care of the community, and promulgated'.[152] Finnis, a leading figure of the modern natural law tradition, describes law in slightly more complex terms as an 'authoritative common ordering of a community, aimed at facilitating the realisation of the common good'.[153] Both these understandings focus attention on the question of what is the aim or purpose of law. Both Aquinas' and Finnis' positions focus on promotion of the common good of the community as being the purpose of law. Finnis identifies seven self-evident 'basic goods' that are universally accepted by all human societies, including knowledge, which comprise the common good of a human community.[154] These basic goods are intrinsically valuable, in that they are valuable to humans in and of themselves. They are hallmarks of human flourishing, regardless of their utilitarian or consequentialist value.[155]

The 'New Haven' or 'policy-oriented' school of international law, founded and developed by McDougal and Lasswell, also focuses on the purpose of law. The New Haven School of international law characterizes law as not simply a system of rules, but as a continual process of authoritative decision-making by which people identify, clarify and secure their common interests.[156] To assist in identifying and pursuing common interests, the

---

151 Vikram Mathur and Aparna Roy, 'Perspectives from India on Geoengineering' (2019) 116 *Current Science* 40, 44–5.

152 St Thomas Aquinas, *Summa Theologiae Part II*, 'Treatise on Law Q90', Art 4, Para F (1273).

153 John Finnis, *Natural Law and Natural Rights* (Oxford University Press 2011) 276–7. See also generally Suri Ratnapala, *Jurisprudence* (Cambridge University Press 2017).

154 Finnis' seven basic goods are: life, knowledge, play, aesthetic experience, friendship, religion and practical reasonableness. See Suri Ratnapala (n 153) 152.

155 Finnis (n 153) 119–20.

156 Myres S McDougal and Harold D Lasswell, 'The Identification and Appraisal of Diverse Systems of Public Order', in MS McDougal and WM Reisman (eds) *International Law Essays: A Supplement to International Law in Contemporary Perspective* (The Foundation Press 1981) 22–4.

New Haven School identifies eight values that all people cherish.[157] These values include enlightenment – the gathering, processing and disseminating of information and knowledge – which aligns with Finnis' basic good of 'knowledge'. Both Finnis' modern natural law tradition and the New Haven School's policy-oriented approach to international law therefore identify the common human good of new knowledge as valuable in and of itself, regardless of any consequential or use value.

Both traditions also recognize that, in practice, decision-makers carry out a pragmatic balancing (that is, using practical reasonableness) of these various human values and goods in constructing an overall conception of the common good or common interest of the relevant society. The key point is that a decision-maker may choose to partially, or even wholly, sacrifice the basic good of knowledge and thereby 'live in ignorance', if such choice is required in an exercise of practical reasonableness to preserve or further a socially desirable mix of the other basic goods of human flourishing. It follows that a choice to sacrifice the basic good of knowledge or enlightenment should only be entertained where it is abundantly clear that it is required to achieve or protect those other basic goods.

From this natural law perspective, the conditions required to justify stifling further human knowledge on SRM appear to be very questionable. We know the underlying biophysical process of human-induced climate change is threatening to undermine several of the basic human goods of large parts of the population of the planet, and this threat will likely increase over the coming decades. The existing scientific knowledge on the course of climate change through the period to 2100 indicates, according to the IPCC, increasing current and future climate change impacts that will likely have significant negative affects upon human health, food security and human physical infrastructure. These impacts will make most of Finnis' basic goods and/or the New Haven School's basic human values much more difficult to pursue. However, further knowledge of SRM R&D has the possibility of opening up policy options for decision-makers (such as limiting short-term global temperature to avoid global temperature overshoot) that could substantially ameliorate the significant risks that climate change will pose to large parts of the human population. In short, there is an absence of strong evidence that *restricting* knowledge of SRM is required to achieve a desirable mix of the other basic goods of human flourishing. On the contrary,

[157] Lung-Chu Chen, *An Introduction to Contemporary International Law: A Policy-Oriented Perspective* (Oxford University Press 2014) 16. See also, for the eight fundamental values of the New-Haven School: human respect, power (that is, in making and influencing community decisions), enlightenment, well-being (safety, health, comfort), wealth, skill, affection (friendship, loyalty) and rectitude.

arguments to restrain SRM R&D, and thereby prevent it from reaching a deployable state, appear to be a plea for humanity to live in ignorance. This approach risks sacrificing the basic human good of human knowledge and at the same time undermining most or all of the other basic goods of human flourishing.

10.5 Conclusion

In this chapter, we summarized the arguments for and against SRM research, to make the case that further knowledge on SRM effects and risks is needed before making a decision on whether or not it should be implemented. We examined the main claims against SRM R&D: the moral hazard concern, the blurred line between research and deployment, the ungovernability of SRM, and the risks of unilateral deployment. We showed that, while these concerns raise complex issues, they point to the need to tailor appropriate arrangements (including legal arrangements) for SRM proposals. We argue that these claims, however valid some of them may be, should not justify a ban of SRM R&D, but encourage and stimulate discussions on the type of governance mechanisms needed to account for SRM risks and uncertainties. Indeed, rigorous SRM research appears to be the best way to determine which options will both be feasible and desirable, and the safeguards needed to ensure that the development of these technologies is done responsibly.

We further argue that a moratorium on SRM would not only freeze R&D of SRM technologies, but also hinder exploration and development of more nuanced governance options. It would assume that SRM governance options could never be good enough, without judging them on their merit. It could discourage scholars from pursuing important research to move the academic conversation forward. From the perspective of natural law and the New Haven approach to international law, to fail to research the potential of SRM, while also likely imperilling other basic human goods through ongoing climate change impacts, would be to live in ignorance – one which effectively sacrifices not only the basic human good of knowledge but also threatens the other basic goods of human flourishing. As Rayner puts it, 'not to do research is to remain ignorant of our ignorance'.[158] Responsible SRM research should, therefore, be allowed to proceed, if not to prevent dangerous climate overshoot, then at least to improve climate predictions and build capacity to monitor and respond to risks of rogue SRM deployment.

158 Steve Rayner, 'To Know or Not to Know? A Note on Ignorance as a Rhetorical Resource in Geoengineering Debates' (2014) Climate Geoengineering Governance Working Paper Series: 010, 1, 9.

11

Whither Law and Regulation on Carbon Capture, Utilization and Storage Technology? A Comparative Analysis

Hao Zhang

11.1 Introduction

The Paris Agreement and carbon neutrality targets adopted by many countries and jurisdictions have highlighted the importance of carbon removal technologies.[1] Carbon capture and storage (CCS) technologies are crucial under carbon-neutral scenarios, providing an alternative for continued use of fossil fuels, especially in carbon-intensive industries like petrochemical, steel and cement production, where complete phase-out of fossil fuels is impractical or expensive.[2] CCS includes technologies that capture carbon dioxide (CO_2) from energy generation, industrial processes, and even the atmosphere, which can be stored permanently in geological formations.[3] The CCS chain, discussed in this chapter, also includes situations where the captured CO_2 is used for enhanced oil recovery (EOR) that leads to permanent storage of the CO_2 used. Essentially, this chapter focuses on carbon removal technologies, excluding situations where CO_2 is used but not permanently stored, and the utilization aspect of CCS in this chapter is

1 International Energy Agency (IEA), 'Energy Technology Perspective 2020. Special Report on Carbon Capture Utilisation and Storage' (September 2020), available at: <www.iea.org/reports/ccus-in-clean-energy-transitions/regional-opportunities>.

2 Mai Bui et al, 'Carbon Capture and Storage (CCS): The Way Forward' (2018) 11 *Energy & Environmental Science* 1062.

3 IEA (n 1).

limited to EOR. With fossil fuels expected to remain the primary source of energy globally for decades, many scholars consider CCS as a viable solution to capture and isolate CO_2 from fossil fuel combustion processes for permanent storage, especially for countries that will continue to rely on fossil fuels as part of their primary energy mix.[4]

Although the technologies underpinning the CCS chain are practically available, various scholars have raised concerns over its large-scale use due to high capital costs and lack of regulatory clarity that encourages investment and responsible operation.[5] Regulatory clarity is essential in facilitating successful CCS project development, and the regulatory framework should establish clear goals and measures to manage risks and stimulate commercialization.[6] Developing a CCS supply chain requires an established regulatory and legal regime at both international and domestic levels to manage risks and support technology investment.[7] The regional and domestic legal and regulatory regimes governing the CCS projects have been identified as a crucial factor that could facilitate or inhibit the creation of a CCS supply chain at a larger scale.[8] Several countries and jurisdictions, representing both the source and sink of CO_2, have supported policy backing for CCS.[9]

Despite general policy support for CCS as a crucial method to remove CO_2 from the atmosphere and industrial processes, the justification and rationale for its widespread adoption remain hotly debated, particularly from ideological and human-rights perspectives. The backing for CCS is fundamentally rooted in scientific findings on the physical threats posed by climate change to humanity and planetary health. However, even though

4 Bui et al (n 2); Kai Jiang et al, 'China's Carbon Capture, Utilization and Storage (CCUS) Policy: A Critical Review' (2020) 119 *Renewable and Sustainable Energy Reviews* 109601; Matthew E Boot-Handford et al, 'Carbon Capture and Storage Update' (2014) 7 *Energy & Environmental Science* 130; Hengwei Liu et al, 'The Role of CO_2 Capture and Storage in Saudi Arabia's Energy Future' (2012) 11 *International Journal of Greenhouse Gas Control* 163.

5 Tim Dixon et al, 'Legal and Regulatory Developments on CCS' (2015) 40 *International Journal of Greenhouse Gas Control* 431; Vivian Scott et al, 'Last Chance for Carbon Capture and Storage' (2013) 3 *Nature Climate Change* 105; Samuela Bassi et al, 'Bridging the Gap: Improving the Economic and Policy Framework for Carbon Capture and Storage in the European Union' (The Centre for Climate Change Economics and Policy and others, June 2015), available at: <www.cccep.ac.uk/wp-content/uploads/2015/10/GRI_LSE_CCS_web.pdf>.

6 Dixon et al (n 5).

7 Zoe Kapetaki and John Scowcroft, 'Overview of Carbon Capture and Storage (CCS) Demonstration Project Business Models: Risks and Enablers on the Two Sides of the Atlantic' (2017) 114 *Energy Procedia* 6623; Tim Dixon et al, 'Update on the London Protocol: Developments on Transboundary CCS and on Geoengineering' (2014) 63 *Energy Procedia* 6623.

8 Kapetaki and Scowcroft (n 7); Dixon et al (n 5).

9 Bui et al (n 2); Jiang et al (n 4).

this support is based on climate science, it faces challenges from scholars and commentators who, while not opposed to climate mitigation, argue that the ambiguous role of CCS in international law and domestic policy and regulatory frameworks could perpetuate fossil fuel consumption.[10] This, they suggest, might exacerbate climate change and delay the transition to cleaner energy. While there is consensus on the importance of CCS in CO_2 removal, the uncertainty surrounding its criticality compared to other mitigation approaches, such as renewable energy, has led some to argue that CCS should be a last resort when other methods are either ineffective or economically unviable.

Support for limited use of CCS is also highlighted in literature concerning human rights and climate mitigation technologies,[11] given the potential implications and ramifications of CCS for human rights, such as the right to food, water, land rights and a healthy environment, the right to access to information and prior consent that is free, informed and readily available.[12] Existing research indicates that a modest CO_2 removal through CCS would require a substantial area of land that may be diverted from other meaningful uses, such as food production and livelihood-related activities.[13] Additionally large-scale deployment of CCS is known to have a significant water footprint, potentially straining fresh water supplies in regions already suffering from scarcity.[14] There are also serious concerns about the risk of underground water contamination from the injection and storage processes used in CCS.[15] Even if CCS does prove effective in slowing or reversing global temperature

[10] Leonie Reins, 'Carbon Capture and Storage' in Sacha Garben and Laurence Gormley (eds), *Oxford Encyclopedia of EU Law* (OUP, 2024, online edn), available at: <https://opil.ouplaw.com/display/10.1093/law-oeeul/law-oeeul-e122?rskey=z7gX5J&result=2&prd=OEEUL>; Haomiao Du and Hao Zhang, 'Climate Neutrality in the EU and China: An Analysis of the Stringency of Targets and the Adaptiveness of the Relevant Legal Frameworks' (2022) 31(3) *Review of European, Comparative & International Environmental Law* 495.

[11] Anne Saab, 'Discourses of Fear on Climate Change in International Human Rights Law' (2023) 34(1) *The European Journal of International Law* 113; 'The Restriction of Geoengineering under International Law – Joint Opinion' (Heinrich Böll Foundation, 2022), available at: <www.ohchr.org/sites/default/files/2022-06/Annex-SubmissionCIEL-ETC-HBF-TWN-Geoengineering-Opinion.pdf>.

[12] Philipp Günther and Felix Ekardt, 'Human Rights and Large-Scale Carbon Dioxide Removal: Potential Limits to BECCS and DACCS Deployment' (2022) 11(12) *Land* 2153; William CG Burns, 'Human Rights Dimensions of Bioenergy With Carbon Capture and Storage: A Framework for Climate Justice in the Realm of Climate Geoengineering' in Randall S Abate (ed) *Climate Justice: Case Studies in Global and Regional Governance Challenges* (Environmental Law Institute, 2017) 149–76.

[13] Burns (n 12).

[14] Günther and Ekardt (n 12).

[15] Ibid.

rise, its impacts on human rights remain poorly understood, particularly through a contextual approach that is rooted in a country's specific and often unique circumstances. The human-rights impacts of CCS are also likely to be geographically uneven, with reinforcement of injustice and negative consequences within and between vulnerable states and communities.[16] Within the current international climate law framework, where national interests often take precedence over global cooperation and meaningful assistance provided to countries in the Global South, the massive adoption of CCS and climate technologies generally will ultimately shift the impacts and responsibilities of climate mitigation actions to vulnerable countries and communities, as well as future generations, with both spatial and temporal consequences.[17]

Aligning regulatory objectives for CCS with ideological and human-rights concerns requires further in-depth, subject-specific and interdisciplinary analysis. This includes regulatory and legal research to understand the current developments and limitations of CCS-related laws and regulations at both international and domestic levels. This chapter examines the international and jurisdictional regulatory frameworks for CCS, highlighting the fragmented nature of the international legal framework governing CCS and the lack of a holistic approach that considers other important aspects, such as human rights. The chapter particularly focuses on jurisdictional and national developments, given that an increasing number of countries and policy makers are keen to advance research and development of CCS in the absence of a comprehensive international legal framework.

To evaluate the legal and regulatory developments underpinning the global matching of CO_2 sources and sinks, this chapter has selected the EU and China, as they represent different stages of regulatory development concerning CCS. This chapter analyses the regulatory development and practices in the EU and China, identifying legal and regulatory obstacles and proposing changes to facilitate the development of a CCS supply chain at both domestic and regional levels. The chapter examines how the regulatory framework stimulates or inhibits CCS development in the selected jurisdictions. Section 11.2 elaborates on the development of a regulatory framework in the EU and China, focusing on the objective, supporting mechanisms and regulatory oversight of CCS. Section 11.3 critically analyses the existing regulatory framework in the selected jurisdictions. Section 11.4 draws broader insights regarding the intersection of climate technology and law. Finally, Section 11.5 concludes the chapter by pinpointing the outstanding legal and regulatory obstacles in the selected jurisdictions.

16 Heinrich Böll Foundation (n 11).

17 Ibid.

It highlights key areas needing focus, and stresses the crucial necessity of establishing a robust legal and regulatory structure for CCS. This framework is vital for enabling upcoming full-scale projects. To successfully bring CCS into widespread use, comprehensive laws and regulations that encompass all fundamental aspects of CCS technology are essential.

11.2 International and jurisdictional regulatory frameworks for CCS

This section delves into legal advancements concerning CCS projects in both the EU and China, aiming to establish domestic regulatory frameworks governing these initiatives and ensuring adherence to international law, particularly in cases involving offshore and cross-border CO_2 transport and storage. Despite technological strides, the proliferation of CCS projects encounters formidable hurdles, primarily due to the substantial capital investment needed and the absence of an encouraging regulatory framework that spurs responsible operation and investment in the international and national context.[18]

Observers stress the criticality of clear domestic legal guidelines for successful project development.[19] They highlight the pivotal role of sustainable funding support, an efficient project permitting process, and robust stakeholder engagement for gaining public acceptance.[20] The symbiotic relationship between finance and regulatory framework development for CCS becomes evident. The substantial costs and long-term planning inherent in CCS projects demand consistent, stable incentives, and regulatory assurance.[21] Simultaneously, a well-defined regulatory framework, encompassing clear objectives and risk management measures, serves to stimulate CCS commercialization and secure sustained, long-term financial commitment from stakeholders.[22]

11.2.1 CCS and international law

Apart from crafting a domestic regulatory framework for CCS projects, international law, notably the London Convention, assumes significance due to the potential transboundary nature of the CCS supply chain involving CO_2 transport and storage.[23] The London Convention, an international

[18] Jiang et al (n 4); Boot-Handford et al (n 4).
[19] Kapetaki and Scowcroft (n 7); Dixon et al (n 5).
[20] Ibid.
[21] Scott et al (n 5); Bassi et al (n 5).
[22] Ibid.
[23] Dixon et al (n 7).

treaty overseeing the disposal of land-based waste at sea, applies when CO_2 storage extends beyond a nation's internal waters or involves transboundary transport and storage, such as in seabed areas beyond national jurisdictions.[24] Under the London Convention, ocean dumping is generally permissible, except for materials explicitly banned and listed in a negative registry. However, the subsequent London Protocol, enacted in March 2006 and a comprehensive amendment to the Convention, adopts a more stringent stance.[25] It restricts sea disposal solely to substances approved for dumping according to the instrument. Initially, CO_2 was not included in the approved materials, potentially hindering offshore CCS projects.

The October 2009 amendments to the London Protocol, more progressive than its parent treaty, removed the CO_2 export restriction.[26] These changes mandated involved nations to create agreements or arrangements mirroring the Protocol's waste-export provisions, establishing a regulatory framework for sea-based CCS projects and outlining guidelines for safe CO_2 storage in sub-seabed formations.[27] In contrast, the Parties to the London Convention agreed in 2005 that no further amendments would be adopted to deal with newly emerging threats to the ocean, including CO_2 storage in the seabed.[28] These amendments require CCS initiatives to adhere to specific guidelines ensuring safe CO_2 storage and minimal environmental impact. Mandates include risk assessments, monitoring, reporting on storage activities, and measures preventing CO_2 leaks. A scientific advisory group was also established to assess project safety and environmental risks while offering guidance to Protocol parties. Encouragement for best practices and information sharing among parties was part of these amendments. Additionally, for CO_2 export to non-parties, agreements must align with Protocol obligations.[29]

Despite these advancements, the amendments did not come into force until a 2019 resolution, allowing ratified countries to transport CO_2 across borders for offshore storage.[30] Pursued by Norway, the Netherlands, and the UK for North Sea CCS potential, this resolution applies to accepting countries,

---

[24] Ibid.

[25] Ibid.

[26] Resolution LP.3(4) on the Amendment to Article 6 of the London Protocol (Adopted on 30 October 2009, by the Fourth Meeting of Contracting Parties to the 1996 Protocol to the Convention on the Prevention of Marine Pollution by Dumping of Wastes and Other Matter, 1972).

[27] Ibid.

[28] Hoon Hong Gi and Joo Lee Young, 'Transitional Measures to Combine Two Global Ocean Dumping Treaties into a Single Treaty' (2015) 55 *Marine Policy* 47.

[29] Resolution LP.3(4) on the Amendment to Article 6 of the London Protocol (n 26).

[30] Dixon et al (n 7).

expanding the legal framework.[31] Although limited in scope, this resolution hints at broader impacts on regional and national regulations. As of January 2023, based on the status of treaties published by the International Marine Organization (IMO), ten countries have accepted the 2009 amendment, marking progress despite unresolved transboundary CCS issues under the London Protocol.[32]

Due to the absence of a universal international regulatory framework for offshore CCS projects extending beyond national waters, the following discussions centre on CCS advancements and regulatory frameworks in the EU and China. This focused analysis aims to grasp the current status of development, laying the groundwork to pinpoint barriers and challenges faced by CCS projects in Section 11.3, and subsequently draw broader insights regarding the intersection of climate technology and law in Section 11.4.

11.2.2 CCS development and regulatory framework in the EU

CCS technology is expected to play a crucial role in achieving the EU's emission-reduction targets at the lowest cost.[33] The European Commission has envisaged CCS in its Energy Roadmap 2050 to mitigate emissions in several emission-reduction scenarios.[34] As of August 2023, the EU boasts 17 ongoing or planned carbon capture projects, envisioned to contribute to a storage capacity of roughly 35 million tons of CO_2 annually by 2030.[35] Broadening this scope to encompass projects in Iceland, Norway and the UK, the collective initiatives surge to 36, forecasting a storage capacity of 110 million tons of CO_2 yearly by 2030.[36] Notably, Norway leads the charge

---

31 IEA (n 1).

32 IMO, 'Status of IMO Treaties' (6 January 2023), available at: <https://wwwcdn.imo.org/localresources/en/About/Conventions/StatusOfConventions/Status%202023.pdf>. These countries are Belgium, Denmark, Estonia, Finland, Iran, the Netherlands, Norway, Korea, Sweden and the United Kingdom.

33 European Commission, 'The European Green Deal' (11 December 2019), available at: <https://eur-lex.europa.eu/legal-content/EN/TXT/HTML/?uri=CELEX:52019DC0640&from=EN>; Du and Zhang (n 10).

34 European Commission, 'Communication from the Commission to the European Parliament, the Council, the European Economic and Social Committee and the Committee of the Regions – A Roadmap for moving to a competitive low carbon economy in 2050' (8 March 2011), available at: <http://eur-lex.europa.eu/resource.html?uri=cellar:5db26ecc-ba4e-4de2-ae08-dba649109d18.0002.03/DOC_1&format=PDF>.

35 International Association of Oil & Gas Producers, 'Map of CO_2 Storage Projects in Europe' (10 October 2023), available at: <https://iogpeurope.org/resource/map-of-eu-ccus-projects/>.

36 Ibid.

with eight projects, closely shadowed by the UK with seven, solidifying their positions as hubs for extensive CCS endeavours – be it operational, in planning stages or under construction.[37]

This robust deployment mirrors a consolidated effort within these regions to bolster CCS capabilities significantly. Norway and the UK, particularly, are emerging as pivotal hubs for advancing CCS technologies on a substantial scale, reflecting a focused commitment toward carbon neutrality. What stands out in these initiatives is the strong collaboration among multiple countries, underscoring the inherently cross-border nature of carbon capture, transport and storage efforts.[38] This signifies a concerted effort toward a more integrated CCS chain in the EU context, fostering collaboration for a comprehensive approach to tackling carbon emissions.[39] As will be discussed later, in contrast, China's current projects primarily revolve around demonstration projects that target specific segments of the CCS chain, such as capture or storage, individually. This approach diverges fundamentally from the EU's strategy, showcasing a different emphasis and approach to CCS deployment – wherein integration and collaboration across various stages of the CCS chain remain central in the European context.

To facilitate the development and uptake of CCS technology, the EU has established a regulatory framework governing CCS development. This framework comprises Directive 2009/31/EC, adopted by the European Parliament and Council, and four Guidance Documents issued by the European Commission to provide more details on its implementation.[40] While the Directive does not generally apply to the utilization of CO_2, except when combined with geological storage of CO_2 emissions (that is, EOR), this chapter focuses on the utilization of captured carbon that also involves permanent storage of CO_2. Therefore, the following section focuses on Directive 2009/31/EC. It is worth noting that the utilization of CO_2 in

[37] Ibid.

[38] Global CCS Institute, '14 CO_2 Storage and Transport Projects Selected Under the EU's Projects of Common Interest and Projects of Mutual Interest' (30 November 2023), available at: <www.globalccsinstitute.com/news-media/latest-news/14-co2-storage-and-transport-projects-selected-under-the-eus-projects-of-common-interest-and-projects-of-mutual-interest/>.

[39] Ibid.

[40] Directive 2009/31/EC of the European Parliament and of the Council of 23 April 2009 on the Geological Storage of Carbon Dioxide and amending Council Directive 85/337/EEC, European Parliament and Council Directives 2000/60/EC, 2001/80/EC, 2004/35/EC, 2006/12/EC, 2008/1/EC and Regulation (EC) No 1013/2006 (referred to Directive 2009/31/EC). See also Alla Shogenova et al, 'Implementation of the EU CCS Directive in Europe: Results and Development in 2013' (2014) 63 *Energy Procedia* 6662.

the EU is regulated by a different set of regulations at the EU and member states levels,[41] which this section does not address.

Directive 2009/31/EC is the principal legislation in the EU governing CCS-related activities and sets out the general principles and rules for geological storage of CO_2 that member states must comply with. The Directive mandates member states to enact domestic laws and regulations to transpose the Directive's provisions. The Directive also lays out detailed procedural and substantive requirements for the entire life cycle of a CCS project. For instance, a storage permit is mandatory for developing and operating a CCS project, and the European Commission reviews the permit application before issuance.[42] The Directive further includes clear provisions on operation, closure and post-closure obligations. To ensure environmental safety, member states and site operators must conduct monitoring, inspection and reporting duties.[43] For CCS project development, the Directive stipulates practical issues related to on-site closure, transfer of responsibility, and the financial obligations of site operators.[44] The Directive clarifies the risk allocation between the site operator and member states and requires operators to provide financial security from the time of application to the commencement of injection.[45] Additionally, operators are obligated to make a financial contribution for monitoring costs by the member states for at least 30 years after the transfer of obligations.[46]

11.2.3 CCS development and regulatory framework in China

China, as the largest greenhouse gas emitter, is facing pressure to transition to a low-carbon economy.[47] The Chinese government recognizes the importance of increasing energy efficiency, reducing emissions and pursuing decarbonization in the energy-intensive sectors. To achieve these goals, China has adopted mid- and long-term climate targets, aiming to reach peak emissions before 2030 and become carbon neutral by 2060.[48] While

---

41 Suren Erkman et al, 'Recycling and Utilisation of Carbon Dioxide in the European Union's Directives' (2017) 26(1) *European Energy and Environmental Law Review* 2–12.

42 Ibid, Arts 6–10.

43 Ibid, Arts 13–15.

44 Ibid, chapter 4.

45 Ibid, Arts 17–20.

46 Ibid, Art 20.

47 Jiang et al (n 4); Yan Gu, 'Carbon Capture & Storage Policy in China' (Center for Climate Change Law White Paper, October 2013), available at: <http://columbiaclimatelaw.com/files/2016/06/Gu-2013-10-CCS-in-China.pdf>.

48 Du and Zhang (n 33). The official document has yet to clarify whether China's climate targets focus on only CO_2 or include other greenhouse gases. China's special envoy for climate change, Mr Xie Zhenhua, said 'China's carbon neutrality targets included greenhouse gas emissions from all sectors of the economy'. Eduardo Baptista, 'China's

emphasizing the development and utilization of clean energy, China's adoption of a carbon neutrality target has reinforced the role of CCS in achieving emission reductions and removal of CO_2 from the atmosphere. Due to the expected dominance of coal as a fuel for the foreseeable future in China, CCS is seen as a crucial technology.[49] An official estimate suggests that the cost of meeting China's long-term climate change mitigation targets for CO_2 will be 25 per cent higher without deploying CCS.[50]

China currently operates or is constructing approximately 40 CCS demonstration projects, boasting a collective carbon capture capacity of three million tons annually.[51] These CCS endeavours span 19 provinces, reflecting diverse capture sources and storage/utilization methods.[52] Notably, the government has offered policy backing for advancing capture technologies, particularly in crucial sectors such as power generation, cement, iron and steel, and petrochemicals, facing limited emission-reduction options. Among these initiatives, there are 13 dedicated capture demonstration projects linked to power and cement plants, capturing a total of 856,500 tons of CO_2 per year.[53]

China's CO_2 capture encompasses a spectrum of technologies: pre-combustion, post-combustion and oxygen-enriched combustion for coal-fired power plants, post-combustion capture for gas-fired plants, CO_2 capture in coal chemical industries, and post-combustion capture in cement production. The robust financial support from the Chinese government and state-owned enterprises has notably bolstered large-scale CCS engineering capabilities.[54] However, there's a gap in technologies for utilization and storage compared to capture and transportation. Emphasizing economic advantages, China's focus within the projects pertaining to CO_2 storage lies in prioritizing EOR, foreseeing benefits from CO_2 storage and increased

2060 Carbon Neutral Goal 'Covers other Greenhouse Gases"' (*South China Morning Post*, 28 July 2021), available at: <www.scmp.com/news/china/science/article/3142771/chinas-2060-carbon-neutral-goal-covers-other-greenhouse-gases>.

[49] Asian Development Bank (ADB), 'Roadmap for Carbon Capture and Storage Demonstration and Deployment in the People's Republic of China' (November 2015), available at: <www.adb.org/sites/default/files/publication/175347/roadmap-ccs-prc.pdf>; Smriti Mallapaty, 'How China Could Be Carbon Neutral by Mid-Century' (2020) 586 *Nature* 482.

[50] ADB (n 49).

[51] Environmental Planning Institute of MEE et al, 'Annual Report on Carbon Capture, Utilization and Storage (CCUS) in China (2021) – Research on the CCUS Pathway in China' (July 2021) (in Chinese).

[52] Ibid.

[53] Ibid.

[54] Ibid.

oil production.[55] The current focus on individual components of the CCS chain in demonstration projects necessitates policy action toward cultivating comprehensive industrial clusters. This imperative involves actively developing full-process clusters that encompass capture, transportation, utilization and storage technologies.

To facilitate the development of CCS demonstration projects, Chinese government authorities have promulgated several policy documents to carry out the key areas of work for technology advancement. These specialized policies provide the basis for regulatory clarity to a limited extent, which is fundamental to further developing the legal and regulatory system governing CCS. The general environmental laws apply when developing CCS projects involves risk and environmental impact assessment. As of now, China has yet to adopt a national law or regulation specifically on CCS.

China's central leadership has long recognized the potential of CCS as part of a portfolio of efforts to mitigate emissions.[56] In 2007, the National Development and Reform Commission (NDRC) issued China's first national programme to address climate change, which explicitly included CCS as an integral part of the technological toolbox to combat climate change.[57] Subsequently, key government stakeholders, including the Ministry of Science and Technology (MOST), NDRC and the Ministry of Environmental Protection (MEP, now superseded by the Ministry of Ecology and Environment), have translated the strategic importance of developing CCS into more specific actions and policy instruments.[58]

To further support CCS development, in February 2013 MOST released a Specialized Plan for Technology Development of CCS in the 12th Five-Year Period.[59] As the first national guideline on CCS development, this specialized plan highlights the strategic significance of CCS and prioritizes research and development for CO_2 capture, transportation and storage. It also outlines the supporting measures for CCS, including agency coordination and demonstration projects. In line with the Specialized Plan for Technology Development of CCS in the 12th Five-Year Period, the NDRC released a Notice on Promoting CCS Pilot and Demonstration Projects in April 2013, which was the first national policy on CCS.[60] The Notice issued by the

55 Lirong Dou et al, 'Trend of Global Carbon Dioxide Capture, Utilization and Storage Industry and Challenges and Countermeasures in China' (2023) 50(5) *Petroleum Exploration and Development* 1246.

56 Jiang et al (n 4).

57 Gu (n 47).

58 Jiang et al (n 4).

59 Specialised Plan for Technology Development of CCS in the 12th Five-Year Period, issued by the MOST, 16 February 2013 (in Chinese).

60 Promoting CCS Pilot and Demonstration Projects, issued by the NDRC, 27 April 2013 (in Chinese).

NDRC in April 2013 sets six major tasks to be carried out for promoting CCS pilot and demonstration projects. These tasks are (1) promoting demonstration projects in coal chemical, oil and gas, and thermal power industries while advancing site assessments in emission-intensive areas; (2) prioritizing the development of cost-effective CCS projects, aiming for a complete development cycle model, including application, approval and management procedures; (3) exploring incentive mechanisms such as tax benefits and financing to diversify investment sources for CCS projects; (4) integrating CCS research into strategic climate, energy and environmental planning, assessing development potential, barriers, risks and impacts; (5) developing country-specific standards and protocols for CCS, establishing safety and environmental supervision systems, including long-term risk assessment and control; (6) strengthening capacity building, international cooperation, utilizing international funds and experiences, and enhancing public awareness and acceptance of CCS in China.[61] Although these tasks resemble the essential regulatory aspects contained in the CCS regulation in the EU, specific regulatory measures in China are yet to be detailed or further developed, leading to the lack of regulatory clarity and a regulatory vacuum for developing CCS projects in China.

The slow progress in regulating CCS largely stems from the absence of an interconnected CCS infrastructure and specialized regulatory framework needed for its technologies. Despite the stalled development of CCS regulation in China, there have been significant efforts to establish a regulatory framework for environmental regulation and risk assessment of CCS.

The MEP has attempted to establish general rules for environmental protection and risk assessment of CCS projects through a Notice on Strengthening the Environmental Protection of CCS Demonstration Projects (2013) and Technical Guidelines for Environmental Risk Assessment for CCS (Trial) (2016).[62] The Notice acknowledges that the existing governance arrangements in China are not yet adapted to CCS projects and faces prominent problems caused by uncertainties in environmental safety.[63] These problems include the potential for secondary pollution resulting from residual waste after the use of adsorbent solvent in the capture process, sudden leakage during transportation and utilization that may damage the local natural environment and threaten public health, and

[61] Ibid.

[62] Notice on Strengthening the Environmental Protection of CCS Demonstration Projects, issued by the MEP (former), 28 October 2013 (in Chinese); Technical Guidelines for Environmental Risk Assessment for CCS (Trial), issued by the MEP (former), 20 June 2016 (in Chinese).

[63] Ibid.

the possibility of CO_2 leakage due to inappropriate site characterization or selection, leading to environmental issues such as groundwater pollution, soil acidification and ecological destruction.[64] Although the Notice requires stringent environmental impact assessment, monitoring and supervision of demonstration projects on a case-by-case basis, it does not aim to address these issues systematically.[65] As stated in the Notice, the MEP plans to consult with relevant stakeholders to improve and develop environmental protection requirements and standards to support CCS project development in China.[66] There has been no further development of CCS regulation in China since the release of the Technical Guidelines in 2016.

The 2016 Technical Guidelines provide detailed descriptions of potential risks that could arise from different stages of CCS project development, referring to Chinese national standards and regulations as well as assessment methods and standards from other jurisdictions. They provide a general framework for conducting environmental impact assessments of CCS demonstration projects in China and supersede other general systems that do not consider the specific characteristics of CCS. While the Technical Guidelines provide guidance to regulators in assessing environmental impact and risks, they lack provisions on key aspects of CCS regulation, such as reporting of irregularities and risk allocation and transfer. Under the current regime, project developers bear all risks related to CCS projects, which creates barriers and challenges in site characterization, resource impact and leakage, leading to increased costs and uncertainty.[67] The Technical Guidelines have limited application, being only applicable to onshore projects, as China's current CCS demonstration projects mainly involve onshore geological utilization and storage, supervised by inland environmental protection departments. Future regulations will need to consider constraints under the London Protocol, to which China is a signatory country, to avoid potential conflicts with international law, and any potential bilateral agreement for transboundary CCS must mirror the Protocol's provisions and requirements.

11.3 Legal and regulatory obstacles for CCS development

The preceding analysis of the legal and regulatory framework governing CCS in the EU and China highlights significant differences in the

64 Ibid.

65 Qi Li et al, 'Application of China's CCUS Environmental Risk Assessment Technical Guidelines (Exposure Draft) to the Shenhua CCS project' (2017) 114 *Energy Procedia* 4270.

66 Notice on Strengthening the Environmental Protection of CCS Demonstration Projects (n 62).

67 Li et al (n 65).

development of laws and regulations to facilitate and regulate CCS activities. To further explore this issue, this section draws upon scholarly discussions regarding the importance of incorporating CCS into climate mitigation strategies, addressing the legal barriers hindering the advancement of the capture, transportation and storage of CO_2, and ensuring consistency with international law. This section discusses specific legal and regulatory obstacles inhibiting CCS uptake in the EU and China. The critical analysis presented in this section forms the basis for the recommendations proposed and elaborated in Section 11.4, which will shed light on possible solutions to overcome these obstacles.

11.3.1 Inadequate incorporation of CCS into climate mitigation strategies

The development of regulations for CCS is not the only incentive for boosting its development. Rather, regulations must be developed in the context of a country's specific climate mitigation plan.[68] The establishment of a clear climate policy regime, with mitigation targets and strategies, will ensure that CCS regulations are enforceable, particularly in terms of acceptable levels of leakage, risk assessment and liability allocation.[69] Mainstreaming CCS into the climate mitigation plan creates an enabling environment that could potentially improve the financial viability of CCS projects by linking carbon pricing with revenue streams.[70] However, high upfront capital contributions and incremental operating expenses required for CCS operation are still significant, and the commercial viability gap faced by CCS developers can only be covered with the help of a supportive policy environment in combination with fiscal and financial support measures provided by the government.[71] Therefore, establishing stringent emission-reduction targets in the near and long term and a strong carbon pricing signal are crucial incentives to support CCS.

Numerous studies emphasize the necessity of fostering a conducive policy landscape to drive CCS advancement, which involves a combination of regulatory frameworks and financial incentives.[72] While early CCS ventures grapple notably with financial hurdles, it is intriguing to consider the observation that the scaling-up impediment might primarily reside in financial challenges rather than legal constraints. This insight, if valid, stands out in the broader context of subsurface technologies, suggesting they

---

[68] Ibid; Bassi et al (n 5).

[69] Kapetaki and Scowcroft (n 7).

[70] Jiang et al (n 4); Gu (n 47).

[71] ADB (n 49).

[72] Jiang et al (n 4); Gu (n 47); Kapetaki and Scowcroft (n 7).

encounter fewer legal barriers but face a financial obstacle to growth. In contrast, above-surface technologies that actively displace fossil fuel usage often contend with an array of legal hindrances. The explicit incorporation of CCS within a comprehensive mitigation strategy establishes a steady policy framework conducive to its progression. This environment, coupled with government-backed fiscal and financial support measures, holds promise in bridging the commercial viability gap for CCS developers. Consequently, the evolution of CCS regulations should align within an overarching climate policy framework, bolstering stringent emission-reduction targets, robust carbon pricing mechanisms, and comprehensive financial aid.

The EU has set medium- and long-term emission-reduction targets and has improved the functionality of the EU's emissions trading scheme (ETS) to support carbon capture, utilization and storage as a key strategy to meet its climate goals.[73] The EU ETS is seen as a significant driver to incentivize low-carbon investments, including CCS, while public funding schemes have been adopted to facilitate its development. However, mobilizing private funding has proven challenging due to the uncertainty of carbon prices in the EU.[74] The Market Stability Reserve was introduced under the EU ETS in 2019 to address this challenge, resulting in an increase in carbon prices (around €85 per ton as of May 2023) that is projected to continue to grow until at least 2030.[75] However, the outcomes of previous funding mechanisms indicate that most of the funding went to projects and technologies with lower costs, creating uncertainties about whether and to what extent CCS projects will be funded by the Innovation Fund despite its larger endowment.[76]

In China, the adoption of a carbon neutrality target and the official recognition of CCS as a mitigation strategy is a strong policy signal to the carbon-intensive industries to explore the use of CCS. However, compared to the more advanced legal and regulatory framework governing CCS projects in the EU, CCS in China is still driven mainly by climate policy and is yet to be incorporated into the national legal framework. Although an increasing number of provinces in China have adopted plans to develop CCS projects, the absence of comprehensive and adequate financial support in the long term has hindered CCS deployment. All existing CCS demonstration projects in China rely on public funding

[73] Cristina Fernanda Alves Rodrigues et al, 'Review of European Energy Policies regarding the recent "Carbon Capture, Utilization and Storage" Technologies Scenario and the Role of Coal Seams' (2015) 74 *Environmental Earth Science* 2553.

[74] Shogenova et al (n 40).

[75] Stig Schjølset, 'The MSR: Impact on Market Balance and Prices' (Point Carbon, 2016), available at: <https://ec.europa.eu/clima/sites/clima/files/docs/0094/thomson_reuters_point_carbon_en.pdf>.

[76] Bassi et al (n 5).

and are incremental and pragmatic, with small projects focusing on the capture or storage individually during the early stages to test various types of technologies.[77] Only a small number of more integrated CCS projects are in operation.[78] Among the various types of CCS projects, CCS for EOR provides some financial relief and revenue opportunities to offset some costs of capture and storage.[79] This explains the concentration of CCS projects for EOR in China, compared to the projects that focus more on integrated technology in the EU. In addition to EOR, revenues can also be sourced from the sale of captured CO_2 where a carbon pricing system exists. However, to date, there is yet to be a strong carbon pricing signal in China as the national ETS is still in an early stage, which entails its limited role in driving low-carbon investment.[80] There is yet to be a linkage between CCS projects and the national ETS.

Despite the growing mentions of CCS in China's national communications and biennial update reports (BURs), its role remains limited to showcasing efforts in promoting demonstration projects, emphasizing the necessity for technology transfer from developed nations, and enhancing standardization processes.[81] However, CCS has yet to gain official recognition as an indispensable mechanism within China's national climate mitigation strategy. This lack of formal recognition, coupled with the absence of robust regulatory tools, significantly impacts the financial feasibility of CCS initiatives and diminishes market confidence in the technology's potential. While China acknowledges CCS in reporting mechanisms, positioning it as a part of climate change mitigation endeavours, the failure to embed it firmly within the national strategy restrains its progression. The absence of comprehensive regulatory frameworks and official endorsement poses significant barriers, hindering both financial investment and broader market trust in CCS's viability as a sustainable and impactful climate mitigation tool within China.

---

77 Jiang et al (n 4); Gu (n 47).

78 Li et al (n 65); Dou et al (n 55).

79 Ibid.

80 IEA, 'China's Emissions Trading Scheme – Analysis' (June 2020), available at: <https://iea.blob.core.windows.net/assets/d21bfabc-ac8a-4c41-bba7-e792cf29945c/China_Emissions_Trading_Scheme.pdf>.

81 For example, China's latest National Communication (NC3), submitted on 25 June 2019, mentions CCS in China's efforts to promote emissions control demonstration projects, the technology needed for mitigation and international collaboration and exchange. See China NC3, available at: <https://unfccc.int/documents/197660>; Similarly, China's latest biennial update report (BUR 2), submitted on 25 June 2019, only mentions CCS once under the section regarding promoting the development of low-carbon pilot and demonstration projects. See China BUR 2, available at: <https://unfccc.int/documents/197666>.

11.3.2 Legal barriers hindering the advancement of CCS

To establish a standard operating framework and offer transparency to investors and operators, the implementation of explicit laws and regulations becomes imperative for the development of CCS projects. Within the EU, Directive 2009/31/EC has been enacted to establish uniform regulatory standards for CCS, thereby fostering a fair environment for project advancement.[82] This Directive serves as a valuable point of reference for other jurisdictions in terms of project permitting, monitoring and reporting requirements, as well as provisions pertaining to liability. The management of liability is a crucial regulatory aspect that necessitates careful deliberation, as it must strike a balance between ensuring responsible project execution and providing developers with a predictable environment, devoid of excessive regulations.[83] Various approaches can be adopted to address liability concerns, including liability transfer, future liability considerations, and contingency liability frameworks.[84] The EU Directive seeks to strike a balance by enabling liability transfer after an agreed-upon timeframe.[85] However, certain provisions within the Directive introduce potential uncertainties and liability exposures, thereby creating barriers to the development of CCS projects.

The first situation mandates that project operators bear the responsibility for all environmental liabilities associated with CCS projects throughout their lifespan. In the event of CO_2 leakage from the storage site, operators are required to acquire emissions allowances to mitigate the leakage.[86] The second situation revolves around the transfer of liability, which is subject to stringent criteria before it can take place.[87] The transfer is contingent upon the competent regulator of the member state being convinced of the extremely low likelihood of leakage. This assurance is upheld through the following provisions: (1) the complete sealing of the storage site and removal of sequestration facilities, with a minimum elapsed period of 20 years, and the provision of compelling evidence that storage will remain permanent indefinitely;[88] (2) the operator's financial contribution to cover monitoring expenses for 30 years following the transfer of liability.[89] These rigorous standards impose significant liabilities and costs on storage site operators, potentially discouraging prospective project developers.

82 Dixon et al (n 5); Shogenova et al (n 40).
83 Ibid.
84 Bassi et al (n 5).
85 Dixon et al (n 5).
86 Shogenova et al (n 40).
87 Ibid.
88 Directive 2009/31/EC, chapter 18.
89 Ibid, chapter 20.

The EU has set up a regulatory structure for CCS, yet China currently lacks a comprehensive policy or regulatory framework to drive CCS development. One of the primary challenges in China is the absence of a formal permitting or authorization process for CCS projects within the domestic system.[90] Given the multifaceted regulatory issues inherent in CCS projects, the absence of an integrated approval process is likely to impede the adoption of CCS technology. This is due to the practical and financial difficulties associated with obtaining all the necessary clearances and permits without a well-established permitting process.[91]

The second challenge stems from the absence of well-defined regulations concerning technical standards and environmental management criteria in China. The establishment of explicit regulations for CCS is imperative due to the extensive deployment of CO_2 capture, transportation and storage, which necessitates specific rules governing each phase of project development.[92] These regulations encompass various aspects, such as the selection and characterization of storage sites, environmental impact assessments, risk assessments, long-term stewardship, and liability considerations. The introduction of explicit regulations is essential to standardize operations, address regulatory concerns related to risk management powers held by regulators, and allocate risks and associated costs.[93]

Last, addressing liability issues related to CCS necessitates the exploration of various approaches in designing liability provisions. These approaches encompass the transfer of liability, future liability arising from storage site closure and monitoring activities, as well as contingency liability pertaining to environmental impact and CO_2 leakage.[94] In the initial stages of CCS projects, liability coverage supported by national laws and regulations is essential. In the context of China, where all CCS projects are currently demonstration projects supported by public funding, they have received distinct treatment concerning liability considerations that do not apply to commercial projects and private investors.[95] Nonetheless, the absence of explicit liability provisions will significantly impede the long-term diffusion of CCS in China, as it introduces ambiguity into the liability framework and potentially exposes storage operators to open-ended liabilities. Thus, there is a pressing need to establish a well-defined liability regime for CCS to facilitate its widespread adoption in China. Section 11.4 provides specific recommendations elaborating on this matter.

---

[90] Gu (n 47); Li et al (n 65).
[91] Ibid.
[92] Bassi et al (n 5); Li et al (n 65).
[93] ADB (n 49).
[94] Bassi et al (n 5).
[95] Li et al (n 65); Dou et al (n 55).

11.3.3 Absence of universally recognized regulatory frameworks underpinned by international law

The London Protocol, which governs offshore CCS activities, demonstrates a more progressive stance compared to its parent treaty, the London Convention. However, the adoption of the 2009 amendment, enabling the transboundary transport and sequestration of CO_2, has encountered challenges. Despite not yet coming into force, a significant step was taken in October 2019 when a resolution was approved under the London Protocol, effectively removing this obstacle. This achievement was the culmination of collective efforts by numerous countries over several years.[96] The resolution allows countries that have ratified the 2009 amendment to transport CO_2 across national borders for offshore storage. Notably, Norway, the Netherlands and the United Kingdom played an active role in pursuing this resolution due to their shared interest in harnessing the offshore storage potential in the North Sea, considered the epicentre of CCS development in Europe.[97] However, it is important to note that the resolution's applicability is limited, exclusively extending to countries that have accepted the 2009 amendment. The legalization of transboundary transport and sequestration of CO_2 carries the potential for significant ramifications on regional and national regulatory frameworks.[98]

Despite significant advancements achieved under the London Protocol toward establishing a common regulatory foundation for CCS, there still remains a substantial journey ahead to ensure widespread acceptance of the Protocol, particularly in countries where CCS holds strategic importance. The London Protocol serves as a valuable regulatory framework for CCS, specifically in Europe and regions where offshore geological storage is necessary, potentially involving the export and storage of CO_2.[99] Given that most European countries are contracting parties to the London Protocol, it is imperative to consider its applicability and relevance when aligning CO_2 sources and sinks within the region. Moreover, the EU's Directive mandates member states to establish bilateral agreements for the export and receipt of CO_2 for sequestration, necessitating compliance with the rules and regulations outlined in the London Protocol.[100]

However, the absence of a common regulatory framework poses a potential barrier to establishing a regional and global CCS supply chain by hindering

---

96 IEA (n 1).

97 Shogenova et al (n 40).

98 IEA (n 1).

99 Shogenova et al (n 40).

100 Ibid.

the matching of CO_2 sources and sinks on a larger scale. This limitation significantly impacts the ability to create a seamless global framework for CCS.[101] Notably, China's ratification of both the London Convention and the London Protocol provides opportunities for potential cooperation with other contracting parties, especially countries that lack adequate storage sites and seek to obtain negative emissions. The London Protocol offers the most practical mechanism for a multilateral cooperative approach in establishing a global CCS supply chain that capitalizes on the vast oil and gas reserves in countries and regions abundant in these resources.

11.4 Insights for climate technology–law interface in the context of climate emergency

The analysis in Section 11.3 underscores a series of intricate intersections between climate technology and law, particularly in the realm of CCS practices within the regulatory frameworks of the EU and China. Within this context, several key challenges and deficiencies surface, shedding light on critical aspects requiring deeper scrutiny and resolution.

An overarching challenge lies in the inadequate integration of CCS technologies into comprehensive climate mitigation strategies. Despite CCS's recognized potential in reducing greenhouse gas emissions, its incorporation into broader climate action plans remains limited in China while the EU has made more explicit references to CCS in its climate mitigation strategies. This limitation emphasizes the necessity for a more coordinated approach to ensure the effective inclusion of CCS within overarching climate change mitigation frameworks. Additionally, the analysis underscores the prevalence of legal barriers hindering CCS technology progression. These barriers manifest in various forms, including complex permitting processes, ambiguous liability frameworks and inconsistent regulations across jurisdictions. Their existence poses significant challenges for stakeholders involved in CCS initiatives, necessitating the development of streamlined and harmonized regulatory mechanisms to facilitate implementation while ensuring legal certainty and minimizing risks.

Furthermore, the absence of universally recognized regulatory frameworks rooted in international law emerges as a prominent challenge. Despite efforts by the EU and China to establish domestic regulations and policies, the lack of a coherent international regulatory framework impedes global CCS deployment, hindering cross-border collaboration, technology transfer and international investments. Addressing these concerns necessitates a meticulous examination of the regulatory landscape surrounding CCS.

[101] Bui et al (n 2).

Resolving these issues would foster an environment conducive to widespread CCS adoption, significantly contributing to climate goals, transitioning to a low-carbon economy, and ensuring global sustainability.

In the EU context, bolstering the carbon pricing signal within the EU ETS and implementing complementary regulatory measures, such as emission performance standards for carbon-intensive sectors, are crucial steps. Similarly, in China, integrating CCS into the low-carbon technology portfolio, establishing specific targets for projects, and implementing robust climate policies are essential strategies. Regarding regulatory clarity, establishing joint liability funds in the EU and adopting an integrated approach involving multiple government agencies in China can enhance certainty and confidence for investors. Learning from the EU's experience in liability provision and project approval can aid China in establishing a fair and transparent framework. Moreover, aligning cooperative efforts with the London Protocol in the EU and considering bilateral agreements among member states can facilitate international collaboration, ensuring consistency, predictability and legality in cross-border CCS projects.

The analysis of the intersection between climate technology and law, particularly within the context of CCS, reveals several overarching observations regarding their interaction, namely interconnected challenges and complexities; regulatory gaps and adaptation; need for harmonization, consistency and clear regulations; and a collaborative and international approach.

First of all, the relationship between climate technology and law presents a nexus of interconnected challenges and complexities. Integrating innovative climate technologies such as CCS into regulatory frameworks is not a linear process. Instead, it involves navigating multifaceted challenges, ranging from technical feasibility to legal compliance. This intricate interplay requires a nuanced understanding of how legal frameworks adapt to accommodate and regulate emerging climate technologies.

Second, there exists a noticeable gap between the pace of technological advancements in climate mitigation and the agility of regulatory frameworks to adapt. Rapid developments in climate technology often outpace the evolution of legal structures designed to govern them. Consequently, this disparity creates regulatory gaps, leaving uncertainties, ambiguities and inconsistencies in the legal landscape. Addressing these gaps necessitates ongoing adaptation and evolution of legal frameworks to keep pace with technological advancements.

Third, the absence of universally recognized and harmonized regulatory frameworks poses a significant hurdle. Diverse regulations across different jurisdictions create barriers for technology deployment, hindering international cooperation, technology transfer and investment. Achieving consistency and alignment within and across legal frameworks becomes crucial to foster collaboration, encourage innovation and ensure a predictable environment for stakeholders involved in climate technology development and deployment. In addition, clarity and certainty in regulations are paramount for fostering a conducive environment for climate technology adoption. Ambiguities, complex permitting processes, and unclear liability frameworks deter investment and hinder the widespread deployment of climate technologies like CCS. Establishing transparent, clear and predictable regulations not only attracts investments but also mitigates risks and uncertainties associated with the implementation of these technologies.

Finally, given the global nature of climate change, a collaborative and international approach is essential. The complexities of climate technology deployment, particularly across borders, necessitate coordinated efforts and cooperation among nations. Harmonizing regulatory standards, aligning legal frameworks and establishing international agreements facilitate cross-border technology deployment, enabling the maximization of the benefits of climate technologies on a global scale.

11.5 Conclusion

The deployment of CCS plays a crucial role in meeting the climate targets established by the Paris Agreement and enjoys support from policy makers worldwide. This chapter conducts a review and analysis of the legal and regulatory arrangements governing CCS in the EU and China. It aims to identify key challenges that these two jurisdictions face. By doing so, this chapter explores insights into the intersection of climate technology and law, focusing on the regulatory frameworks concerning CCS in the EU and China, within the context of climate emergencies.

The effective implementation of CCS projects within the EU is hindered by various obstacles encountered by Directive 2009/31/EU. These obstacles primarily revolve around liability provisions, uncertainties related to remediation issues, and the transfer of liability from storage site operators to competent government authorities. Addressing these challenges is crucial to ensure the smooth progress of CCS initiatives in the EU. On the other hand, China faces a different set of challenges in the development of CCS projects. Specifically, the country lacks specific regulatory frameworks that can provide much-needed clarity on project approval and liability provisions. These gaps in the regulatory landscape impede the advancement of CCS development in China.

The mainstreaming of CCS in climate mitigation strategies varies across regions. In the EU, there has been a significant increase in public funding and efforts to enhance the functionality of the EU ETS, which demonstrates the EU's commitment to integrating CCS technologies into its climate goals. In contrast, China's focus on CCS is primarily driven by climate policy, and the country is still in the early stages of research and project demonstration. This suggests that China is yet to fully establish a comprehensive framework and support system for CCS implementation. Both the EU and China face distinct challenges in promoting CCS projects. By addressing the issues surrounding liability provisions, uncertainties and regulatory frameworks, these regions can accelerate the deployment of CCS technologies and contribute to global efforts in combating climate change.

The London Protocol emerges as a valuable common regulatory denominator for offshore CCS projects. The provisional removal of restrictions on offshore CCS projects involving transboundary issues represents a significant step forward for the EU. China's ratification of the London Convention and London Protocol creates potential opportunities for collaboration with other contracting parties. However, due to the complex geopolitical situations in China's coastal regions, the future of offshore CCS development remains uncertain, with a shift toward exploring onshore project developments in the coming years.

This review and analysis provide an opportunity to address the challenges and weaknesses within the existing institutional and regulatory systems by accommodating necessary changes. It identifies priority areas that require attention and emphasizes the imperative need to develop an effective legal and regulatory framework for CCS to facilitate full-scale projects in the near future. The commercialization and large-scale deployment of CCS will necessitate a more holistic approach to law and regulation, fully integrating all core elements of CCS technology. The interaction between climate technology and law through the case of developing a CCS regulatory framework unveils a landscape riddled with intricacies and challenges. Bridging the gap between evolving technologies and regulatory frameworks, establishing harmonized and transparent regulations, and fostering international collaboration stand as pivotal factors in realizing the full potential of climate technologies and combatting climate change effectively. Future research should explore the means and extent to which the development and improvement of domestic legal and regulatory frameworks can be achieved within a government-led approach that incentivizes CCS deployment.

Acknowledgement

This chapter is an updated version of an earlier journal article by the author, 'Regulations for carbon capture, utilization and storage: Comparative analysis of development in Europe, China and the Middle East' (2021) 157 *Resources, Conservation and Recycling* 105722. The author would like to acknowledge funding support from The Chinese University of Hong Kong Direct Grant for Research 2021–22 (4059051).

12

Evidence-Based Climate Law? The Case of Carbon Capture, Use and Storage

Bettina Lange

12.1 A significant expansion of carbon capture, use and storage as part of climate law and regulation?

Carbon capture, use and storage (CCUS) is one of the technological options for reaching targets enshrined in law for reducing the concentration of carbon dioxide (CO_2) in the atmosphere. Carbon capture and storage (CCS) entails capturing CO_2 emissions from gas- or oil-fuelled power plants, from hydrogen production or from industrial plants with significant combustion, such as those producing steel, cement, aluminium, plastics and glass. Once captured, the CO_2 is transported, usually through pipelines, to an onshore or offshore storage location, such as an underground decommissioned gas or oil field, deep-saline aquifer, the ocean,[1] or coal seams that are not being mined.[2] CC*U*S goes further and re-uses the captured CO_2, for instance in the manufacture of plastics[3] or carbonated drinks.

What evidence is there in favour of deploying CCUS as a climate-mitigation technology? CCUS is controversial for various reasons. The

---

[1] In the UK, storage of CO_2 directly in the ocean is prohibited by Reg 2 of The Storage of Carbon Dioxide (Licensing etc.) Regulations 2010, SI 2010/2221. It is considered to increase the salinity of ocean water and thus pose a risk to aquatic ecosystems.

[2] Barbara Praetorius and Katja Schumacher, 'Greenhouse gas mitigation in a carbon constrained world: The role of carbon capture and storage', (2009) *Energy Policy* 37, 5081, 5082.

[3] Laura-Jane Tiley, 'Carbon Capture Usage and Storage' (Briefing Paper, Number CBP 8841, House of Commons Library, 2020) 3.

Advisory Committee of the UN Human Rights Council considers that the potential benefits of CCS do not outweigh the risk that it may infringe international human rights, such as a right to life as well as a right to information and public participation, as set out in the International Covenant on Civil and Political Rights. CCS may also infringe a right to health[4] and a clean, healthy and sustainable environment.[5]

Against this backdrop of legal controversy, this chapter is focusing on one specific aspect surrounding debate about CCUS: its high costs, such as its capital and operational costs,[6] compared to renewables such as wind and solar.[7] There are currently very few commercial-scale CCUS plants in operation worldwide.[8] Some of these benefit from large

4 Art 25 Universal Declaration of Human Rights. Report of the UN Human Rights Council Advisory Committee, 'Impact of New Technologies Intended for Climate Protection on the Enjoyment of Human Rights', 10 August 2023, A/HRC/54/47, para 51.

5 In addition, the Advisory Committee to the UN Human Rights Council considers CCS as potentially limiting sustainable development, and not to be a negative emission technology given the fact that the construction and operation of CCS plant also generates new CO_2 emissions. The most severe criticism of the Advisory Committee is reserved for the use of CCS that does not involve its deployment for energy generation or the production of goods, such as Direct Air Capture of CO_2 (Report of the Human Rights Council Advisory Committee (n 4) paras 19, 10 and 7). This has to be balanced with CCS potentially supporting just transition for communities that have relied in the past on the fossil fuel industry, for example, for employment (Report of the Special Rapporteur on the promotion and protection of human rights in the context of climate change, Ian Fry, 'Initial Planning and Vision for the Mandate', 24 June 2022, A/HRC/50/39, para 48). A Report of the Special Rapporteur on the issue of human rights obligations relating to the enjoyment of a safe, clean, healthy and sustainable environment recommends fitting CCS technology to fossil fuel fired power plants (UN Report of the Special Rapporteur on the issue of human rights obligations relating to the enjoyment of a safe, clean, healthy and sustainable environment, A/74/161, 15 July 2019, para 77).

6 Conor Hickey, Sam Frankhauser, Stephen M. Smith and Myles Allen, 'A review of commercialisation mechanisms for carbon dioxide removal' (2023) *Frontiers in Climate*, 8.

7 Stuart Haszeldine, 'Chapter 18: Getting CO_2 Storage Right – Arithmetically and Politically', in Mai Bui and Niall Mac Dowell (eds) *Carbon Capture and Storage* (The Royal Society of Chemistry 2020), 564.

8 Figures vary as to how many plants exist. Twenty-eight *CCS* plants are referred to by S. Garcia Freites and C. Jones, 'A Review of the Role of Fossil Fuel-Based Carbon Capture and Storage in the Energy System' (Tyndall Centre, The University of Manchester Research, 2021) <https://research.manchester.ac.uk/en/publications/a-review-of-the-role-of-fossil-fuel-based-carbon-capture-and-stor> 12. Twenty-one *CCS* plants are referred to by the then Minister for BEIS in House of Commons, Business, Energy and Industrial Strategy Committee, Oral Evidence: Carbon Capture, Usage and Storage (CCUS) HC 1094 and Clean Growth Strategy (2018), HC 871, Q. 148. Around 40 *CCUS* plants are referred to by the International Energy Authority <www.iea.org/energy-system/carbon-capture-utilisation-and-storage>.

subsidies.[9] The UK government pledged £800 million in its March 2020 Budget through an infrastructure fund for developing CCUS for, at least, two sites.[10] According to a previous UK government minister, CCUS 'is more of an economic problem than a technical problem'.[11]

This has sparked debate about whether states should intervene in private business activity to create economic incentives for commercial-scale deployment of CCUS. In the United Kingdom, the government has taken the view that state intervention is necessary. It has established an economic regulatory framework in Parts 1 and 2 of the Energy Act 2023, which also seeks to enhance the economic evidence base in favour of the commercial viability of CCUS. This chapter examines key features of this framework. It argues, first, that scientific, including economic, evidence for or against technologies is co-produced with political actors through legal rules that act as 'boundary objects'. Hence, law does not merely prescribe specific actions in relation to the use of climate-mitigation technologies but it acts as a bridge between economic evidence and political choices for or against particular climate-mitigation technologies.

The chapter develops this argument in five sections. Section 12.2 suggests that evidence about the economics of different climate-mitigation technologies is a distinct form of 'second-level' evidence which can become linked to politics through boundary objects. Section 12.3 sets out key features of Parts 1 and 2 of the Energy Act 2023 which seek to establish an economic regulatory framework for CO_2 transport and storage networks and the capture of CO_2 from low-carbon hydrogen and industrial production in the UK. Section 12.4 argues that these legal rules are 'boundary objects' which link evidence about the economics of CCUS to policy choices. Section 12.5 draws attention to other policy choices, such as promotion of the circular economy and democratically governed local energy projects which may be sidelined through the distinct links between economic evidence and policy created by Parts 1 and 2 of the Energy Act 2023.

9 According to the then Minister for BEIS giving evidence to the House of Commons BEIS Committee, five of the 21 CCS plants she is referring to receive large subsidies (n 8).

10 BEIS, 'Carbon Capture, Usage and Storage: A Government Response on Potential Business Models for Carbon Capture, Usage and Storage' (August 2020)<chrome-extension://efaidnbmnnnibpcajpcglclefindmkaj/https://assets.publishing.service.gov.uk/media/5f3a447ee90e0732dafc03af/CCUS-government-response-business-models.pdf> 7.

11 BEIS (n 8) Q. 157.

12.2 Knowledge resources for climate law

12.2.1 Why focus on evidence about the economics of climate-mitigation technologies?

Climate law can be broadly defined as a distinct field of ideally coherent legal norms which is evolving incrementally to limit dangerous anthropogenic climate change.[12] Various knowledge resources are harnessed for understanding what amounts to dangerous anthropogenic climate change, and how it can be prevented and mitigated. Hence, the idea of evidence-based climate law is well established. But the focus has been initially on natural-science evidence[13] for targets for reducing CO_2 concentrations in the atmosphere and thus to flesh out the meaning of preventing '*dangerous* anthropogenic interference with the climate system'.[14] For instance, natural-science evidence about the rapid pace of global warming has been a driver of the European Union (EU) emission-trading scheme.[15] At a range of governance levels, spanning the international level of the United Nations Framework Convention on Climate Change[16] and Paris Agreement,[17] EU law[18] and domestic legislation, climate law now requires legislatures and executives to consider advice based on scientific evidence when exercising their legal powers. In the United Kingdom, the secretary of state for the Department for Energy Security and Net Zero (DESNZ) must consider, but is not bound by, the independent advice of the Committee for Climate Change – which is based on scientific expertise – when he/she sets a national carbon budget.[19] Hence, climate law provides procedural routes for generating and considering this 'first level' of natural-science evidence. That first level of evidence also informs targets for reducing the concentration of greenhouse gases in the atmosphere. It can also include information about the emission reductions that various climate-mitigation technologies can achieve. For instance, the Intergovernmental

[12] Building on Benoit Mayer and Alexander Zahar, *Debating Climate Law* (CUP 2021), 4.

[13] Report of the Human Rights Council Advisory Committee (n 4) para 21.

[14] Art 2 of the UN Framework Convention on Climate Change.

[15] For example, Preamble 3 to Directive 2008/101/EC of the European Parliament and of the Council of 19 November 2008 amending Directive 2003/87/EC so as to include aviation activities in the scheme for greenhouse gas emission allowance trading within the Community [2009] OJ L 8/3.

[16] Art 4 (2) (c) and (d), Arts 5 and 9 of the UN FCCC.

[17] Arts 4 (1), 7 (5) and 14 (1) of the Paris Agreement.

[18] Regulation (EU) 2021/1119 of the European Parliament and of the Council of 30 June 2021 establishing the framework for achieving climate neutrality and amending Regulations (EC) No 401/2009 and (EU) 2018/1999 ('European Climate Law') [2021] L 243/1, art 4 (4).

[19] Climate Change Act 2008 s 9 (1) (a) and 9 (4).

Panel on Climate Change (IPCC) considers CCUS to be part of the mix of climate-mitigation technologies needed to achieve net zero by 2050.[20]

There is also a 'second level' of evidence which addresses a subsequent question about climate-mitigation technologies. Are particular mitigation technologies likely to achieve in practice specific reductions in the concentration of CO_2 in the atmosphere, and thus compliance with climate targets, such as the 1.5 °C warming limit? Second-level evidence consists of social science evidence, for example about the cost-effectiveness of specific climate technologies and whether they are accepted by citizens. Such evidence may be even more complex to produce than evidence at the first level because the impact of a specific climate technology on concentrations of CO_2 in the atmosphere depends on a range of contextual factors. These factors are difficult to predict, such as rates of economic growth that increase emission of greenhouse gases which may offset reductions in CO_2 emissions from plants with CCUS, as well as policy preferences of specific governments. In the United Kingdom, for instance, the Conservative government relaxed only in 2023 its planning-policy restrictions for onshore wind farms. The restrictions had prompted a member of the House of Commons Committee on Business, Energy and Industrial Strategy (BEIS) to state that support for CCUS but not onshore wind is 'a political ideological decision'.[21]

Factors that shape second-level evidence vary between countries, given their different mixes of energy sources[22] and production and employment profiles (and thus greenhouse gas emissions), as well as variation in how regulatory frameworks shape carbon prices and whether government subsidies for innovative technologies are already a feature of the national political culture. Some factors that shape evidence about the economics of CCUS are unrelated to politics, such as a country's geology. For instance, costs of CO_2 transport and storage are reduced if former gas and oil fields – one storage option – exist close to industrial plants that emit CO_2.

Procedures for bringing second-level evidence to the attention of public policy makers and legislators are less well established given its complexity. Such procedures need to address that second-level evidence may be more likely to be contested since it is often produced in-house, for example, as a regulatory impact assessment prepared by government civil servants or reports of government-commissioned consultancies. Moreover, second-level

[20] Intergovernmental Panel on Climate Change, 'Global Warming of 1.5°C, Summary for Policy Makers' (2018) <www.ipcc.ch/report/sr15/> 15.

[21] House of Commons, Business, Energy and Industrial Strategy Committee (n 8) Q. 107.

[22] In the UK, it is uncertain what proportion of energy will be provided by gas-fired power plants with CCUS, as a form of 'mid-merit' power, in the context of a commitment to increase the share of energy generated from renewable sources. That uncertainty may affect confidence of potential private CCUS investors (BEIS (n 10) 32).

evidence may involve less clear-cut distinctions between assessing the evidence and a policy choice, since economic evidence that favours, for example, a particular climate technology will have an opportunity cost, that is, investment foregone for other climate technologies. Finally, evidence about the economics of climate technologies can be uncertain and in flux. The UK government's renewed commitment to deploy CCUS at a commercial scale by 2030 has been subject to the qualification that the costs for CCUS will come down.[23]

Despite these difficulties of generating and bringing to the attention of policy makers and legislators second-level evidence, courts increasingly require consideration of such evidence. For instance, the Administrative Court in the UK declared in recent cases[24] the UK government's net-zero strategy as unlawful since it did not spell out in sufficient detail *emissions reduction pathways.* Such pathways specify how particular policies will lead to a particular percentage reduction in greenhouse gas emissions.[25] So, what do we know from previous academic scholarship about how evidence, including economic evidence, may influence public policy making and regulation?

12.2.2 Insights from the literature on evidence-based public policy making and regulation

Some of the literature on evidence-based public policy making and regulation starts from the assumption that 'evidence' and 'policy' correspond to two domains distinguished both institutionally and epistemologically.[26] It has, therefore, set out typologies of how evidence may exert influence on policies. It also seeks to identify factors that facilitate or hinder interactions between evidence and policy, and to argue for ways in which they could relate better. A frequent, often implicit, assumption is that by improving

---

23 HM government, 'Clean Growth: The UK Carbon Capture Usage and Storage Deployment Pathway: An Action Plan' (2018) <chrome-extension://efaidnbmnnnibpcajpcglclefindmkaj/https://assets.publishing.service.gov.uk/media/5bfd760bed915d118adbb940/beis-ccus-action-plan.pdf> 9.

24 *The Queen, on the application of FoE, Client Earth, Good Law Project and Joanna Wheatley* v *Sec. of State for BEIS* [2022] EWHC 1841 (Admin); *Friends of the Earth, Clientearth, Good Law Project* v *Secretary of State for Energy Security and Net Zero* [2024] EWHC 995 (Admin).

25 *The Queen, on the application of FoE, Client Earth, Good Law Project and Joanna Wheatley* v *Sec. of State for BEIS* [2022] EWHC 1841 (Admin), paras 237–9. HM Government 'Powering Up Britain' (2023) sets out in further detail emission pathways in response to the Hight Court ruling <www.gov.uk/government/publications/powering-up-britain>.

26 See for example, Arie Freiberg and Wesley Carson, 'The limits to evidence-based policy: Evidence, emotion and criminal justice' (2010) *The Australian Journal of Public Administration* 69(2), 152–64, 160.

institutional and epistemological links between the two domains, policy making will be enhanced.[27]

Typologies of the impact of evidence on policy have identified the instrumental, conceptual and symbolic influence of evidence. In the case of *instrumental* influence, the evidence directly steers the specific policy option chosen. Here science provides the insights needed to reach a decision. This can be facilitated by the evidence provider understanding how political decision-making for crafting policy works.[28] Evidence has *conceptual* influence when it does not directly steer a specific policy choice, but nevertheless shapes how policy makers come to view and frame the public policy issue for which they want to develop policy. For instance, evidence may indicate that CCUS cannot be developed without government support. But the choice of the specific policy option is left open, that is, whether government will provide financial subsidies to companies operating CO_2 transport and storage networks through direct payments, or through a process of companies bidding in a competitive process for subsidy payments. Evidence can also have *symbolic* influence on public policy.[29] In this case it does not shape the policy, but is used with hindsight in order to justify a particular policy which has been chosen on the basis of political considerations. This can be particularly effective if the evidence relied upon is perceived as 'unbiased'.[30]

The literature has further nuanced this typology. Weible, for instance, differentiates between impacts according to what type of public policy making process is at stake.[31] He argues that our understanding of the use of expert-based information is shaped by how we conceptualize the public policy making process. 'Punctuated equilibrium theory' is one way for understanding a public policy making process. It suggests that policy change happens not only in incremental steps but also in major strides at certain points in time. This accords with UK government policy making for CCS. A crucial point in time was 2007, when a UK Labour government set up

27 John Holmes and Rebecca Clark, 'Enhancing the use of science in environmental policy-making and regulation' (2008) *Environmental Science & Policy* 11(8), 702–11.

28 Brian Head, 'Three lenses of evidence-based policy' (2008) *Australian Journal of Public Administration* 67(1), 1–11, 1, 5.

29 Nabil Amara, Mathieu Ouimet and Réjean Landry, 'New evidence on instrumental, conceptual, and symbolic utilization of university research in government agencies' (2004) *Science Communication*, 26(1), 75–106, 78, 91.

30 Sheila Jasanoff, 'The Practices of Objectivity in Regulatory Science', in C. Camic, N. Gross and M. Lamont (eds) *Social Knowledge in the Making* (University of Chicago Press 2011), 307–38, 308.

31 Christopher M. Weible, 'Expert-based information and policy subsystems: A review and synthesis' (2008) *The Policy Studies Journal*, 36(4), 615–35.

a competition for CCS technology. This was, however, cancelled in 2011 by a Conservative-Liberal Democrat government due to the high capital costs of installing CCS at a gas or coal-fired power station, estimated then to be between £1 billion and £2.5 billion.[32] In 2012, the government revived the competition idea, though this was again cancelled by the then Conservative government in 2016.[33] A National Audit Office (NAO) report contributed to the UK Treasury's decision to withdraw support for the second competition. The Treasury considered the capital expenditure for the projects to be too high, taking into account that the cost was ultimately to be borne by energy consumers. According to the Treasury the money could be better spent on other policies.[34] In a somewhat circular argument, the NAO also suggested that the competition was premature: the UK government was seeking to deliver CCS before it was cost-efficient to do so.[35] In 2022, the UK's Conservative government revived its support, now for CCUS, by proposing an economic regulatory framework, enshrined in Parts 1–2 of the Energy Act 2023.

To summarize, literature on evidence-based public policy making and regulation has identified instrumental, symbolic and conceptual impacts of evidence on policy and regulatory frameworks. It conceives of evidence as a discrete body of knowledge that, initially, can be clearly distinguished from political decision-taking. Science and technology studies (STS), however, work with a less clear-cut distinction between evidence and policy. They analyse how boundaries between science (including social science) and policy are actively constituted. They argue that sometimes the boundaries between science and policy, as distinct epistemological fields and institutional practices, can become blurred. Hence, STS provides further nuance and a micro-level account of interactions between science and policy. This can shed light on

32 Tiley (n 3) 7.

33 Ibid 15.

34 National Audit Office (NAO), 'Carbon Capture and Storage: The Second Competition for Government Support' (2017) <www.nao.org.uk/reports/carbon-capture-and-storage-the-second-competition-for-government-support/#downloads> 6.

35 Ibid. The demonstration projects, which would have been financially supported through the competition, should have made CCS commercially more viable by reducing its costs. According to the NAO report, the UK government spent £168 million on the two CCS competitions. Cancelling CCS projects is not unique to the UK. In 2003 the 'FutureGen' project was started in the US for installing CCS at a coal-fired electricity generating plant. The US government promised $1.1 billion support. Construction began in 2014, in 2015 the project was cancelled by the federal government. Another CCS project was suspended in Mississippi in 2017 (Rachel M. Cohen, 'The Environmental Left Is Softening on Carbon Capture Technology. Maybe That's OK', The Intercept, 20 September 2019 <https://theintercept.com/2019/09/20/carbon-capture-technology-unions-labor/>.

how economic evidence either for or against state intervention in private business activity – including technology development – is constructed.

12.2.3 Two insights from STS: boundary objects and a non-instrumental conception of technology

STS literature has developed the idea that nature is co-produced with society, science with politics, and facts with values.[36] Sheila Jasanoff and several others argue that we should not see either side of these binaries as determining the other, but as influencing one another, especially in the case of risk controversies.[37] Hence, STS literature suggests that boundaries around epistemological fields and institutional practices, such as science and politics, have to be actively constructed through 'boundary work', but this does not lead to rigid demarcations.[38] Negotiators in international treaty deliberations, for instance, may be 'speaking simultaneously in the name of government and of science'.[39] Thus, how much distance there is as a result of boundary work, for example, between economic evidence and politics, can vary.[40]

Analysing boundary-drawing enables one to go beyond the more static and schematic snapshots of 'symbolic', 'conceptual', or 'instrumental' links between science, policy and regulation discussed in the literature on evidence-based public policy making. Boundary-drawing can throw light on how fields of social practices are *represented* in distinct ways, for example, as science or politics. Boundary work means not just 'to create and shape', but also to 'disrupt boundaries'.[41] Hence, there can be 'boundary closure' as well as 'connecting', and 'breaching' of boundaries.[42] Boundary work is political since contesting boundaries in 'boundary conflicts'[43] can lead to

[36] Sheila Jasanoff, 'The Idiom of Co-Production', in S. Jasanoff (ed) *States of Knowledge: The Co-Production of Science and Social Order* (Routledge 2004), 1–12; Sheila Jasanoff, 'Ordering Knowledge, Ordering Society', in S. Jasanoff (ed) *States of Knowledge: The Co Production of Science and Social Order* (Routledge 2004), 13–45. Sheila Jasanoff, 'Serviceable truths: Science for action in law and policy' (2015) *Texas Law Review* 93(7), 1723–49.

[37] James Fairhead and Melissa Leach, *Science, Society and Power: Environmental Knowledge and Policy in West Africa and the Caribbean* (CUP 2003), 14; Sheila Jasanoff, 'Contested boundaries in policy-relevant science' (1987) *Social Studies of Science* 17(2), 195–230, 195.

[38] Sheila Jasanoff, *The Fifth Branch: Science Advisers as Policy Makers* (Harvard University Press 1990), 236.

[39] Fairhead and Leach (n 37) 1.

[40] Susan Owens, *Knowledge, Policy, and Expertise: The UK Royal Commission on Environmental Pollution 1970–2011* (OUP 2015) 13.

[41] Charlene Zietsma and Thomas Lawrence, 'Institutional work in the transformation of an organizational field: The interplay of boundary work and practice work' (2010) *Administrative Science Quarterly* 55(2), 189–221, 195.

[42] Ibid 194.

[43] Jasanoff, 'Ordering Knowledge, Ordering Society' (n 36) 13, 19.

questioning existing systems of social stratification – for instance, between lay and expert knowledge.[44] Boundary objects can breach boundaries and act as bridges, for example, between the field of science and the field of politics.[45] Hence, boundary objects are 'entities that enhance the capacity of an idea, theory or practice to translate across culturally defined boundaries, for example, between communities of knowledge or practice'.[46]

Boundary objects have three characteristics. First, they are flexible in their interpretation.[47] This enables various actors with different interests to attribute their preferred meaning to the idea or practice.[48] Second, facilitative boundary objects are characterized by their coordination function.[49] They bring together people and ideas where there is no consensus, to enable working together.[50] Third, boundary objects coordinate because they create links between the ill-structured, indeterminate parts of the boundary object and its more locally tailored, specific uses.[51] Various material and linguistic artefacts, such as concepts ('sustainability', 'resilience', and so on), legal provisions,[52] reports and models[53] can be boundary objects. A multitude of boundary objects together constitute 'boundary infrastructures'.[54]

Section 12.3 argues that Parts 1 and 2 of the Energy Act 2023 are a boundary object that links the UK government's policy choice for CCUS with economic evidence through the economic regulatory framework of the two Parts. Hence, CCUS as a climate-mitigation technology does not automatically reduce CO_2. From a UK perspective it needs to be embedded in a network that links the hardware of CO_2 capture to infrastructure and human actors, including economic regulators. This accords with STS's rejection of the idea that technology can determine human action. CCUS

---

44 Zietsma and Lawrence (n 41) 205.

45 David Guston, 'Stabilizing the boundary between US politics and science: The rôle of the Office of Technology Transfer as a boundary organization' (1999) *Social Studies of Science* 29(1), 87–111.

46 Nick J. Fox, 'Boundary objects, social meanings and the success of new technologies' (2011) *Sociology* 45(1), 70–85, 70.

47 Susan Leigh Star, 'This is not a boundary object: Reflections on the origin of a concept' (2010) *Science, Technology, & Human Values* 35(5), 601–17, 601.

48 Fox (n 46) 72; Star (n 47) 602.

49 Fox (n 46) 80.

50 Star (n 47) 602.

51 Ibid 601, 603.

52 Nicholas Haagensen, 'Boundary objects in complex governance systems: collective action clauses in European sovereign debt governance' (2022) *Journal of European Public Policy* <https://research.cbs.dk/en/publications/boundary-objects-in-complex-governance-systems-collective-action->.

53 Sarah Michaels, 'Matching knowledge brokering strategies to environmental policy problems and settings' (2009) *Environmental Science & Policy* 12(7), 994–1011, 1001.

54 Star (n 47) 602.

is a climate-mitigation technology not just in the instrumental sense of 'any tool or technique, any product or process, any physical equipment or method of doing or making, by which human capability is extended'.[55] Human capability is required to turn CCUS into a technology that can reduce CO_2 concentrations in the atmosphere at a significant scale. But there is debate about what human capability should be expanded to in order to turn CCUS into a climate-mitigation technology. Should states intervene in private business activity to promote commercial-scale application of CCUS? If so, how?

12.3 Should states intervene in private business activity to promote CCUS?

12.3.1 Is CCUS necessary but too costly for the private sector to implement it alone?

Some argue that state intervention is necessary to achieve commercial-scale application of CCUS given its significant costs.[56] The UK Committee for Climate Change considers CCUS to be 'a necessity not an option' in order to reduce greenhouse gas emissions by 78 per cent from 1990 levels by 2035, a target set in the United Kingdom's Sixth Carbon Budget.[57] Moreover, it has been suggested in the UK context that 'overall system costs are lower when CCUS is deployed in the power sector', compared to scenarios where there is no CCUS deployment for achieving climate targets.[58]

Worldwide, so far, there has been little application of CCUS. One of the reasons for this are the financial risks that are distinct to the CCUS value chain. They are akin to a 'chicken and egg' dilemma: without assurance that there will be sufficient capture of CO_2 from industrial or low-carbon hydrogen production, operators of CO_2 transport networks and storage sites

---

55 Lyria Bennett Moses, 'Regulating in the Face of Sociotechnical Change', in R. Brownsword, E. Scotford and K. Yeung (eds) *The Oxford Handbook of Law, Regulation and Technology* (OUP 2016), 573–96, 576 referring to Donald Schön, *Technology and Change* (Pergamon Press 1967), 1.

56 Hickey et al (n 6) 1.

57 Press Release, 'UK Enshrines New Target in Law to Slash Emissions by 78% by 2035' <www.gov.uk/government/news/uk-enshrines-new-target-in-law-to-slash-emissions-by-78-by-2035>.

58 BEIS, 'Energy Bill – Impact Assessments, Annex 1.1. CO_2 Transport and Storage Regulatory Investment (TRI) Impact Assessment, Section 2 Rationale for Intervention' (2022) <https://Bills.parliament.uk/Bills/3311/publications> 9.

are unlikely to invest in appraising potential storage sites and laying pipelines.[59] Without enough capacity for CO_2 transport and storage, industrial emitters and low-carbon hydrogen producers will not invest in CO_2 capture at their plants. Moreover, it is not clear who in the value chain will bear the financial risk if one of the components in the chain fails.[60] For instance, if the CO_2 output of a cement or steel factory is reduced, should the factory bear the cost through higher prices for CO_2 transport and storage, or should the transport and storage company bear it through reduced revenue? CCUS value chains pose the challenge of how to coordinate investment decisions with long lead times made by different industrial operators.

Legal regulatory frameworks are, therefore, considered necessary to create incentives for private investment for commercial-scale application of CCUS.[61] How much difference such frameworks make is debatable. During the 14 years that the EU Directive on geological storage of CO_2 has been implemented in EU member states it has not led to significant commercial deployment, even though the Directive creates some economic incentives, such as the transfer of liability from private operators to the state for CO_2 storage, usually when at least 20 years have elapsed after closure of a site.[62]

The economic argument for state intervention varies from country to country depending on, for example, the mix of energy sources a country relies on. In the UK, the economic case for CCUS has been strengthened through the government's commitment to low-carbon hydrogen as a new energy source, including for heating and for difficult-to-decarbonize sectors such as transport.[63] Blue hydrogen production splits a fossil fuel, such as gas or oil, into hydrogen and CO_2.[64] A legal requirement to capture CO_2 from this process will create new demand not only for CO_2-capture technology, but also for transport and storage services. But through what regulatory tools should the government intervene? How do such tools create evidence

59 BEIS, 'Energy Bill – Impact Assessments, Annex 1.4. Hydrogen and Industrial Carbon Capture (ICC) Business Models Impact Assessment' (2022) <https://Bills.parliament.uk/Bills/3311/publications>, 5.

60 Ibid.

61 Aileen McHarg and Mark Poustie, 'Risk, Regulation, and Carbon Capture Storage: The United Kingdom Experience', in Donald Zillman, Aileen McHarg, Lila Barrera-Hernández and Adrian Bradbrook (eds) *The Law of Energy Underground* (OUP 2014), 249–74, 249.

62 As implemented in the UK through The Storage of Carbon Dioxide (Termination of Licences) Regulations 2011, SI 2011/1483, reg 8 (b).

63 European Academies Science Advisory Council, 'Decarbonisation of Transport: Options and Challenges' (2019), EASAC Policy Report, <www.easac.eu> 5.

64 UK Parliament Post Note, 'Low Carbon Hydrogen Supply' (2021) Number 645, <https://post.parliament.uk/research-briefings/post-pn-0645/#:~:text=The%20UK%20Government%20is%20expected,low%2Dcarbon%20hydrogen%20in%202021> 1.

in favour of the economics of CCUS? This is of interest also for countries that are relying on fossil fuels for power generation in order to promote economic development, and which are still considering whether and to what extent to deploy CCUS.[65]

12.3.2 How does the UK government propose to intervene in private economic activity to promote CCUS?

The UK government seeks to promote the deployment of CCUS at commercial scale through two main means. Part 1 of the Energy Act 2023 sets out various ways of providing financial support for companies that transport and store CO_2. Part 2 of the Act establishes procedures for paying revenue support to installations that capture CO_2 from industrial and hydrogen production.

12.3.2.1 Providing financial support for CO_2 transport and storage companies

Market failure and 'income gap' justifications for state intervention

The Regulatory Impact Assessment (RIA) produced by the then UK Department for Business, Enterprise and Industrial Strategy (BEIS)[66] relies on market failure for CO_2 transport and storage services as a justification for state intervention. Some of these failures are significant, others may not be. For instance, the price regulation of Part 1 of the Energy Act 2023 seems necessary to ensure efficient returns on capital investment, while also ensuring, through a price cap, that regional monopoly operators of CO_2 transport and storage services do not abuse their market power. BEIS anticipated that there will be three operators of CO_2 transport and storage networks in the UK,[67] which will act as regional monopolies. Upfront capital expenditure costs for those who provide CO_2 transport infrastructure through pipelines is high, compared to CO_2 transport by ship, rail or road.[68]

---

65 DACCS and CCS for coal-fired power plants can involve higher water consumption, while BECCS can entail both increased water consumption and changed land use which may increase food prices. The latter are particularly significant in developing countries (IPCC, 'Climate Change 2022: Impacts, Adaptation and Vulnerability, Working Group II Contribution to the Sixth Assessment Report of the IPCC' (CUP 2022) <www.ipcc.ch/report/ar6/wg2/> 654.

66 In February 2023, the Department for Business, Energy and Industrial Strategy was dissolved and separated into two new UK central government departments: the Department for Energy Security and Net Zero and the Department for Science, Innovation and Technology.

67 BEIS (n 58) para 65.

68 Ibid para 24.

Hence, price regulation seeks to make income for service providers more predictable. It also seeks to remove barriers to entry into the market for CO_2 transport services. It thereby attempts to increase competition in the provision of these services and thus reduce their price, which can also limit the costs for governments to provide financial support for CCUS.

Economic evidence for some of the other market failures that the UK government seeks to rely on to justify its intervention is more questionable. For instance, the RIA for Part 1 of the Energy Bill 2022 refers to an 'early-mover risk', that is, the risk that initial providers of CCUS services face higher costs than later adopters[69] since fewer participants in a less competitive market may be associated with higher prices for CCUS services. But this scenario may be limited, since the UK government considers CCUS technologies as an opportunity for UK companies to export to other countries.[70] Such export opportunities may arise where developing countries have included low-carbon fossil-fuel energy into their Nationally Determined Contribution, as part of the fulfilment of their Paris Agreement obligations.[71] Moreover, the UK government is considering opening up the UK CO_2 transport and storage network to CO_2 captured in other countries.[72] This points to early-mover *advantages* rather than risks. In the RIA itself, export opportunities are projected to create 46,000 jobs by 2030 within the CCUS sector, and to add £3.5 billion in gross value per annum (GVA) to the UK economy.[73] There may also be early-mover advantages from financial returns from intellectual property protection of innovative CCUS technology. Research about industrial-pollution control technologies has found first-mover advantages for companies that reap efficiency gains from innovative production processes.[74]

---

69 Ibid para 8.

70 Ibid para 89.

71 Such as, in the case of Indonesia, a commitment to 'clean coal'. -

72 BEIS, 'Carbon Capture, Usage and Storage: An Update on the Business Model for Transport and Storage' (2022) <chrome-extension://efaidnbmnnnibpcajpcglclefindmkaj/https://assets.publishing.service.gov.uk/government/uploads/system/uploads/attachment_data/file/1045066/ccus-transport-storage-business-model-jan-2022.pdf> 11. It is unclear why the box in the RIA that asks whether the measures 'will have an impact on international trade' is not ticked (BEIS (n 58) 2).

73 BEIS (n 58) para 90 referencing an Energy Innovation Needs Assessment.

74 Johan Albrecht, 'Working Paper: Environmental Regulation, Comparative Advantage and the Porter Hypothesis' (1998), Nota di Lavoro, No. 59, Fondazione Eni Enrico Mattei (FEEM), Milano <chrome-extension://efaidnbmnnnibpcajpcglclefindmkaj/https://www.econstor.eu/bitstream/10419/154917/1/NDL1998-059.pdf>; Michael Porter and Claas van der Linde, 'Toward a new conception of the environment-competitiveness relationship' (1995) *Journal of Economic Perspectives* 9(4), 97–118. Mads Greaker, 'Spillovers in the development of new pollution abatement technology: A new look at the Porter-hypothesis', (2006) *Journal of Environmental Economics and Management* 52, 411–20.

The UK government also justifies its intervention in private business activity on the ground that companies that store CO_2 face the risk of closure of transport and storage sites, due to defects, including CO_2 leakage.[75] The government will therefore indemnify CO_2 transport and storage companies against CO_2 leakage, for which there is currently no commercial insurance.[76] But the technical literature on CCUS considers the risk of CO_2 leakage to be small if appropriate risk management standards are complied with.[77]

To summarize, an economic regulatory framework for CO_2 transport and storage services can be justified with reference to some market failures. Significant transfers of risks and liabilities from the private sector to the state may, however, create a moral hazard, that is, incentives for companies to limit their efforts to reduce risks and liabilities.[78]

More fundamentally, some NGOs,[79] as well as businesses and universities involved in the development of CCS, question the preference for and economic value of competition. The UK government justifies its intervention into private business activity in relation to CCUS with reference to market failure, that is, a failure of markets to be sufficiently competitive. More collaboration and knowledge exchange, from research-and-development projects among CCS technology developers for example,[80] may reduce the costs of progressing CCS. This may entail limited applicability of intellectual property rights[81] and a requirement that publicly funded CCS projects are to be configured in such a way that they can support other projects.[82] This can occur, for example, through CO_2 transport infrastructure having to serve industrial installations that capture CO_2, in not just one but several areas, and to transfer CO_2-capture technology between installations.

While a market-failure analysis addresses the wider picture of the welfare of the UK economy as a whole, economic incentives have also been justified with reference to the 'income gap' theory. This suggests that there is insufficient income for economic actors, such as investors and

75 BEIS (n 58) para 94 e.

76 Ibid para 42.

77 Health and Safety Executive (2011) 'Assessment of the Major Hazard Potential of Carbon Dioxide (CO_2)' <www.hse.gov.uk/carboncapture/assets/docs/major-hazard-potential-carbon-dioxide.pdf> 2. Stuart Haszeldine (n 7) 565.

78 DESNZ, 'Carbon Capture, Usage and Storage, Industrial Carbon Capture Business Models Update for Track-1 Expansion and Track-2' (2024) - 9.

79 Jason Anderson and Joana Chiavari, 'Understanding and improving NGO position on CCS' (2009) *Energy Procedia* 1, 4811–17, 4815.

80 Subject to the provisions of the UK Subsidy Control Act 2022. House of Commons, Business, Energy and Industrial Strategy Committee (n 8) Q. 225.

81 Ibid Q. 218.

82 Ibid Q. 252.

operators, to deploy CCUS at a commercial scale.[83] This theory has been questioned by research that shows that what will determine whether CCUS is commercially applied is not just the finance available, but also a range of political-economy factors, such as: (1) whether government departments for energy, climate change and (in the United Kingdom) the Treasury are aligned in their support for CCUS, (2) whether legal requirements for reducing CO_2 concentrations in the atmosphere are stringent enough to create sufficient pressure for capturing CO_2, (3) whether there are other significantly cheaper energy sources, such as renewables, available, (4) how much public opposition there is to the continued use of fossil fuels (for example, when hydrogen is produced from natural gas), and (5) whether the price for carbon is high enough, due to CO_2 emission trading or a carbon border adjustment mechanism, to create incentives for CCUS.[84] Moreover, exogenous shocks to the energy-supply system, such as those generated by the war in Ukraine, have generated increased profits for energy providers. All of these factors raise questions about the 'income gap' theory as a good justification for state intervention in support of the commercial deployment of CCUS.[85]

These limits to market failure and the income gap justifications should be borne in mind when considering whether and to what extent the state should intervene to promote CCUS. How, then, does the UK government seek to advance CCUS?

Licensing operators that provide CO_2 transport and storage services

According to Part 2 of the Energy Act 2023, the lawful operation of CO_2 transport and storage networks requires a licence. Initially, the secretary of state for DESNZ will issue these licences. Once 'the sector has sufficiently matured', the economic regulator for gas and electricity services in Great Britain will be issuing, enforcing and modifying the licences.[86] The legislation enables the Gas and Electricity Markets Authority (GEMA) to allocate licences on a competitive basis. Licences would be allocated to those operators that provide best value for money.[87] Further upfront

[83] For examples of countries providing subsidies and tax credits in order to address such income gaps, see Hickey et al (n 6) 5–6.

[84] Florian Kern, James Gaede, James Meadowcroft and Jim Watson, 'The political economy of carbon capture and storage: An analysis of two demonstration projects' (2016) *Technological Forecasting & Social Change* 102, 250–60, 257–8.

[85] Alex Lawson, 'Calls for Bigger Windfall Tax after Shell Makes "Obscene" $40 bn Profit', *The Guardian* (London, 2 February 2023) <www.theguardian.com/business/2023/feb/02/shell-profits-2022-surging-oil-prices-gas-ukraine>.

[86] BEIS (n 58) paras 9 and 25.

[87] Ibid.

financial incentives, to entice companies to enter the market for CO_2 transport and storage services, may be necessary in order to have more competitors. This will enable the allocation of licences via auction, which can enhance efficiency in the provision of transport and storage services. This, in turn, can reduce the costs of financial support. Sections 11(1) and (3) Energy Act 2023 also empower, initially the secretary of state for DESNZ, and then GEMA, to regulate the service price and provide revenue support.

Regulating prices and providing revenue support for CO_2 transport and storage

Price regulation seeks to enable companies providing CO_2 transport and storage services to recover the costs of their expenditure on capital and operations over a set period of time, so that only 'necessary' and 'reasonable' costs are charged.[88] The details of this will be fleshed out through secondary legislation.[89] Section 57 of the Energy Act 2023 provides for further financial assistance for operators of CO_2 transport and storage networks through revenue-support contracts. The secretary of state for DESNZ can appoint a person to be a counterparty for a revenue-support contract with a licensed provider of CO_2 transport and storage services.[90] Through this contract, revenue-support payments, from users to operators of the CO_2 transport and storage network, will be administered. This will reduce the financial risk for CO_2 transport and storage companies that the operation of the network may be delayed in the case of a slow start to capturing CO_2 at industries and hydrogen-production facilities.[91]

Economic evidence about the effectiveness of price regulation and revenue support is limited since there is insufficient information to date to conduct a full cost-benefit analysis. It is not yet known what level of price GEMA will allow operators of CO_2 transport and storage networks to charge, or how much money will be spent on revenue support for these operators. The details of the network delivery are still being negotiated between the government and future transport and storage companies.[92] The benefits of the regime have also not been quantified yet. These include emission reductions achieved,[93] potentially reduced costs for energy-intensive businesses, and the protection of jobs, especially in the former industrial heartlands of the UK (including the North-East, North-West, and South Wales). Protecting

88 Ibid 2.
89 Ibid.
90 Section 59 (1) Energy Act 2023.
91 BEIS (n 58) para 15 b.
92 Ibid para 58.
93 Ibid para 84.

employment in these areas will contribute to the UK government's 'levelling up' policy. This is referred to by some as 'carbon justice' for regions that have become socio-economically deprived due to de-industrialization.[94] The government estimates that using CCUS in the oil and gas, chemicals, and other combustion-intensive industries will prevent job-loss for about 15 per cent of UK workers.[95] Moreover, the economic regulatory framework will deliver financial benefits for CO_2 emitters by protecting them from excessive monopolistic pricing by a limited number of CO_2 transport and storage companies.[96]

The evidence about the economics of CO_2 transport and storage services may be strengthened if other sources of CO_2 – beyond industry and hydrogen production – are deployed, such as bioenergy carbon capture and storage (BECCS) and direct air carbon capture and storage (DACCS).[97] BECCS involves absorption of CO_2 from the atmosphere through photosynthesis by plants. The resulting biomass is turned into biofuel. The CO_2 generated when biofuels are produced (or the biomass is burned) is captured and stored.[98] BECCS, which involves the transformation of land use through planting forests on a large scale which are then cut down for the production of biofuels, has been controversial since it is considered to pose a risk of negative carbon sinks, loss of biodiversity and potential loss of livelihoods for previous land users.[99] BECCS, however, can contribute significantly to decarbonizing transport. According to the Committee for Climate Change, biofuels could reduce the UK's consumption of traditional jet fuel by about 10 per cent by 2050.[100] Whether BECCS itself can be considered cost-effective in a given national context depends on a range of factors, such as

94 House of Commons, Business, Energy and Industrial Strategy Committee (n 8) Q. 1.

95 Ibid. A BECCS trial plant at the UK Drax power station is considered to support 6,000 jobs in the area (House of Commons, Business, Energy and Industrial Strategy Committee (n 8) Q. 212).

96 BEIS (n 58) 3.

97 Such further evidence is currently not contained in the UK government's RIA for CO_2 transport and storage services. S 67(7) Energy Act 2023 provides a legal power to provide revenue support for BECCS and DACCS.

98 Tiley (n 3) 11.

99 IPCC, 'Climate Change 2022: Impacts, Adaptation and Vulnerability', Contribution of Working Group II to the Sixth Assessment Report of the IPCC (CUP 2022) <www.ipcc.ch/report/ar6/wg2/> 163. Concerns about BECCs potentially infringing a right to life due to potential loss of land otherwise used for agricultural production have also been expressed by the Advisory Committee to the UN Human Rights Council (n 4) p 22, fn 6.

100 Climate Change Committee, 'The Sixth Carbon Budget: Aviation' <chrome-extension://efaidnbmnnnibpcajpcglclefindmkaj/https://www.theccc.org.uk/wp-content/uploads/2020/12/Sector-summary-Aviation.pdf>10, 21.

availability of land for biomass generation and the costs of BECCS compared to other available energy sources.

Limiting financial risks when CO_2 transport and storage networks stop operating

Chapter 4 of Part 1 of the Energy Act 2023 seeks to limit financial risks that arise when CO_2 transport and storage networks cease to operate. The proposed rules strengthen existing provisions for building up funds for the decommissioning of networks. A Special Administration Regime and Transfer Scheme will limit the risk of financial liability for the UK state and – in a yet unknown distribution – for investors and users of CO_2 transport and storage sites.

Existing rules on decommissioning[101] are strengthened in order to avoid insufficient, or too many, funds being accumulated. Prospective operators of CO_2 transport and storage networks will be required, through conditions in their licences, to set up a decommissioning fund to which network users will contribute, with the details of this regime to be fleshed out in secondary legislation.[102] Operators of transport and storage companies will be responsible for the investment of the funds in appropriate security trusts in order to maximize them.[103] The development of CO_2 transport and storage infrastructure has already been incentivized in the UK by the secretary of state heading DESNZ having powers to limit the decommissioning obligations for owners of oil and gas facilities in case they sell their installations for reuse in CCUS. The Energy Act 2023 provides powers to limit the taxpayer contribution to this reduction in expenditure on oil and gas assets.[104]

In addition, Part 1, Chapter 4, of the Energy Act 2023 seeks to reduce the costs of CCUS by limiting – for tax payers, investors and users of CO_2 transport and storage services – the risk of transport and storage companies becoming insolvent and ceasing operations. Under the Special Administration Regime and Transfer Scheme, the secretary of state for DESNZ has the power to appoint a special administrator to a transport and storage company in case of a risk of insolvency. The administrator would continue to run the company[105] or would transfer it to another company.[106] The administrator's overall objective would be to keep the CO_2 transport and storage network

[101] Storage of Carbon Dioxide (Licensing etc.) Regulations 2010/2221 as amended, regs 13 and 14; and Storage of Carbon Dioxide (Termination of Licences) Regulations 2011/1483 as amended.

[102] BEIS (n 58) para 29.

[103] Ibid paras 43 and 79.

[104] Section 299 Energy Act 2023.

[105] BEIS (n 58) para 30.

[106] Section 43(3) and (4) Energy Act 2023.

running.[107] In contrast to other insolvency regimes, liquidation of the CO_2 transport and storage network should be avoided, as far as possible, which may limit protection of creditors.[108]

12.3.2.2 Revenue support for CO_2 capture from industry and hydrogen production

UK state intervention, justified on the basis of market failure and income gaps, also extends to revenue support for capturing CO_2 from industrial production, such as steel and chemicals production, waste incineration and fossil-fuel power plants,[109] and hydrogen production.[110] All these are considered of strategic importance to the UK economy.[111] Revenue support not only provides a financial subsidy for operators of CO_2 capture plants, but should also stimulate demand for CO_2 transport and storage services, thereby rendering the latter economically more viable.

Sections 65 and 67 of the Energy Act 2023 provide powers for the secretary of state for DESNZ to designate a counterparty, such as the Low Carbon Contracts Company,[112] to enter into private-law contracts with hydrogen producers or commercial or industrial facilities capturing their CO_2 emissions. The counterparty will administer the payment of the financial support.[113] Similar to the provision of revenue support for CO_2 transport and storage companies, these contracts will be negotiated initially between the secretary of state for DESNZ and the particular industrial or hydrogen producer, with reference to a set of criteria. Once markets in CO_2 capture have matured, revenue support can be allocated on a more competitive basis, for example, through auctions.[114] This builds on the 'contracts for difference' auctions for low-carbon-electricity generation in the United Kingdom under Chapter 2 of the Energy Act 2013. These contracts have contributed to reducing costs of electricity generated from UK offshore windfarms.[115]

For low-carbon hydrogen production, revenue support will cover the difference between a so-called *strike price*, which reflects the cost of producing hydrogen, and a *reference price*, which is the market value

---

107 Section 43(1) Energy Act 2023. BEIS (n 58) para 44.

108 Ibid.

109 Revenue support can also be extended to capturing CO_2 from the air and CO_2 dissolved in sea water. Section 67(7) Energy Act 2023.

110 Section 65 Energy Act 2023.

111 BEIS (n 58) para 92.

112 <www.lowcarboncontracts.uk/>.

113 BEIS (n 59) para 15.

114 Section 76(5)(a) Energy Act 2023.

115 BEIS (n 59) para 17.

of hydrogen. If demand for hydrogen falls, the strike price will be higher compared to the reference price. The revenue support will thus compensate for low market demand.[116] This regime seeks to reduce the costs of companies starting out in low-carbon hydrogen production and may develop markets for this new energy source. If paid over longer periods, without auctioning revenue support to the most efficient low-carbon hydrogen producers, revenue support may create a moral hazard of inefficient hydrogen producers passing some of their production costs on to third parties, such as tax payers or other energy companies. Section 70 of the Energy Act 2023 makes further provision for financial support for hydrogen production, transport and storage through levy payments to a low-carbon hydrogen administrator. The levy will be paid by gas shippers.[117] Its purpose is to finance payments under revenue-support contracts at a later stage.[118] The section now clarifies that the hydrogen levy will not be financed by gas or electricity suppliers nor by domestic energy bill payers. The then UK government had to make this change in order to ensure that its own Conservative Party back-benchers and the Labour Party's members of parliament would support the passage of the Energy Bill 2022 through parliament.[119]

It remains to be seen what exactly the costs and the quantified benefits of the proposed economic regulatory framework in Parts 1 and 2 of the Energy Act 2023 will be. These costs and benefits also raise important *distributional* issues. Who in the CCUS commercial chain – energy-supply companies, users of CO_2 transport and storage companies, operators of networks, or business and domestic energy bill payers – will contribute what, if any, proportion of the financial revenue support for carbon capture at low-carbon hydrogen-production facilities and at industrial plants? What, if any, contribution will be made by the public, that is, general tax payers? The primary legislation of the Energy Act 2023 does not clearly determine how much of the cost of the state intervention through revenue-support contracts for CCUS operators will be borne by each of these different potential contributors.[120] Hence, parliamentary approval of these provisions confers limited democratic legitimacy. In addition, legal rules for steering the distribution of the costs of financing

[116] Ibid para 23.

[117] Section 70(8) Energy Act 2023.

[118] BEIS (n 59) para 14.

[119] Miranda Bryant, 'Shapps Signals U-turn on Planned Hydrogen Levy for Energy Bills', *The Guardian* (24 June 2023) <www.theguardian.com/politics/2023/jun/24/shapps-signals-u-turn-planned-hydrogen-levy-energy-Bills>.

[120] BEIS (n 58) para 78.

the revenue support for CCUS are limited. Revenue-support contracts are negotiated private-law contracts. Funding out of general taxation for revenue support[121] will be authorized by existing tax legislation. Other funding provided by commercial entities in the CCUS chain will have to comply with the polluter pays[122] and proportionality principles, so that potential state interference with Article 1 to the First Protocol of the European Convention on Human Rights private property right is lawful. The amount of finance to be provided by a commercial entity through a revenue-support contract for CCUS will also have to be necessary and reasonable, and not excessive,[123] for achieving the specific ends of the commercial viability of the carbon capture at a low-carbon hydrogen-production or industrial plant.

Moreover, revenue support has to satisfy general criteria, such as whether 'value for money' has been achieved in spending taxpayers' money. Given the limits to quantified cost-benefit analysis of CCUS, this can be difficult to determine. Hence, public-sector costs are often only known with hindsight, for example, when audits are carried out in the UK by the NAO[124] or when Parliamentary Committee reports are compiled, such as those of the Public Accounts Committee.[125]

Having considered the main legal provisions in the Energy Act 2023 through which the UK government seeks to provide economic incentives for commercial-scale application of CCUS, and thus to improve the economic evidence for CCUS, it is necessary to further analyse how the rules may achieve this by being 'boundary objects'.

[121] BEIS (n 72) 19.

[122] Environment Act 2021 Section 17(5)(e).

[123] Philip Baker, 'Taxation and the European Convention on Human Rights', British Tax Review (2000) <chrome-extension://efaidnbmnnnibpcajpcglclefindmkaj/https://www.fieldtax.com/wp-content/uploads/2015/04/BTR-2000-4-211-377-Taxation-and-the-European-Convention-on-Human-Rights.pdf>.

[124] For example, National Audit Office, 'Delivering Major Projects in Government: A Briefing for the Committee of Public Accounts' (2016) <www.nao.org.uk/briefings/delivering-major-projects-in-government-a-briefing-for-the-committee-of-public-accounts/>.

[125] UK Public Accounts Committee, 'Carbon Capture and Storage Inquiry' (2017) www.parliament.uk/business/committees/committees-a-z/commons-select/publicaccountscommittee/inquiries/parliament-2015/carbon-capture-storage-16-17/. The RIA for the economic regulatory framework for the UK CO_2 transport and storage network refers to 'impact evaluations' and 'value for money' assessments of this framework once it has been applied (BEIS (n 58) para 108 c). That includes an assessment of whether the fees charged for network users (that is, CO_2 emitters) for decommissioning are not too high.

12.4 Parts 1 and 2 of the Energy Act 2023: boundary objects?

Parts 1 and 2 of the Energy Act 2023 seek to strengthen economic evidence for CCUS by reducing its costs and balancing financial risks between those who capture CO_2 and those who transport and store it. The legal provisions bring into dialogue social science economic evidence with politics, that is, the UK's policy choice for CCUS. The legal rules facilitate this by being boundary objects that – as discussed in Section 12.2.3 – provide interpretative flexibility, a coordination function and links between the different meanings that the legal rules can have and their specific local applications.

12.4.1 Interpretative flexibility

Parts 1 and 2 of the Energy Act 2023 provide for interpretative flexibility in relation to the politically contentious issue of how the costs for financial support of CCUS in the UK should be distributed between tax payers, energy firms and business and domestic energy consumers.[126] This interpretative flexibility also arises from the complexity of legal rules for energy markets. How the rules for CCUS will interact with other rules of the Energy Act 2023 will affect whether financing support for CCUS will lead to higher bills for business and domestic energy consumers. For instance, sections 203 and 215 of the Energy Act 2023 seek to promote greater competition between energy network companies and higher uptake of smart meters among energy consumers. Smart meters enable shifting energy use to times of the day when energy is cheaper, for example, from renewables.[127] More competition and smart meters may reduce customer bills. This may offset any increase in bills that may arise from contributing to revenue support for CCUS.

Finally, interpretative flexibility seems to be a structural and thus inevitable feature of Parts 1 and 2, since the meaning of these rules depends on their interaction with uncertain carbon market dynamics. The price of carbon will have an impact on the cost of CCUS. What that price will be is uncertain, because this depends on how the new post-Brexit UK Emissions Trading Scheme is working. Some industrial plants are provided with free

---

[126] See, for instance, the speeches of Baroness Sheehan and Lord Moylan during the Second Reading of the Energy Bill 2022, and Lord West of Spithead and Viscount Trenchard during the Committee Stage in the House of Lords <https://Bills.parliament.uk/Bills/3311/publications>.

[127] BEIS, 'Energy Bill Summary Impact Assessment' (2022) <chrome-extension://efaidnbmnnnibpcajpcglclefindmkaj/https://Bills.parliament.uk/publications/47255/documents/2> 23.

emission allowances. If these plants use CCUS[128] they will have surplus emission allowances which they can sell in the UK carbon market, thereby potentially lowering the price for carbon.[129] A low price for carbon may lower commercial demand for CCUS. The price for carbon will also depend on how much the UK will charge carbon-intensive products being imported into its markets through a new carbon border adjustment mechanism.[130] Hence, the rules for CCUS are a boundary object since they sit at the interface between specific legal prescriptions and carbon market dynamics which are in flux.[131]

12.4.2 Linking unstructured features to locally tailored uses

A further characteristic of a boundary object is that it creates a link between its more unstructured features and its locally tailored uses.[132] In the case of legal rules, this is a link between their open-textured language and their local implementation in specific instances. For instance, the primary legislation – Parts 1 and 2 of the Energy Act 2023 – provide a procedural mechanism for revenue support for CCUS. This is couched in open-textured language. The local application of these rules occurs when the secretary of state or an allocation body provides specific sums of revenue support to particular CO_2-capture installations and operators of CO_2 transport and storage services. Hence, there is a link between the legal rules in Chapter 1 of Part 2 of the Energy Act 2023 that set out the *overarching procedure* for providing revenue support for CCUS and the *local, specific expenditure decisions*, that are only taken subsequently by politicians and regulators in relation to specific CCUS projects. This particular link, that is, leaving specific expenditure decisions for later, limited the risk of insufficient political support for the Bill during the law-making process. A constant 'forth and back' over time between the procedural provision,[133] which is largely unstructured, and its local application, that is, a specific expenditure decision, is a further feature of the boundary object. It can lead to the standardization of legal decision-making by clarifying what are routine and what are exceptional cases.[134]

[128] BEIS (n 72).

[129] A risk recognized as existing also on an international scale in the Report of the UN Human Rights Council Advisory Committee (n 4) Annex, para 5.

[130] DESNZ, 'Factsheet: UK Carbon Border Adjustment Mechanism' (2023) <www.gov.uk/government/consultations/addressing-carbon-leakage-risk-to-support-decarbonisation/outcome/factsheet-uk-carbon-border-adjustment-mechanism>.

[131] Star (n 47) 604.

[132] Ibid 614.

[133] Such as Energy Act 2023 s 57.

[134] Star (n 47) 613–14.

Finally, Parts 1 and 2 of the Energy Act 2023 are boundary objects because they coordinate various actors in the absence of agreement between them.

12.4.3 Coordination in the absence of agreement between actors

Various sections in Parts 1 and 2 of the Energy Act 2023 promote coordination between actors in the absence of agreement. For instance, Section 99 of the Act coordinates policy makers and independent economic regulators, such as GEMA, who may not necessarily agree on the details of the implementation of the economic regulatory framework. The section suggests that the secretary of state will establish the strategic priorities and policy outcomes for CO_2 transport and storage, while the economic regulator will take day-to-day decisions with reference to its statutory duties.[135] Here the legal rules provide a procedural mechanism for coordinating politicians and economic experts, who may disagree. Such disagreement may be promoted by some overlap in practice in the decision-making of politicians, such as the secretary of state, and an economic regulator, operating at arm's length from the government. There is no clear-cut distinction between making policy and assessing uncertain economic evidence.

Finally, Parts 1 and 2 of the Energy Act 2023 coordinate different approaches to reducing concentrations of CO_2 in the atmosphere, in the absence of consensus on how CO_2 is best reduced. On the one hand, CO_2 can be cut through stringent statutory emission limits that curb how much CO_2 can be emitted by industrial operators (through conditions in licences), such as power plants or cement, steel, glass, plastics, and pulp and paper production. Stringent emission limits can be technology forcing.[136] They can incentivize more *efficient production and combustion processes* that *prevent* CO_2 from being generated. On the other hand, CO_2 emissions can be reduced by capturing them 'end of pipe' – whatever amount of CO_2 an installation generates. A competitive market in CCS, with reduced costs for capture, transport and storage, may create economic incentives for not preventing CO_2 emissions. Hence, 'end of pipe' CCS may undermine regulators' efforts to promote innovative and efficient production and combustion techniques, for example in the industrial and power sector.[137] For instance, post-combustion CCUS for power plants is associated with some loss of

---

[135] BEIS (n 58) para 26.

[136] Andries Nentjes, Frans P. de Vries and Doede Wiersma, 'Technology-forcing through environmental regulation' (2007) *European Journal of Political Economy* 23(4), 903–16.

[137] The risk of CCUS undermining efforts to prevent CO_2 emissions in the first instance from arising in industrial production or in power generation has also been recognized by the Report of the UN Human Rights Council Advisory Committee (n 4) para 4.

efficiency in energy generation.[138] By avoiding directly addressing this potential tension between different approaches to reducing CO_2 emissions, the provisions of the Energy Bill 2022 provided a tacit coordination function.

12.5 CCUS as limiting radical transformation of energy and production systems?

Parts 1 and 2 of the Energy Act 2023 link norms, that is, the UK government's policy commitment to CCUS, in distinct ways to facts, such as reduced costs for investors in CCUS in order to promote its commercial-scale application. But there is nothing inevitable about the distinct link between norms and facts that the provisions of the Energy Act 2023 create. In fact, this co-production of norms and facts has been criticized by some for holding back a more radical transformation of energy and production *systems* at speed and scale in order to reduce long-term concentrations of CO_2 in the atmosphere. This argument against state intervention in private business activity is based on three claims. The first suggests that CCUS not only 'locks in' fossil fuels in energy generation, but also *unlocks* new fossil fuels, such as tar oil.

12.5.1 Does CCUS promote fossil fuel 'lock-in' and 'unlocking'?

Deploying CCUS at commercial scale may reduce political pressure to phase out energy generation from fossil fuels.[139] One of the oldest and so far commercially best established CCUS techniques is enhanced oil recovery (EOR). It has been used since the 1970s,[140] in particular in the US, to increase the amount of oil that is pumped out of a field. It involves injection of CO_2 into a field in order to increase pressure within it. This helps to pump out residual oil. The CO_2 injection can be repeated several times, and thus is a form of CO_2 reuse.[141] It can be applied to both oil and gas fields. The CO_2

138 Environment Agency, 'Guidance: Post-Combustion Carbon Dioxide Capture: Best Available Techniques (BAT)' (2021), last updated November 2022 <www.gov.uk/guidance/post-combustion-carbon-dioxide-capture-best-available-techniques-bat> para 3.1.

139 UN Human Rights Council, 'The Toxic Impacts of Some Proposed Climate Change Solutions', A/HRC/54/25, para 70. Stephen Melzer, 'Carbon Dioxide Enhanced Oil Recovery (CO_2 EOR): Factors involved in adding Carbon Capture, Utilization and Storage (CCUS) to Enhanced Oil Recovery' (2012) <chrome-extension://efaidnbmnnnibpcajpcglclefindmkaj/https://carboncapturecoalition.org/wp-content/uploads/2018/01/Melzer_CO2EOR_CCUS_Feb2012.pdf> 2.

140 Stuart Haszeldine and Navraj Singh Ghaleigh, 'Geological Factors for Legislation to Enable and Regulate Storage of Carbon Dioxide in the Deep Surface', in Ian Havercroft, Richard Macrory and Richard Stewart (eds) *Carbon Capture and Storage: Emerging Legal and Regulatory Issues* (Bloomsbury 2018), 5–31, 23.

141 Melzer (n 139) 11.

for EOR or enhanced gas recovery (EGR) can come from different sources. It can be generated by separating it from the natural gas that is pumped out of a gas field.[142] Or it can be CO_2 that has been captured from industrial emitters or hydrogen producers and is then transported to the gas or oil field. The CO_2 used for EOR or EGR will finally be stored and contained within the field.[143] EOR is costly and may be cost-effective only when the price of oil is high.[144] The cost of EOR and EGR can be reduced by allocating carbon credits for the final storage of the CO_2 in the gas or oil field. Such a subsidy may prolong reliance on fossil fuels and limit pressure to switch to renewable-energy sources, which have become cheaper.[145] Some estimate that 'lock-in' of coal to electricity can last for about 40–100 years.[146] The capital investment costs for power generation with CCS are higher than for those without. Hence, plants with CCS need to be running for a significant amount of time to recoup these capital costs.[147] 'Locked-in' CO_2 emissions will be even higher if the energy needed for operating CCS itself is generated from burning fossil fuels[148] and it should be borne in mind that CCS at power plants does not remove CO_2 (let alone methane emissions) that arise when fossil fuels are extracted and transported.[149]

While major environmental NGOs oppose this potential fossil fuel lock-in of CCUS,[150] the UK government justifies it as providing security of energy supply: energy supply from 'traditional flexible carbon-intensive electricity generators' should not be priced out of the market, because they can address the problem of intermittency of renewables. If the wind does not blow and the sun does not shine, renewables may not meet short-term and unexpected high demand. Such disruption to energy supply would lead to loss of output, with detrimental impacts for both consumers and businesses.[151] Without quantified economic evidence for this view, it appears that security of energy supply may be an objective pursued at very significant costs, including to the interests of national security.

---

142 Ibid 4.

143 Ibid 8.

144 Ibid 5.

145 Haszeldine (n 7) 564.

146 Praetorius and Schumacher (n 2) 5091.

147 Ibid 5090.

148 CIEL, 'Confronting the Myth of Carbon-Free Fossil Fuels' <www.ciel.org/reports/carbon-capture-is-not-a-climate-solution/#:~:text=real%20climate%20solutions.-,Confronting%20the%20Myth%20of%20Carbon-Free%20Fossil%20Fuels%3A%20Why%20Carbon,industry%20with%20a%20license%20to> 7.

149 BEIS (n 10) 32.

150 Such as Greenpeace and Global Witness and Friends of the Earth.

151 BEIS (n 58) para 14 a.

CCUS is also being used for 'unlocking' new oil resources. For instance, CO_2 is injected into oil sands in Alberta, Canada, in order to push oil out of the sands. Revenue from these oil exports provides about a third of Alberta's total gross domestic product.[152] A second argument against the kind of state intervention for CCUS enabled by the UK Energy Act 2023 is that it may undermine a circular-economy approach to industrial production.

12.5.2 Limiting the circular economy and democratically controlled renewable-energy projects?

CCS technologies, and especially CO_2 transport and storage, may increase demand for CO_2 including CO_2 captured from industrial production or waste incineration. Profitable CO_2 transport and storage businesses may thus limit incentives for more fundamental and long-term decarbonization strategies, such as reuse and recycling of industrial products and waste prevention as part of degrowth strategies[153] or a 'circular economy'.[154] For instance, cement kilns generate significant CO_2 emissions which can be captured through CCUS. But a circular-economy approach may be cheaper, and also save other resources used for cement production, such as water and iron ore. The European Commission's Circular Economy Action Plan identifies construction and buildings as a target for increased recycling and reuse of materials. This will be economically incentivized through legal rules requiring a minimum of recycled content in building products and quality standards for such products. This should promote consumer confidence in these goods and thus develop markets for them.[155] A third claim suggests that by promoting climate-mitigation technologies, often deployed by large corporate actors, the development of small-scale democratically controlled renewable-energy community projects may be impeded.

12.5.3 Limiting the development of small-scale democratically controlled renewable-energy community projects?

CCUS is a technology linked to large corporations operating fossil-fuel power plants and industrial installations. This may be at odds with promoting

[152] Florian Kern et al (n 84) 257.

[153] IPCC, 'Climate Change 2022: Impacts, Adaptation and Vulnerability', Contribution of Working Group II to the Sixth Assessment Report of the IPCC (CUP 2022) <www.ipcc.ch/report/ar6/wg2/> 173.

[154] Report of the Human Rights Council Advisory Committee (n 4), para 12.

[155] European Commission, Communication from the Commission to the European Parliament, the Council, the European Economic and Social Committee and the Committee of the Regions, 'A New Circular Economy Action Plan for a Cleaner and More Competitive Europe' (COM (2020) 98 final, 11, 14).

small-scale, democratically controlled community projects that generate renewable energy. A significant expansion of local community renewable-energy projects may limit demand for large-scale gas-fuelled power plants with CCS. Some large-scale power plants, however, would still be necessary, because generation from renewable-energy sources can be intermittent. With less demand for large-scale generated power, the cost of subsidizing such plants and CCUS may increase.[156]

Whether this potential tension between large-scale power plants with CCUS and community renewable-energy projects will arise will also depend on how the Independent System Operator and Planner (ISOP) for energy – to be established through Section 161 of the Energy Act 2023 – will use its powers. The ISOP has to promote the net-zero,[157] 'security of supply'[158] and 'the efficiency and economy' objectives.[159] In carrying out its functions, the ISOP should consider the energy *system* as a whole for the UK in an integrated manner.[160] This could include promotion of both CCUS, as long as this meets the 'efficiency and economy' objective, *and* small renewable-energy community projects. The latter may contribute to promoting resilience and local security of supply by diversifying energy generation.

12.6 Conclusion: Legal rules as boundary objects mitigating evidence deficits about the costs of CCUS?

This chapter has examined a key debate about climate law that seeks to rely on climate-mitigation technologies to achieve targets for reducing greenhouse gas emissions. The debate addresses to what extent and how the state should intervene in private business activity in order to reduce the costs of CCUS and thus enable its deployment at a commercial scale. This debate is not just informed by normative preferences about the right balance between state-market interactions for tackling the climate crisis, but also by the well-established idea that climate law should be evidence-based.[161]

156 See, for instance, the speech of Baroness Sheehan during the Second Reading of the Energy Bill 2022 in the House of Lords https://Bills.parliament.uk/Bills/3311/publications. Victoria Johnson and Stephen Hall, 'Community energy and equity: The distributional implications of a transition to a decentralised electricity system' (2014) *People, Place and Policy* 8(3), 149–67, 149, 157.

157 Energy Act 2023, s 163(1)(a).

158 Energy Act 2023, s 163(1)(b).

159 Energy Act 2023, s 163(1)(c).

160 Energy Act 2023, s 164(1)(c) requires the system operator to consider 'the whole-system impact of a relevant activity' when carrying out its functions.

161 Dan Tarlock, 'Is there a there there in environmental law?' (2004) *Journal of Land Use and Environmental Law* 213, 239–40.

Given that the costs of CCUS can limit support for this climate-mitigation technology, the chapter focused on evidence about the economics of CCUS. It has analysed how the economic regulatory framework of Parts 1 and 2 of the UK Energy Act 2023 seek to construct economic evidence for CCUS, in particular through price regulation and revenue support. The chapter has argued that the legal rules that create this economic regulatory framework are 'boundary objects' that bridge the UK government's policy commitment to CCUS with the science – here, economics, a branch of social science which generates evidence about different public policy choices, including the costs of CCUS. The distinct, but not inevitable link that Parts 1 and 2 of the Energy Act 2023 create between 'norms' and 'facts' can transform what different actors consider as the economic evidence for CCUS. The legal provisions of the Energy Act 2023 also create what appears to be an unquestioned transition from natural-science evidence for CO_2 reduction targets to economic evidence for particular CO_2 mitigation technologies. By acting as a boundary object, Parts 1 and 2 seek to make CCUS – a controversial climate-mitigation technology – more acceptable. Rendering transparent how legal rules as boundary objects construct economic evidence for climate-mitigation technologies should thus contribute to debates and further research about evidence-based climate law.

13

Nature-Based Climate Solutions and the Rights of Nature: A Conundrum of Values

Felicity Deane and Justine Bell-James

13.1 Introduction

This chapter explores nature-based climate solutions (NBCS). While nature functions as a balanced ecosystem independent of human intervention, the ecosystem can be enhanced or manipulated to deliver economically valuable 'services' over time with human input. Active nurturing of particular ecosystem functions – including carbon sequestration – is the premise of NBCS. While NBCS is a critical technology that can be employed in the mitigation of climate change, it is also a double-edged sword, for human intervention designed specifically to support carbon-sequestration services may have negative ramifications, such as inadvertent impacts on biodiversity values.[1] Anthropocentric approaches to environmental problems, such as NBCS, are therefore not universally popular. The need to protect biodiversity in climate change responses is powerfully expressed in the literature.[2]

We will proceed on the basis that NBCS is a promising approach to addressing the climate crisis provided that biodiversity is front and centre in deliberations. In this sense NBCS is a compromise and a midway point between ecocentric, nature-focused approaches (for example, rights of

[1] 'The eco–climate nexus', (2022) 12(7) *Nature Climate Change* 12, 595–95.

[2] Brett R. Scheffers and Gretta Pecl, 'Persecuting, protecting or ignoring biodiversity under climate change' (2019) 9(8) *Nature Climate Change* 581–6; Nathalie Pettorelli et al, 'Time to integrate global climate change and biodiversity science-policy agendas' (2021) 58(11) *Journal of Applied Ecology* 2384–93.

nature), and anthropocentric approaches that may mitigate climate change and prioritize the survival of humanity at all cost.[3] While these ecocentric versus anthropocentric – or intrinsic versus instrumental – conceptions of nature are often viewed as inherently irreconcilable,[4] we argue that an NBCS approach can draw upon aspects of the Rights of Nature (RoN) paradigm through exploring the competing concerns of the 'use' of nature by humans, as a tool to mitigate climate change, against the intrinsic value of the environment and the needs of non-human species – all of this while retaining control of climate change.

Pursuing this middle ground leads to a conundrum: can NBCS only be available where these solutions align with the natural functions and services of the existing ecosystem? In order to support good outcomes for both climate and biodiversity, it will be necessary to evaluate the impacts of particular actions in a way that leads to an understanding of the impacts of human intervention. Hence, we argue that legal intervention to encourage or permit landholders to develop NBCS projects should be accompanied by a requirement to measure the short- and long-term impacts of intervention. Further, where those measurements indicate that actions impact adversely on critical ecosystem functions, proponents must be required to change their approach and engage in restoration where needed.

In Section 13.2 we explain the connection between NBCS, natural capital (NC), and ecosystem services. We include discussion about the importance of measuring ecosystem impacts where NBCS is employed. NC accounting may provide a method for doing this. Section 13.3 addresses some of the theoretical objections to concepts like NBCS. In this part, we consider the RoN paradigm. Section 13.4 aims to determine if and where there is a balance to be struck. In this regard, the section identifies that RoN and the ecosystem services paradigm are not necessarily mutually exclusive. We conclude that employing a measurement approach can and should do nothing to devalue the intrinsic value of nature, while allowing NBCS to be used as a tool to confront the climate crisis.

13.2 NBCS, NC and ecosystem services

The uptake of mitigation technology must increase dramatically if we are to keep global temperature increases to well below 2 °C of warming.

3 Nathalie Pettorelli et al, 'Time to integrate global climate change and biodiversity science-policy agendas' (2021) 58(11) *Journal of Applied Ecology* 2384.

4 Justine Bell-James, 'Ecosystem services as a metaphor in environmental law: Balancing intrinsic and instrumental values' (2020) 39(3) *University of Queensland Law Journal* 525.

NBCS is one such tool. NBCS is encapsulated under the broader concept of 'nature-based solutions' or NbS (and we refer to both throughout this chapter). They are an essential technology,[5] with the ability to provide up to 30 per cent of the amount of mitigation necessary by 2030 to alleviate climate change and restrict global warming to below 2 °C.[6] According to the Intergovernmental Panel on Climate Change (IPCC), 'NbS provide adaptation and mitigation benefits in ways that support wild species and habitats, often contributing to other sustainable development goals'.[7] NBCS emerged as a concept in 2008,[8] and has been the subject of further attempts at conceptual clarification ever since. It builds upon other anthropocentric concepts of nature, including ecosystem services and NC.[9] The European Commission defines NbS as a way to address societal challenges with 'solutions that are inspired and supported by nature, which are cost-effective, [and] simultaneously provide environmental, social and economic benefits and help build resilience'.[10] The International Union for Conservation of Nature and Natural Resources (IUCN) defines NbS as 'actions to protect, sustainably manage and restore natural or modified ecosystems, which address societal challenges … while simultaneously providing human well-being and biodiversity benefits'.[11]

---

5 NbS are a technology insofar as they require human ingenuity and action in order to be effective.

6 Nathalie Seddon et al, 'Understanding the value and limits of nature-based solutions to climate change and other global challenges' (2020) *Philosophical Transactions of the Royal Society B: Biological Sciences* 375(1794) 20190120.

7 Intergovernmental Panel On Climate Change (IPCC), 'Climate Change 2022 – Impacts, Adaptation and Vulnerability: Working Group II Contribution to the Sixth Assessment Report of the Intergovernmental Panel on Climate Change' (IPCC 2022) 303 <www.cambridge.org/core/product/identifier/9781009325844/type/book> accessed 2 September 2023.

8 Kathy MacKinnon, Claudia Sobrevila and Valerie Hickey, 'Biodiversity, Climate Change, and Adaptation: Nature-Based Solutions from the World Bank Portfolio' (World Bank Group, Washington, 2008) <https://documents.worldbank.org/en/publication/documents-reports/documentdetail/149141468320661795/biodiversity-climate-change-and-adaptation-nature-based-solutions-from-the-world-bank-portfolio> accessed 12 February 2023.

9 Carsten Nesshöver et al, 'The science, policy and practice of nature-based solutions: An interdisciplinary perspective' (2017) 579 *Science of The Total Environment* 1215.

10 European Commission, 'Communication from the Commission to the European Parliament, the Council, the European Economic and Social Committee and the Committee of The Regions: Next Steps for a Sustainable European Future – European Action for Sustainability' (Strasbourg, 22 November 2016) COM/2016/0739.

11 Emmanuelle Cohen-Shacham and others, 'Nature-Based Solutions to Address Global Societal Challenges' (International Union for Conservation of Nature, Gland, 2016) 6.

These definitions are high-level and broad,[12] which has meant that categorizations of the approaches that fall under the umbrella of NbS are sometimes contested. However, NbS as a category will generally include actions with any *natural* aspect, designed to enhance nature and increase 'natural features and processes' into urban environments particularly.[13] It follows that NbS will generally involve the protection, conservation, restoration, management or enhancement of natural ecosystems, with the purpose of improving an environmental (human caused) problem such as climate change.[14]

NBCS is a sub-category of NbS and has been heralded as an essential mechanism for mitigating and adapting to climate change.[15] NBCS is NbS deployed for climate change mitigation and adaptation purposes. Nature can provide 'a cost-effective way of [mitigating or] adapting to climate change whilst also protecting the ecosystem services on which we depend'.[16] The most common form of NBCS is the conservation of terrestrial forests or woodlands for carbon sequestration.[17] However, NBCS can support mitigation and adaptation goals for all countries. For example, restoration of soil organic carbon (SOC) 'represents 25% of the potential of natural climate solutions'.[18] Further, the restoration of shorelines can support waterway health and potentially remove risks associated with erosion and flooding in some communities.[19] These solutions can therefore serve many human purposes, but at the same time it is critical that the impact on the ecosystem is not narrowly evaluated.

The commitment to NBCS increased significantly during the Paris Agreement negotiations, when 66 per cent of nation state signatories committed to incorporate NBCS into their climate pledges and domestic responses to climate change.[20] NBCS also permeate the Paris Agreement

---

12 Barbara Sowińska-Świerkosz and Joan García, 'What are Nature-Based Solutions (NBS)? Setting core ideas for concept clarification' (2022) 2 *Nature-Based Solutions* 100009.

13 European Commission (n 9); Judy Bush and Andréanne Doyon, 'Building urban resilience with nature-based solutions: How can urban planning contribute?' (2019) 95 *Cities* 102483.

14 Shannon Osaka, Rob Bellamy and Noel Castree, 'Framing "nature-based" solutions to climate change' (2021) 12(5) *WIREs Climate Change* e729, 1.

15 Nathalie Seddon et al, 'Grounding nature-based climate solutions in sound biodiversity science' (2019) 9 *Nature Climate Change* 84; Bronson W Griscom et al, 'Natural climate solutions' (2017) 114 *Proceedings of the National Academy of Sciences* 11645. IPCC (n 7).

16 Seddon et al (n 6).

17 Ibid.

18 DA Bossio and others, 'The role of soil carbon in natural climate solutions' (2020) 3 *Nature Sustainability* 391, 391.

19 Nathalie Seddon et al, 'Getting the message right on nature-based solutions to climate change' (2021) 27(8) *Global Change Biology* 1518, 1519.

20 Nathalie Seddon et al (n 15) 84.

itself; for example, the preamble to the Agreement recognizes the 'importance of the conservation and enhancement … of sinks and reservoirs' of greenhouse gases and notes the need to ensure the integrity of all ecosystems.[21] During the Paris negotiations, the actions identified as critical predominately included conserving and enhancing forests as carbon sinks and increasing the resilience of ecological systems.[22] These actions leave much of the detail to the project developers themselves, and as a result there are concerns that these references in the Paris Agreement, could in fact have some perverse consequences.

The popularity of NBCS stems in part from the idea that it may be possible to address climate change through environmental interventions that may not require dramatic societal changes.[23] However, NBCS also require that natural resources are manipulated and *used*.[24] Indeed, the broader category of NbS focuses on the benefits that humans derive from nature (rather than considering the needs of nature independently from humans), and this builds upon the earlier concepts of ecosystem services and stocks of NC.[25] 'Ecosystem services' is a term used to identify the benefits that nature provides to humans,[26] related to the idea of nature as a capital asset, and includes the 'ecological characteristics, functions, or processes that directly or indirectly contribute to sustainable wellbeing'.[27] Certain functions are necessary for

---

21 Paris Agreement (entered into force 4 November 2016, opened for signature 22 April 2016) [2016] ATS 24, 1 (Preamble).

22 Himangana Gupta and Lokesh Chandra Dube, 'Benefits of Evaluating Ecosystem Services for Implementation of Nature-Based Solutions Under the Paris Agreement', in Mohamed Behnassi et al (eds), *Social-Ecological Systems (SES): From Risks and Insecurity to Viability and Resilience* (Springer International Publishing, 2021) 39.

23 Nathalie Seddon et al (n 19) 1519.

24 Nicolas Faivre et al, 'Nature-based solutions in the EU: Innovating with nature to address social, economic and environmental challenges' (2017) 159 *Environmental Research* 509, 510.

25 Carsten Nesshöver et al, 'The science, policy and practice of nature-based solutions: An interdisciplinary perspective' (2017) 579 *Science of The Total Environment* 1215.

26 Millennium Ecosystem Assessment, *Ecosystems and Human Well-Being: Synthesis* (Island Press, 2005) 40.

27 Robert Costanza, 'Valuing natural capital and ecosystem services toward the goals of efficiency, fairness, and sustainability' (2020) 43 *Ecosystem Services* 101096, 1. This aligns with the System of Environmental-Economic Accounting (SEEA), which defined ecosystem services as: 'the contributions of ecosystems to benefits used in economic and other human activity': United Nations et al, *System of Environmental-Economic Accounting 2012: Central Framework* (United Nations, February 2014) 2.2. See also the definition of ecosystem services as the conditions and processes from natural ecosystems which help to 'sustain and fulfil human life' in Gretchen C Daily, 'Introduction: What Are Ecosystem Services?' in Gretchen C Daily (ed), *Nature's Services: Societal Dependence on Natural Ecosystems* (Island Press, 1997), 1–10. Daily has made the point elsewhere that human beings depend 'utterly on the continuation of natural cycles for their very existence'.

basic human rights and the production of goods or benefits, such as 'food, drinking water, fibre, building and manufacturing materials', and may or may not require human intervention.[28] Ecological functions and processes also include 'carbon storage, air and water filtration, pollination, protection from storms and floods, and places for rest and recreation'.[29] Escobedo and colleagues conceptualize ecosystem services as the components of ecosystems that are directly enjoyed, consumed or used to produce measurable human benefits.[30] However, the long-term derivation of benefits is dependent on responsible use.

NbS also relate to NC, a phrase and a paradigm that is connected with the concept of ecosystem services and that was first defined over 100 years ago in reference to renewable and non-renewable 'natural assets'.[31] The definition of NC has since expanded and is now considered to include the intrinsic value that nature provides.[32] Over time, ecological economists have recognized that natural assets could be valued in a way similar to the valuation of other stocks of capital assets. Contemporary economists acknowledge three categories of capital assets: produced capital, human capital (education and health) and NC.[33] There are critical differences between each category, however NC (on which humans can have an impact) is not dependent on human production. NC is therefore broadly defined as the living and non-living components of ecosystems on which humans rely for goods and services, including biodiversity.[34] While it is increasingly acknowledged that NC and the services it provides (ecosystem services) are essential to human existence and human rights to food, water and shelter, it is typically unpriced or undervalued.[35]

See Gretchen C Daily, 'What Are Ecosystem Services?' in David E Lorey (ed), *Global Environmental Challenges of the Twenty-First Century: Resources, Consumption, and Sustainable Solutions* (SR Books, 2003) 227, 229.

28 *Australia's Strategy for Nature 2019–2030* (Commonwealth of Australia, 2019) 5.

29 Ibid.

30 Francisco J Escobedo, Timm Kroeger and John E Wagner, 'Urban forests and pollution mitigation: Analyzing ecosystem services and disservices' (2011) 159(8–9) *Environmental Pollution* 2078, 2080.

31 Carter Brandon et al, 'Integrating natural capital into national accounts: Three decades of promise and challenge' (2021) 15(1) *Review of Environmental Economics and Policy* 134, 135.

32 Ibid.

33 Partha Dasgupta, 'The Economics of Biodiversity: The Dasgupta Review' (HM Treasury, London, 2021) 38. See also the discussion of other forms of capital (financial, intellectual, social, and so on) in John B Ruhl, Steven E Kraft and Christopher L Lant, *The Law and Policy of Ecosystem Services* (Island Press, 2007) 41.

34 Stephen Polasky and Gretchen Daily, 'An introduction to the economics of natural capital' (2021) 15(1) *Review of Environmental Economics and Policy* 87, 88.

35 Robert Costanza et al, 'The value of the world's ecosystem services and natural capital' (1998) 25(1) *Ecological Economics* 3.

An assumption that often underlies approaches involving NbS, conserving NC and supporting sustainability of ecosystem services, is that co-benefits will necessarily flow. Indeed, NC is intricately connected with long-term productivity and effective land-management strategies. As noted by Brady and colleagues:

> Soil biodiversity underpins agricultural productivity through interactions that generate functions and ultimately supporting ecosystem services. … To the extent that soil biodiversity benefits farmers and these benefits or supporting ecosystem *services* are recurring then soil biodiversity is, from an economic perspective, equivalent to other assets (e.g., breeding livestock), and should be valued and managed from a long-term perspective.[36]

However, this should be taken subject to two caveats. First, win/win scenarios are not guaranteed in these approaches, and it may be the case that an NbS intervention aimed at restoring or enhancing delivery of a particular ecological function operates to the detriment of another. For example, restoring a forest purely for carbon sequestration may result in a monoculture plantation, with no benefits for biodiversity beyond climate change mitigation.[37] The second caveat to supporting an NC approach is the risk of socio-ecological trade-offs. In particular, implementation of an approach of payment for ecosystem services can result in financial benefits flowing to some community members only, to the exclusion of others.[38]

NC accounting has been proffered as a means to achieve better alignment in local, state and national policies on biodiversity.[39] NC accounting refers to 'reporting systematically on NC stocks and flows'.[40] As the name suggests, NC accounting borrows from traditional accounting methods to calculate the total stocks and flows of natural resources and their associated services in a given ecosystem. Although the use of NC accounting to inform and evaluate policy has been limited to date, the 'mainstreaming' of NC accounting is one way policy makers and corporate entities have sought to

[36] Mark V Brady et al, 'Valuing supporting soil ecosystem services in agriculture: A natural capital approach' (2015) 107(5) *Agronomy Journal* 1809.

[37] Anastasia Telesetsky, 'Ecoscapes: The future of place-based ecological restoration laws' (2013) 14(4) *Vermont Journal of Environmental Law* 493.

[38] See, for example, Amrit Melissa Dencer-Brown et al, 'Integrating blue: How do we make Nationally Determined Contributions work for both blue carbon and local coastal communities?' (2022) 51(9) *Ambio* 1978.

[39] Brandon et al (n 31) 134.

[40] Ibid. 135.

incorporate the environment and ecosystem condition into their decision-making.[41] According to the Australian CSIRO, 'a guiding principle of NC accounting is that the valuation of nature can help better conserve it'.[42] In this respect, Rujis and colleagues argue that NC accounting 'can help to answer a broad range of natural resources and environment policy questions during all phases of the policy cycle'.[43]

To explain the basics of these concepts, in NC accounting, NC stocks refer to 'physical accounts' which aim to capture the extent and condition of the ecosystem.[44] These stocks then result in a 'flow' of services (referred to as ecosystem services).[45] From a monetary accounts' perspective, the flow (supply and use) of services arising from the ecosystem can be translated into a monetary figure which amounts to the net present value of the future services. However, NC accounting may become ethically problematic when we consider the ease of quantifying some assets and services in financial terms vis-à-vis others. Natural assets which are used to produce goods, such as fossil fuels, minerals, agricultural land and timber, more readily lend themselves to valuation in monetary terms because 'standard principles of productive asset valuation can be used'.[46] Indeed, many of these components of NC have existing market values.[47] However, there are many forms of NC for which there is little data as it has not yet been collected systematically,[48] and others which are difficult to reduce to monetary form (for example, cultural value). This could mean that some natural assets are favoured over others, despite the fact that all have value. We make these observations for a reason: for NBCS to form part of the climate solution, an NC accounting methodology will need to be utilized. Understanding exactly how NBCS impact on the environment will be essential, particularly if NBCS are used to support up to 30 per cent of the necessary mitigation required by 2030 if we are to stay on a pathway of limiting warming to below 2 °C. Accounting for nature is not without criticism, and certainly the concept of applying a monetary value to nature is completely foreign in many indigenous cultures.

41 United Nations Department of Economic and Social Affairs, *Natural Capital Accounting for Integrated Biodiversity Policies* (United Nations, 2020) <www.un-ilibrary.org/content/books/9789210052528> accessed 30 November 2021.

42 CSIRO, 'Natural capital accounting in Australia's primary industries' (Fisheries Research and Development Corporation, 2017) <www.frdc.com.au/sites/default/files/products/2017-175-DLD-A1.pdf>.

43 Arjan Ruijs et al, 'Natural capital accounting for better policy' (2018) 48(7) *Ambio* 714.

44 Ibid.

45 Unai Pascual et al, 'On the value of soil biodiversity and ecosystem services' (2015) 15 *Ecosystem Services* 11; Brady et al (n 36).

46 Brandon et al (n 31).

47 Polasky and Daily (n 34) 89.

48 Ibid.

The uncertainty surrounding the impacts of NBCS coupled with a distaste for assigning a value to nature means that these approaches attract critics. The debate between the importance of recognizing the intrinsic value of nature instead of assigning value to only the instrumental components, feature frequently in the ecosystem services literature.[49] Indeed, the concept of ecosystem services has been riddled with controversy and there have been many criticisms of an ecosystem services paradigm and NBCS in the literature.[50] McCauley's 2006 commentary in *Nature* made four arguments against ecosystem services which have formed the basis for subsequent critiques. First, he argued that the paradigm of ecosystem services requires an assumption that the 'biosphere is benevolent'.[51] McCauley suggests that this logic overlooks the critical truth that ecosystems do not act for the benefit of any species in particular. Rather, just as there are ecosystem services, there are also disservices.[52] Because of the existence and potential for ecosystem disservices there is a need to ensure that ecosystems are not altered in a way that could lead to an imbalance in favour of those disservices.[53] In order to ensure any alteration is not unfavourable, this first and foremost requires recognition of the complexity of ecological systems. We address this in more detail in Section 13.4.

McCauley's second point is that markets fluctuate. This observation is of course correct, however, the point in relation to ecosystem services is a criticism of *payments for ecosystem services* (PESs) that are made through a market-based instrument. It is the mechanism of the market, rather than the recognition of ecosystem services, that this criticism is directed toward. Hahn and colleagues identify seven different degrees of commodification within ecosystem services policy, ranging from no commodification and appreciation of ecosystems for their intrinsic value, all the way through to complete commodification and trade in nature.[54] Indeed, the ecosystem services framing may be used merely as a metaphor or a communication tool, without any market interaction.[55]

49 See, for example, Justine Bell-James (n 4) 525; Marc D Davidson, 'On the relation between ecosystem services, intrinsic value, existence value and economic valuation' (2013) 95 *Ecological Economics* 171.

50 Escobedo, Kroeger and Wagner (n 29); Osaka, Bellamy and Castree (n 14); Seddon et al (n 15); Telesetsky (n 36).

51 Douglas J McCauley, 'Selling out on nature' (2006) 443(7107) *Nature* 27.

52 Ibid.

53 Jari Lyytimäki, 'Ecosystem disservices: Embrace the catchword' (2015) 12 *Ecosystem Services* 136.

54 Thomas Hahn et al, 'Purposes and degrees of commodification: Economic instruments for biodiversity and ecosystem services need not rely on markets or monetary valuation' (2015) 16 *Ecosystem Services* 74.

55 Bell-James (n 4).

McCauley's third and fourth arguments are more relevant to this chapter. His third point is that NbS, and the resulting reliance on ecosystem services, 'underestimates the technological prowess of humans'.[56] Put differently, if we prioritize NBCS, then we are likely overlooking or discounting the value of other approaches. Although this is a common criticism of NBCS,[57] we suggest that there should not be an *either-or* approach to mitigating climate change. Indeed, to keep global temperatures to well below 2 °C of warming, efforts well in excess of a modern-day race to the Moon will be required.[58] In support of this argument, many commentators warn against the overreliance on the promise of carbon sequestration through tree planting when there is an urgent need to phase out combustion of fossil fuels and releases of greenhouse gases into the atmosphere from other activities, such as farming and mining.[59] This is a point that we agree with, as the urgency for *all* action is increasing daily. The metaphorical toolbox should include NBCS.

McCauley's fourth point is that through its aesthetic beauty, cultural significance and evolutionary necessity, nature has infinite value to humans and other living creatures. As a result, reducing it to a limited number of human benefits, or (worse) using it to profit from conservation efforts, does little to recognize this value. McCauley argues that protecting nature for nature's sake has always produced the most 'impressive' 'track record of achievements'.[60] We pause here to suggest that this may have been the case in times prior to the climate emergency, but one could argue that this view is outdated where drastic intervention is required to address the climate crisis. Despite this, McCauley's view aligns with concerns that NBCS may lead to over-emphasis on monoculture tree planting rather than ensuring protection of biodiversity and conservation of diverse ecosystems (thus the need to account for ecosystem changes),[61] and that the reliance on forests as carbon sinks will 'come at the cost of carbon rich and biodiverse native ecosystems and local resource rights'.[62] This may also interfere with some cultural practices of different indigenous peoples. We will now explore this in more detail, although from a slightly different angle.

---

56 McCauley (n 51) 28.

57 Osaka, Bellamy and Castree (n 14).

58 Barbara A Cosens et al, 'Designing Law to Enable Adaptive Governance of Modern Wicked Problems' (2020) 73(6) *Vanderbilt Law Review* 1687.

59 Seddon et al (n 19) 1518.

60 McCauley (n 51) 28.

61 Seddon et al (n 19) 1529.

62 Ibid 1518.

13.3 Posthumanism and RoN

Legal tradition remains rooted in concepts of commodification and human dominion over nature, and thus we must remain conscious of this when we consider how nature can (and whether it should) be used as a part of the solution for climate change mitigation. Indeed, there is growing awareness of the impact that NBCS may have on biodiversity.[63] In this section we address the fact that NBCS may not be universally embraced by all (which of course is not a necessary prerequisite to any policy approach), and consider the different perspectives associated with the RoN movement as a philosophical approach that modifies a traditional human-nature relationship. We consider NBCS to be a departure from the RoN approach. Indeed, rather than recognizing nature's rights to exist without modification, an NBCS directly modifies the landscape to either mitigate or assist humans to adapt to climate change. Despite recognizing this, one of our key arguments is that it is possible to reconcile an RoN approach with the landscape modification that NBCS requires.

The RoN movement is based on a posthumanist perspective that disputes the anthropocentric view of human beings at the centre of existence, which in turn incorporates a lack of emphasis on the importance of a flourishing environment. There is no shortage of commentators who argue that a human-centric ethic is problematic for a number of reasons,[64] most prominently the resulting ethical superiority of particular groups of people. This causes disadvantage for other persons, beings or nature through the unfairness that results from this superiority.[65] For this reason alone it is critical to challenge the viewpoint of the human as the most important (and at times only) universal concern.

Posthuman ethics dismantles 'the dominating centrality of the human for society and theory',[66] constituting humans as '(a) physically, chemically, and biologically enmeshed and dependent on the environment; (b) moved to action through interactions that generate affects, habits, and reason; and

[63] Osaka, Bellamy and Castree (n 14) 12.

[64] See, for example, Bennett Gilbert, 'The problem of Anthropocentrism and the human kind of personhood' [2022] *Philosophy & Social Criticism* 01914537221110900.

[65] Roderick Frazier Nash, *The Rights of Nature: A History of Environmental Ethics* (The University of Wisconsin Press, 1988) 8.

[66] Jannice Käll, *Posthuman Property and Law: Commodification and Control Through Information, Smart Spaces and Artificial Intelligence* (Taylor & Francis Group, 2022) 5 <http://ebookcentral.proquest.com/lib/qut/detail.action?docID=7026906>.

(c) possessing no attribute that is uniquely human but is instead made up of a larger evolving ecosystem'.[67]

Posthumanist philosophy can serve many purposes, but most relevantly here it underpins different theories for regulating the human–environment relationship. This includes theories of Earth jurisprudence,[68] deep ecology[69] and the RoN perspective (as opposed to rights over, or to, nature).[70] If these theories are adopted as a paradigm to address global issues, then questions must necessarily be raised about how nature can be utilized by humans. A posthuman ethic will shift the question from 'what are my rights over nature' and 'how can nature be harnessed to help resolve problems' to 'what are my duties toward it'. Indeed, humans are in a unique position of power in terms of environmental decision-making, but this position brings with it a great deal of responsibility. The following section identifies one approach to dealing with this responsibility – that is, RoN.

The importance of valuing the environment and the responsibility to care for it has deep historical and cultural roots. For example, it has long been understood by indigenous populations all over the world that ecosystems and unmodified landscapes are an inseparable part of human existence.[71] Although different cultures have unique belief systems, there is a common central tenet associated with the 'interdependence of humans and nature'.[72] The perspective of some of Australia's First Nations peoples offers an example. Many First Nations peoples in Australia align their connection to nature with duties rather than individual rights.[73] Indeed, the law for

67 Diane Marie Keeling and Marguerite Nguyen Lehman, 'Posthumanism', *Oxford Research Encyclopedia of Communication* (26 April 2018) <https://oxfordre.com/communication/view/10.1093/acrefore/9780190228613.001.0001/acrefore-9780190228613-e-627>.

68 Thomas Berry, *The Great Work: Our Way into the Future* (Crown, 2011).

69 George Sessions, 'The Deep Ecology Movement: A review' (1987) 11(2) *Environmental Review* 105.

70 Laura Forlano, 'Posthumanism and design' (2017) 3(1) She Ji: *The Journal of Design, Economics, and Innovation* 16, 17.

71 Alexis Bunten, 'What do the rights of nature have to do with indigeneity?', *Bioneers* (27 March 2017) <https://bioneers.org/rights-nature-indigeneity/>.

72 Anne Haluska, 'Restorative justice and the rights of nature: Using indigenous legal traditions to influence cultural change and promote environmental protection' (2023) 49(1) *Mitchell Hamline Law Review* 93, 96.

73 Marcia Langton, 'The Estate as Duration: "Being in Place" and Aboriginal Property Relations in Areas of Cape York Peninsula in North Australia', in Lee Godden and Maureen Tehan (eds), *Comparative Perspectives on Communal Lands and Individual Ownership* (Routledge-Cavendish, 1st ed, 2010) 75, 75, 76.

these peoples encompasses an obligation to 'care for country', which existed long before postcolonial legal ideals.[74] Referred to as 'First Law' by some First Nation writers, it 'is the way of living on Country handed down through countless generations'.[75] Some indigenous scholars have also argued, however, that the cultural obligations (and indeed the human right to practise those cultural obligations) of First Nations peoples are 'antithetical to the concept of legal personhood' for nature.[76] As Virginia Marshall explains: 'The concept of personhood of river systems is foreign within Aboriginal ideology, where there is no separation between land, water and Aboriginal cultural/legal obligations that derive from the fundamental Aboriginal belonging to the land and waters.'[77] Therefore, although the RoN movement is supported by indigenous peoples in some jurisdictions,[78] it remains controversial in others.

The historical roots extended beyond the culture of indigenous peoples. In 1867, John Muir observed the extremes of human selfishness,[79] and wrote: 'How narrow we selfish, conceited creatures are in our sympathies! How blind to the rights of all creation!'[80] It was Muir's belief that humans and nature could co-exist as equals, rather than one serving the other. These ideas have, of course, been contested, even within the discipline of environmental philosophy;[81] however, Muir's arguments have been echoed in the past century in many influential works.[82] Indeed, there are many commentators all over the world who have argued that popular theories formed in post-industrial society will be inadequate to reverse the rate of environmental degradation that has caused climate change and will be made worse by it.[83]

---

[74] Virginia Marshall, 'Removing the veil from the "Rights of Nature": The dichotomy between First Nations customary rights and environmental legal personhood' (2019) 45(2) *Australian Feminist Law Journal* 233, 236.

[75] Martuwarra River Of Life et al, 'Recognizing the Martuwarra's First Law right to life as a living ancestral being' (2020) 9(3) *Transnational Environmental Law* 541, 544.

[76] Marshall (n 74) 238.

[77] Virginia Marshall, 'First Nations' Perspective on Legal Rights for Rivers' (Exploring the Legal Status of Nature, University of Melbourne, 11 August 2017); cited by Marshall (n 72) 238.

[78] Craig M Kauffman and Pamela L Martin, 'Constructing Rights of Nature norms in the US, Ecuador, and New Zealand' (2018) 18(4) *Global Environmental Politics* 43, 44.

[79] Linda C Forbes, 'An introduction to John Muir's "Our National Parks"' (2002) 15(1) *Organization & Environment* 59, 59.

[80] Nash (n 65) 3.

[81] Forbes (n 79) 59.

[82] Berry (n 68); Aldo Leopold, *A Sand County Almanac, and Sketches Here and There* (Oxford University Press, 1989); Christopher D Stone, *Should Trees Have Standing? Law, Morality, and the Environment* (Oxford University Press, 2010).

[83] Haluska (n 72) 93.

In *Should Trees Have Standing?*, Christopher Stone argued that 'until a rightless thing receives its rights, we cannot see it as anything but a thing for the use of us – those who are holding the rights at the time'.[84] As such, as nature does not have rights in most nations, NBCS are not seen as exploitation. However, when we consider the impacts on the individual ecosystems, and adopt a viewpoint that accepts nature has rights, there may be questions of whether the ecosystem as it exists has been sufficiently considered and whether nature's rights should override the potential benefits that the NBCS may provide. In the jurisdictions where rights to nature have been recognized, these questions may need to be answered sooner rather than later. This is because NBCS may be inconsistent with some forms of life that nature supports.

As we demonstrate below, RoN can be initiated either through legal measures to promote particular rights or through recognition of legal personhood for the environment. There are only subtle differences between these two means to achieve RoN, but these differences can be critical depending on desired outcomes. For instance, legal rights for nature will mean that any actions that impact on nature must not encroach on those particular rights. For instance, a particular legal framework could grant an ecosystem (or nature generally) the right to restoration.[85] In these instances, there would be a requirement placed on any persons altering the landscape to restore it to its previous condition (even though systems theory argues that restoration to original conditions is not possible).[86] The second method in recognizing RoN is through granting legal personhood. This will mean that 'an entity exists for its own interests and not for the value it contains for others'.[87] It will allow nature (through representation) to sue for damages where there are breaches of its rights.

Between 2006 and 2014, RoN were recognized in a number of places, including Brazil, Bolivia, Colombia, Ecuador, India, Mexico, New Zealand and parts of the United States.[88] The different social, environmental and institutional arrangements across jurisdictions mean that the recognition was

---

84 Stone (n 82) 8.

85 Allison Katherine Athens, 'An indivisible and living whole: Do we value nature enough to grant it personhood?' (2018) 45(2) *Ecology Law Quarterly* 187.

86 Heriberto Cabezas et al, 'Sustainable Systems Theory: Ecological and other aspects' (2005) 13(5) *Journal of Cleaner Production* 455. Similar arguments are explored in the ecological restoration literature: Benjamin J Richardson, 'It is about time: Understanding the textures of time in Australian environmental law' (2018) 35 *Environmental and Planning Law Journal* 299.

87 Athens (n 85) 206.

88 Kauffman and Martin (n 78) 44.

unique in each instance.[89] The first national constitutional recognition of rights to the environment was in 2008, when Ecuador passed a constitutional amendment to acknowledge RoN, as follows:[90] 'Nature, or Pacha Mama, where life is reproduced and occurs, has the right to integral respect for its existence and for the maintenance and regeneration of its life cycles, structure, functions and evolutionary processes.'[91]

This constitutional recognition was broad in scope (applying to all of nature, including that beyond Ecuadorian borders),[92] and it granted extensive rights through the highest legal power of the nation. Since 2008, there have been more than a dozen lawsuits brought on the basis of protecting the RoN in Ecuador, and in the majority of cases the court has decided in favour of the environment.[93]

New Zealand's parliament elected to protect two distinct ecosystems, rather than nature generally, when it passed the Te Urewera Act in 2014 and the Te Awa Tupua Act in 2017. These Acts embody traditional Māori values. They recognize legal personhood for the Te Urewera national park and the Whanganui River. Although they define clear boundaries for the rights (thus limiting their application),[94] the recognition is extensive and provides all the 'rights, powers, duties and liabilities of a legal person' to these ecosystems.[95] These laws support the 'interests, well-being and rights' of these ecosystems, effectively limiting the allowable impacts of human actions.[96] There are nevertheless limits to the New Zealand approach, such as the categories of persons who can represent nature and the extent of protection provided. For instance, although the laws grant these ecosystems procedural access to the legal system, they do not assure 'the ecosystems' right to maintain their integrity or be restored, much less to flourish'.[97]

Although we have offered just a brief overview of the RoN paradigm and how it has been implemented domestically, our discussion demonstrates that an RoN approach is no longer confined to theory. These rights exist, they are enforceable, and regardless of whether one agrees with the framing, it challenges uses of the environment for human purposes – which is what NBCS require. Although there is a global push toward NBCS as a mitigation

89 Ibid.
90 Athens (n 85) 204.
91 *Constitution of the Republic of Ecuador 2008*, Article 71 (translation by the Comparative Constitutions Project).
92 Kauffman and Martin (n 78) 49.
93 Ibid.
94 Ibid.
95 Te Awa Tupua (Whanganui River Claims Settlement) Act 2017 (New Zealand) s 14(1).
96 Kauffman and Martin (n 78) 51.
97 Ibid 50.

technology, the compatibility of this approach with the RoN will no doubt be questioned, even though mitigation of climate change will be needed to support ecosystem functions. This challenge is considered in more detail in the next section, and we conclude that a balance must be struck to ensure NBCS can be utilized.

13.4 Integrating aspects of RoN into NBCS

The post-Industrial Revolution view that 'nature' is a commodity available to benefit humans has led to the sustainable development paradigm, which focuses on how the environment can be sustainably used, rather than the best way to conserve its value.[98] The assignment of property rights over land alongside this paradigm has led to an unbalanced emphasis on economic outcomes,[99] changed landscapes and biodiversity destruction. There is concern that extensive use of NBCS as a technology to address climate change will exacerbate these impacts, particularly given the speed with which actions to mitigate climate change must be taken. Further, the widespread adoption of NBCS may be viewed as incoherent with an RoN approach, which requires that nature is protected for its own sake, rather than that of humans for human purposes. However, as we rapidly approach dangerous rising temperatures, NBCS will be necessary as *one response* to climate change. We say this even though the literature supports further research into the precision and effectiveness of different methods for NBCS.[100]

If we accept that NBCS must be deployed, then one important challenge to resolve is the 'intrinsic versus instrumental value' debate.[101] There must also be a balance between the human rights that support the use of nature for food, water and shelter and those associated with conserving and respecting nature, such as cultural practices require. On the one hand, applying a human economic lens to nature will potentially amplify and perpetuate the era of anthropocentrism. On the other hand, failing to factor ecosystem services and NC conservation into decision-making through valuation is folly,[102] particularly when doing

[98] Haluska (n 72) 93.

[99] Sophie Riley, 'Sustainable development and the United Nations dialogues' (2019) 9(1) *Victoria University Law and Justice Journal* 31.

[100] Kimberly A Novick et al, 'Informing nature-based climate solutions for the United States with the best-available science' (2022) 28(12) *Global Change Biology* 3778; Tom Bradfer-Lawrence et al, 'The potential contribution of terrestrial nature-based solutions to a national "net zero" climate target' (2021) 58(11) *Journal of Applied Ecology* 2349.

[101] Davidson (n 49).

[102] Justine Bell-James, 'Integrating the ecosystem services paradigm into environmental law: A mechanism to protect mangrove ecosystems?' (2019) 31(2) *Journal of Environmental Law* 291, 295.

so may lead to better climate change mitigation. The common ground is, perhaps, that the current state of the environment and the corresponding failure of environmental policies necessitates a change in approach away from treating nature 'as if it were free and as if it were limitless'.[103]

The ecosystem services paradigm shifts natural resources from being viewed as a free commodity to something that must be valued and conserved. However, measurement is essential to ensure that environmental decision-making takes environmental losses into account, which may not be reflected in a financial value. We posit that taking environmental losses into account will lead to better decision-making by landholders and policy makers alike, as well as ensure a real consideration of the costs of action on the environment. Making 'visible the environmental costs via monitoring, measuring and reporting in terms of various environmental indicators forms a much more direct and complete articulation of environmental rationality'.[104] The provision of information on environmental reform can potentially be transformative, if that information is reliable, understandable and transparent.[105]

We also argue that labelling nature as 'capital' does not necessarily lead to the commodification of nature, although we do recognize that this label is distasteful to many people.[106] In contrast to this view, Chenoweth and colleagues suggest that NC and the ecosystem services framing can serve as a metaphor that will require nature is at least considered in economic decision-making.[107] Ultimately, decisions will consider economic returns, but an NC paradigm will mean that those returns will be properly calculated. In this respect, recognizing NC in economic decision-making can assist to determine quantitative and qualitative changes in resource use, allowing for the measurement and reporting of changes in ecosystem conditions.[108]

---

[103] United Nations, 'Reflect Nature's "True Value" in Economic Policies and Decisions, UN Chief Urges', *UN News* (2 March 2021) <https://news.un.org/en/story/2021/03/1086102>.

[104] Arthur PJ Mol, 'Environmental governance in the information age: The emergence of informational governance' (2006) 24(4) *Environment and Planning C: Government and Policy* 497, 504.

[105] Ibid 501–2. Note the Craik Review indicated that there would be some public policy reasons for not providing some information publicly but that these situations would be limited.

[106] See, for example, Emma Terama et al, 'Accounting for the environment as an economic asset: Global progress and realizing the 2030 Agenda for Sustainable Development' (2016) 11(6) *Sustainability Science* 945.

[107] Jonathan Chenoweth et al, 'The interrelationship of green infrastructure and natural capital' (2018) 75 *Land Use Policy* 137.

[108] United Nations, 'Frequently asked questions', *System of Environmental Economic Accounting* <https://seea.un.org/content/frequently-asked-questions>; United Nations and SEEA, *System of Environmental-Economic Accounting—Ecosystem Accounting: Final Draft* (Background Document, United Nations Department of Economic and Social Affairs Statistic Division, 5 February 2021).

We accept that the ecosystem services paradigm requires that an anthropocentric lens is applied to the conservation of nature,[109] but we maintain that this lens will not necessarily result in negative environmental impacts. Nor does this lens override the intrinsic value of nature and its importance for its own sake. Further, these concepts were not developed exclusively to commodify and profit from nature,[110] although we accept that some frameworks for PESs do have this effect. Rather, ecosystem services and NC were conceptualized to lead to a better understanding of the value of nature to all life.[111] Further, recognition allows one to consider long-term impacts of the 'use' of nature for human purposes while at the same time underscoring its value for sustaining life. The language of 'capital' is valuable in linking the human economy with its ecological surroundings,[112] and provides a useful concept for mainstreaming policy thinking about the interconnectedness between humans, economics and the natural world. For instance, Potschin emphasizes that 'much of the work in the field of ecosystem services is directed towards measuring and communicating the importance of nature to people'.[113] Thus, the identification, measuring and labelling of the services provided by NC will not necessarily reduce nature to a mere asset to be privatized, marketed and commodified, as built capital frequently is,[114] with the result that the intrinsic value of nature is overlooked.[115]

It is on this point that we suggest that the frameworks in support of the RoN could be combined with NBCS – but this should be carefully considered. The broad constitutional recognition, as per the Ecuadorian example, could have widespread consequences. Without specificity, this form of nature protection could preclude any human interference that is not seen as directly complementary to the ecosystem as it exists. Further, when combined with the precautionary principle, this broad recognition may ultimately pose a barrier to NBCS. This could be considered a barrier to addressing climate change (or at least a barrier to one critical tool for

[109] Bell-James (n 102) 295.

[110] See, for example, Hahn et al (n 54).

[111] Bell-James (n 102) 295.

[112] Robert Costanza and Herman E Daly, 'Natural capital and sustainable development' (1992) 6(1) *Conservation Biology* 37, 43–4.

[113] Marion B Potschin et al, 'Have ecosystem services been oversold? A response to Silvertown' (2016) 31(5) *Trends in Ecology & Evolution* 334, 335.

[114] Robert Costanza and Shuang Liu, 'Ecosystem services and environmental governance: Comparing China and the U.S.' (2014) 1(1) *Asia & the Pacific Policy Studies* 160, 162. R Costanza et al, 'The value of the world's ecosystem services and natural capital' (1998) 25 *Ecological Economics* 3.

[115] Bell-James (n 4).

addressing it). As an alternative, the protection of specific ecosystems, in the form of protected areas, would allow NBCS in appropriate spaces. Although identifying specific areas for protection, as opposed to nature more broadly, may be considered a compromise, it is one that can also be effective for biodiversity outcomes. Studies have indicated that identifying protected areas with accompanying management plans will stabilize critical ecosystems. As noted by Dasgupta: 'Analysis of the global impacts of terrestrial Protected Areas and Marine Protected Areas on conservation, agriculture, forestry, and fisheries sectors showed that benefits of protecting 30% of land and ocean would exceed costs, and would provide better financial outcomes and higher non-monetary benefits than currently.'[116]

Using protected areas or NBCS both require evaluation. To accept this is to accept that nature can and perhaps should be measured and monitored (particularly when the alternative mainstream view for the past centuries has been to use nature as a free resource). Most socio-ecological systems are highly complex in nature,[117] and theories have progressed over time to understand the systematic changes that can be identified in environmental systems.[118] Environmental systems have a variety of elements with both known and unknown interdependencies, and the removal of any one element may modify or even destroy elements of the entire system.[119] Hence, if one is to argue in support of NBCS, it is important to frame nature in the context of broader ecological systems.[120] This 'systems theory' approach recognizes that the individual components of any environment interact and entangle,[121] so that a single action may reap different results depending on the state of the broader ecosystem.[122] Even where similar species reside, the different traits that can feature in the same species in different locations can result in

[116] Dasgupta (n 33) 438.

[117] See the discussion of complexity, including in relation to ecosystem services law and policy, in Ruhl, Kraft and Lant (n 33).

[118] Cosens et al (n 58) 1694.

[119] Ibid 1695.

[120] Much of the literature frames this 'systems theory' approach through the framework of adaptive ecosystem management. See, for example, Jennifer N Solomon, Michael C Gavin and Meredith L Gore, 'Detecting and understanding non-compliance with conservation rules' (2015) 189 *Biological Conservation* 1, 2. See also John B Ruhl, 'Regulation by adaptive management – is it possible?' (2005) 7(1) *Minnesota Journal of Law, Science & Technology* 21.

[121] See, for example, Lael Parrott, 'Complexity and the limits of ecological engineering' (2002) 45(5) *Transactions of the ASAE* 1697, 1699; Lael Parrott, 'Measuring ecological complexity' (2010) 10(6) *Ecological Indicators* 1069, 1069; Danièle Bourcier and Pierre Mazzega, 'Toward measures of complexity in legal systems' (ICAIL07: 11th International Conference on Artificial Intelligence and Law, Stanford, June 2007) 4–8.

[122] John B Ruhl and Daniel M Katz, 'Measuring, monitoring, and managing legal complexity' (2015) 101(1) *Iowa Law Review* 191, 212.

different outcomes of the same actions.[123] Hence, having complete (or near complete) information is a necessary step in understanding the benefits or costs of a particular action,[124] and understanding how systems respond to change is essential to 'navigate that change'.[125]

Irrespective of the method used, commentators argue that environmental decision-making ultimately needs more timely data and analytics.[126] In this respect, NC accounting has the potential to inform government, corporate, landholder and consumer decision-making around the use or consumption of natural resources.[127] It is not an easy task, as the development of comprehensive NC accounts is a multidisciplinary effort requiring input from economists, ecologists, accountants, statisticians and data scientists.[128] The complexity of NC accounting, the relationship of ecosystem services to biodiversity, and the methodological and data challenges all present difficulties in the use of these accounts.[129] However, where it is done well it will likely support sustainable behaviour and potentially benefit the agricultural sector, both financially and socially.[130] Hence, the development of methodologies and tools to enable the measurement of NC at a national scale, but also on a farm scale, is recognized as essential by government, and by the sector generally.[131] This will naturally mean better outcomes for biodiversity, which of course is a critical component of NC.[132]

In arguing in support of metrics for nature we recognize that it is vital that nature is valued for its own sake and that its intrinsic properties are not overlooked. The recognition of these intrinsic values was behind advocacy for the recognition of nature as a subject of the law rather than an object of property for human exploitation.[133] However, we argue that the question

---

123 Colin T Reid and Walters Nsoh, *The Privatisation of Biodiversity? New Approaches to Conservation Law* (Edward Elgar Pub, 2016) 4.

124 Elinor Ostrom, 'Elinor Ostrom speaks about property rights' (2020) 35(3) *Journal of Private Enterprise* 7.

125 Cosens et al (n 58) 1692.

126 Brandon et al (n 31).

127 Ibid 144–7.

128 Geoff Bright, Emily Connors and Joe Grice, 'Measuring natural capital: Towards accounts for the UK and a basis for improved decision-making' (2019) 35(1) *Oxford Review of Economic Policy* 88, 91.

129 Brandon et al (n 31) 146.

130 Price Waterhouse Coopers, 'Farming for the Future' (Macdoch Foundation, 2021) Discussion Paper; Brady et al (n 36); Ruijs et al (n 43).

131 Price Waterhouse Coopers (n 130).

132 Ben Groom and Zachary Turk, 'Reflections on the Dasgupta Review on the economics of biodiversity' (2021) 79(1) *Environmental and Resource Economics* 1.

133 See, generally, Julien Bétaille, 'Rights of nature: Why it might not save the entire world' (2019) 16(1) *Journal for European Environmental & Planning Law* 35.

of whether to extend rights to nature is a matter for each jurisdiction and remains separate to recognition of its intrinsic value.[134] It is also a separate question as to whether a particular ecosystem should be altered for NBCS. The answer to this will not always be yes, nor will it always be no. Some specificity is needed. However, where RoN have been legally recognized, there will be limits to the types of actions humans can initiate (even if those actions are in support of climate change mitigation). This is not to say that NBCS should not be used, but whether they are appropriate to use must be based on place-based analysis.

13.5 Conclusion

The rapidly worsening climate crisis demands urgent solutions to complex problems. NBCS offer an approach that can be used to restore nature and preserve and deliver ecosystem services, including the sequestration of carbon for climate mitigation. NBCS can (and we argue should) be underpinned by an NC accounting framework. An NC accounting framework alongside the evolution of data science and technology has meant that it is possible to predict how humans can impact the environment. As such, it is important to understand how particular actions will affect ecosystems when it is possible to do so. The purpose of an NC framework and accounting for nature is not to undermine any environmental philosophies, such as deep ecology, Earth jurisprudence, or an RoN approach. Rather this paradigm can support the ends desired by any and all of them, depending on how the metrics are used. This chapter suggests that the metaphors of ecosystem services and NC can be useful, and that measuring environmental impacts is essential, particularly where NBCS are widely adopted.

NBCS, NC and ecosystem-service approaches are each criticized on the basis that they continue the anthropocentric tradition of promoting human needs over environmental concerns. There is some truth in this. If inadequately regulated, these methods could be used as a way to continue to commodify nature and for select groups to benefit from it. We argue that this is why these tools should be appropriately utilized, and that the benefits are spread to all life, rather than only to select groups of humans.

In this regard, systems theory suggests that ecosystems are complex, with a variety of elements that interact and interweave.[135] Modification of one

[134] Ibid 45.

[135] See, for example, Parrott, 'Complexity and the limits of ecological engineering' (n 121) 1699; Parrott, 'Measuring ecological complexity' (n 121) 1069; Bourcier and Mazzega (n 121) 4–8.

of those elements may lead to unintended and perverse consequences. This could mean considering already modified landscapes and accepting that there could be economic trade-offs in doing so. Hence, employing robust systems of measuring and evaluation may be useful to determine whether particular places are appropriate for NBCS. Ultimately, there will always be conflicts in land use, with many competing concerns, and a posthuman ethic would suggest that species should all be considered equally in balancing those concerns, rather than preferencing human interests over all others. Measuring and valuing nature should not compromise this. In fact, if used appropriately, it could align with this outcome.

Recognizing the intrinsic value of nature requires that one accepts the argument that protecting the environment is an important end in itself, irrespective of any ecosystem services that an ecosystem provides. However, as the urgency for controlling climate change increases, new approaches need to be deployed, which will inevitably result in some opposition to the methods used. NBCS will be one such method.[136] In this chapter, we have argued that the ecosystem services paradigm has an important place in both recognizing environmental values and protecting them. However, it should not be the only way we accept that nature has value. It is merely one way of seeing nature and assigning value to it. It does not necessarily lead to commodification, nor does it support human interference. However, our final point is that controlling climate change will require human intervention through (among other things) NBCS. Whatever the intervention, what is most important is that the intervention must be evaluated on the basis of impacts and benefits. This may not be an easy task, but it will be absolutely essential in coming decades.

[136] Although we recognize that there is a wide variety of different methods that come under the banner of NBCS.

14

Ban on Coal: Rationale and Legality for Restrictions on Coal Production, Sale and Use

Daria Shapovalova

14.1 Introduction

Coal was the first fossil fuel to be commercially exploited, due to its relatively shallow location.[1] The use of coal for steam engines in manufacturing and shipping propelled economic development in the 19th century, with coal representing 95 per cent of aggregate global use of commercial energy through the first half of the 20th century.[2] Coal is also the fossil fuel with the greatest carbon content, releasing about 30 per cent more carbon dioxide per unit of energy than refined oil, and almost 80 per cent more than natural gas.[3] Today, coal accounts for over a quarter of the global energy demand, representing a large share of the energy mix even in economically developed states.[4] The International Energy Agency (IEA) predicts 2023 to be the year of peak coal demand, with growth in India and South-East Asian countries offsetting declines in the European Union (EU) and the

1 Vaclav Smil, *Energy: A Beginner's Guide* (Oneworld Publications 2006) 93–95. See also Barbara Freese, *Coal: A Human History* (Basic Books 2016).

2 Smil (n 1) 98.

3 Ibid 101; Marquita K Hill, *Understanding Environmental Pollution* (4th edn, Cambridge University Press 2020) 138.

4 IEA, 'World Energy Balances: Overviews' (2021) <www.iea.org/reports/world-energy-balances-overview> accessed 18 December 2023.

United States (US).[5] China remains the 'decisive player' in setting the trends for coal in the coming decade.[6]

As a fossil fuel with the greatest carbon content, coal is first on the phase-out list in the climate goals-compliant scenarios. Most analyses on the climate and energy outlook agree on the need for the use of coal to be drastically reduced, and in the longer term to only be allowed with carbon capture technologies.[7] About 80 per cent of coal reserves (as well as 50 per cent of gas and 30 per cent of oil reserves) cannot be 'burned and emitted' if global warming is limited to 2 °C.[8]

The move away from coal will cause social and economic issues, as closing mines and power plants is often associated with environmental and social fallout in the host communities.[9] Furthermore, coal is a staple source of base-load power, and in developing states it can be a feasible answer to energy security. The phase-out of coal will need to be supplemented by a sufficient amount of a backstop energy source, and the choice of the alternative fuel will have a significant impact not just on climate, but on socio-economic and environmental indicators as well.[10] There are further concerns about the impact of coal bans on investment protection regimes. In 2021–22, proceedings were initiated against the Netherlands by RWE and Uniper, alleging that enacting the Coal Ban Law amounted to expropriation under the Energy Charter Treaty and international law.[11] Regulating the phase-out of coal, therefore, is riddled with complexity.

---

5 IEA, 'Coal 2023: Analysis and Forecast to 2026' (2023) 11 <https://iea.blob.core.windows.net/assets/a72a7ffa-c5f2-4ed8-a2bf-eb035931d95c/Coal_2023.pdf> accessed 18 December 2023.

6 Ibid.

7 See, for example, IEA, 'World Energy Outlook 2023' (2023) 140 <https://iea.blob.core.windows.net/assets/42b23c45-78bc-4482-b0f9-eb826ae2da3d/WorldEnergyOutlook2023.pdf> accessed 18 December 2023; IPCC, 'An IPCC Special Report on the Impacts of Global Warming of 1.5°C' (2019) <www.ipcc.ch/sr15/download/#full> accessed 18 December 2023.

8 IPCC, 'Synthesis Report of the IPCC Sixth Assessment Report (AR6)' (2023) 24 <https://report.ipcc.ch/ar6syr/pdf/IPCC_AR6_SYR_LongerReport.pdf> accessed 18 December 2023.

9 Guri Bang, Knut Einar Rosendahl and Christoph Böhringer, 'Balancing Cost and Justice Concerns in the Energy Transition: Comparing Coal Phase-out Policies in Germany and the UK' (2022) 22 *Climate Policy* 1000; Thomas Spencer et al, 'The 1.5°C Target and Coal Sector Transition: At the Limits of Societal Feasibility' (2018) 18 *Climate Policy* 335.

10 See IPCC's comparison in IPCC, 'Climate Change 2014: Synthesis Report. Contribution of Working Groups I, II and III to the Fifth Assessment Report of the Intergovernmental Panel on Climate Change' (2014) 469. On the importance of just transition in coal phase-out see Michael Jakob et al, 'The Future of Coal in a Carbon-Constrained Climate' (2020) 10 *Nature Climate Change* 704.

11 Wet verbod op kolen bij elektriciteitsproductie (Law Banning Coal in Electricity Production) 11 December 2019. *WE AG and RWE Eemshaven Holding II BV v. Kingdom of*

The Netherlands is not alone in adopting policies and laws on coal restrictions. Many countries have adopted a variety of regulatory measures directed at coal phase-out: from strict air emissions standards, to subsidy withdrawal, license conditions and regulated phase-out timelines.[12] International initiatives and proposals to ban coal have also been presented in recent years, including proposals for a coal ban treaty.[13] Such initiatives, however, need to be carefully considered for compatibility with wider regulatory frameworks including international economic law.[14]

This chapter builds on this discourse by developing a framework to systematically examine existing and potential future policies and laws on restrictions of coal use, sale and production. It categorizes and examines international and domestic measures to provide a critical outlook in light of concerns over the climate emergency and the need for a just transition. It argues that despite the challenges in regulating coal bans at both international and national levels, the variety of regulatory and policy tools available provide sufficient design choices to address specific concerns in a tailored manner. It starts with the discussion of coal as an energy source and natural resource, providing background on its impact on the climate and the geographical distribution of coal reserves, consumption and trade. It then examines international legal and governance frameworks for coal phase-out in Section 14.3. Section 14.4 focuses on domestic policy and regulatory restrictions on coal, critically examining the existing and proposed instruments. It categorizes these measures as indirect (through energy planning and climate law) and direct (restrictions on coal use, sale and production) before outlining the growing

the Netherlands, ICSID Case No. ARB/21/4; *Uniper SE, Uniper Benelux Holding B.V. and Uniper Benelux N.V. v. Kingdom of the Netherlands*, ICSID Case No. ARB/21/22. Both cases were withdrawn following the German Supreme Court ruling prohibiting intra-EU investment arbitration. See Georg Scherpf and Simon Hermes, 'Another "Bleak House"—the Decision of the German Federal Supreme Court on the Inadmissibility of Intra-EU ICSID Arbitrations' [2023] Arbitration International 1.

12 See, for example, Katharina Baudisch and Dörte Fouquet, 'Germany's Coal Exit Plan: A Legal and Political Outline on How Germany Plans to Phase out Coal Fired Power Stations by 2038' (2019) 9 *Renewable Energy Law and Policy Review* 45; Melissa Harris, Marisa Beck and Ivetta Gerasimchuk, 'The End of Coal: Ontario's Coal Phase-Out' (IISD 2015) <www.iisd.org/system/files/publications/end-of-coal-ontario-coal-phase-out.pdf> accessed 18 December 2023.

13 See, for example, Anthony Burke and Stefanie Fishel, 'A Coal Elimination Treaty 2030: Fast Tracking Climate Change Mitigation, Global Health and Security' (2020) 3 *Earth System Governance* 100046; Mathieu Blondeel, Thijs Van de Graaf and Tim Haesebrouck, 'Moving beyond Coal: Exploring and Explaining the Powering Past Coal Alliance' (2020) 59 *Energy Research & Social Science* 101304; Peter Newell and Andrew Simms, 'Towards a Fossil Fuel Non-Proliferation Treaty' (2020) 20 *Climate Policy* 1043.

14 Anatole Boute, 'Phasing Out Coal through Electricity Market Regulation' (2022) 59(4) *Common Market Law Review* 1007.

body of just-transition measures aiming to address the socio-economic fallout. Section 14.5 contains conclusions and identifies further research needs.

14.2 Coal as an energy source

To examine the legal and policy framework around coal phase-out, it is first important to establish the rationale and parameters for such a phase-out. This section reviews the role of coal as an energy source, its main use and deposit locations, as well as the climate impacts of its mining and combustion.

Based on carbon and heat content, coal is divided into four types: lignite or brown coal (25–35 per cent carbon), sub-bituminous (35–45 per cent), bituminous (45–86 per cent), and anthracite (86–97 per cent).[15] Around the world, coal is primarily used for electricity production and heating, while higher quality coal can also be converted to coke, used for steelmaking. Depending on the geology of the deposit, coal can be developed by either surface or underground mining. Coal mining itself contributes significantly to greenhouse gas emissions as one of the largest sources of methane.[16] The exact contribution of coal mining to methane emissions is not certain. Conservative estimates place it at 11 per cent of global anthropogenic methane emissions, but this could be a significant underestimation.[17] Abatement technologies are available to reduce emissions in underground mining operations, such as recovery and use or flaring of methane.[18] In addition to carbon dioxide (CO_2), combustion of coal also contributes significantly to air pollution, producing sulfur dioxide, nitrogen oxide, particulates, mercury and other heavy metals.[19]

Meeting the goal of the Paris Agreement to limit warming to 2 °C or below requires a significant reduction in coal use and mining. In the medium to long term, using coal to produce power is only viable with carbon capture technologies.[20] About 80 per cent of coal in the ground

---

15 US Energy Information Administration, 'Coal Explained' <www.eia.gov/energyexplained/coal/> accessed 18 December 2023.

16 IPCC WGIII, 'Climate Change 2022: Mitigation of Climate Change' <www.ipcc.ch/report/ar6/wg3/downloads/report/IPCC_AR6_WGIII_FullReport.pdf> accessed 18 December 2023; Nazar Kholod et al, 'Global Methane Emissions from Coal Mining to Continue Growing Even with Declining Coal Production' (2020) 256 *Journal of Cleaner Production* 120489.

17 Kholod et al (n 16); Stefan Schwietzke et al, 'Upward Revision of Global Fossil Fuel Methane Emissions Based on Isotope Database' (2016) 538 *Nature* 88.

18 US EPA, 'Global Non-CO_2 Greenhouse Gas Emission Projections & Mitigation 2015–2050' (2019) 14–15 <www.epa.gov/sites/default/files/2019-09/documents/epa_non-co2_greenhouse_gases_rpt-epa430r19010.pdf> accessed 18 December 2023.

19 US Energy Information Administration, 'Coal and the Environment' <www.eia.gov/energyexplained/coal/coal-and-the-environment.php> accessed 18 December 2023.

20 See, for example, IEA (n 7) 140; IPCC (n 7).

must remain unused between 2010 and 2050, meaning some states will need to leave their reserves unexploited.[21] Furthermore, some of the existing coal mines and power plants would need to cease operations before their economic lifespan ends.[22]

The IPCC acknowledges that limiting global warming to 2 °C or below 'could strand considerable fossil fuel infrastructure' with coal assets projected to be at risk of being stranded before 2030.[23] Coal assets are expected to represent three-quarters of power sector fossil fuels stranded assets in G20 countries, and China would account for 45 per cent of the total.[24] The stranded assets in the upstream sector require further consideration of location and ownership of coal reserves.[25]

Reserves of coal are estimated at about 1,070 billion tonnes, and are heavily concentrated in the US (23.2 per cent), Russia (15.1 per cent), Australia (14 per cent), China (13.3 per cent) and Indonesia (10.3 per cent).[26] In Europe, Germany, Ukraine and Poland are the top three coal reserve holders and producers, and consumers.[27] Consumption of coal traditionally occurs in the production regions, but the volume of internationally traded coal has been steadily rising to represent about 16 per cent of total coal demand

21 IPCC (n 8) 24; Christophe McGlade and Paul Ekins, 'The Geographical Distribution of Fossil Fuels Unused When Limiting Global Warming to 2 °C' (2015) 517 *Nature* 187.

22 IPCC (n 16) 625; IEA, 'Net Zero Roadmap: A Global Pathway to Keep the 1.5 °C Goal in Reach' (2023) 13, 150 <https://iea.blob.core.windows.net/assets/9a698da4-4002-4e53-8ef3-631d8971bf84/NetZeroRoadmap_AGlobalPathwaytoKeepthe1.5CGoalinReach-2023Update.pdf> accessed 18 December 2023; Jessica Jewell et al, 'Prospects for Powering Past Coal' (2019) 9 *Nature Climate Change* 592.

23 IPCC (n 16) 28.

24 Deger Saygin et al, 'Power Sector Asset Stranding Effects of Climate Policies' (2019) 14 *Energy Sources, Part B: Economics, Planning, and Policy* 99.

25 See, for example, Ben Caldecott, James Tilbury and Yuge Ma, 'Stranded Down Under? Environment-Related Factors Changing China's Demand for Coal and What This Means for Australian Coal Assets' (Smith School of Enterprise and the Environment 2013) <www.smithschool.ox.ac.uk/sites/default/files/2022-03/stranded-down-under-report.pdf> accessed 18 December 2023.

26 For anthracite, bituminous, sub-bituminous and lignite coal. Total proved reserves of coal are 'those quantities that geological and engineering information indicates with reasonable certainly can be recovered in the future from known deposits under existing economic and operating conditions'. Energy Institute, '2023 Statistical Review of World Energy' (2023) data sheet <www.energyinst.org/statistical-review> accessed 18 December 2023.

27 Energy Institute (n 26). Eurostat, 'Coal Production and Consumption Statistics' <https://ec.europa.eu/eurostat/statistics-explained/index.php?title=Coal_production_and_consumption_statistics#Consumption_and_production_of_hard_coal> accessed 18 December 2023. See also, Euracoal Statistics (2022) <https://euracoal.eu/info/euracoal-eu-statistics/> accessed 18 December 2023.

in 2022.[28] Australia and Indonesia are the largest exporters of coal, while China and India are the two biggest importers and two of the world's biggest consumers.[29]

North America and Europe represent around 6 per cent share of the world's coal demand each but, in both regions, the demand has been falling in the last decade.[30] At the same time, the Asian market for coal has been steadily growing[31] due to the growing overall energy demand and 'infrastructure inertia'.[32] China alone accounts for more than half the world's coal demand.[33] Coal is an abundant resource – the reserve to production ratio estimated that the 2023 coal reserves accounted for 139 years,[34] which is well beyond the climate goal-compliant scenarios, even with carbon capture technologies applied.

This section reviewed the role of coal in the international energy market and its role in climate change. It is clear that reducing coal use requires a collective approach and, importantly, focused efforts from coal reserve holders, producers and large (and growing) consumers. This would require policy and regulatory interventions at both international and national levels. In the absence of significant climate-related policy shifts on coal, its market and international trade is expected to continue to grow.[35] The next two sections map out and examine the existing international and national regulatory and policy restrictions on coal.

14.3 Climate restrictions on coal in international law and governance

As noted earlier, restricting the use of coal requires a collective international effort. The need for cooperation is amplified as coal use in some regions produces climate change-induced impacts in others. International law already

[28] IEA (n 5) 60. See Franziska Holz et al, 'The COALMOD-World Model: Coal Markets until 2030', in Mark C Thurber and Richard K Morse (eds), *The Global Coal Market* (Cambridge University Press 2015).

[29] Energy Institute (n 26) data sheet.

[30] Ibid.

[31] Ibid; IEA (n 5) 11. See also Mark C Thurber and Richard K Morse, 'The Asia-Centric Coal Era', in Mark C Thurber and Richard K Morse (eds), *The Global Coal Market* (Cambridge University Press 2015); Holz et al (n 28) 441–7.

[32] Gareth AS Edwards, 'Coal and Climate Change' (2019) 10 *WIREs Climate Change* 1, 3.

[33] IEA (n 5) 11.

[34] 'Reserves-to-production (R/P) ratios represent the length of time that those remaining reserves would last if production were to continue at the previous year's rate. They are calculated by dividing remaining reserves at the end of the year by the production in that year.' Energy Institute (n 26) data sheet.

[35] Holz et al (n 28) 454–6.

indirectly regulates the use of coal, through air pollution conventions and climate change treaties imposing emission limitations. Any restrictions or outright ban on mining or use of coal imposed from the international level would interfere with states' sovereignty over natural resources. However, there is space for international cooperation and law-making to develop a regime for sealing coal deposits. Informal international governance also has a role to play with Powering Past Coal Alliance gaining prominence in coal phase-out globally.

14.3.1 International air pollution regulation

Despite air pollution regulation targeting emissions associated with coal use, international and regional air quality regulations have not been seen as driving coal phase-out. Burning coal is one of the largest sources of sulfur dioxide, and the Convention on Long-Range Transboundary Air Pollution (CLRTAP)[36] was adopted in 1979 primarily to address the European acid rain problem caused by coal-fired power plants.[37] The CLRTAP is a regional framework treaty adopted under the auspices of the United Nations Economic Commission for Europe (UNECE) and supplemented by eight protocols.[38] The CLRTAP framework imposed emission limitations but despite the wide uptake of the agreement and protocols among the UNECE members, it is doubtful that the CLRTAP compelled states to transition from coal to natural gas, as domestic processes are believed to have played the main role in that transition.[39]

The EU air pollution legislation and, in particular, the Industrial Emissions Directive,[40] could be instrumental in reducing the use of coal in Europe. It requires the use of Best Available Techniques (BAT) to set emission limits for large combustion plants including sulfur dioxide and nitrogen oxides, attributable to coal combustion. When first adopted, the targets under the Directive were too modest to effect meaningful change in the energy

36 (Signed 3 November 1979, entered into force 16 March 1983) 1302 UNTS 217.

37 Jorge E Viñuales, *The International Law of Energy* (Cambridge University Press 2022) 162–3.

38 See the Protocols at <www.unece.org/env/lrtap/status/lrtap_s.html> accessed 18 December 2023.

39 Jørgen Wettestad, *Clearing the Air: European Advances in Tackling Acid Rain and Atmospheric Pollution* (Routledge 2002) 164–76; Loren R Cass, 'Acid Pollution and Acid Rain', in Paul G Harris (ed), *Routledge Handbook of Global Environmental Politics* (Routledge 2014) 395.

40 Directive 2010/75/EU of the European Parliament and the Council on Industrial Emissions (Integrated Pollution Prevention and Control) of 24 November 2010, OJ L334/17.

sector, especially given the exceptions for existing plants.[41] However, the 2017 BAT conclusions, with specific provisions on coal, would require coal power plants to either undergo significant modernization or 'disappear from the energy map of Europe'.[42] As time progresses, the derogations and exceptions allowable under the Directive become unavailable. For example, a 'limited lifetime derogation' allowing states to permit operators to keep production with shorter operating hours, has not been available since 31 December 2023.[43]

In addition, although the EU emission trading scheme or climate targets did not specifically address coal, the increase in price due to emission payments can make coal economically unattractive.[44] Air pollution and climate change law, even if not directly addressing coal, can thus be instrumental in its phase-out if they set out ambitious limits and targets.

14.3.2 International climate change law

The international climate change regime was quite late to address the fossil fuels problem. The United Nations Framework Convention on Climate Change (UNFCCC) and the Paris Agreement do not place specific obligations directly regulating fossil fuel production.[45] At COP26 (Conference of the Parties to the United Nations Framework Convention of Climate Change) in 2021, the outcome document for the first time 'broke the taboo'[46] and referred specifically to parties to the Paris Agreement 'accelerating efforts towards the phasedown of unabated coal power and phase-out of inefficient fossil fuel subsidies' while emphasizing the need for a just transition.[47] The synthesis report of Nationally Determined

---

[41] Giuseppe Bellantuono, 'Legal Pathways of Decarbonisation in the EU: The Case of Coal Phase-Out' (2019) 17 *Oil, Gas & Energy Law* 13–15.

[42] Maciej M Sokołowski, 'Burning out Coal Power Plants with the Industrial Emissions Directive' (2018) 11 *The Journal of World Energy Law & Business* 260, 266. See also Gerard Wynn and Paolo Coghe, 'Europe's Coal-Fired Power Plants: Rough Times Ahead: Analysis of the Impact of a New Round of Pollution Controls' (IEEFA 2017) <https://ieefa.org/wp-content/uploads/2017/05/Europe-Coal-Fired-Plants_Rough-Times-Ahead_May-2017.pdf> accessed 18 December 2023.

[43] Industrial Emissions Directive, Art 33.

[44] But see Boute (n 14) on how this approach underperformed in the energy crisis. See also Bellantuono (n 41). Gerrit Rentier, Herman Lelieveldt and Gert Jan Kramer, 'Varieties of Coal-Fired Power Phase-out across Europe' (2019) 132 *Energy Policy* 620.

[45] See Daria Shapovalova, 'Climate Change and Oil and Gas Production Regulation: An Impossible Reconciliation?' (2023) 26 *Journal of International Economic Law*.

[46] Harro van Asselt, 'Breaking a Taboo: Fossil Fuels at COP26' (*EJIL: Talk!*, 26 November 2021) <www.ejiltalk.org/breaking-a-taboo-fossil-fuels-at-cop26/> accessed 2 January 2024.

[47] Glasgow Climate Pact, para 36, 1/CMA.3, FCCC/PA/CMA/2021/10/Add.1.

Contributions (NDCs) under the Paris Agreement indicates that 9 per cent of Parties, accounting for 58 per cent of total global electricity generation from coal in 2020, indicated measures to phase down unabated coal power, such as replacing coal-fired power plants with combined-cycle gas power plants.[48]

So far, the NDCs of top fossil fuel producing countries have not systematically engaged with coal phase-down.[49] However, the latest, COP28, took a broader approach, aiming at fossil fuels collectively, but again emphasized the need to accelerate unabated coal phase-down.[50] Given that this language made it into the first Global Stocktake document, the parties will now have to explain how their NDC was informed by it.[51]

14.3.3 Informal coal phase-out governance

In the absence of direct action on fossil fuels in climate treaty regimes, more informal intergovernmental cooperation emerged to phasing out the use of coal, in the form of the Powering Past Coal Alliance (PPCA). PPCA, launched during COP23 by Canada and Denmark, does not emphasize coal production, calling instead for a phase-out of 'existing unabated coal power generation'.[52] The Alliance attracted the support of 48 national governments, 49 sub-national governments and 71 organizations.[53] However, in a trend similar to the Beyond Oil and Gas Alliance[54] or the nuclear weapons prohibition treaty,[55] it is countries that 'have no coal in their electricity mix and that have adopted a phase-out plan'[56] that are most likely to join

---

48 'Nationally Determined Contributions under the Paris Agreement. Synthesis Report by the Secretariat (26 October 2022) FCCC/PA/CMA/2022/4', para 190.

49 Natalie Jones and Paola Yanguas Parra, 'How the Transition Away From Fossil Fuel Production Can Be Included in New Climate Commitments and Plans' (IISD, 2024) <www.iisd.org/publications/report/fossil-fuel-transition-new-climate-commitments> accessed 10 August 2024.

50 First Global Stocktake, para 28, FCCC/PA/CMA/2023/L.17.

51 Paris Agreement Art 4(9). See also Harro van Asselt, ' "Historic" or "Historic Failure"? Fossil Fuels at COP28' (*EJIL: Talk!*, 28 December 2023) <www.ejiltalk.org/historic-or-historic-failure-fossil-fuels-at-cop28/> accessed 2 January 2024.

52 Powering Past Coal Alliance Declaration <www.poweringpastcoal.org/about/declaration> accessed 18 December 2023.

53 Powering Past Coal Alliance Members <www.poweringpastcoal.org/members> accessed 18 December 2023.

54 See Shapovalova (n 45).

55 Treaty on the Prohibition of Nuclear Weapons (signed 7 July 2017, entered into force 22 January 2021) 3379 UNTS.

56 Blondeel, Van de Graaf and Haesebrouck (n 13).

the Alliance. PPCA does not place any obligations on states to phase out coal but rather allows them to 'cooperate without giving up sovereignty in sensitive areas'.[57]

14.3.4 *Would an internationally imposed ban be possible?*

Proposals and initiatives to adopt a treaty placing bans on coal mining and binding timelines for coal phase-out have been put forward since the adoption of the Paris Agreement. For example, Tuvalu and Vanuatu spearheaded the initiative to develop a Fossil Fuel Non-Proliferation Treaty.[58] Proponents of this proposal call for fossil fuel production and infrastructure phase-outs, reporting regime, and acceleration of energy transition through directing subsidies to renewables and sharing low-carbon technologies.[59] Newell and Simms note that coal's carbon intensity, diminishing financial viability and its substitutability in most cases, might make it 'an appropriate focus for the first wave of negotiations'.[60]

Burke and Fishel made a case for a 'Coal Elimination Treaty' that would provide for the 'complete phasing out and prohibition of coal's mining, sale and burning by 2030'.[61] Exploring three potential models, and acknowledging challenges in climate diplomacy, they conclude that such a treaty would place 'additional moral and normative pressure on the burning of fossil fuels'.[62]

Questions therefore arise around the feasibility of such treaties and their interaction with other international obligations and national legal frameworks. International law around energy development revolves around the notion of national sovereignty over natural resources, which confirms states' right to develop their resources as well as their own energy policies and relevant legal frameworks without external coercion.[63] Rights to develop and use natural sources do not come without duties, including respect of indigenous peoples' rights, environmental obligations, fair and equitable treatment of foreign investments.[64] Nevertheless, regulation of energy activities remains

---

57 Felicity Vabulas and Duncan Snidal, 'Cooperation under Autonomy: Building and Analyzing the Informal Intergovernmental Organizations 2.0 Dataset' (2021) 58 *Journal of Peace Research* 859.

58 See <https://fossilfueltreaty.org/> accessed 18 December 2023.

59 Newell and Simms (n 13).

60 Ibid 1048.

61 Burke and Fishel (n 13).

62 Ibid.

63 UNGA Res 1803 (XVII) (14 December 1962); Viñuales (n 37) 45.

64 Nico Schrijver, *Sovereignty over Natural Resources: Balancing Rights and Duties* (Cambridge University Press 1997) 306–67.

largely a domestic matter.[65] Thus, the importance of sovereignty and consent in international law would make any treaty-based coal ban or phase-out binding only on consenting states.[66]

There are further concerns about the impact of coal bans on investment protection regimes. Some national measures to phase out coal have already caused investors to initiate investment arbitration proceedings.[67] Under bilateral and multilateral investment treaties, investors are protected from direct and indirect expropriation. If a foreign company owns rights or infrastructure for coal mining or coal power plants, climate regulation which could strand those assets might amount to such expropriation.[68] Investment arbitration may thus be seen as a barrier to the energy transition, risking the 'regulatory chill' for fossil fuel phase-out.[69] Scholars propose some solutions to address this issue, such as adjusting the calculation of damages to avoid large compensations,[70] 'carving-out' climate change-related measures for investment treaties,[71] or outright investment treaty termination.[72]

It is not only foreign investors who might require compensation. Some states-holders of coal reserves might be more amenable to agreeing to an internationally imposed ban if compensation was forthcoming. Suggestions have been put forward to develop a coal deposit trading scheme or a global fund as ways to leave some reserves undeveloped while compensating reserve holders or supporting communities and workers.[73] Any such model would be very complex to develop and implement. For example, the resources could be sealed

65 See Viñuales (n 37) 32. See also Shapovalova (n 45).

66 Vienna Convention on the Law of Treaties (signed 23 May 1969, entered into force 27 January 1980) 1155 UNTS 331, Arts 34 and 35.

67 See, for example, *WE AG and RWE Eemshaven Holding II BV v. Kingdom of the Netherlands*, ICSID Case No. ARB/21/4; *Uniper SE, Uniper Benelux Holding B.V. and Uniper Benelux N.V. v. Kingdom of the Netherlands*, ICSID Case No. ARB/21/22.

68 'Expropriation Regime under the Energy Charter Treaty' (Energy Charter Secretariat 2012) 9–12 <www.energycharter.org/fileadmin/DocumentsMedia/Thematic/Expropriation_2012_en.pdf> accessed 18 December 2023.

69 Anatole Boute, 'Investor Compensation for Oil and Gas Phase out Decisions: Aligning Valuation Methods to Decarbonization' (2023) 23 *Climate Policy* 1087. See also Kyla Tienhaara et al, 'Investor-State Dispute Settlement: Obstructing a Just Energy Transition' (2023) 23 *Climate Policy* 1197.

70 Boute (n 69).

71 Joshua Paine and Elizabeth Sheargold, 'A Climate Change Carve-Out for Investment Treaties' (2023) 26 *Journal of International Economic Law* 285.

72 Tienhaara et al (n 69).

73 Bård Harstad, 'Buy Coal! A Case for Supply-Side Environmental Policy' (2012) 120 *Journal of Political Economy* 77; Paul Collier and Anthony J Venables, 'Closing Coal: Economic and Moral Incentives' (2014) 30 *Oxford Review of Economic Policy* 492. See also Global Transition Fund in Newell and Simms (n 13).

off temporarily or completely.[74] If the latter is adopted – such commitments would not likely survive political changes in states, especially as the resource prices would be hiking up due to scarcity. Such a scheme could further be contrary to many states' constitutions and wider legal frameworks preventing alienation of natural resources or requiring their development.[75] Some further research is needed into the interaction of a potential deposit trade/compensation scheme with international law and specific national legal frameworks.

Finally, any international measures on coal ban or phase-out would need to take into consideration the notion of just transition. Just transition refers to a fair distribution of burdens and benefits as we transition to a low-carbon economy. A narrower definition of just transition focuses on workers, owing to the term's origin from US trade unions in the 1980s.[76] The Solidarity and Just Transition Silesia Declaration, adopted at COP24, is an example of a workforce-centred just transition document.[77] The Paris Agreement preamble mentions 'taking into account the imperative of a Just Transition of the workforce and the creation of decent work and quality jobs in accordance with nationally defined development priorities'.[78] For coal phase-out this is imperative as, globally, the coal mining industry employs around 11 million people.[79] Over time, a wider approach emerged, particularly in academia, bringing together all elements of society in transition, and encompassing energy justice, climate justice and environmental justice.[80] Beyond impact on workers, phase-out of coal mining or use will have an impact on energy price and availability, disproportionately affecting states relying on coal for access to energy. The Global Stocktake, adopted at COP28, reaffirms that 'sustainable

[74] Hans-Werner Sinn, *The Green Paradox: A Supply-Side Approach to Global Warming* (MIT Press 2012) 173–4, 176–8.

[75] For example, in the United Kingdom, the regulator responsible for oil and gas licensing is legally required to follow the objective to 'maximise the economic recovery' from the North Sea. UK Infrastructure Act 2015, section 41 amending section 9A of the UK Petroleum Act 1998. The Constitution of Ecuador in Art 1 establishes that 'Non-renewable natural resources of the State's territory belong to its inalienable and absolute assets, which are not subject to a statute of limitations'.

[76] Peter Newell and Dustin Mulvaney, 'The Political Economy of the "Just Transition"' (2013) 179 *The Geographical Journal* 132.

[77] Solidarity and Just Transition Silesia Declaration <www.ioe-emp.org/index.php?eID=dumpFile&t=f&f=134978&token=91237abd5b4e38c1e7c2e4364b2b8e7095d8e0fd> accessed 18 December 2023.

[78] Paris Agreement preamble.

[79] IBIS World, 'Global Coal Mining – Market Size, Industry Analysis, Trends and Forecasts (2023–2028)' <www.ibisworld.com/global/market-research-reports/global-coal-mining-industry/#TableOfContents> accessed 18 December 2023.

[80] Kirsten EH Jenkins et al, 'Politicising the Just Transition: Linking Global Climate Policy, Nationally Determined Contributions and Targeted Research Agendas' (2020) 115 *Geoforum* 138.

and just solutions to the climate crisis must be founded on meaningful and effective social dialogue and participation of all stakeholders'.[81] It further links action on fossil fuels to just transition by requiring the transition away from fossil fuels in energy systems to be conducted 'in a just, orderly and equitable manner' and the phase-out of 'inefficient fossil fuel subsidies that do not address energy poverty or just transitions'.[82] At present only very few parties to the Paris Agreement (3 per cent) mentioned the need and/or measures for a just transition for communities and workers dependent on coal.[83]

To conclude, international law on air pollution and climate change, so far, has had limited success in inducing coal phase-out at the international level. Informal initiatives and coal treaty ban proposals have been put forward, but in addition to the lack of political will, there are some further challenges that would need to be addressed in any such arrangement. These include sovereignty over natural resources, investment protection and compensation schemes, and just transition. The next section maps out and examines the national legislation and policy on restrictions.

14.4 Restrictions on coal production, sale and use in national legislation and policy

As energy production and use are primarily regulated at the national level, it is national legislation and policy that contain more specific and targeted interventions on coal restriction. Some of these, similar to international regulations, affect the coal industry indirectly through, for example, air pollution and climate regulation.[84] However, the adoption of climate targets has clearly accelerated coal phase-out in some states and led to regulations directly restricting coal use. Broadly speaking the policy/regulatory measures on coal restrictions could be divided into five types depending on activity: (1) energy planning and political pledges, (2) climate legislation, (3) restrictions of coal use, (4) restrictions on coal sale/export, and (5) restrictions on coal mining. These are examined in the following subsections.

14.4.1 Energy planning and political pledges

The adoption of binding coal phase-out or ban is often preceded by information action and energy planning policy indicating the vector of

---

81 First Global Stocktake, para 28, FCCC/PA/CMA/2023/L.17 para 9.

82 Ibid para 28.

83 'Nationally Determined Contributions under the Paris Agreement. Synthesis Report by the Secretariat (26 October 2022) FCCC/PA/CMA/2022/4' (n 48) para 190.

84 US 42 U.S.C. §7401 et seq. (1970). See Viñuales (n 37) 128.

progress. It is thus important not to discount non-binding pledges made at COPs or domestic forums. Political commitments to phase out coal use or production, as well as national energy planning, often precede the adoption of binding coal bans or phase-outs, or they provide a framework for accountability. However, as such commitments do not create binding obligations, governments can, and do, delay action without legal consequences. For example, the president of Hungary announced that the country would stop using coal by 2030 during the United Nations Climate Action Summit in 2019.[85] Hungary since joined the PPCA, pledging to phase out coal at its last remaining lignite power plant Matrai Eromu by 2025.[86] In November 2023, the Hungarian government extended the coal power plant's closure date until 2027.[87]

Italy's National Energy Strategy, adopted in 2017, pledged the phase-out of coal in electricity generation by 2025.[88] Although the government remains committed to the phase-out date,[89] there are concerns that the extension of a coal power plant in Sardinia might jeopardize the target.[90]

The UK has made significant progress toward the goal of unabated coal power phase-out by October 2024 with no direct coal restricting legislation. While the UK has ambitious climate change legislation, discussed earlier, the coal phase-out declarations remained political, despite the government's commitment to table the relevant legislation 'at the earliest opportunity' in 2021.[91] Two years later, with no bill proposed, the government reportedly

---

85 See Beyond Fossil Fuels, 'Europe's Coal Exit' <https://beyondfossilfuels.org/europes-coal-exit> accessed 18 December 2023.

86 Government of Hungary, 'Elkezdődött a mátrai erőmű átalakítása, a munkahelyek nincsenek veszélyben' (The Transformation of the Mátra Power Plant Has Started, Jobs Are Not in Danger) <https://kormany.hu/hirek/elkezdodott-a-matrai-eromu-atalakitasa-a-munkahelyek-nincsenek-veszelyben> accessed 18 December 2023.

87 'Hungary Plans to Keep Lignite-Fuelled Matrai Eromu Running until 2027' (*Newsbase*, 21 November 2023) <https://newsbase.com/story/hungary-plans-to-keep-lignite-fuelled-matrai-eromu-running-until-2027-302145> accessed 2 January 2024.

88 Italy's National Energy Strategy 2017 <www.mimit.gov.it/images/stories/documenti/BROCHURE_ENG_SEN.PDF> accessed 18 December 2023.

89 IEA, 'Italy 2023: Energy Policy Review' (2023) 128 <https://iea.blob.core.windows.net/assets/71b328b3-3e5b-4c04-8a22-3ead575b3a9a/Italy_2023_EnergyPolicyReview.pdf> accessed 18 December 2023.

90 Beyond Fossil Fuels, 'European Countries Delay Coal Phase-Outs and Gamble on Fossil Gas amidst Push for Decarbonised Power' (*Beyond Fossil Fuels*, 7 December 2023) <https://beyondfossilfuels.org/2023/12/07/european-countries-delay-coal-phase-outs-and-gamble-on-fossil-gas-amidst-push-for-decarbonised-power/> accessed 2 January 2024.

91 UK Government, 'End to Coal Power Brought Forward to October 2024' (20 June 2021) <www.gov.uk/government/news/end-to-coal-power-brought-forward-to-october-2024> accessed 18 December 2023.

considers legislation not necessary as 'coal generators have already committed to phasing out their coal operations'.[92]

Political pledges can thus be an important precursor to coal phase-out measures or an instrument for accountability. However, without legally binding obligations specific to the coal phase-out, their applicability might be limited.

14.4.2 Climate legislation and coal phase-out

National climate legislation is an important driver of coal phase-out albeit not always directly. Even where the emission targets in the legislation are non-sectoral, for countries with a coal sector, phase-out is often the only option for achieving them. For example, the UK's 2008 Climate Change Act was the first act of law establishing a legally binding climate change target.[93] The target was updated in 2019 to achieve net zero by 2050. The government must develop carbon budgets as well as proposals and policies to meet these budgets.[94] The Act further establishes the Committee on Climate Change, an independent, non-departmental public body, to advise the UK and devolved governments and parliaments.[95] While the committee does not provide the government with specific policy recommendations, its scientific advice has important weight in policy and law-making, including on coal phase-out.[96] However, the letter from the committee's chair did not dissuade UK ministers from approving a deep coal mine in Cumbria in 2021.[97] Ultimately, the way to achieve the binding target is up to government, and plans, incorporating coal, might still comply with the target.

Climate regulation can be used more directly to address coal phase-out. For example, in Canada, climate regulation is in federal jurisdiction while energy is in provincial. In 2012, the Canadian government adopted regulations to reduce CO_2 emissions from coal-fired power stations, establishing emission performance standards.[98] Such regulatory design allows operators to choose

---

92 Dimitris Mavrokefalidis, '"UK Abandons Proposed Law to Ban Coal Power Generation"' (*Energy Live News*, 26 June 2023) <www.energylivenews.com/2023/06/26/uk-abandons-proposed-law-to-ban-coal-power-generation/> accessed 3 January 2024.

93 For an overview of the Act see Thomas L Muinzer, *Climate and Energy Governance for the UK Low Carbon Transition: The Climate Change Act 2008* (Springer 2019).

94 Sections 1, 4–10 and 12–15.

95 Sections 32–42.

96 See discussion in Paula Walk and Nora Stognief, 'From Coal Phase-out to Net Zero: Driving Factors of UK Climate Policy' (2022) 138 *Environmental Science & Policy* 76.

97 'Climate Change: Minister Rapped for Allowing Cumbria Coal Mine' (*BBC News*, 30 January 2021) <www.bbc.com/news/science-environment-55871503> accessed 2 January 2024.

98 Reduction of Carbon Dioxide Emissions from Coal-fired Generation of Electricity Regulations SOR/2012-167 (30 August 2012).

to deploy carbon capture or switch to less polluting generating fuels or renewables. The performance standard applied to all new units from 2015, with staggered application dates for existing units based on their date of commission. In 2018, these regulations were updated with an accelerated timeline bringing forward the compliance dates.[99] The regulations were supplemented with provincial regulations and policies, with Ontario leading the coal phase-out in 2014.[100]

A similar approach emerged in the US, where a number of states legislated emission performance standards for power plants, effectively banning new coal-fired power plants. California pioneered the process in 2006, establishing emission standards for local publicly owned electric utilities to the level of emissions of a gas-fired plant.[101]

Using climate legislation to phase out coal power can thus be an effective regulatory tool, and allows a tailored approach based on the features of specific jurisdictions and energy systems. However, it also can leave an option to use carbon capture with continued coal use or, where targets are non-sectoral, avoid coal-related measures altogether.

14.4.3 Coal use restrictions

Legally binding measures to ban or phase out the use of coal directly are not new. As early as 1257, in England, the burning of 'sea coal' (bituminous coal) was banned to address air pollution in London.[102] Today, countries around the world pass legislation to end or limit the use of coal for electricity production, or close existing and prevent the opening of new coal power plants. The local context and structure of the energy system dictate the appropriate regulatory design in each case.

For example, in Europe, Finland, Germany and the Netherlands passed legislation to place a legally binding obligation to phase out the use of coal in electricity production by 2029, 2038 and 2025, respectively.[103] The

---

99 Regulations Amending the Reduction of Carbon Dioxide Emissions from Coal-fired Generation of Electricity Regulations: SOR/2018-263.

100 Harris, Beck and Ivetta Gerasimchuk (n 12).

101 California Emission Performance Standard – SB 1368 (Perata, Chapter 598, Statutes of 2006).

102 Cass (n 39) 389.

103 Germany – Kohleausstiegsgesetz (Law to Reduce and End Coal-fired Power Generation and to Amend Other Laws (Coal Phase-out Act) 8 August 2020, hereinafter German Coal Law; the Netherlands – Wet verbod op kolen bij elektriciteitsproductie (Law Banning Coal in Electricity Production) 11 December 2019, hereinafter Dutch Coal Law; Finland – Laki hiilen energiakäytön kieltämisestä (Act on Banning the Use of Coal as Energy) 29 March 2019, hereinafter Finnish Coal Law.

approach in all three Acts is similar in that they set a general prohibition with a cut-off date, a set of exceptions, monitoring and compliance authority, as well as penalties for violations. Exceptions to the ban in all three cases are rather limited and include phase-out extensions for certain installations, for example based on efficiency, amount of CO_2 produced,[104] or security of supply in emergencies.[105]

The issue of compensation for power plant operators is raised in the German and Dutch Acts. The Dutch law allows operators to request compensation if they are disproportionately affected compared to other operators of coal-fired production installations.[106] The compensation may be granted if it does not raise state aid rules concerns in the European Commission.[107] The German law develops a tender procedure, whereby operators of hard coal power plants may participate in annual tenders in which they offer to shut down their plant voluntarily in exchange for compensation.[108]

In Germany, the coal phase-out legislation was particularly challenging to adopt due to the high reliance on coal following the closure of nuclear power stations. Coal legislation was adopted following a lengthy policy development and public debate process,[109] and in addition to developing an elaborate compensation regime also differentiates between hard and brown coal, establishes a phase-out date for coal mining, a structural support framework, which is discussed later.[110]

Hawaii closed its last power plant in September 2022, as the state's contract with the utility was set to expire after 30 years in operation.[111] Two years prior, the state legislature adopted the law requiring no new coal permits to be awarded after 31 December 2022.[112] Legislation, in this case, is relatively simplistic and straightforward given the absence of disgruntled investors, but it does not address the coal ban's impact on the state's electricity mix which now heavily relies on oil, raising prices for consumers.[113] Nevertheless,

[104] Dutch Coal Law, Art 3.

[105] Finnish Coal Law, Art 6. Emergency exceptions are limited to three months, renewable up to three months at a time by government decree.

[106] Dutch Coal Law, Art 4.1.

[107] Ibid, Art 4.3.

[108] German Coal Law, sections 10–26. See also Louisa Raitbaur, 'The New German Coal Laws: A Difficult Balancing Act' (2021) 11 *Climate Law* 176, 179–80.

[109] See Baudisch and Fouquet (n 12); Raitbaur (n 108).

[110] Strukturstärkungsgesetz Kohleregionen (Structural Support Act for Coal Regions) 8 August 2020.

[111] Hallie Golden, 'Hawaii to Close Its Only Coal Power Plant in a Step toward Renewable Energy' (*The Guardian*, 1 September 2022) <www.theguardian.com/us-news/2022/aug/31/hawaii-close-coal-power-plant-renewable-energy> accessed 2 January 2024.

[112] Hawaii Bill for an Act Relating to the Environment SB2629 SD2 HD1 (2020).

[113] Golden (n 111).

elsewhere long-term contracts can place significant barriers to the accelerated energy transition.[114]

Although coal phase-out legislation is by no means a universal phenomenon, it is a useful tool to create legal commitments on coal able to withstand political changes in the era of polarization of climate discourse. The examples of coal use phase-out legislation in this section demonstrate how regulation can be tailored to address the legal culture and structure of the energy system in different jurisdictions.

14.4.4 Coal sale regulation

Legal restrictions on coal sales can be deployed for a number of reasons. Large producer-exporter states might restrict sales in times of domestic power crises. For example, Indonesian Domestic Market Obligation decrees require coal mining companies to sell a share of produced coal to the national utility at a price capped below market level.[115] The mechanism was used in September 2022, to ban coal exports in light of the global energy crisis.[116] However, especially in light of COP28's fossil fuel focus, large producers might consider phasing out exports on climate grounds.[117]

Climate concerns can also motivate coal sale bans. Despite reluctance to legislate on electricity generation coal phase-out, the UK did adopt regulations restricting the sale of traditional house coal.[118] On a larger scale, the state of Washington in 2019 passed a law which prohibits utilities from buying electricity generated from coal after 2025, and in Hawaii, the Coal Ban Law prohibited the coal power purchase agreements.[119]. These

[114] Anatile Boute, 'Environmental Force Majeure: Relief from Fossil Energy Contracts in the Decarbonisation Era' (2021) 33(2) *Journal of Environmental Law* 339.

[115] Richard Bridle, Anissa Suharsono and Mostafa Mostafa, 'Indonesia's Coal Price Cap: A Barrier to Renewable Energy Deployment' (IISD 2019) <www.iisd.org/system/files/publications/indonesia-coal-price-cap.pdf> accessed 18 December 2023.

[116] Christina Bernadette, 'Indonesia Bans Coal Exports in January on Domestic Power Worries' (*Reuters*, 1 January 2022) <www.reuters.com/markets/commodities/indonesia-bans-coal-exports-january-domestic-supply-worries-media-2022-01-01/> accessed 3 January 2024.

[117] Fergus Green, 'From Laggard to Leader? Why Australia Must Phase out Fossil Fuel Exports, Starting Now' (*The Conversation*, 18 December 2023) <http://theconversation.com/from-laggard-to-leader-why-australia-must-phase-out-fossil-fuel-exports-starting-now-219912> accessed 3 January 2024.

[118] UK Government, 'Restrictions on Sale of Coal and Wet Wood for Home Burning Begin' (1 May 2021) <www.gov.uk/government/news/restrictions-on-sale-of-coal-and-wet-wood-for-home-burning-begin> accessed 18 December 2023.

[119] Senate Bill SB 5116 Supporting Washington's Clean Energy Economy and Transitioning to a Clean, Affordable, and Reliable Energy Future (2019) section 3; Hawaii Bill for an Act Relating to the Environment para 342B.

regulations would be appropriate in specific energy market structures as they can be used instead of, or in addition to, coal use phase-outs to place obligations on energy suppliers.

14.4.5 Coal production restrictions and just transition

As discussed in Section 14.2, coal mining produces a significant contribution to methane emissions, in addition to developing coal for further consumption. It is therefore important to develop legally binding commitments to prevent awards of new coal mining permits and consider whether the existing permits could be revoked on climate grounds.

Preventing the award of new coal mining permits is fairly straightforward with appropriate regulation, but this has not been a popular regulatory tool so far.[120] The UK Energy Act 2023 nearly included a clause prohibiting the opening of new coal mines and the licensing of new coal mines by the Coal Authority or its successors.[121] If included, this clause would effectively prohibit future licensing of all new coal extraction, including area extensions to existing coal mining in the UK. Before the Bill was passed, it was amended to remove the clause.[122]

Litigation can be useful to challenge new coal permits on climate grounds, as was the case in New South Wales Rocky Hill case.[123] The court rejected the coal mining company's appeal against the refusal of planning permission based on environmental and social impact assessment.[124] Nevertheless, the potential contribution of the mine to greenhouse gas emissions was an important consideration in the court's deliberations.[125] The proposed Cumbria coal mine in the UK, mentioned previously, has a pending challenge in the High Court on the grounds of application of market substitution argument,[126] and failures to consider downstream emissions and international impacts of the project.[127]

---

[120] See, for example, Jo Clarke, 'Australia Will Not Ban Coal, Gas Developments' (*Argus*, 27 February 2023) <www.argusmedia.com/en//news/2423773-australia-will-not-ban-coal-gas-developments> accessed 3 January 2024.

[121] Energy Bill (17 April 2023) HL Bill 130, section 270.

[122] House of Commons, Energy Bill [HL] Amendment Paper (10 May 2023).

[123] *Gloucester Resources Limited v. Minister for Planning*, [2019] NSWLEC 7.

[124] Ibid, para 556.

[125] Ibid, para 515, paras 555–6.

[126] See Guy Dwyer, '"Market Substitution" in the Context of Climate Litigation' (2022) 12 *Climate Law* 1; Justine Bell-James and Brianna Collins, '"If We Don't Mine Coal, Someone Else Will": Debunking the "Market Substitution Assumption" in Queensland Climate Change Litigation' (2020) 37 *Environmental and Planning Law Journal* 167.

[127] *Friends of the Earth v. Secretary of State for Levelling Up, Housing and Communities; and South Lakeland Action on Climate Change v. SSLUHC (Whitehaven coalmine)*. See complaint in

Refusing a permit in the presence of a license or revoking the existing permits on climate grounds presents significant complications in most jurisdictions. In addition to the threat of investment arbitration, there might not be the legal basis for permit withdrawal. In some cases, where the industry heavily relies on government support, withdrawal of subsidies might be enough to close production.[128] In 2007, Germany ended subsidies for hard coal production, which led to the closure of mines in the Saarland region in 2012, and in the Ruhr in 2018.[129]

Normally, when a licence for a mine is granted, there is an exhaustive list of grounds for refusing a permit. Similarly, for an already operating mine, the conditions for permit revocation are usually restricted in legislation. These challenges came to the fore in Germany during the Hambach forest mine extension discourse. The German Federal Mining Act lists the grounds for rejection of an extraction licence or approvals.[130] The grounds include the mining location 'overriding public interest' prohibiting exploration in the relevant field.[131]

Whether climate emergency and the need to reduce emissions would constitute 'harm to society' would depend on judicial interpretation. Withholding and revoking coal mining permits without a well-developed appropriate legal framework would result in violations and high compensations. However, withdrawal of government support in some cases should be more actively pursued. Currently, despite EU state aid rules restricting coal subsidies, fiscal support for coal mining is declining but not eliminated.[132]

14.4.6 Just transition

Closures of coal power plants and coal mines are associated with job losses and wider socio-economic impacts on the regions. In this context, a number of

Climate Case Chart database <https://climatecasechart.com/non-us-case/friends-of-the-earth-v-secretary-of-state-for-levelling-up-housing-and-communities-and-south-lakeland-action-on-climate-change-v-ssluhc-whitehaven-coalmine/> accessed 18 December 2023.

[128] Karl Storchmann, 'The Rise and Fall of German Hard Coal Subsidies' (2005) 33 *Energy Policy* 1469.

[129] Pao-Yu Oei, Hanna Brauers and Philipp Herpich, 'Lessons from Germany's Hard Coal Mining Phase-out: Policies and Transition from 1950 to 2018' (2020) 20 *Climate Policy* 963.

[130] German Federal Mining Act (BBergG) sections 11–13 and 55. See Ryan Rafaty, Sugandha Srivastav and Björn Hoops, 'Revoking Coal Mining Permits: An Economic and Legal Analysis' (2020) 20 *Climate Policy* 980, 986.

[131] German Federal Mining Act (BBergG) sections 48(2), 11, 12(1). Rafaty, Srivastav and Hoops (n 130).

[132] Ipek Gençsü et al, 'Phasing out Public Financial Flows to Fossil Fuel Production in Europe' (2020) 20 *Climate Policy* 1010.

just transition-oriented legal and policy measures can be adopted to support regions highly dependent on coal production or use. These include detailed transition planning at the policy level; community-led place-based planning; establishment of ringfenced funding for retraining and reskilling of workers as well as re-purposing and reusing of the existing infrastructure; and monetary compensation to laid-off workers and resource holders. As part of its Just Transition Mechanism, the EU has established a EUR7.5 billion fund to support regions in the transition toward climate neutrality by 2050.[133] In Germany, the coal phase-out law was accompanied by a law on structural support for coal regions.[134] The law provides for financial support for coal workers of pre-retirement age, and subsidies for the energy grid and heating transition to partially offset higher prices for electricity.[135] The Dutch coal exit law established a fund from which payments can be made to employees of a production installation as well as for retraining.[136]

These processes are usually associated with creating independent commissions with participation from trade unions, NGOs, scientists and industry representatives.[137] Such bodies are tasked with planning and contributing to the achievement of social consensus through public participation and co-development of outcomes.[138] For example, in Canada, the proposed Sustainable Jobs Bill would see the establishment of a Sustainable Jobs Partnership Council to engage with stakeholders and provide the government with independent advice.[139] Overall, there is a growing body of just transition policies and regulations around climate, labour and public participation, at both the national and international levels.

In conclusion, there are a number of legislation and policy designs available to parliaments and regulators at the national level. Regulation on coal phase-out must be tailored to national circumstances, including the status as producer or consumer, legal framework and obligations, energy market design and just transition considerations.

---

[133] Regulation (EU) 2021/1056 of the European Parliament and of the Council of 24 June 2021.

[134] German Structural Support Act. See Raitbaur (n 108).

[135] Raitbaur (n 108). German Coal Law, section 57–58. German Structural Support Act section 26.

[136] Dutch Coal Exit Law Art 4(a).

[137] Raphael J Heffron, *Achieving a Just Transition to a Low-Carbon Economy* (Palgrave Macmillan 2021) 43–85.

[138] See, for example, Scottish Just Transition Commission <www.justtransition.scot/the-commission/> accessed 18 December 2023.

[139] House of Commons Canada, Bill C-50 An Act Respecting Accountability, Transparency and Engagement to Support the Creation of Sustainable Jobs for Workers and Economic Growth in a Net-zero Economy (11 December 2023).

14.5 Conclusion

The climate rationale of rapid coal phase-out is not in doubt in the energy and climate change mitigation analysis and modelling. Despite coal's abundance and large share in today's power production, achieving global climate goals requires both mining and use of coal to rapidly decline over the next few decades. Achieving this necessitates both a collective international approach and tailored regional and national regulatory interventions. Importantly, the burdens and benefits of these regulations will not fall fairly on countries and communities, as more focused efforts are needed from coal reserve holders, producers and large (and growing) consumers.

This chapter mapped out and examined policies and laws on restrictions of coal use, sale and production. At the international level, existing air and climate change regulations have a role to play, especially with increased attention to fossil fuels since COP26. An internationally imposed ban would be challenging to adopt, owing to the lack of political will as well as legal limitations placed by sovereignty, but also concerns of investment law and equity. Nevertheless, informal governance and potential frameworks of reserve, holder compensation offer some innovative legal solutions.

At the national level, many countries have adopted policies and legislation targeting specific activities in the coal sector, allowing them to tailor regulations to specific local concerns. Such regulations need to take into account the wider legal framework and consider potential conflict with, for example, state-aid limitations or constitutional and energy law.

Finally, just transition is an increasingly important notion, at the intersections of public, climate and labour law, calling for targeted support for the most affected regions, workers and communities in coal phase-out regulation and policy.

PART 2

The Broader Normative Context

15

Should Owners of Carbon-Intensive Assets Be Compensated if Required to Cease Operation Due to the Energy Transition?

Ignacio Herrera Anchustegui and Gunnar S. Eskeland

15.1 Introduction

Businesses come and go, and so do individual assets, practices and technologies. From the general point of view of economic development, these transitions may involve significant adjustment costs but also reflect and utilize improvements in productivity and changes in priorities at both the individual and collective levels.

In an economy that includes individuals, firms and contracts, usually when assets have been stranded by economic development – for example, yesterday's steam engines – no compensation is offered to their owners. Often, the mere prospect of such compensation would risk impeding the process of development. For instance, it might make engine owners and their bankers less concerned about the future and less dynamic in their investment in new technologies.

In the context of the energy transition, illustrative examples of long-lived assets that risk being stranded by the energy and technology transformation include coal mines, oil wells, diesel locomotives and internal combustion vehicles, as well as some industrial processes related to heavy industrial activity such as steel production, electricity generation and/or cement plants. Climate-friendly technologies to help mitigate adverse climate effects, such as renewable energy, including onshore, offshore wind and solar panels and 'green' hydrogen, are being promoted and benefit from public subsidies. In this scenario, investors are redirected to shift their portfolio from carbon-intensive assets to these new energy projects.

The current climate emergency, starting with the Rio Convention of 1992 and continuing with the Kyoto Protocol of 1997, has compelled states to commit to reducing greenhouse gas (GHG) emissions so that they do not exceed global warming of 2 °C (or preferably 1.5 °C) above preindustrial levels.[1] A consequence of this is government action that may bring about profound changes in government regulation and our economic system.[2]

In particular, government policies may play a role in prematurely closing emission-intensive activities or retiring assets as their operation conflicts with a need to mitigate GHG emissions. Some conflicts have already been adjudicated in international arbitration tribunals or national courts, as discussed in this chapter. Similarly, broader discussions revolve around protecting investments under the 1994 Energy Charter Treaty, under renewed scrutiny as a result of the withdrawal of several European countries.[3] In the area of oil and gas, some European states have banned the exploration for, or extraction of, oil and gas, and some contracts have been terminated, leading to compensation claims.[4] Litigation cases have also emerged regarding electricity-generating plants, such as the compensation paid to owners of coal- and lignite-fired electricity-producing plants in Germany and the Netherlands due to the termination of their licences (discussed in Sections 15.3 and 15.4). Anticipated closures or the revocation of licences for some polluting activities can be linked to states' obligations to reduce their GHG emissions. These can be demanded by law and courts vis-à-vis states or companies. Examples of the former are the *Urgenda* Dutch Supreme Court case,[5] or the 2023 *Klimaatzaak* case where the Brussels Court of Appeal declared that the Belgian state must reduce GHG emissions by at least 55 per cent by 2030 in comparison to

1 Paris Agreement to the United Nations Framework Convention on Climate Change, 12 December 2015.

2 David Khachvan, 'Non-compensable regulation versus regulatory expropriation: are climate change regulations compensable?' [2020] 35(1–2) *ICSID Review – Foreign Investment Law Journal*, 154.

3 European Commission, European Commission proposes a coordinated EU withdrawal from the Energy Charter Treaty (7 July 2023); Martin Dietrich Brauch, 'Should the European Union fix, leave or kill the Energy Charter Treaty?' [2021], Columbia Center on Sustainable Investment; J Robert Basedow, 'The Achmea judgment and the applicability of the Energy Charter Treaty in intra-EU investment arbitration' [2020] 23 *Journal of International Economic Law*, 271.

4 Anatole Boute, 'Investor compensation for oil and gas phase out decisions: aligning valuation methods to decarbonization' [2023] 23 *Climate Policy*, 1087.

5 *De Staat der Nederlanden v. Stichting Urgenda*, Hoge Raad, ECLI:NL:HR:2019:2007, 13 January 2020.

1991.[6] An example of the latter is the 2022 *Milieudefensie et al v. Royal Dutch Shell plc* case in the Netherlands.[7]

In this chapter, we examine instances in which states directly demand the cessation of operations in carbon-intensive industrial sectors before the end of their expected economic or authorized operational lifespan and discuss whether compensation is due. To explore this, we draw from legal and economic arguments. We find that the answer depends on the protection afforded to assets, including property and contractual rights. Such rights are themselves limited by the rights of others. The protection of property rights typically has limitations, including those related to the exercise of state powers or the promotion of public policies. One crucial constraint on government action against asset owners is protection against expropriation. To be lawful, expropriation, as we shall show, must usually satisfy several criteria, including that of owners receiving adequate compensation.

In general, the legitimate scope for the government to shape policy in protecting the public interest has been deemed confined to environmental protection.[8] This includes policies encouraging the energy transition, such as emission regulation and taxes, quotas, and support for renewable energy and green research and development. Government measures play a significant role in the envisioned energy transition, such as Europe's Emissions Trading System (ETS), potentially phasing out or stranding fossil fuel-related assets.

A compensation-free regime entails that each incremental tightening of expected climate policies reduces investment in fossil fuels. This potentially saves money and reduces resistance to the advancing transition. However, such a policy affects the security of property rights, the flow of investment, and legal predictability, and is often deemed illegal, potentially leading to financial instability.[9] If large industries have to close due to their carbon footprint, it could result in job losses, unpaid loans and, possibly, material damage and energy scarcity.[10] Regarding the social, or total, cost of the

[6] *VZW Klimaatzaak v. Kingdom of Belgium & Others*, Cour d'appel Bruxelles – 2021/AR/1589, 2022/AR/737 et 2022/AR/891, 30 November 2023.

[7] *Milieudefensie et al. v. Royal Dutch Shell plc*, Rechtbank Den Haag C/09/571932 / HA ZA 19–379, 26 May 2021. This was reversed on appeal, in 2024.

[8] See dealing with this issue from a European perspective, inter alia: Andrew J Jordan and Duncan Liefferink (eds), *Environmental Policy in Europe: The Europeanization of National Environmental Policy* (Routledge 2004); Geert Van Calster and Leonie Reins, *EU Environmental Law* (Edward Elgar Publishing 2017); Mariachiara Alberton, 'Environmental protection in the EU member states: changing institutional scenarios and trends' [2012] 363 *L'Europe en Formation*, 287.

[9] Achim Hagen, Niko Jaakkola and Angelika Vogt, 'The interplay between expectations and climate policy: compensation for stranded assets' [2019], IAEE Energy Forum, 29.

[10] See generally on 'carbon bubble': JF Mercure et al, 'Macroeconomic impact of stranded fossil fuel assets' [2018] 8 *Nature Climate Change*, 588; Christophe McGlade and Paul

transition, much therefore hinges on policies not being introduced with undue haste and not being discriminatory. The same factors are important when deciding whether or not policy-induced closures justify compensation to owners.

Our approach to presenting arguments for and against compensation is as follows. In Section 15.2, we explain what is meant by our focus: states requiring assets to be closed down. In Section 15.3, we link our discussion to the role and nature of government action and to when and in what circumstances the rights of owners of property rights or contractual rights demand compensation from the government. We conclude that there are strong arguments both in law and in practice to show that while permits, licences and concessions may be revoked, compensation is owed when the action would otherwise represent a government breach of the right of property or contractual stipulations. Section 15.4 explains the opposing case that compensation is not justified when government action is found to be within its general policy making powers, reflects the public interest, and is non-discriminatory and lawful. In Section 15.5, we conclude with two main observations. First, such cases are likely to arise in the future. Second, both lines of argumentation lead to plausible legal conclusions, and courts tasked with resolving disputes are likely to strike a balance between the two. While governments may pursue public interest goals along general lines that are non-discriminatory and can be anticipated, they must also be prepared to compensate owners if and when policies change in a way that is found to be unduly abrupt and discriminatory.

15.2 What do we mean by states requiring a carbon-intensive asset to cease operating before the end of its expected operational life?

Climate change mitigation necessitates government policy introducing forceful measures. Internationally, most states have committed to implementing actions to reduce their GHG emissions as part of their Nationally Determined Contributions to the Paris Agreement Framework. Others have gone further and set binding targets to mitigate and adapt to

Ekins, 'The geographical distribution of fossil fuels unused when limiting global warming to 2 °C' [2015] 517 *Nature*, 187; Editorial, 'Carbon bubble toil and trouble' (2014) 4 *Nature Climate Change* 229; Forbes, *We're living through a carbon bubble—here's what we can do about it* (26 March 2021), www.forbes.com/sites/ashoka/2021/03/26/were-living-through-a-carbon-bubble- -heres-what-we-can-do-about-it/?sh=205d6e491b62; *The Guardian* and Fiona Harvey, *What is the carbon bubble and what will happen if it bursts?* (4 June 2018), www.theguardian.com/environment/2018/jun/04/what-is-the-carbon-bubble-and-what-will-happen-if-it-bursts.

climate change.[11] Both imply at least efforts – and sometimes commitments – to develop tools and regulatory frameworks to reduce GHG emissions, involving public interventions and citizen engagement. Policies may provide incentives to alter consumption patterns, promote technological solutions, or even prohibit highly polluting activities.[12]

Some European states have adopted stringent measures that directly restrict the use or enjoyment of industrial assets. Some of these tackle carbon-intensive activities and technologies, as we discuss in this chapter in relation to electricity production in Germany and the Netherlands, where electricity-producing plants fuelled by coal have been required to be phased out, and/or to limit their production capacity to 35 per cent or less.[13] Other measures apply to non-carbon-intensive facilities which are perceived as too dangerous, such as in Germany where nuclear power plants had their licences revoked by a decision of the German Bundestag in 2011.[14] In common, these measures involve the cessation of certain activities, either immediately or within a specified timeframe, possibly infringing on the right to enjoy property and results of economic ventures.

While other climate mitigation interventions may curtail the operative lifetime of emission-intensive assets, we analyse cases in which the government intervenes directly in a way that requires ceasing the operation

---

11 Sweden, the United Kingdom, France, New Zealand, Denmark and Hungary are examples of countries with national binding net-zero targets. The European Union has set out 'a binding objective of climate neutrality in the Union by 2050 in pursuit of the long-term temperature goal set out in point (a) of Article 2(1) of the Paris Agreement' in Article 1 of the Regulation (EU) 2021/1119 establishing the framework for achieving climate neutrality (OJ [2021] L 243/1).

12 Illustrative examples are CO_2 taxes on fuels or use of certain polluting activities, discouraging use of gasoline and diesel cars, and electric car tax exemptions (both in use in Norway and many other countries, see more on this: Shiyu Yan and Gunnar S Eskeland, 'Greening the vehicle fleet: Norway's CO_2-differentiated registration tax' [2018] 91 *Journal of Environmental Economics and Management* 247; Ignacio Herrera Anchustegui and Aleksander Glapiak, 'Wind of change: A Scandinavian perspective on energy transition and the "greenification" of the oil and gas sector', in Katarzyna Gromek-Broc (ed), *Regional Approaches to the Energy Transition: A Multidisciplinary Perspective* (Springer 2023). In Europe, several countries are now discussing bans on the sale of new fossil-fuelled cars around 2035.

13 Article 3 of Wet van 11 December 2019, houdende Regels voor het produceren van elektriciteit met behulp van kolen (Wet verbod op kolen bij elektriciteitsproductie [Prohibition of Coal in Electricity Production]).

14 For a detailed discussion of the nuclear phase-out, see: Federal Office for the Safety of Nuclear Waste Management, *The nuclear phase-out in Germany* (19 April 2023). While these examples do not involve fossil fuels, their legal processes are of relevance as they, too, relate to government policing powers, public interest, and whether there is undue haste or discrimination.

of carbon-intensive assets or processes. These interventions are primarily driven by the need to mitigate or adapt to the effects of climate change. The nature of these interventions can vary significantly, but they hinder the free enjoyment of possessions or property. Consequently, their owners or possessors may not utilize them as intended or for the foreseen duration. In some instances, the order may limit or even revoke a granted right, as, for instance, where a concession or licence awarded for a lignite power plant to operate for 50 years is reduced to just 25 years or revoked altogether.

These restrictions on property or contractual rights will reduce the value of the asset for the owner, in some cases resulting in a total loss of the value.[15] The assets are, therefore, stranded. Stranded assets can be defined as assets that lose value well before their anticipated useful life due to changes in legislation, regulation, market forces, disruptive innovation, societal norms or environmental shocks.[16] Rendering an asset stranded by state intervention could involve the physical taking of property (a taking or direct expropriation of an asset) or an action that prohibits or limits an activity but does not involve actually removing the property from a person. Both entail a loss of asset value.[17] In addition, such government interference may lead to an erosion of rights.[18] This can affect both nationals and foreigners, triggering the application of national or international law and, typically, of provisions protecting foreign investment.

Our discussion aims to cover any state intervention that has the effect of demanding the cessation or imposition of a limit on the operation of a carbon-intensive asset. This can take several forms: a law prohibiting an activity, revoking a licence or concession through an administrative decree, or any other expression of the state's intent with a similar outcome. These actions may be classified as direct or indirect expropriation – a topic well debated in the literature of investment law.[19] Expropriation is almost always

---

15 Frederick Van der Ploeg and Armon Rezai, 'Stranded assets in the transition to a carbon-free economy' [2020] 12 *Annual Review of Resource Economics*, 281, p 289.

16 Emily Hammond and Jim Rossi, 'Stranded costs and grid decarbonization' [2016] 82 *Brooklyn Law Review*, 645.

17 See, for example: Christoph Schreuer, 'The concept of expropriation under the ETC and other investment protection treaties' [2005] 2 *Transnational Dispute Management*, 1, p 5; Wil D Verwey and Nico J Schrijver, 'The taking of foreign property under international law: a new legal perspective?' [1984] 15 *Netherlands Yearbook of International Law*, 3; UNCTAD, *Series on Issues in International Investment Agreements: Taking of Property* (2000), pp 4, 41; W Kühn and U Wiegel, 'The application of international law and treaty provisions by arbitrators' [2003] 4 *The Journal of World Investment*, 451.

18 UNCTAD, *Series on Issues in International Investment Agreements: Taking of Property* (2000), p 20.

19 For some literature dealing with the issue of expropriation see, inter alia, Jörg Kammerhofer, *International Investment Law and Legal Theory: Expropriation and the Fragmentation of Sources* (Cambridge University Press 2021); V Lowe, 'Regulation or expropriation?' [2002] 55

compensated, whether it is direct or indirect, lawful or unlawful. Our focus is on situations where the government requires the cessation of an activity, the utilization of a technology or the operation of a facility. Cases in which a plant shuts down because of more general policies, such as emission taxes (or other taxes or input costs), rendering operations unprofitable, can be argued to have a similar effect in the long run but are not part of our concern here.[20]

Two important notes of caution should be mentioned. First, while compensation from the government may be a legal obligation, there are instances where compensation is provided primarily due to political or business considerations to prevent a dispute, such as when a case is settled between parties without establishing or recognizing guilt or responsibility. Second, literature and case law, both domestic and international, allow significant leeway for government actions, including tax changes, without necessarily incurring an obligation to compensate. We cover this in some detail in what follows.

15.3 The case for compensation reflects legal systems that protect rights and discipline governments in the pursuit of public interest

15.3.1 Protection of property: the relationship between an asset and its owner

Legal principles, including ancient systems such as Hammurabi's Code, Roman Law and Mosaic Law, support tranquillity and predictability by defining rights and ascribing liability to those who violate these rights. This applies to cases involving another private party and, ultimately, the state. Relevant to our discussion are rights of property, possession and use related to productive assets, whether it is a truck, a power plant, an oil well or a firm. These principles, which are related to the protection of rights in rem (or 'goods'), also establish penalties for the violator, and may require compensation to be paid by the transgressor to the complainant.[21] In the

Current Legal Problems, 447; Steven Ratner, 'Compensation for expropriations in a world of investment treaties: beyond the lawful/unlawful distinction' [2017] 111 *American Journal of International Law*, 7; Peter D Isakoff, 'Defining the scope of indirect expropriation for international investments' [2012] 3 *The Global Business Law Review*, 189.

20 As will be seen in our analysis, however, such broader policy changes have often been met (and will be) successfully by findings that the policies are within the broad regulatory ('policing') powers of government, thus not expropriatory, thus not commanding compensation to be legal.

21 For instance, in public international (investment) law, when private property is taken or affected by the state it is an accepted principle of law that compensation is due. See: David Collins, *An Introduction to International Investment Law* (2nd edn, Cambridge University Press 2023), p 29.

past, rules akin to 'an eye for an eye' have reflected violation of rights and compensation as part of a general principle of proportionality and restorative justice.[22]

On the side of an owner whose rights to the peaceful enjoyment of their asset have been interfered with (as in reduced, or taken), compensation makes them financially sound or unharmed, even though they have been deprived of an asset or its service flow. On the side of the person or entity affecting the rights, compensation may serve more broadly to discourage or prevent such interference, as they can expect to be sanctioned and/or required to compensate the affected party.

Two other elements are incorporated into these introductory statements. First, legal systems have granted the state the power to enforce the law and compel compensation from the affecting party. Second, over time, it has become an accepted legal principle that whenever the state deprives someone of property or interferes with an owner's ability to enjoy property, there will usually be compensation. In other words, the state is not free to 'take'.

Generally, rights, such as that to property, can provide peace and predictability, facilitating, but also directing, asset creation. Nevertheless, in any given situation, such protection and respect for property is most directly in the owner's interest. This is well reflected in the work of early commentators, such as Adam Smith.[23] Between the protection of private property (as opposed to it being taken or having its use restricted) and the interest of the collective, there may be potential tension and conflict.[24] Such tension, which may be between a property owner and a government, raises the question of whether state action that harms the interests of an owner can be delimited, and if so how.

---

22 The term 'eye for an eye', in the earlier Hammurabi Code as well as in Mosaic law is quite literally proportional: When there is punishment: 'Anyone who injures their neighbour is to be injured in the same manner', and similarly when 'damages' are metered out to be paid by the offender to an owner whose asset has been harmed.

23 'Civil government, in so far as it is instituted for the security of property, is in reality instituted for the defence of the rich against the poor, or of those who have some property against those who have none at all', Adam Smith (1775–76), *An Inquiry into the Nature and Causes of the Wealth of Nations, Book V, Part II, Of the Expense of Justice* (Liberty Classics, Oxford University Press 1976, 1979), p 715.

24 An interesting literature narrates and analyses how rights – acknowledged by others – emerge in frontier conflict settings, such as in the North American west (land rights and mining claims) and the Amazon frontier. Mining claims illustrate well how the rights and their protection can come about as collective action (and emerging government), in the sense that individual miners would find themselves without protection. A collection of essays is found in Michelle R Garfinkel and Stergios Skaperdas, *The Political Economy of Conflict and Appropriation* (Cambridge University Press 1996).

Consequently, there is always tension around rights seen as political and historic constructs: if there is a tendency to protect rights to convey peace and confidence in owners, it may also serve to safeguard current values and policy choices and thereby suppress innovation, new technologies, disruption, youth and potential. Through his idea of creative destruction, Schumpeter emphasizes that such oppression of the new is a deterrent to innovation and technological change.[25] Patents were historically monopoly positions that a state power – a king – could sell to gain revenue, but with consequences that were detrimental to dynamics, competition and innovation. Today's patents, or the protection of intellectual property more generally, can be seen as an attempt to strike a balance: the patent holder is rewarded through a potential monopoly position, but it is limited, not the least in its duration.

15.3.2 Expropriation of assets by the state and protection of owners' rights

Legal provisions that explicitly impose constraints on takings by the state or the sovereign are typically found in modern systems as part of basic principles of the rule of law, respect for private property and separation of powers. Protection of private property and a duty of compensation when the state hinders property (or the ability to enjoy it peacefully) are found in most constitutions across the globe. Such provisions protect asset owners against arbitrary interference and expropriation and are central to most regulatory frameworks. These protections are anchored on the limitation of power of the state, the rule of law and the idea of primacy of individual rights – all these concepts connected to Western liberal ideas.[26]

An example of this is the Fifth Amendment to the United States Constitution, which, as part of the 1791 Bill of Rights, safeguards property from government abuses: 'No person shall be … deprived of life, liberty or property, without due process of law; nor shall private property be taken for public use without just compensation.'

More recently, in December 2021, the South African Congress rejected a proposed modification of Section 25 of the South African Constitution, which would have allowed land expropriation to occur with compensation being 'nil'. The amendment was rejected, even though the argument was

[25] A modern contribution that emphasizes this problem of liberalization, especially through financial channels is Raghuram G Rajan and Luigi Zingales, *Saving Capitalism from the Capitalists: Unleashing the Power of Financial Markets to Create Wealth and Spread Opportunity* (New York: Crown Business 2003).

[26] See, for example, highlighting the link between individualism and western liberalism: Kenneth Pennington, *The Prince and the Law, 1200–1600: Sovereignty and Rights in the Western Legal Tradition* (University of California Press 2023).

that land is the common heritage of all citizens and must be safeguarded for future generations.[27]

Hard-law sources in international law, such as investment protection agreements and human rights instruments, extensively recognize the right to property. International investment treaties provide for investor protection against expropriation, whether direct or indirect, and create obligations of compensation by the state in cases where it interferes with property. Similar protections are found in human rights instruments, as will be discussed later. These protections limit government action to ensure predictability and fairness, and to promote investment that would lead to economic development.

A way of looking at provisions that obligate compensation (say, in the case of expropriation or an order to close a source of GHG emissions) is that such provisions serve as commitment devices. They bestow credibility on government actions by law and afford predictability to investors and creditors. In simple terms, the idea is that asset creation could be subdued, made impossible or only made possible for some if the government were able to take property without compensation. The separation of powers and the rule of law can thus safeguard property and the peaceful enjoyment thereof. Such arrangements, or commitment devices, can be attractive to society at large, including the government, which is thereby weakened in a narrow sense. The image of Ulysses, who benefits from having his men tie his hands to the mast in order not to be tempted to change course, is often invoked to illustrate such a problem in the social sphere: a government that is limited in its ability to expropriate can be more robust in inducing asset creation.

The recognition of property rights as human rights and the state's duty to compensate when these rights are affected is established in several international law instruments, including declarations and conventions. These provisions can be found in Article 17 of the Universal Declaration of Human Rights (1948), the European Convention on Human Rights (1952), the American Convention on Human Rights (1969) and the African Charter on Human and Peoples' Rights (1981).[28]

These instruments recognize two elements: the right to own property and the right to the peaceful enjoyment of one's possessions. They also stress and

[27] Parliament of the Republic of South Africa, Amendment of Section 25 of the Constitution, www.parliament.gov.za/project-event-details/285; Reuters and Tim Cocks, *S. African law change plan to allow land expropriation fails to pass* (7 December 2021), www.reuters.com/markets/commodities/safrican-law-change-plan-allow-land-expropriation-fails-pass-2021-12-07/.

[28] There are, however, international treaties, conventions and declarations that do not include property as a human right as the result of Cold War politics.

create guarantees to prevent the arbitrary deprivation of property rights. If property rights are arbitrarily deprived, compensation is required, which should be based on a fair market valuation.[29]

When dealing with cases involving breach of property rights by states through interference or expropriation, international tribunals have repeatedly found that compensation is due, even if there is no explicit reference to a right to compensation in the taking of property or other forms of interference, such as in the case of Protocol 1 (Article 1) of the European Convention on Human Rights.[30] In European Human Rights Law, it is settled that in cases of expropriation 'there must be a procedure that ensures an overall assessment of the consequences of the expropriation, including the granting of compensation in relation to the value of the expropriated property'.[31] Compensation is accrued as part of a balancing of the interests at stake: that of the property owner, and those legitimate interests protected and pursued with the state measure. In the *Holy Monasteries v. Greece* case, the Greek government was ordered to pay compensation after property held by the monasteries was transferred to the state. The European Court of Human Rights found that the taking of property without compensation was in breach of the European Convention on Human Rights and Protocol 1 thereto, even though the lands were confiscated to be granted to landless people.[32]

Furthermore, a central tenet of investment protection law and customary public international law is the protection of property against direct and indirect expropriation. Almost all international treaties protecting investors in a host state recognize a guarantee of compensation in cases of property being affected.[33] These guarantees are meant, as already discussed, to ensure stability and respect for the investment, which is particularly relevant regarding large infrastructure projects, such as energy-production facilities.[34]

---

[29] Helen Mountfield, 'Regulatory expropriations in Europe: the approach of the European Court of Human Rights' [2002–03] *New York University Environmental Law Journal*, 136, p 140. We do not go into detail on the literature of what constitutes a fair or sufficient compensation.

[30] Aida Grgic et al, *The Right to Property under the European Convention on Human Rights*, Council of Europe (2007), pp 5–6.

[31] European Court of Human Rights, Guide on Article 1 of Protocol No. 1 to the European Convention on Human Rights (31 August 2022), p 32. See cases: *Alfa Glass Anonymi Emboriki Etairia Yalopinakon v. Greece*, § 36; *Pálka and Others v. the Czech Republic*, § 62.

[32] *The Holy Monasteries v. Greece European*, Court of Human Rights Application Nos 13092/87 and 13984/88 Judgment of 9 December 1994.

[33] David Collins, *An Introduction to International Investment Law* (2nd edn, Cambridge University Press 2023), p 39.

[34] Peter Cameron, *International Energy Investment Law: The Pursuit of Stability* (Oxford University Press 2021), para 1.32.

Compensation is due both when expropriation is lawful (done according to the provisions in the treaty) and when it is not – unlawful expropriation. In the latter case, it is considered arbitrary and a breach of public international law, in addition to a breach of the investment protection vis-à-vis the international investor. Examples of this are found, for instance, in Article 6 of the Norwegian Bilateral Investment Treaty Model from 2015 or, more broadly, in Article 13(1) of the Energy Charter Treaty. International investment protection treaties subject expropriation (direct or indirect) to four requirements: conducted for a public purpose; under due process of law; non-discriminatory; and with prompt payment of adequate and effective compensation.

When expropriation is lawful, the state ought to pay compensation equal to the incurred losses (*damnum emergens*). In the case of unlawful expropriation, there is a right to full reparation, as well as room for loss of profit (*lucrum cessans*).[35]

Both nationally and internationally, legal systems, including numerous human rights conventions and international investment treaties, recognize the value of private property as a fundamental societal right. Yes, states may interfere with or expropriate property in the public interest (such as over carbon-intensive assets in the interest of climate change). However, basic principles, such as the rule of law, human rights protection, constitutional provisions and investment law, subject such takings of property to a series of requirements, including the payment of fair compensation. These restrictions aim to ensure that state actions are conducted fairly and are accountable, protect individuals' rights and interests, and provide a legal framework for exercising governmental authority.[36]

Last, the protection of property and the limitation of state powers, ensuring that the taking of carbon-intensive assets is done under restrictive terms and with compensation, are necessary due to the need to strengthen government in a more general sense. More assets can be created when property rights are protected. In other words, a government that is unlikely to order the cessation of operations for carbon-intensive assets unless certain conditions apply would display a discipline that is favourable to investors, as it would resemble the discipline of a government that does not engage in arbitrary expropriation.

---

35 *Factory at Chorzów*, Permanent Court of International Justice (1927). In this chapter, we do not discuss these differences regarding law and unlawful expropriation.

36 Christoph Schreuer, 'The concept of expropriation under the ETC and other investment protection treaties' [2005] 2 *Transnational Dispute Management*, 14, p 2.

15.3.3 Owner's rights and government action that orders the cessation of operations but does not 'take' property

Closely linked to the earlier discussion on expropriation, a state order for cessation of the operation of a carbon-intensive asset may, in terms of value to the owner, be equivalent to a physical taking of the property. Through an interference that causes the cessation of operations, the state is preventing the 'peaceful enjoyment' of the property: productive operation, the production of goods, the offering of services or its net income.

Both human rights law and international investment law, in addition to national constitutional provisions, create, as a matter of principle, hard-law limits for the state, be it in ordering the cessation of operations or revoking an authorization or licence. Carbon-intensive assets, licences, and concessions often serve both as administrative authorizations to conduct specific business activities and as instruments specifying the rights and obligations of the parties involved. If an authorization to operate is revoked before the agreed-upon timeframe, it can be seen as an expropriation or a breach of contractual rights.[37]

European human rights case law stresses that Article 1 of Protocol 1 of the European Convention on Human Rights protects not only the possession of goods but also legitimate expectations, including the expectation of using an asset.[38] The European Court of Human Rights has consistently held that holding a licence or authorization to operate a business constitutes a form of possession, and that the right to peaceful enjoyment of possessions, by both individuals and legal persons, is protected under Article 1 of Protocol 1 of the European Convention on Human Rights.[39] In *Sporrong and Lönnroth v. Sweden*, the Court recognized the protection of property (and the right to compensation) in cases without a clearly identifiable deprivation or limitation. This includes situations where there is a gradual interference with property rights.[40]

---

[37] *Parkerings-Compagniet AS v. Republic of Lithuania*, ICSID Case No. ARB/05/8; *Compañiá de Aguas del Aconquija S.A. and Vivendi Universal S.A. v. Argentine Republic*, ICSID Case No. ARB/97/3 (formerly *Compañía de Aguas del Aconquija, S.A. and Compagnie Générale des Eaux v. Argentine Republic*).

[38] Ledi Bianku et al, 'Right to Property in the Context of Article 1 Protocol 1 to the European Convention on Human Rights and Freedoms' (June 2021), p 17.

[39] See the case law of the European Court of Human Rights, for example, *Megadat.com SRL v. Moldova* (2008), §§ 62–63; *Bimer S.A. v. Moldova* (2007), § 49; *Rosenzweig and Bonded Warehouses Ltd. v. Poland* (2005), § 49; *Capital Bank AD v. Bulgaria* (2005), § 130; *Tre Traktörer Aktiebolag v. Sweden* (1989), § 53. See also: Aida Grgic et al, *The Right to Property under the European Convention on Human Rights*, Council of Europe (2007), p 5.

[40] *Sporrong and Lönnroth v. Sweden* (1982).

In international investment law, indirect expropriation or measures equivalent or tantamount to dispossession are also protected, even if they tend not to be clearly defined. These measures are usually termed 'indirect', 'creeping' or 'de facto' expropriation. An order to stop an activity, or the revocation of a licence to conduct it, can fit such a definition. Tribunals have, for instance, included contractual rights within the scope of expropriation, including wrongful terminations, a breach of an agreement and the revocation of licences.[35]

Protection of property and compensation are warranted when it can be argued that the owner of the asset experiences a significant restriction or complete elimination of its right to 'use, enjoy the benefits of, or dispose of' its property[41] and the owner loses control over its business operations,[42] suffering a substantial deprivation.[43] Revocation of rights can infringe investors' legitimate expectations, for instance, longer-term operations necessary to recoup the investment.[44] As such action may result in the complete loss of the investment value in the future, compensation is due.[45]

15.3.4 European case law confirms that compensation for carbon-intensive assets must be made: German and Dutch examples

15.3.4.1 Introduction

Examples from Germany and the Netherlands show that owners of carbon-intensive assets, particularly plants producing electricity, receive compensation when their rights over an asset are affected. Compensation is due when either their infrastructure is subject to a taking of property or when there is an anticipated termination of a licence or permit that prevents them from utilizing these assets for economic activity, whether wholly or partially. These cases simply reflect basic legal principles related to the protection of the rights of investors, the rule of law, the limitation of state powers, and, as we have discussed, economic- and industrial-policy grounds. These cases also show how compensation to owners or users of existing technology may create barriers to the phasing out of these, as states may become wary of the sums that would have to be paid as a result of these actions.

---

[41] Peter Cameron, *International Energy Investment Law: The Pursuit of Stability* (Oxford University Press 2021), para 6.77.

[42] *Sempra Energy International v. The Argentine Republic*, ICSID Case No. ARB/02/16.

[43] *BayWa r.e. Renewable Energy GmbH and BayWa r.e. Asset Holding GmbH v. Spain*, ICSID Case No. ARB/15/16.

[44] *Tecnicas Medioambientales Tecmed, S.A. v. The United Mexican States*, ICSID Case No. ARB (AF)/00/2.

[45] *Metalclad Corporation v. The United Mexican States*, ICSID Case No. ARB(AF)/97/1.

15.3.4.2 The German coal and lignite cases

Recent cases in Germany illustrate how courts have upheld the rights of carbon-intensive assets to receive compensation if their property or possessions are affected before the end of their planned operational lifetime. A series of compensation claims were brought in Germany arising from measures and suggestions arising from the work by the Commission on Growth, Structural Change, and Employment (the 'Coal Commission'). The Coal Commission had the objective of developing strategies to transition away from coal and attain Germany's climate targets.[46] One of the Commission's recommendations was to gradually phase out the generation of electricity from coal-fired power plants, including anthracite (hard coal) and lignite (brown coal), no later than the end of 2038.[47] The German state took these recommendations on board, and a policy and regulatory process was initiated to accelerate this phase-out. The policy was transformed into the Act to Reduce and End Coal-Fired Power Generation (August 2020),[48] which required the phasing out of hard coal and lignite plants in 2022, 2030 and 2038.[49] The Act and its ban on the use of anthracite and lignite effectively forced some operators to stop operations before the end of their anticipated life.[50]

To address the detrimental effect on private property and rights arising from the granting of a licence to produce electricity, compensation schemes were implemented, depending on the asset involved. In the case of lignite-fired power plants, for example, specific closing dates and compensation resulted in compensation for the 'socially acceptable decommissioning of lignite plants by December 31, 2029', with RWE receiving €2.6 billion and Lausitz Energie Kraftwerk AG €1.75 billion in compensation. For hard coal power plants, the compensation is determined through a tender system that

46 German Federal Government, *Ending coal-generated power*, www.bundesregierung.de/breg-en/service/archive/kohleausstiegsgesetz-1717014.

47 Clean Energy Wire and Carolina Kyllmann, *German coal region brings phase-out forward to 2030 but refires lignite short-term* (4 October 2022), www.cleanenergywire.org/news/german-coal-region-brings-coal-phase-out-forward-2030-refires-lignite-short-term; Reuters, *Germany's cabinet approves accelerated coal exit by 2030 in western state* (2 November 2022), www.reuters.com/business/energy/germanys-cabinet-approves-accelerated-coal-exit-by-2030-western-state-2022-11-02/.

48 Gesetz zur Reduzierung und zur Beendigung der Kohleverstromung (Kohleverstromun gsbeendigungsgesetz – KVBG) 8 August 2020.

49 For a discussion of the three stages of phasing out see: Clean Energy Wire and Julian Wettengel, *Spelling out the coal exit – Germany's phase-out plan*, www.cleanenergywire.org/factsheets/spelling-out-coal-phase-out-germanys-exit-law-draft.

50 Achim Hagen, Niko Jaakkola and Angelika Vogt, 'The interplay between expectations and climate policy: compensation for stranded assets' [2019], IAEE Energy Forum, p 30.

initially allows for voluntary reductions and later imposes legally required reductions. The compensation follows a decreasing scale, with the maximum payment per decommissioned megawatt offered in auctions sliding from €165,000 in 2020 to €89,000 in 2026. As of 2027, plant operators will no longer receive financial compensation.

The anticipated revocation of licences for the German nuclear reactors, non-carbon-intensive assets, has triggered similar legal arguments and results worth commenting on. Pursuant to a political decision to close these plants based on security matters, the German Federal Constitutional Court reviewed the measure's constitutionality in the Atomic Energy Act. In 2016, the Court declared that the phase-out of nuclear reactors was largely compatible with the Constitution, but that nuclear asset owners were entitled to adequate compensation.[51] As a result of subsequent negotiations and legal processes, the federal government settled to pay €2.4 billion to EnBW, E.ON/PreussenElektra, RWE and Vattenfall. That ended a series of legal actions, including a lawsuit before the International Centre for Settlement of Investment Disputes (ICSID).[52]

15.3.4.3 Dutch coal cases

Like Germany, the Dutch state has recognized the right to compensation for the anticipated closure of carbon-intensive assets. In December 2019, the Netherlands implemented the Prohibition of Coal in Electricity Act, with a prohibition after 2030.[53] While four power plants were granted a transition period of 5–10 years, the Hemweg plant was required to cease operations before 1 January 2020, just days after the law was enacted. In this section, we will focus on the Hemweg plant case, while the other four cases will be discussed in Section 15.4.

Article 4 of the Act includes a provision for operators of coal-fired power plants to claim compensation if they can demonstrate that the measure disproportionately affects them. The compensation was introduced, according to the law's preamble, as these measures would prevent owners from recouping their investment, thereby violating property rights under Dutch law.[54] Additionally, the law recognized

---

51 BVerfG, Judgment of the First Senate of 6 December 2016 – 1 BvR 2821/11.

52 BVerfG, Order of the First Senate of 29 September 2020 – 1 BvR 1550/19.

53 Wet van 11 December 2019, houdende Regels voor het produceren van elektriciteit met behulp van kolen (Wet verbod op kolen bij elektriciteitsproductie [Prohibition of coal in electricity production]), Staatsblad van het Koninkrijk der Nederlanden, Jaargang 2019, 493 (19 December 2019).

54 European Commission, State Aid SA.54537 (2020/NN) – Netherlands: Prohibition of coal for the production of electricity in the Netherlands [2020], para 13.

its impact 'rules on property within the meaning of Article 1 of the First Protocol to the European Convention on Human Rights and Fundamental Freedoms'.[55]

Based on these compensation requirements, Vattenfall, the owner of the Hemweg power plant, was awarded a payment of €52.2 million as compensation for lost profits and the costs associated with the abrupt dismissal of staff.[56]

15.3.4.4 Takeaways from the Netherlands and Germany

The German and Dutch cases just discussed provide examples by states and ratified by courts confirming the potential for compensation when asset owners are compelled to terminate activities before the expected end of the asset's lifespan. These cases demonstrate how regulatory climate policies, in general, may have the support of law and courts, even when they deprive owners of value. However, requiring the cessation of operation of carbon-intensive assets is only lawful provided that such measures are justified in terms of promoting the public interest, conducted according to the rule of law, and against the payment of adequate compensation.

A close examination of the cases involving the closure of coal-powered plants in Germany and the Netherlands shows that the compensation payment was linked to the negative effect the measure had on the right to property, as well as the owner's legitimate expectations to operate these assets. Additionally, the compensation was connected to an event that forced a closure in a short time frame, leaving little room for the asset owners to react, a topic which we discuss again in Section 15.4.4. The closures were abrupt and imposed within a short time frame, and thus would result in the denial of the owners' peaceful enjoyment of property and render their assets unsellable. As a result of these case-specific observations, it is not surprising that compensation was provided to affected power plant owners in Germany and the Netherlands.

---

55 Ibid, para 13.

56 Ibid. In a state-aid twist, the General Court overturned this decision, but it did not address whether the payment constituted state aid or not in Judgment of 16 November 2022, *Netherlands v Commission*, T-469/20, EU:T:2022:713 and pending before the Court of Justice in C-40/23 P – *Commission v. Netherlands* (*Appréciation de compatibilité d'une mesure non qualifiée d'aide d'État*).

15.4 The case against: decarbonization policies are in the public interest and non-compensable if they are non-discriminatory

15.4.1 Property rights are limited by the rights of others

Property rights, or other rights of use and peaceful enjoyment and possession, are delimited by the rights of others. These include the rights of government when acting in the public interest. In contrast to a system which prioritizes individual rights, under certain circumstances and within specific requirements, the state may interfere and affect private individual rights to promote the public interest.[57] In these cases, owners of carbon-intensive assets may see their property affected and compensation might not be due.

This is highlighted, for example, by Article 1 of Protocol 1 to the European Convention on Human Rights, which indicates that the right to enjoy possessions does not 'impair the right of a State to enforce such laws as it deems necessary to control the use of property in accordance with the general interest or to secure the payment of taxes or other contributions or penalties'.

Interfering with the right to peaceful enjoyment of possessions is legitimate and is to be allowed if: (1) prescribed by law; (2) in the public interest; and (3) necessary in a democratic society, these being cumulative conditions.[58] Similarly, the European Court of Human Rights accepted that a limitation of rights based on the rights of others and public policy is justified when it is under due process, pursues a legitimate aim and the measure is proportionate.[59] With regard to the democratic element, this recognizes that a democratic process must be permitted to make and change policy without being, in effect, held to ransom by a small group of persons who have invested while relying and based on an earlier policy.[60]

---

[57] See more on this interplay in an international investment setting and stressing the importance of proportionality: Benedict Kingsbury and Stephan W Schill, 'Public law concepts to balance investors' rights with state regulatory actions in the public interest: The concept of proportionality', in Stephan W Schill (ed), *International Investment Law and Comparative Public Law* (Oxford University Press 2010).

[58] Aida Grgic et al, *The Right to Property under the European Convention on Human Rights*, Council of Europe (2007), p 12.

[59] *Beyeler v. Italy* [GC], §§ 108–114.

[60] Helen Mountfield, 'Regulatory expropriations in Europe: the approach of the European Court of Human Rights' [2002–03] *New York University Environmental Law Journal*, 146.

15.4.2 Compared to expropriation, 'interference' makes less of a case for compensation

Limits to interference with the rights of private entities owning or operating carbon-intensive assets (prescribed by law; in the public interest; and necessary in a democratic society) bear a strong similarity to the conditions delimiting legal expropriation (Section 15.3). Nevertheless, 'interference with the right to peaceful enjoyment' differs from 'expropriation' in two ways: in terms of the consequence for the value to the owner and the result related to the public purpose.

For example, the government may act in the public interest by reducing harmful noise in a neighbourhood (or air pollution as discussed in Section 15.4.3) and interfere in a factory's operations by ordering it to cease operating during night hours, acting in the exercise of 'police powers'. The consequence in value to the owner will likely be smaller than that of expropriation or revoking a factory operation permit, since daytime hours of operation are still available for the owner's 'peaceful enjoyment'. The consequence is also, to a greater extent, related to and driven by the public interest because the public purpose of less noise is not 'mixed' with – or contaminated by – a possible desire of the government to possess or control the valuable factory or its outputs.

As lawful expropriation necessitates compensation, whereas 'interference with peaceful enjoyment' does not, the economic logic can be as follows: Under expropriation, the government typically takes the asset (or the authorization to operate), sometimes because the government wants the asset. The requirement of compensation under expropriation serves as a deterrent to government expropriation over and above what is needed to deter undue interference. In the case of interference, the government would not typically gain command or peaceful enjoyment of the asset or its output. Thus, a compensation requirement is not needed to deter that temptation.

15.4.3 Not all limitations on property grant compensation; interference based on the state's regulatory powers never does

Section 15.3 showed how compensation can be warranted in the case of property or rights being affected by the state when requiring a carbon-intensive activity to be stopped. However, this argument overlooks a crucial issue: states have the right to regulate and exert powers to benefit their citizens in a democratic society.[61] An increasing number of examples show

[61] OECD, *International Investment Law: A Changing Landscape*, a companion volume to *International Investment Perspectives*, Chapter 2.

that measures taken by the state affecting private parties' rights do not lead to compensation.[62]

While it is reasonable and in line with the limitation that state powers that impact property rights without compensation should be restricted and based on special requirements, the term 'police powers' may be misleading. States have not only the right but also the duty to regulate and make strategic decisions for the benefit of society as part of their regulatory function. This may include the authority to order the cessation of an activity or the closure of a carbon-intensive asset.

Under public international law, and in many national legal systems, it is permissible for the state to interfere with the property rights and other rights of investors and private parties in connection with the right to regulate. This power to affect property or other rights of individuals or entities without compensation is typically linked to the concept of 'police powers'.[63] These powers enable the government to act in the public interest, even if they have an impact on private property or contractual rights.

It is widely accepted that measures such as taxation, trade restrictions, antitrust rules, environmental protection, land-planning orders, and carbon taxes to mitigate climate change are permissible without triggering any compensatory payment.[64] Requiring the cessation of carbon-intensive activities is a decision justified by the protection of others (right to health, for instance) and the environment (rights of nature and the right to a healthy environment), and it serves as an example of this, provided it is done in a non-discriminatory manner and in accordance with due process of law.[65]

Let us turn to carbon trading systems as an example of such interference. For instance, a European system of tradable CO_2 emission quotas, the ETS, spreads these policy shocks and their responses across sectors and

[62] Peter Cameron, *International Energy Investment Law: The Pursuit of Stability* (Oxford University Press 2021), para 6.75.

[63] For some literature dealing with this concept of the state's police powers and the concept of expropriation or taking, see inter alia, D Benjamin Barros, 'The police power and the takings clause' [2003] 58 *University of Miami Law Review*, 471; Joseph L Sax, 'Takings and the police power', in Richard A Epstein (ed), *Constitutional Protection of Private Property and Freedom of Contract* (Routledge 2013); Ben Mostafa, 'The sole effects doctrine, police powers and indirect expropriation under international law' [2008] *Australian International Law Journal*, 267; Noam Zamir, 'The Police Powers doctrine in international investment law' [2017] 14 *Manchester Journal of International Economic Law*, 318.

[64] See more on this: OECD, *International Investment Law: A Changing Landscape*, a companion volume to *International Investment Perspectives*, p 47; Ian Brownlie, *Public International Law* (6th ed, Oxford University Press 2003), p 509; M Sornarajah, *The International Law on Foreign Investment* (5th ed, Cambridge University Press 2021).

[65] Peter Cameron, *International Energy Investment Law: The Pursuit of Stability* (Oxford University Press 2021), para 6.89.

over decades. An additional element of the ETS, which has softened its expropriatory content, has been distributing free quotas to existing firms. Thus, firm assets that might lose value due to government regulations (emission quotas) are buffered (if not compensated) by the fact that European competitors face the same constraints and cost increases and may either buy quotas or benefit from selling quotas if they reduce emissions.[66] Within the EU, the ETS scheme qualifies for non-compensation as it is based on the public interest (addressing GHG emissions directly); non-discriminatory (cross-sectoral, and allowing trade in quotas to soften differences in compliance costs); designed (and modified) so that quota transfers over time (banking) can protect against untimely tightening of policy; introduced in a lawful and democratic setting.

More than a handful of investment law cases illustrate that state intervention may not warrant compensation even if property is significantly affected. For example, *The Occidental Exploration and Production Company v. Republic of Ecuador*,[67] *Too v. Greater Modesto Insurance Associates*,[68] *Lauder v. Czech Republic*,[69] *Técnicas Medioambientales Tecmed S.A. v. United Mexican States* arbitration cases, and the European Court of Human Rights case *Jahn and others v. Germany*,[70] stress that measures that affect individual rights that are non-discriminatory are bona fide, are for the benefit of the collective and are under due process do not lead to a compensation payment. Even in liberal states such as Canada and the United States, the state has been absolved of paying compensation (or settled with a much smaller sum) when ordering the ban of additives on gasoline for environmental reasons. In *Methanex Corp. v. United States of America*, compensation claims were dismissed by the arbitration tribunal in full,[71] and similarly in *Ethyl Corporation v. Government of Canada* (settled with a payment out of court of $13 million out of the $215 million claimed).[72]

National constitutional or statutory provisions also reflect similar cases for non-compensation.[73] For example, in Germany, the Federal Constitution distinguishes between expropriation, which always requires compensation,

[66] Bard Harstad and Gunnar Eskeland, 'Trading for the future: signaling in permit markets' [2010] 94(9–10) *Journal of Public Economics*, 749.

[67] *Occidental Exploration and Production Co. v. Ecuador, Award*, 1 July 2004, at 80–92.

[68] Award, 29 December 1989, *23 Iran-United States Cl.* Trib. Rep.378.

[69] *Lauder (U.S.) v. Czech Republic* (Final Award) (3 September 2002).

[70] European Court of Human Rights, *Jahn and others v. Germany* (2004, Grand Chamber).

[71] *Methanex Corp. v. United States of America*, final award 3 August 2005.

[72] *Ethyl Corporation v. The Government of Canada* (1997).

[73] Ulrike Deutsch, 'Expropriation without compensation: the European Court of Human Rights sanctions German legislation expropriating the heirs of "new farmers"' [2019] *German Law Journal*.

and decisions that limit ownership and do not need compensation, as discussed by the Federal Constitutional Court in the Groundwater Case.[74]

In summary, there is a range of state measures that are considered strategic and of the utmost importance for guiding society. These measures may interfere with the rights of owners and operators of carbon-intensive assets. However, measures and intervention powers are non-compensable as they are justified in the public interest, non-discriminatory and conducted under the rule of law. Thus, when there is a requirement to cease the operation of carbon-intensive assets due to climate concerns, which is done generally and as part of the legitimate democratic process, states can be found to simply be acting in the public interest and not be liable to compensate.

15.4.4 In the public interest of climate policy, 'stranding' of some assets is to be expected

The challenge of climate change necessitates state intervention in the economy: an emission trading scheme like EU ETS is only one example. A combination of such government policies (and treaties) that 'interfere with the peaceful enjoyment of possessions' can certainly satisfy the requirements (for non-compensation) above if they are: prescribed by law; in the promotion of public interest (climate mitigation, energy transition); and necessary in a democracy.

Such policies will be, should be and are expected to result in some stranding of assets, as when cars, ships or power plants are scrapped, closed or modified earlier in their lifetime than they otherwise would be. It is a consequence to be expected, not only of established climate science but also of international agreements and national targets to mitigate climate change and drastically cut GHG emissions. These obligations and goals inevitably compel states to act, including in ways that limit the rights of private individuals and firms. Consequently, car owners and car manufacturers are subjected to tightening emission standards, fuel sellers to blend-in mandates, power plants and steel mills to cutting emission quotas, and certain technologies are not allowed to operate due to their impact on climate, health or the environment.

Given the current climate ambitions and policy targets, it is unsurprising that states would begin to demand the cessation of certain polluting activities. It is for debate whether this is sufficiently and legitimately driven by the public interest of climate mitigation, lawful and required in a democratic society. These measures and related issues are the subject of intense public

[74] Federal Constitutional Court, *The Groundwater Case*, Judgment of 15 July 1981, 58 BVerfGE 300, paras 330–331.

debate.[75] In terms of legality, essential questions have been shown to be whether the action taken by the government is in line with the public interest and discriminatory.

15.4.4.1 Non-compensation has important efficiency implications

Policies that do not provide compensation have positive implications in terms of efficiency since compensation can be harmful to efficiency. First, as a central benchmark, households and firms that anticipate climate policy tightening early on and take measures to mitigate climate change directly save societal resources, reduce the costs of such tightening and demonstrate the feasibility of mitigation. Such anticipation, cost reductions and leadership may be hindered if owners of carbon-intensive assets expect or demand compensation. Second, significant investment in carbon-intensive assets that could be eligible for compensation might discourage the tightening of policies, and prospects of compensation could incentivize such investment. Both the rent-seeking literature, initiated by Anne Krueger in 1974,[76] and the literature on commitment problems,[77] deal with how some unproductive and costly actions influence government decisions. In environmental economics, the 'new source bias' is a prominent example of the government imposing more lenient policies for existing facilities and equipment (such as cars) than for new ones.[78] While such differentiation can be a sensible part of ramping up mitigation ambitions, typically it has the adverse effect of accelerating investment in older and more polluting assets when it is expected. Hans Werner Sinn's 'green paradox' in 2008 is one example of this, as he points out that fossil resource owners who anticipate tighter climate policies may accelerate fossil fuel extraction and combustion.[79]

Thus, if compensation is expected in light of climate action, investors may be more likely to develop even more carbon-intensive assets. This effect is akin to the exemption from stricter climate policies for established assets,

[75] Rechtbank Den Haag, *RWE and Uniper v. the Netherlands (Ministry of Climate and Energy)*, ECLI:NL:RBDHA:2022:12628 (RWE), para 5.16.3.

[76] Anne O Krueger, 'The political economy of the rent-seeking society' [1974] 64(3) *The American Economic Review*, 291.

[77] Finn E Kydland, and C Prescott Edward, 'Rules rather than discretion: the inconsistency of optimal plans' [1977] 85(3) *Journal of Political Economy*, 473; Bard Harstad and Gunnar Eskeland, 'Trading for the future: signaling in permit markets' [2010] 94(9–10) *Journal of Public Economics*, 749.

[78] Gunnar S. Eskeland, 'A presumptive pigovian tax: complementing regulation to mimic an emissions fee' [1994] 8(3) *The World Bank Economic Review*, 373.

[79] Hans-Werner Sinn, 'Public policies against global warming: a supply side approach' [2008] 15 *International Tax and Public Finance*, 360.

which creates a new source bias. Similar stimulating effects on polluting investments, reinvestments and the continuation of such activities have been assessed in the literature in connection with free emission quotas (grandfathering).[80] These allocations are often seen as the price one has to pay to reduce resistance from existing industries to environmental reforms, which often also overcompensate the firms affected.[81]

15.4.4.2 Non-compensation and expectations: 'If you knew it, you won't be compensated'

Legal certainty and legitimate expectations remain intact when a diligent investor can anticipate the need to act to align with climate change objectives. Entities engaged in polluting and carbon-intensive activities bear a societal responsibility to contribute to emissions reduction. They should be 'extra alert to the possibility of regulatory measures by the government and adapt to this',[53] by adjusting their line of activity and business model. As the Hague District Court clearly stated in November 2022 in *RWE and Uniper v. the Netherlands*, '[i]f a company does not do this or does so insufficiently, it is at its own risk' and ought not to receive any compensation if there are other balancing measures in place.[82]

Let us review this case in relation to legitimate expectations and the cessation of carbon-intensive industries. In November 2022, the Hague District Court dismissed compensation claims by RWE and Uniper regarding the premature closure of three coal-fired plants. These projects, initiated in 2009 and completed in 2015 and 2016, respectively, were originally anticipated to have a minimum operational lifespan of 40 years.[83]

However, and as discussed in Section 15.3.4.3, in 2019, the Dutch government enacted the Prohibition of Coal in Electricity Act, reducing the operational lifespan of carbon-intensive industries by about 25 years. The basis for the claim for compensation was the protection of property rights under Protocol 1 of the European Convention on Human Rights (ECHR) and Article 17 of the Charter of Fundamental Rights of the EU, closely aligning with the arguments in Section 15.3.4 of this chapter

[80] Bard Harstad and Gunnar Eskeland, 'Trading for the future: signaling in permit markets' [2010] 94(9–10) *Journal of Public Economics*, 749.

[81] Dallas Burtraw et al, 'The effect on asset values of the allocation of carbon dioxide emission allowances' [2002] 15(5) *The Electricity Journal*, 51.

[82] Rechtbank Den Haag, *RWE and Uniper v. the Netherlands (Ministry of Climate and Energy)*, ECLI:NL:RBDHA:2022:12628 (RWE), para 5.17.3.

[83] Ibid; and Rechtbank Den Haag, *Uniper and Utilities Center Maasvlakte v. the Netherlands*, ECLI:NL:RBDHA:2022:12653.

when reviewing the coal cases in Germany and the Netherlands. The measure negatively impacts the peaceful enjoyment of property and fundamental rights. However, the measures were foreseeable and struck a fair balance between the climate objectives being pursued and the respect for individual rights.[84]

The judgment by the Hague District Court highlights that a foreseeability test is key[85] and that companies engaged in 'controversial industries' are 'expected to be aware of national and international developments that may (in due course) lead to government measures that could affect the activities of its company'.[86] The exact measure may not be expressly foreseen; what matters is whether it falls within what could reasonably be expected based on the context of national climate plans and public interest. If it could have been legitimately expected and the state acts in a bona fide manner, no compensation can be claimed. In this balancing exercise, the Court assessed the public interest, the measure itself, and the alternatives and possibilities provided to the property holder whose rights have been impacted.[87] In the context of climate change, compensation becomes 'only one of many factors' when evaluating foreseeable measures that promote public interests and balance.[88]

If this judgment is upheld by the Hague Appeals Court, the argument in this section would be strengthened. That is, in cases where a state revokes a licence or permit as part of reasonable and expected climate policy, shortening the expected lifespan of a carbon-intensive industry, compensation should not be granted automatically (or at all), even if such an asset would become stranded.

15.4.5 Stranded assets, financial implications and the carbon crash

Apart from the critical questions about incentives and mitigation associated with compensation, discussion and analysis also need to be had concerning stranded assets and financial risks, including those related to financial system stability. When Mark Carney, who was the Governor of the Bank

---

[84] Rechtbank Den Haag, *RWE and Uniper v. the Netherlands (Ministry of Climate and Energy)*, ECLI:NL:RBDHA:2022:12628 (RWE), para 5.16.

[85] Ibid, para 5.19.3.

[86] Ibid, para 5.17.3.

[87] Ibid, para 5.20.

[88] Ibid, para 5.19.1. As we write these lines, the legal strength and/or the extent of this argument for non-compensation is being tested. RWE and Uniper appealled this judgment by the Hague District Court. On January 2025, the Hague Court of Appeal heard both parties and a decision is expected in 2025 or 2026.

of England, highlighted in a landmark speech in 2015 that financial systems are vulnerable, his main emphasis was on climate risk, such as assets lost due to flooding, storms or drought. Later, the term 'climate risk' was also used to cover the risks that climate policy may pose for carbon-intensive assets and activities.[89]

When there are significant and unforeseen changes in regulatory or market conditions, asset operators may be unable to recover their investments[90] due to redundancy or obsolescence,[91] rendering market values lower than book values.[92] The assets are then stranded if it becomes impossible or prohibitively expensive to transition to alternative uses of the assets.[93]

Historically, discussions about stranded costs arose when electricity markets were transitioning as a result of reregulation from monopolies to competitive structures, as to whether investors were entitled to compensation to recover their costs in markets of regulated rates for monopolies.[94]

There are important questions to be asked revolving around whether climate risks and the stranding of assets pose a bigger threat to financial systems or to financial stability. Van der Ploeg and Rezai point out that redirecting investment from carbon-intensive to carbon-free industries under a socially optimal policy leads to the optimal and forward-looking composition of capital. Some have proclaimed an ongoing bubble in carbon-intensive sectors (the so-called 'carbon bubble') that bursts once climate policy hits, but Van der Ploeg and Rezai emphasize that the scope for surprises has two sides to it.[95]

Compared to the role of real estate and mortgages, which were at the root of the financial crisis of 2007–08, the energy transition affects a narrower asset base in terms of value, a lower degree of leverage, and has greater predictability in the institutional underpinning of the energy transition.

---

[89] Mark Carney and Bank of England, *Breaking the tragedy of the horizon – climate change and financial stability*, Lloyd's of London (29 September 2015).

[90] Body of Knowledge on Infrastructure Regulation, *Stranded Costs*, https://regulationbodyofknowledge.org/glossary/s/stranded-costs/.

[91] *What's Stranded Cost?* (3 February 1999), https://web.archive.org/web/20100104092419/http://home.earthlink.net/~tpco/stranded.html.

[92] Emily Hammond and Jim Rossi, 'Stranded costs and grid decarbonization' [2016] 82 *Brooklyn Law Review*, p 646.

[93] Frederick Van der Ploeg and Armon Rezai, 'Stranded assets in the transition to a carbon-free economy' [2020] 12 *Annual Review of Resource Economics*, 281, p 289.

[94] Timothy J Brennan and James Boyd, 'Stranded costs, takings, and the law and economics of implicit contracts' [1997] *Journal of Regulatory Economics*, 11, 41; Emily Hammond and Jim Rossi, 'Stranded costs and grid decarbonization' [2016] 82 *Brooklyn Law Review*, p 647.

[95] Frederick Van der Ploeg and Armon Rezai, 'Stranded assets in the transition to a carbon-free economy' [2020] 12 *Annual Review of Resource Economics*, 284.

Unlike mortgages and the banking sector, the fossil fuel industry is large enough to envisage broader financial difficulty if the transition to renewable energy is disorderly. Nevertheless, in terms of quantitative impacts focusing on real assets exposed to the risk of stranding, Van der Ploeg and Rezai write: 'Stranding of assets mostly affects the 20 biggest oil, gas, and coal companies ... but it also affects carbon-intensive industries such as steel, aluminum, cement, plastics, and greenhouse horticulture.'[96] A disorderly transition to a carbon-free economy will lead to stranded assets and legal claims.

Van der Ploeg and Rezai also highlight two conditions regarding stranding: first, an unanticipated future change in conditions affecting the profitability of fossil fuel assets and, second, adjustment costs. One cause of stranding carbon-intensive assets is when the government suddenly intensifies climate action to limit the total amount of cumulative carbon emissions, catching the private sector unaware of the potential policy change.[97]

The differences between fossil-based assets and real estate are also significant. Van der Ploeg and Rezai point out that the mere size of carbon assets that could possibly be stranded,[98] while large, is more modest than in real estate: fossil fuel companies constitute about $5 trillion out of the $70 trillion of world stock markets, and out of total financial assets of $300 trillion.[99] This does not rule out 'bubbles', but financial regulators are increasingly aware of, and monitor, these risks.[100] McGlade and Ekins assess that 80 per cent of all coal reserves must be stranded if the 2 °C Paris limitation on warming is not to be exceeded.[101] However, Harstad's 'Buy Coal' paper highlights that coal reserves are heterogeneous, implying that reserves with a high carbon content and mining costs can be taken 'off the table' at moderate costs.[102]

---

[96] Ibid.

[97] Ibid, 282.

[98] Ibid, 283.

[99] BloombergNEF, *Fossil fuel divestment: a $5 trillion challenge* (26 August 2014); MarketWatch and Wallace Witkowski, *Global stock market cap has doubled since QE's start* (12 February 2015), www.marketwatch.com/story/global-stock-market-cap-has-doubled-since-qes-start-2015-02-12.

[100] Mark Carney and Bank of England, *Breaking the tragedy of the horizon – climate change and financial stability*, Lloyd's of London (29 September 2015).

[101] Christophe McGlade and Paul Ekins, 'The geographical distribution of fossil fuels unused when limiting global warming to 2 °C' [2015] 517 *Nature*, 187.

[102] Bård Harstad, 'Buy Coal! A supply side approach to climate policy' [2012] 120(1) *Journal of Political Economy*, 77.

Efforts such as the Carbon Tracker,[103] and many other initiatives (the EU's taxonomy included),[104] share a common goal of creating awareness and conveying information to optimally reallocate financial flows and exposures for the climate transition, subject to the commitment of policy makers. Karydas and Xepapadeas deconstruct disaster risk into that which is temperature-related and that which is not, showing the effects on asset pricing and asset holdings.[105] Following Carney's alert in 2015, central bank governors have highlighted the risks of a 'carbon bubble' and emphasized the need for capital market regulators to ensure oversight and corporate disclosure.[106] Thus, if national policies lag behind – or when treaties lag behind – so will financial flows and real asset composition and exposures. What banking regulators could do in addition, perhaps, is to force highly leveraged financial sectors to lean away from highly correlated risks, such as surprising climate change damage or unanticipated moves away from carbon emissions in political decisions or market movements.

One conclusion that can be drawn regarding 'stranded assets' is that some financial disturbances can be anticipated. However, the predictability in a long-term transition and the narrower asset base in energy, compared to real estate, suggest that financiers should focus on measuring, monitoring and regulating risk-taking.

Crucially, the recognized risks could be effectively addressed by the discipline outlined earlier: resist compensating owners and lenders for stranded assets when closures adhere to the law and public interest, are non-discriminatory and proportionate, reflect democratic procedures, and can be anticipated.

In financial crises, a recurring issue is the bailout of the banking industry and its leaders and owners by the government and taxpayers – a form of compensation for owners and managers. When banks, including many large ones, incur significant losses, they are often assisted, with managers retained and owners buffered, rather than banks facing dissolution. This

103 Frederick Van der Ploeg and Armon Rezai, 'Stranded assets in the transition to a carbon-free economy' [2020] 12 *Annual Review of Resource Economics*.

104 Regulation (EU) 2020/852 on the establishment of a framework to facilitate sustainable investment, and amending Regulation (EU) 2019/2088 (OJ [2020] L 198/13).

105 Christos Karydas and Anastasios Xepapadeas, 'Pricing climate change risks: CAPM with rare disasters and stochastic probabilities' [2019], CER-ETH Economics working paper series 19/311.

106 See, for example, Mark Carney and Bank of England, *Breaking the tragedy of the horizon – climate change and financial stability*, Lloyd's of London (29 September 2015).

practice fosters excessive risk-taking in banking. One interpretation of this government and regulatory leniency is that some banks are considered too big and essential to be allowed to dissolve. Therefore, efforts to prevent excessive size and monitor and regulate risk-taking have been sought in banking regulation and oversight. Notably, climate risk has been incorporated into monitoring and stress-testing in the financial sector since Carney's Bank of England speech in 2015.

Energy sectors, such as electricity generation, may also be essential because governments could face significant challenges if the sector were threatened with a fundamental or financial breakdown. However, as in banking, such risks would not be mitigated by compensation, or the expectation of compensation, for stranded assets. On the contrary, compensation practices could exacerbate the challenges. In bankruptcy, banks and other firms are either dissolved or assets are put in the control – administration – of others, such as creditors, courts or government agencies. In the energy sector, too, the government will be part of such administration of assets and industries, if only temporarily.

15.5 Summary and conclusion

Basic principles of law establish rights and obligations, thereby reducing conflict and enabling investment and development. Among these are property rights, or owners' rights to the 'peaceful enjoyment of their property'. When respected, such rights are crucial in facilitating asset creation (investment), peace and productivity. However, property rights are not absolute. They are delimited by the rights of others, as illustrated by the government's authority to implement policies in the public interest, such as climate policy.

A government action that requires the cessation of an activity or forbids the utilization of a fossil-intensive technology may be considered to be of an expropriatory nature if it is found to be 'discriminatory'. In such cases, there is a legal argument that the owner is entitled to compensation. Reaching the opposite conclusion, policies in the public interest (such as climate change mitigation) may fall within the ordinary policy-making authority of the government and, therefore, not be compensable, even if they lead to closures, as long as they are not discriminatory. This constitutes the legal case for non-compensation.

From an economic perspective, secure property rights have positive efficiency implications as they facilitate saving, investment and development. However, the government possesses powers, such as taxation, which are employed in pursuing public interest, including welfare goals such as wealth redistribution. The use of these powers can change in response to the evolving needs and interests, as seen in the rising prominence of climate policy. A compensation requirement can discourage government from appropriating specific assets arbitrarily, and thus benefit investor confidence and asset creation. In contrast, non-compensation can serve the purpose of sensitizing investors to anticipate changing needs in line with government policies.

In our specific case, examining the legitimate public interest behind government actions leading to cessation in the use of fossil-intensive equipment, technologies or processes, there is, a priori, a tension between two desirable 'disciplining effects' in the 'public interest' of climate policy: The first, property rights protection, aims to discourage government actions that are expropriatory in nature – actions that are unduly specific, discriminatory or hasty, relative to the public interest goal of climate policy. Undue haste or discrimination relative to the public interest goal should be met with a compensation requirement. The second, encouraging alertness and responsibility in investors and asset creation, aims to encourage investors to be observant of emerging public-interest goals, such as climate policy, and to recognize the risks in public-interest areas that the government might act upon. Compensation should not be expected when government actions are not unduly hasty or discriminatory.

Our review of arguments, examples and case law regarding compensation in policy examples reveals a consistent pattern. Compensation is typically deemed necessary when the policy change is considered excessively rapid or specific – disproportionate to the policy objective at hand. In contrast, policies that are gradually and universally tightened over time are less likely to warrant compensation.

We have identified two key qualities that government policies should possess to avoid compensation:

1. *Universality*: Treating all emitting sources equally or similarly, such as by implementing tradable quotas or emission taxes. From the point of view of environmental economics, such universality is commendable, as GHGs, especially CO_2, are globally distributed, making any given emission reduction equally valuable wherever it may happen. From a legal perspective, this equal treatment also upholds the principle of equality before the law and the principle of proportionality.
2. *Predictability*: Implementing emission quotas in a general and stringent manner over time, reflecting the evolution of the environmental problem at hand. From the perspective of environmental economics, such predictability allows for the early detection and direction of investment and divestment opportunities. For owners and operators and investors, this implies actively following the policy process and anticipating and understanding the interaction between industrial activities and environmental challenges.

Ideally, these qualities of universality, proportionality and predictability, akin to the rule of law and equality under the law, not only serve to prevent compensation payments but also deter against investment in assets that might be stranded, thus avoiding costs and conflict. One of the essential goals of government policy, including legislation, is to ensure that the government and the law remain firm and impervious to manipulation by private-sector behaviours.

However, there are cases in which compensation is justifiable and necessary, not only in situations involving expropriation but also when specific activities or practices need to be targeted by the government. An instructive example reflecting this and the impact climate change policy has on technology development (its introduction but also phasing out) might be the EU's current consideration of a ban on the sale of new internal combustion engine cars starting in 2035. While such a policy would not lead to stranded assets in the form of existing cars (as car lifespans are relatively short and existing cars might still be allowed some use), it could potentially devalue and strand existing car manufacturing factories in Europe. Compensation in this scenario is less likely to be associated with a legal obligation, because the policy is both part of broader climate policies and outlined explicitly in legislation 12 years in advance (2023), allowing for anticipation and adjustment of planned investment over a long time frame. Practically speaking, European car manufacturers are likely to focus on the challenges ahead and support the industry's transition, which would have more significant positive implications than the prospects of, or legal requirements for, compensation.

Acknowledgement

This work has been made possible by the generous funding granted by the Research Council of Norway to the Norwegian Centre for Hydrogen Research – HyValue (Project: 333151).

16

Rising Tides, Rising Justice: Navigating the Intersection of Climate Mitigation and Energy Justice

Laura Kaschny

16.1 Introduction

The energy transition, driven in large part by the political consensus reached by the Paris Agreement and the need to decarbonize the energy sector to mitigate climate change, demands a thorough reevaluation of how to address the challenges posed by the Anthropocene. The necessary large-scale disruptive changes of the energy transition extend beyond the decarbonization of energy services. They are characterized by a sociocultural shift, driven by increased environmental awareness, technological progress and gradually more liberalized markets. This modernization of the energy sector entails a transformation toward more sustainable, distributed and overall smarter energy systems that place a greater emphasis on demand-side actors and technological solutions.[1] Consequently, the energy transition encompasses

[1] Derk Loorbach, Niki Frantzeskaki and Flor Avelino, 'Sustainability Transitions Research: Transforming Science and Practice for Societal Change' (2017) *Annual Review of Environment and Resources* 599, 601; Saskia ACM Lavrijssen, 'The Right to Participation for Consumers in the Energy Transition' (2016) 25 *European Energy and Environmental Law Review* 152, 153; Ruven Fleming, Kaisa Huhta and Leonie Reins, 'What Is Sustainable Energy Democracy in Law?', in Ruven Fleming, Kaisa Huhta and Leonie Reins (eds), *Sustainable Energy Democracy and the Law* (Brill 2021) 3, 6.

the simultaneous decarbonization,[2] decentralization,[3] digitalization[4] and democratization[5] of energy services.[6]

Such a broad understanding of the energy transition, which extends well beyond the narrow focus on decarbonizing energy technologies, is echoed in the European Union's (EU) ambition for a fair, competitive and green transition to climate neutrality by 2050 and beyond.[7] This approach not only recognizes energy technologies as essential tools for decarbonization but also critically examines their potential to either reinforce existing inequalities or distribute benefits more equitably, especially to marginalized communities.

After an extensive 'predevelopment' and 'take-off' period,[8] the energy transition in the EU has entered the phase of 'acceleration', which has been described as a breakthrough in structural developments and an accumulation of interrelated sociocultural, economic, ecological and technological changes.[9] However, the question of how the transitional phase of 'acceleration' can be governed and supported by law remains largely unanswered and open for scholarly discussion.[10] A significant threat to delaying or impeding the EU's energy transition stems from its disruptive magnitude, as it seeks to decarbonize a sector traditionally heavily reliant on

[2] Decarbonization refers to the reduction or elimination of carbon dioxide emissions from various sectors, including energy and transport, to mitigate climate change.

[3] Decentralization in the context of the energy transition describes a shift away from traditional, centralized generation of energy to a more distributed system where energy is generated close to where it will be used. Examples include the installation of solar panels.

[4] Digitalization in the energy transition refers to the integration of technological solutions such as smart grids and data analytics to optimize the real-time management of the electricity grid.

[5] Democratization in the context of the energy transition describes the empowerment of consumers and communities to actively participate in energy production, ownership and decision-making.

[6] Marga E Edens, 'Public Value Tensions for Dutch DSOs in Times of Energy Transition: A Legal Approach' (2017) 18(1–2) *Competition and Regulation in Network Industries* 132, 134.

[7] Commission, '"Fit for 55": Delivering the EU's 2030 Climate Target on the Way to Climate Neutrality' COM (2021) 550 final, 1–3.

[8] 'Predevelopment' here refers to the emergence of novel sustainable practices that do not yet lead to visible social change, while the 'take-off' period describes the first shifts toward a new dynamic equilibrium of social development including the adoption of new practices and common visions and expectations.

[9] Jan Rotmans, René Kemp and Marjolein van Asselt, 'More Evolution than Revolution: Transition Management in Public Policy' (2001) 3(1) *Foresight: The Journal of Future Studies, Strategic Thinking and Policy* 15, 17; only followed by the final phase of 'stabilization'.

[10] Jonathan Köhler et al, 'An Agenda for Sustainability Transitions Research: State of the Art and Future Directions' (2019) 31 *Environmental Innovation and Societal Transitions* 1, 9, 21.

fossil fuels. This transformation poses a risk, potentially exacerbating existing deficiencies and introducing new challenges as the transition unfolds.[11]

The European Commission aims to alleviate these existing and emerging adverse societal impacts by pursuing a 'gradual',[12] 'just',[13] 'inclusive'[14] and 'socially fair'[15] transitional approach. However, the Commission's strategy falls short in articulating how it intends to achieve the ambitious and pathbreaking climate targets essential for the energy transition and effective climate change mitigation, while simultaneously ensuring that 'no one is left behind'.[16]

The concept of 'energy justice' aims to bridge this gap by guiding the energy transition toward more just and socially accepted outcomes. Energy justice has received increasing attention worldwide and has penetrated into a multitude of disciplines, including social sciences, policy and law.[17] This chapter sets out to explore the concept of energy justice within the context of the far-reaching requirements of the Anthropocene. It seeks to assess whether the notion of justice can effectively guide the energy transition, or whether profound compromises and trade-offs are unavoidable, potentially undermining justice in the process.

The chapter is structured as follows. Section 16.2 provides an overview of the current conceptualization approaches of energy justice in the social sciences. This is conducted through a review of scientific literature on the origin and definition of energy justice and its implications for the governance of technological solutions in the energy sector. Section 16.3 analyses the capacity of energy justice in steering and accelerating the energy transition toward socially accepted energy services. This analysis delves deeper into the role of economic efficiency, technological solutions, and the overall need for an overhaul of EU energy regulation. Section 16.4 explores the inherent trade-offs that energy justice encounters in the context of the energy transition, including how to navigate unavoidable injustices.

---

11 Ruven Fleming, Kaisa Huhta and Leonie Reins, 'What Is Sustainable Energy Democracy in Law?', in Ruven Fleming, Kaisa Huhta and Leonie Reins (eds), *Sustainable Energy Democracy and the Law* (Brill 2021) 3.

12 Commission, 'Clean Energy for All Europeans' (2019), 1 <https://energy.ec.europa.eu/topics/energy-strategy/clean-energy-all-europeans-package_en>.

13 Commission, 'The European Green Deal' COM (2019) 640 final, 2.

14 Ibid.

15 Commission, '"Fit for 55": Delivering the EU's 2030 Climate Target on the Way to Climate Neutrality' COM (2021) 550 final, 4.

16 Commission, 'Clean Energy for All Europeans' (2019), 5 <https://energy.ec.europa.eu/topics/energy-strategy/clean-energy-all-europeans-package_en>.

17 Raphael J Heffron and Darren McCauley, 'The Concept of Energy Justice across the Disciplines' (2017) 105 *Energy Policy* 658, 666.

16.2 Defining energy justice and its role in the Anthropocene

While environmental and climate justice are rooted in the anti-establishment and activist movements of the United States through the 1970s, energy justice emerged as a predominantly academic concept in the 2010s.[18] While the aforementioned concepts of justice share the common ambition of bridging the climate change discourse with human rights and a wider array of justice concerns, energy justice was intentionally separated to address justice matters pertaining to energy technologies in a more strategically impactful manner and with greater conceptual rigour.[19] As such, energy justice also arose as a critique of environmental and climate justice and their failure to be translated into policy.[20] Today, scholarship describes energy justice as the 'primary driver',[21] 'overall *raison d'être*',[22] and 'main trigger of energy law in the 21st century'.[23] Related concepts such as 'energy democracy'[24] and 'sustainable energy democracy'[25] are also emerging from the energy justice debate in academia.

Sovacool and colleagues describe energy justice as 'a global energy system that fairly distributes both the benefits and burdens of energy services, and

18 Raphael J Heffron, Darren McCauley and Benjamin K Sovacool, 'Resolving Society's Energy Trilemma through the Energy Justice Metric' (2015) 87 *Energy Policy* 168, 169; Kristen Jenkins, 'Setting Energy Justice Apart from the Crowd: Lessons from Environmental and Climate Justice' (2018) 39 *Energy Research and Social Science* 117, 120; Mariëlle Feenstra and Gül Özerol, 'Energy Justice as a Search Light for Gender-energy Nexus: Towards a Conceptual Framework' (2021) 138 *Renewable and Sustainable Energy Reviews* 2.

19 Kristen Jenkins, 'Setting Energy Justice Apart from the Crowd: Lessons from Environmental and Climate Justice' (2018) 39 *Energy Research and Social Science* 117, 119–20.

20 Raphael J Heffron, Darren McCauley and Benjamin K Sovacool, 'Resolving Society's Energy Trilemma through the Energy Justice Metric' (2015) 87 *Energy Policy* 168, 175.

21 Raphael J Heffron and Kim Talus, 'The Evolution of Energy Law and Energy Jurisprudence: Insights for Energy Analysts and Researchers' (2016) 19 *Energy Research & Social Science* 1, 9.

22 Raphael J Heffron, 'Applying Energy Justice into the Energy Transition' (2022) 156 *Renewable and Sustainable Energy Reviews* 1.

23 Imelda Maher and Oana Stefan, 'Delegation of Powers and the Rule of Law: Energy Justice in EU Energy Regulation' (2019) 128 *Energy Policy* 84, 84.

24 See, for example, Matthew Burke and Jennie Stephens, 'Energy Democracy: Goals and Policy Instruments for Sociotechnical Transitions' (2017) 33 *Energy Research & Social Science* 35; Eleanor Stein, 'Energy Democracy: Power to the People? An Introduction', in Raya Salter, G Carmen Gonzalez and A Elizabeth Kronk Warner, *Energy Justice* (Edward Elgar Publishing 2018) 258.

25 See Ruven Fleming, Kaisa Huhta and Leonie Reins (eds), *Sustainable Energy Democracy and the Law* (Brill 2021).

one that contributes to more representative and inclusive energy decision-making'.[26] As an emerging concept, energy justice still lacks a uniformly accepted definition – both conceptually and substantively. Energy justice has been described in various ways, including as a conceptual, analytical or decision-making tool,[27] a theory,[28] an idea,[29] an objective,[30] an agenda[31] or a theoretical instrument.[32] Three possibilities have emerged for conceptualizing the notion of energy justice: (1) framing it as a triumvirate of tenets;[33] (2) defining ten 'energy justice principles';[34] and (3) abandoning pursuit of a fixed definition altogether.[35]

1. *Justice tenets:* Most commonly referred to in literature is the triumvirate of tenets.[36] While there is an ongoing debate as to how many tenets are inherent to energy justice,[37] the three predominantly agreed upon are: distributive justice,[38] recognition justice and procedural justice.[39] Distributive justice

26 Benjamin K Sovacool et al, 'New Frontiers and Conceptual Frameworks for Energy Justice' (2017) 105 *Energy Policy* 677, 677.

27 Benjamin K Sovacool and Michael H Dworkin, 'Energy Justice: Conceptual Insights and Practical Application' (2015) 142 *Applied Energy* 435, 436–40.

28 Raphael J Heffron and Darren McCauley, 'The Concept of Energy Justice across the Disciplines' (2017) 105 *Energy Policy* 658, 666.

29 Benjamin K Sovacool et al, 'New Frontiers and Conceptual Frameworks for Energy Justice' (2017) 105 *Energy Policy* 677.

30 Ruven Fleming, Kaisa Huhta and Leonie Reins, 'What Is Sustainable Energy Democracy in Law?', in Ruven Fleming, Kaisa Huhta and Leonie Reins (eds), *Sustainable Energy Democracy and the Law* (Brill 2021) 3, 17.

31 Kirsten EH Jenkins et al, 'Energy Justice: A Conceptual Review' (2016) 11 *Energy Research & Social Science* 174.

32 Max Lacey-Barnacle, Rosie Robison and Chris Foulds, 'Energy Justice in the Developing World: A Review of Theoretical Frameworks, Key Research Themes and Policy Implications' (2020) 55 *Energy for Sustainable Development* 122.

33 Kerstin Jenkins et al, 'Energy Justice: A Conceptual Review' (2016) 11 *Energy Research & Social Science* 174.

34 Benjamin K Sovacool et al, 'New Frontiers and Conceptual Frameworks for Energy Justice' (2017) 105 *Energy Policy* 677.

35 Michael C LaBelle, 'In Pursuit of Energy Justice' (2017) 107 *Energy Policy* 615; Kristen EH Jenkins et al, 'Towards Impactful Energy Justice Research: Transforming the Power of Academic Engagement' (2020) 67 *Energy Research & Social Science* 2.

36 Raphael J Heffron and Darren McCauley, 'The Concept of Energy Justice across the Disciplines' (2017) 105 *Energy Policy* 658; Darren A McCauley et al, 'Advancing Energy Justice: the Triumvirate of Tenets' (2013) 32(3) *International Energy Law Review* 107.

37 Some scholars further refer to additional tenets such as cosmopolitan justice, that is, the consideration of cross-border effects of energy activities, and restorative justice, that is, the rectification of injustices caused by the energy sector.

38 Often also referred to as 'distributional justice', despite a lack of clear differentiation between both concepts.

39 See Raphael J Heffron, Darren McCauley and Benjamin K Sovacool, 'Resolving Society's Energy Trilemma through the Energy Justice Metric' (2015) 87 *Energy Policy* 168, 168;

examines the fair allocation of burdens and benefits within the energy system, focusing on 'who receives what'.[40] Recognition justice aims to ensure a fair representation of those affected, questioning 'who is misrepresented'.[41] Procedural justice emphasizes equal access to decision-making processes and is commonly linked to public participation, asking 'who decides'.[42] Conceptually, the triumvirate of tenets has been considered particularly useful as an analytical tool, which has the potential to aid legal decision-making in highlighting injustices and gaining clarity on how energy justice issues are framed.[43] The tenets critically evaluate energy policy and existing legal frameworks by offering an analytical framework, with justice dilemmas at the forefront.[44]

2. *Energy justice principles*: This conceptualization defines energy justice according to so-called energy justice principles. Initially, eight principles were introduced,[45] which were later extended. Today, they encompass the following ten principles: availability, affordability, due process, good governance,[46] sustainability, intragenerational equity, intergenerational equity, responsibility, resistance and intersectionality.[47] Despite not being principles in a legal sense, this framework enlists the values that are inherent to the

---

Kirsten EH Jenkins et al, 'Energy Justice: A Conceptual Review' (2016) 11 *Energy Research & Social Science* 174, 176–7.

40 Kirsten EH Jenkins et al, 'Energy Justice: A Conceptual Review' (2016) 11 *Energy Research & Social Science* 174, 176–7; Blanka Vitéz and Saskia ACM Lavrijssen, 'The Energy Transition: Democracy, Justice and Good Regulation in the Heat Market' (2020) 13(5) *Energies* 4.

41 See Raphael J Heffron, Darren McCauley and Benjamin K Sovacool, 'Resolving Society's Energy Trilemma through the Energy Justice Metric' (2015) 87 *Energy Policy* 168, 168; Kirsten EH Jenkins et al, 'Energy Justice: A Conceptual Review' (2016) 11 *Energy Research & Social Science* 174, 176–7.

42 Kirsten EH Jenkins et al, 'Energy Justice: A Conceptual Review' (2016) 11 *Energy Research & Social Science* 174, 178; Ruven Fleming, Kaisa Huhta and Leonie Reins, 'What Is Sustainable Energy Democracy in Law?', in Ruven Fleming, Kaisa Huhta and Leonie Reins (eds), *Sustainable Energy Democracy and the Law* (Brill 2021) 3, 10.

43 Benjamin K Sovacool and Michael H Dworkin, 'Energy Justice: Conceptual Insights and Practical Application' (2015) 142 *Applied Energy* 435, 437.

44 Mariëlle Feenstra and Gül Özerol, 'Energy Justice as a Search Light for Gender-energy Nexus: Towards a Conceptual Framework' (2021) 138 *Renewable and Sustainable Energy Reviews* 2

45 The eight initial principles introduced in Benjamin K Sovacool and Michael H Dworkin, 'Energy Justice: Conceptual Insights and Practical Applications' (2015) 142 *Applied Energy* 435 were later extended to ten in Benjamin K Sovacool et al, 'New Frontiers and Conceptual Frameworks for Energy Justice' (2017) 105 *Energy Policy* 677.

46 Later also referred to as transparency and accountability.

47 Benjamin K Sovacool and Michael H Dworkin, 'Energy Justice: Conceptual Insights and Practical Applications' (2015) 142 *Applied Energy* 435; Benjamin K Sovacool et al, 'New Frontiers and Conceptual Frameworks for Energy Justice' (2017) 105 *Energy Policy* 677.

energy sector and highlights public interests that legal decision-making should take into account. The framework's conceptual understanding as a decision-making tool can assist policy makers, regulators and the judiciary, but also consumers and businesses, to make more informed and equitable energy choices.[48]

3. *Abandoning the pursuit of a fixed definition*: Some scholars consider that energy justice, as a workable notion in practice, must be capable of answering to local particularities and differences.[49] Given that states greatly vary in their historical fossil-fuel dependencies, infrastructures, energy mix and socio-economic circumstances,[50] energy justice responses require a tailored approach that takes the contemporary societal environment and political-economic landscape into account.[51] Hence, beyond a universal energy justice framework, more nuanced interpretations must exist that respond to justice issues on all legal and regulatory scales of the energy sector to ensure that the multifaceted character or energy regulation and its societal consequences are sufficiently addressed.[52]

While none of the conceptualization approaches have avoided criticism,[53] many of them are complementary and mutually reinforcing.[54] This synergy becomes particularly relevant as climate change has raised increasing interest in socio-technical solutions to existing and continually evolving energy problems. However, research has demonstrated that these technological solutions are marked by the possibility of unfairness and injustice.[55] Low-carbon technologies are often seen as essential for accelerating decarbonization and addressing energy challenges, yet there are concerns that their roll-out and integration could lead to newly arising injustices and perpetuate existing inequalities. Hence, it is crucial to fully recognize and assess the intricate

48 Benjamin K Sovacool and Michael H Dworkin, 'Energy Justice: Conceptual Insights and Practical Applications' (2015) 142 *Applied Energy* 435, 436.

49 Annabel Pinker, 'Just Transition: A Comparative Approach – A Report Prepared for the Just Transition Commission' (The James Hutton Institute & SEFARI Gateway 2020), 4.

50 Ibid, 64.

51 Ibid.

52 Michael C LaBelle, 'In Pursuit of Energy Justice' (2017) 107 *Energy Policy* 615.

53 Romain Mauger, 'Making Sense of Changing Concepts for the Energy Transition: An Energy Transition Concepts Nexus for Policy and Law', in Ruven Fleming, Kaisa Huhta and Leonie Reins (eds), *Sustainable Energy Democracy and the Law* (Brill 2021) 28, 35.

54 Ross Gillard, Carolyn Snell and Mark Bevan, 'Advancing an Energy Justice Perspective of Fuel Poverty: Household Vulnerability and Domestic Retrofit Policy in the United Kingdom' (2017) 29 *Energy Research and Social Science* 53, 54.

55 Karen Bickerstaff, Gordon Walker and Harriet Bulkeley, 'Introduction: Making Sense of Energy Justice', in Karen Bickerstaff, Gordon Walker and Harriet Bulkeley (eds), *Energy Justice in a Changing Climate: Social Equity and Low-Carbon Energy* (Bloomsbury Publishing 2013), 6.

relationship between energy technologies and their societal impacts when navigating the path toward decarbonizing energy services. This includes considering the role of low-carbon technologies within the context of other existing and emerging infrastructures, spatial considerations, socio-economic underpinnings and (democratic) decision-making processes.[56] Ultimately, technological systems, such as energy systems, must always be viewed as intrinsic parts of broader social systems.[57]

The urgency of the energy transition in the era of the Anthropocene, driven by the imperative for extensive decarbonization and socially accepted energy services, presents an important opportunity to delve deeper into the potential and shortcomings of the interrelation of justice and energy technologies, as well as their capacity to drive transformative change.[58] This leads to the fundamental question of whether energy justice can guide energy regulation in aligning energy technologies with societal values, while also considering the constraints posed by the challenges of the Anthropocene when striving for justice. This question will be explored in the following sections, first examining the guidance the energy justice can provide to energy regulation, before turning to the boundaries and limitations of a just transition.

16.3 Energy justice: exploring opportunities and imperatives

In the last decade, the field of energy law has been described as having undergone an 'ethical turn',[59] generating a fresh impetus for the pursuit of energy justice on a global scale. Within the context of EU law, three notable developments exemplify the growing social consciousness of EU energy policy and the need of energy justice to inform the energy transition and the management of energy technologies. These developments not only highlight the evolving nature of EU energy regulation but also emphasize the increasing recognition of social concerns shaping energy policies within the EU. By exploring these developments, a deeper understanding can be

56 Duncan McLaren, Kristian Krieger and Karen Bickerstaff, 'Justice in Energy System Transitions: The Case of Carbon Capture and Storage', in Karen Bickerstaff, Gordon Walker and Harriet Bulkeley (eds), *Energy Justice in a Changing Climate: Social Equity and Low-Carbon Energy* (Bloomsbury Publishing 2013) 159, 163.

57 Ibid.

58 Aparajita Banerjee et al. 'Renewable, Ethical? Assessing the Energy Justice Potential of Renewable Electricity' (2017) 5(5) *AIMS Energy* 768.

59 Aileen McHarg, 'Energy Justice: Understanding the "Ethical Turn" in Energy Law and Policy', in Iñigo del Guayo et al (eds), *Energy Justice and Energy Law* (Oxford University Press 2020) 15.

gained of the potential that energy justice holds in aligning the regulation of energy technologies with the needs of the Anthropocene and broader societal values.

16.3.1 The role of energy justice in resource scarcity

As a sector historically rooted in the use of fossil fuels, economic regulation has been traditionally relied upon to guide energy regulation in its efforts to manage energy as a scarce resource.[60] Today, the prevailing view of the European Commission is that economic efficiency serves as a primary mechanism for maximizing consumer welfare and 'the most efficient means of ensuring affordable energy prices'.[61] Since the 1990s, EU energy regulation has therefore been predominantly concerned with economic integration and market liberalization to move a state-controlled energy sector toward an increasingly competitive and integrated internal energy market.[62] Market-based mechanisms have been heavily relied upon as a method to facilitate the most economically efficient allocation and distribution of energy resources. After two centuries of industrialization, the conversion of this economic system into a climate-neutral market economy poses a major challenge.[63]

Historically, a narrow understanding of economic efficiency as an optimal allocation of available resources to maximize welfare has resulted in the prevalence of fossil fuels, with a low-carbon economy and renewable energy technologies being a secondary concern for energy regulation.[64] Fossil fuels are considered low-cost, and, because of that, economically efficient, as negative externalities such as CO_2 emissions and climate impacts are not accounted for as part of the price.[65] Moreover, network operators, as managers of natural monopolies, are bound by economic efficiency.[66]

---

60 Louis-Gaëtan Giraudet and Antoine Missemer, 'The History of Energy Efficiency in Economics: Breakpoints and Regularities' (2023) 97 *Energy Research & Social Science* 2.

61 Commission, '2020 report on the State of the Energy Union pursuant to Regulation (EU) 2018/1999 on Governance of the Energy Union and Climate Action' COM (2020) 950 final, 8.

62 Kim Talus, 'Introduction: Development of EU Energy Law and Policy', in Kim Talus (ed), *Introduction to EU Energy Law* (Oxford University Press 2016) 1, 3.

63 Paul JJ Welfens, 'Climate Policy Problems: The Concept of a Sustainable Social Market Economy', in Paul JJ Welfens (ed), *Global Climate Change Policy: Analysis, Economic Efficiency Issues and International Cooperation* (Palgrave Macmillan 2022) 341.

64 Raphael J Heffron and Darren McCauley, 'The Concept of Energy Justice across the Disciplines' (2017) 105 *Energy Policy* 658, 664.

65 Ibid, 665.

66 See, for example, Directive (EU) 2019/944 of the European Parliament and of the Council of 5 June 2019 on Common Rules for the Internal Market for Electricity and amending Directive 2012/27/EU (recast) [2019] OJ L158/125 ('EU Electricity Directive'), Art 31.

Investments in the expansion of energy infrastructure is often considered economically inefficient and are closely monitored by the natural regulatory authorities.[67] This has led to substantial shortcomings in infrastructure investment and grid capacity in many member states such as the Netherlands, Spain, Germany, Poland, Bulgaria and Ireland.[68] As a consequence, electricity networks across the EU are failing to serve the integration of renewable energy sources and other low-carbon technologies today.[69] Despite the focus on economic efficiency and the acceptance that it leads to affordable energy services, the increasing number of vulnerable EU citizens that live in energy poverty further contribute to the disillusionment of the promise of neo-liberal energy policies.[70]

However, when a social market economy, such as the EU, transitions toward climate neutrality, the objective must be a sustainable social market economy, where economic growth and environmental quality are high, and where governmental institutions can finance redistributive and social policies.[71]

By embracing the concept of energy justice, the approach to energy regulation as a means of managing resource scarcity can be reevaluated and reframed. Energy justice centres energy regulation around the notion of 'justice', treating energy as a good and service.[72] The contemporary understanding of the concept of 'justice' is synonymous with 'fairness', and particularly fair social structures that lead to a fair distribution of goods and services.[73] As a 'merit good', that is a good or service that society agrees should be accessible to all regardless of income, energy is closely linked to

---

[67] See, for example, Electricity Directive, Art 59(1)(k).

[68] Eurelectric, 'Power System of the Future: Keys to Delivering Capacity on the Distribution Grid' (2023), 5 <www.eurelectric.org/publications/power-system-of-the-future-keys-to-deloivering-capacity-on-the-distribution-grid/>

[69] Laura Kaschny and Saskia Lavrijssen, 'The Independence of National Regulatory Authorities and the European Union Energy Transition' (2023) 72(3) *International & Comparative Law Quarterly* 715; Renee van Hest and Jan Kleinnijenhuis, 'Reconstruction: How Regulators Contributed to Power Grid Shortages' (*NOS*, 3 October 2022) <https://nos.nl/nieuwsuur/artikel/2446934-reconstructie-hoe-toezichthouder-tekorten-op-het-elektriciteitsnet-mede-veroorzaakte>.

[70] Often described as a situation in which households are unable to access essential energy services. Data from Eurostat, SILC accessible at <https://ec.europa.eu/eurostat/databrowser/view/ILC_MDES01__custom_1485289/default/line?lang=en>.

[71] Paul JJ Welfens, 'Climate Policy Problems: The Concept of a Sustainable Social Market Economy', in Paul JJ Welfens (ed), *Global Climate Change Policy: Analysis, Economic Efficiency Issues and International Cooperation* (Palgrave Macmillan 2022) 341, 342.

[72] Benjamin K Sovacool and Michael H Dworkin, 'Energy Justice: Conceptual Insights and Practical Applications' (2015) 142 *Applied Energy* 435, 436.

[73] Ibid.

the ideals of social justice and economic equity.[74] According to social justice theory, such a 'primary good' defines questions of justice as those relating to the distribution of rights and liberties, powers and opportunities, and income and wealth.[75] Social and economic justice, therefore, relate to rights of access and affordability, with a heightened responsibility of the state to guarantee fair distribution of energy as a merit or primary good and service.[76]

While economic regulation is not entirely contradictory to this approach, energy justice allows the negative externalities of a narrow application of economic efficiency to be highlighted. As a framework aimed at addressing a fair energy distribution, energy justice becomes a crucial tool for legislators and the judiciary in reevaluating both the physical allocation of energy as a scarce resource and the equitable share of benefits and responsibilities. The concept also goes further by outlining the complexities of justice that go beyond material inequalities and highlighting inequalities in process and representation.[77] Energy justice, therefore, transcends and complements economic efficiency by encompassing concepts that have historically been difficult to quantify within economic frameworks and market-based regulation, including sustainability and social justice.

16.3.2 Energy justice as an accelerator of the energy transition

While EU law promotes the integration of renewable and new-energy sources, there is a risk of new injustices arising from the challenge of decarbonizing the energy sector.[78] Often, these injustices and risks can be associated with both the deployment and prohibition of energy technologies.[79] Both responses are characterized by risks of environmental harm: While the importance of renewable energy sources for climate

[74] Stéphanie de Somer, *Autonomous Public Bodies and the Law – A European Perspective* (Edward Elgar Publishing 2017), 33.

[75] Benjamin K Sovacool and Michael H Dworkin, 'Energy Justice: Conceptual Insights and Practical Applications' (2015) 142 *Applied Energy* 435, 437.

[76] Stéphanie de Somer, *Autonomous Public Bodies and the Law: A European Perspective* (Edward Elgar Publishing 2017), 33.

[77] Chiara Armeni, 'What Justice? The Scope for Public Participation in the European Union Just Transition' (2023) 60(4) *Common Market Law Review* 1027, 1028.

[78] Consolidated Version of the Treaty on the Functioning of the European Union [2012] OJ C326/47, art 194(1)(c) TFEU; Benjamin K Sovacool, *Energy & Ethics: Justice and the Global Energy Challenge* (Palgrave Macmillan 2013) 1–2; Romain Mauger, 'Making Sense of Changing Concepts for the Energy Transition: An Energy Transition Concepts Nexus for Policy and Law', in Ruven Fleming, Kaisa Huhta and Leonie Reins (eds), *Sustainable Energy Democracy and the Law* (Brill 2021) 28, 37.

[79] Jesse L Reynolds and Floor Fleurke, 'Climate Engineering Research: A Precautionary Response to Climate Change' (2013) *CCLR* 101, 105.

change mitigation is undisputed, it is also important to recognize the diverse ecological risks associated with their implementation. For instance, both onshore and offshore wind-energy systems have exhibited adverse ecological effects, including noise pollution, shift of species migration patterns and harm to wildlife.[80] Hydropower has been repeatedly called out to be a major contributor to the deterioration of aquatic ecosystems in EU member states,[81] while carbon capture and storage technologies raise concerns regarding soil and ocean acidification.[82] Many of these harms are spatially and socially unevenly located with marginalized and vulnerable populations at higher risks.[83] Additionally, these climate considerations often come with social trade-offs such as high unemployment rates in coal-intensive regions, which once were the backbone of national economic growth.[84]

Therefore, the energy transition presents a complex risk-risk trade-off between climate change mitigation, environmental harm and societal considerations. On the one hand, there is a looming threat of catastrophic environmental damage due to global warming, while on the other hand, the widespread adoption of renewable energy technologies, aimed at managing climate risks, also creates significant environmental and societal challenges. Striking a balance between these risks is a crucial aspect of sustainable energy and climate policy.

The Anthropocene era highlights a significant shift in this trade-off, placing greater emphasis on climate change mitigation, including justifying intervention on massive scales in global physical, chemical and biological Earth systems to counterbalance climate change.[85] The role of regulation

---

80 See Y Teff-Seker et al, 'Noise Pollution from Wind Turbines and its Effects on Wildlife: A Cross-National Analysis of Current Policies and Planning Regulations' (2022) 168 *Renewable and Sustainable Energy Reviews*; Josep Lloret et al 'Unravelling the Ecological Impacts of Large-scale Offshore Wind Farms in the Mediterranean Sea' (2022) 824 *Science of the Total Environment*.

81 Alban Kuriqi et al, 'Ecological Impacts of Run-of-River Hydropower Plants: Current Status and Future Prospects on the Brink of Energy Transition' (2021) 142(C) *Renewable and Sustainable Energy Reviews* 4–11; Åsa Össbo and Patrik Lantto, 'Colonial Tutelage and Industrial Colonialism: Reindeer Husbandry and Early 20th-Century Hydroelectric Development in Sweden' (2011) 36(3) *Scandinavian Journal of History* 324, 336–40.

82 MN Anwar et al, 'CO_2 Capture and Storage: A Way forward for Sustainable Environment' (2018) 226 *Journal of Environmental Management* 131.

83 Jason Byrne and Chloe Portanger, 'Climate Change, Energy Policy and Justice: A Systematic Review' (2014) 36(2) *Analyse & Kritik* 315, 316.

84 See, for example, Hanna Brauers, Philipp Herpich and Pao-Yu Oei, 'The Transformation of the German Coal Sector from 1950 to 2017: An Historical Overview', in Christian von Hirschhausen et al (eds), *Energiewende 'Made in Germany'* (Springer 2018) 45–78.

85 Jesse L Reynolds and Floor Fleurke, 'Climate Engineering Research: A Precautionary Response to Climate Change' (2013) *CCLR* 101, 101.

is to guide technology and innovation toward outcomes that are aligned with societal needs, as unchecked innovation – or innovation for the sake of innovation – can carry significant or unknown costs to the environment and human health.[86] Therefore, while prioritizing technological advancements is crucial to address the challenges of the Anthropocene, innovation must ensure an acceptable level of environmental and public health risks.

In this context, energy justice provides an overarching framework that extends beyond the narrow regulation of specific energy technologies or network industries. In its pluralist approach, it promotes cross-sectoral governance and connects economic, societal, technical and environmental considerations. Energy justice assesses energy technologies within a system-wide focus that allows them to be valued at full cost.[87] Technology is always mediated by social practices, institutions and systems, and must therefore be guided by regulatory frameworks that align energy technologies with societal needs.[88]

Energy justice plays an important role in enabling a more informed assessment and effective management of the risk-risk trade-off between climate and environmental concerns. It highlights the economic, societal and environmental costs associated with energy technologies and decision-making processes in a broader sense. It acts as a safeguard to prevent the emergence of new and unnecessary costs for the environment and society. Energy justice emphasizes transparency and accountability within legislative and judicial decision-making processes and strives to achieve a more equitable distribution of environmental costs. Moreover, the transition toward low-carbon energy systems can also be viewed as an embodiment of energy justice, whether as an act of fairness toward future generations or as a recognition of fundamental spatial and historical injustices underlying the climate change issue.[89]

---

86 See for instance, transboundary implications of weather modification programmes; Manon Simon, Jan McDonald and Kerryn Brent, 'Transboundary Implications of China's Weather Modification Programme' (2023) *Transnational Environmental Law*.

87 Raphael J Heffron and Darren McCauley, 'Achieving Sustainable Supply Chains through Energy Justice' (2014) 123 *Applied Energy* 435, 435; Kirsten EH Jenkins et al, 'Energy Justice: A Conceptual Review' (2016) 11 *Energy Research & Social Science* 174, 180.

88 Siddharth Sareen and Håvard Haarstad, 'Bridging Socio-technical and Justice Aspects of Sustainable Energy Transitions' (2018) 228 *Applied Energy* 624, 625.

89 Duncan McLaren, Kristian Krieger and Karen Bickerstaff, 'Justice in Energy System Transitions: The Case of Carbon Capture and Storage', in Karen Bickerstaff, Gordon Walker and Harriet Bulkeley (eds), *Energy Justice in a Changing Climate: Social Equity and Low-carbon Energy* (Bloomsbury Publishing 2013) 158.

16.3.3 The overhaul of energy regulation as an opportunity for justice

The energy transition requires a complete review and overhaul of energy regulation. It has become clear that the envisioned low-carbon energy sector, relying on renewable sources and new technological solutions, is largely incompatible with the traditional market structures of large, centralized producers supplying energy to passive end-consumers.[90] Rather, the energy transition encompasses the simultaneous decarbonization, decentralization, digitalization and democratization of the energy sector, all of which require significant technological, structural and social transformations of the current market structure. Such a transition brings about substantial changes to the roles and responsibilities of market actors and encourages increasing market participation of consumers.

The European Commission's Fit-for-55 (2020) initiative is a driving force behind the EU's energy transition by setting increasingly tight climate targets to achieve net-zero emissions in the EU by 2050.[91] These objectives are reflected in far-reaching legislative packages of the European Commission introducing extensive amendments to EU energy law to reflect the new realities of a decarbonized, digitalized and decentralized energy sector.[92]

However, these transformative developments and the disruptive magnitude of the energy transition threaten to create new legal inconsistencies and exacerbate existing ones.[93] If left unmitigated, these regulatory tensions will ultimately result in adverse societal consequences and a reduced social acceptability of the transition, leading to delays in the energy transition by causing setbacks in the integration of renewable energy sources and the phase-out of fossil fuels.

The EU Electricity Regulation codifies the significance of a 'just transition' by recognizing the 'social and economic impacts of the clean

---

90 Matthew Burke and Jennie Stephens, 'Political Power and Renewable Energy Futures: A Critical Review' (2018) 35 *Energy Research & Social Science* 78.

91 Commission, '"Fit for 55": Delivering the EU's 2030 Climate Target on the way to Climate Neutrality' COM (2021) 550 final; see most recently Directive (EU) 2023/2413 of the European Parliament and of the Council of 18 October 2023 amending Directive (EU) 2018/2001, Regulation (EU) 2018/1999 and Directive 98/70/EC as regards the promotion of energy from renewable sources, and repealing Council Directive (EU) 2015/652 [2023].

92 The fourth revision of the EU energy law framework. For electricity, European Commission, 'Clean Energy for All Europeans Package' (2019) <https://energy.ec.europa.eu/topics/energy-strategy/clean-energy-all-europeans-package_en>; for gas, European Commission, 'Hydrogen and Decarbonised Gas Market' (2021) <https://energy.ec.europa.eu/topics/markets-and-consumers/hydrogen-and-decarbonised-gas-market_en>.

93 Romain Mauger, 'Making Sense of Changing Concepts for the Energy Transition: An Energy Transition Concepts Nexus for Policy and Law', in Ruven Fleming, Kaisa Huhta and Leonie Reins (eds), *Sustainable Energy Democracy and the Law* (Brill 2021) 28.

energy transition'.[94] This understanding is mirrored by the Commission's acknowledgement of the modernization of the energy sector as a 'unique opportunity to reduce systemic inequality'.[95] The European Commission emphasizes the importance of a 'socially fair'[96] transition that ensures that 'no one is left behind'[97] and guarantees 'access to energy for all, tackling energy poverty at its roots and protecting vulnerable consumers'.[98]

However, the amendments to EU energy law do not fully align with the level of inclusiveness envisioned by the European Commission. It remains unclear how a 'gradual' and 'progressive'[99] transition aligns with the increasingly tighter and ambitious EU climate objectives and the need to accelerate decarbonization. Both the EU Electricity Directive and Regulation are important regulatory instruments governing the EU energy transition.

However, while the EU Electricity Regulation emphasizes the importance of a just transition, it constrains the concept to the challenges arising within current 'coal and carbon-intensive regions' and the reskilling of the workforce in line with the European Social Model.[100] Moreover, the energy justice focus of the EU Electricity Directive is predominantly restricted to the protection of vulnerable consumers living in energy poverty.[101] This demonstrates that the EU's approach to energy justice is often limited to seeking a minimum standard for combatting social exclusion and marginalization but does not encompass a broader whole-of-economy approach to a just transition.

Justice concerns related to the EU energy transition are extensive and complex. In the context of technology, justice concerns include exclusionary attributes of existing and emerging technological solutions. For instance, the integration of electric vehicles (EVs) and solar panels, which, despite their environmental and social importance, can exacerbate socio-economic

---

94 Regulation (EU) 2019/943 of the European Parliament and of the Council of 5 June 2019 on the Internal Market for Electricity (recast) [2019] OJ L158/54 ('EU Electricity Regulation'), Art 4.

95 Commission, '"Fit for 55": Delivering the EU's 2030 Climate Target on the Way to Climate Neutrality' COM (2021) 550 final, 4.

96 Ibid.

97 Commission, 'Clean Energy for All Europeans' (2019) 5 <https://energy.ec.europa.eu/topics/energy-strategy/clean-energy-all-europeans-package_en>.

98 Ibid, 12.

99 EU Electricity Regulation, Art 4.

100 Ibid; Commission, 'Clean Energy for All Europeans' (2019) 12 <https://energy.ec.europa.eu/topics/energy-strategy/clean-energy-all-europeans-package_en>.

101 Electricity Directive, Arts 28 and 29.

disparities.[102] These technologies create a threat of shifting costs from more affluent households, which can readily adopt and benefit from these technologies, to poorer ones, which cross-subsidize the resulting additional costs imposed on the energy system as a whole.[103] In those situations, less affluent households find themselves more often in passive consumer roles, that is consumers who do not actively participate in and benefit from decentralized energy systems, further widening the gap between the wealthy and the poor.

The energy transition extends beyond questions of hardware choices for energy technologies. As I have discussed, economic efficiency alone provides little insight into how the costs and benefits of the energy transition should be distributed in society. Shifting to a low-carbon economy requires regulation to ensure that the societal costs, risks and benefits of this transition are distributed in a just and equitable manner.[104] Energy justice allows for a normative engagement with these factors from the perspective of citizens and consumers.[105] As an overarching concept, energy justice centres energy regulation around the notion of 'justice', offering a unifying framework to help resolve regulatory tensions around a common objective. Serving as an instrument for legal and judicial decision-making, energy justice supports a fair energy transition by fostering progress toward more inclusive and equitable energy services.

16.4 Navigating the boundaries of energy justice

While holding significant promise, from a legal perspective, energy justice does not exist without criticism. The following section delves into the notion of energy justice as a guiding concept for the energy transition, exploring the boundaries of its promises and the inherent trade-offs it must navigate. It also assesses the constraints of energy justice in terms of its potential to

---

[102] Kirsten Jenkins, Benjamin K Sovacool and Darren McCauley, 'Humanizing Sociotechnical Transitions through Energy Justice: An Ethical Framework for Global Transformative Change' (2018) 117 *Energy Policy* 66, 68.

[103] Ibid. In these cases, active customers benefit from technological advancements (for example, solar panels) to reduce their overall energy bill. However, in most EU member states, the electricity grid is not traditionally built for the rapid electrification of the energy sector and experiences capacity constraints. This requires substantial investments into the grid, which are socialized costs that are carried by all network users in a flat tariff structure. As a result, both passive and active consumers are facing higher network charges. However, passive consumers are not benefiting from the advantages of these technological advancements, unlike active consumers, and therefore cross-subsidize the latter.

[104] Siddharth Sareen and Håvard Haarstad, 'Bridging Socio-technical and Justice Aspects of Sustainable Energy Transitions' (2018) 228 *Applied Energy* 624, 624.

[105] Ibid.

instigate positive change and the possible extent to which society might have to tolerate injustices.

16.4.1 Energy justice as a value trade-off

While energy justice aims to offer answers to regulatory tensions, energy justice is not free from tensions within itself. The various justice tenets and values carry inherent trade-offs. For instance, intragenerational equity (justice within a generation) and intergenerational equity (justice between generations) do not always harmoniously coexist, but carry difficult trade-offs.[106] The previously highlighted risk-risk trade-off between climate change mitigation, environmental protection and justice is a complex challenge to reconcile. This includes the risk of turning the energy transition into a 'justice transition' without sufficient regard for the drastic measures required for effective climate change mitigation.

This is exacerbated by the fact that energy injustices are not always characterized by clear-cut scenarios with immediate 'winners' and 'losers'.[107] Instead, they exist in a complex web of interconnected issues, wherein addressing one injustice may inadvertently give rise to another. This contributes to the intricacy of identifying individual justice concerns and effectively addressing and regulating them in practice.

The EU and national governments often face the responsibility of navigating highly complex and often obscure trade-offs, as they aim to promote and safeguard energy justice. Questions include whether and how should a single (or a few) generation(s) of energy consumers bear the entire costs of the energy transition? It also creates a new discourse around accountability and responsibility of the state, market actors and consumers. It raises fundamental questions about the alignment of energy regulation with public interests, prompting a debate of who benefits from the energy transition and who bears the burdens of the negative externalities and social costs of decarbonizing energy services. There is a risk that this framing may centre energy justice predominantly around discussions of injustice and unsustainability rather than emphasizing a positive conception of justice and sustainability.[108] This, in turn, can trigger more radical political opposition

[106] Aileen McHarg, 'Energy Justice: Understanding the "Ethical Turn" in Energy Law and Policy', in Iñigo del Guayo et al (eds), *Energy Justice and Energy Law* (Oxford University Press 2020) 15, 22.

[107] Benjamin K Sovacool et al, 'New Frontiers and Conceptual Frameworks for Energy Justice' (2017) 105 *Energy Policy* 677, 684.

[108] Noel Healy and John Barry, 'Politicizing Energy Justice and Energy System Transitions: Fossil Fuel Divestment and a "Just Transition"' (2017) 108 *Energy Policy* 451, 452.

and instrumentalization of injustices, rather than a meaningful debate around how to create future-proof energy services.[109] It becomes evident, then, that the implementation and discourse of energy justice is intensely political.[110] Simultaneously, due to its abstract legal character, energy justice provides few safeguards against politicization for short-term political agendas – as will be further discussed later. As such, the effectiveness of energy justice, as a guiding legal instrument for the energy transition, invariably reflects the values and norms of the society within which it is applied.

16.4.2 Energy justice as an obligation to act?

Reflecting on energy justice within the context of the Anthropocene also questions the extent to which it demands not just the removal of injustices but also proactive efforts to establish justice. In other words, does energy justice encompass an obligation to act? Energy justice and related concepts, such as environmental justice, emerged from the discussion of how individual action can positively contribute to climate change mitigation,[111] and how resource depletion can be redefined as an injustice to future generations, carrying a duty to address it.[112] However, this discussion is ultimately a legal one.

Governmental mitigation responsibilities have been repeatedly challenged before national courts of the EU with increasing success.[113] In several member states, the debate is gradually shifting to discuss the role of fossil-fuel producers and their obligations to mitigate climate change.[114] However, at the consumer level, the prevailing policy interventions primarily involve 'nudging',[115] such as providing incentives for the implementation

[109] Ibid. An example of this is the debate around 'net metering', which is a billing system where solar panel owners offset purchased energy (in kWh) with generated energy and often net out at zero kWh billed. As such, solar panel owners can drastically reduce their energy bill but also do not contribute to the overall costs of maintaining the network upon which they rely. This results in higher overall costs for non-metering customers. However, the debate around abandoning net metering is often framed as an injustice that solar panel owners have to face, rather than a broader discussion on how to approach network costs more fairly across society and in light of the increasing costs of the energy transition.

[110] Ibid.

[111] Raphael J Heffron and Darren McCauley, 'The Concept of Energy Justice across the Disciplines' (2017) 105 *Energy Policy* 658, 663.

[112] Benjamin K Sovacool and Michael H Dworkin, 'Energy Justice: Conceptual Insights and Practical Application' (2015) 142 *Applied Energy* 435, 437.

[113] See, for example, *State of the Netherlands v. Urgenda Foundation* ECLI:NL:HR:2019:2006; German Federal Constitutional Court, Order of the First Senate of 23 March 2022, 1 BvR 1187/17.

[114] See, for example, *Milieudefensie et al v. Royal Dutch Shell plc*, 21 January 2021; *Lliuya v. RWE AG* (pending), *ClientEarth v. Polska Grupa Energetyczna*, 22 September 2020.

[115] A policy intervention, which aims to align consumer choices with their own will.

of energy-efficiency measures or the integration of renewable energy technologies. These policy measures promote certain societal norms without imposing legal obligations.[116]

As a concept emerging from the social sciences, energy justice largely lacks explicit legal implementation and application in the EU. In comparison, the related concept of a 'just transition' has been integrated into secondary legislation[117] and is now embodied in Article 4 of the EU Electricity Regulation.[118] However, this inclusion confines its applicability solely to the realm of the Regulation itself, limiting its integration as an overarching concept of EU energy law. It can therefore not inform more integrated cross-sectoral governance that encompasses economic, technological, and environmental decision-making more broadly. Consequently, for energy justice to fully realize its potential as a guiding concept, there is a need for its formal inclusion into overarching EU Treaty law.[119] However, in such cases, energy justice must be understood as a concept of general orientation for legislative or judicial decision-making. It can therefore, by definition, not reach a state of fulfilment.[120] As an abstraction lacking absolute implications, its justiciability is greatly limited – except in situations where governmental actions grossly neglect considerations of justice.[121]

Consequently, energy justice, as a legal concept of EU law, cannot be relied upon to impose legal obligations to demand positive change in response to climate change. As an interpretative framework for legislative and judicial decision-making, it requires further concretization through more specific legislation and case law to give rise to legal obligations in narrower contexts.

16.4.3 How much injustice is acceptable?

As much as energy justice has the potential to guide the energy transition toward more equitable and socially acceptable energy decision-making, certain injustices are unavoidable. This is not merely a result of the intricate nature of identifying and rectifying energy injustices without inadvertently

116 Louis-Gaëtan Giraudet and Antoine Missemer, 'The History of Energy Efficiency in Economics: Breakpoints and Regularities' (2023) 97 *Energy Research & Social Science* 4.

117 Secondary legislation in EU law constitutes the body of law (such as Regulations and Directives) that is derived from the principles and objectives set out in the EU treaties.

118 EU Electricity Regulation, Art 4.

119 Primary law refers to the EU treaties that lay down the legal framework of the EU.

120 Ruven Fleming, 'The "Trias": A New Methodology for Energy Law' (2019) 28(5) *European Energy and Environmental Law Review* 164, 166.

121 See for more detailed discussion, Laura Kaschny, 'Energy Justice and the Principles of Article 194(1) TFEU Governing EU Energy Policy' (2023) 12(2) *Transnational Environmental Law* 270.

giving rise to new ones; it is also attributable to the unavoidable prioritization of climate change mitigation over other societal and environmental costs.

That certain injustices must be tolerated was also put forward in the *Opal* judgment of the Court of Justice of the EU (CJEU).[122] In that case, the Court held that, while EU member states are required to take each other's and the interests of the EU into account in energy decision-making, this consideration does not confer an absolute right.[123] In other words, certain negative consequences for other member states and the EU through national energy decision-making must be accepted in line with national competences. The judgment applies on an institutional level governing interactions between the EU and its member states. However, it raises the question of where the line is drawn between having to tolerate injustices and the risk of infringing upon each other's interests.

While being strongly rooted in economic efficiency, EU energy regulation has increasingly focused on the protection of vulnerable consumers to mitigate the negative externalities of market-based mechanisms. With 41 million EU citizens living in energy poverty, and a concerning upward trend, the 2019 amendments to the EU Electricity Directive have imposed on member states the requirement to monitor and address their vulnerable citizens.[124] Although EU law refrains from defining energy poverty, it delegates this task to member states, allowing them to delineate vulnerable citizens based on the intricacies of their national circumstances, energy technologies and infrastructure.[125] Despite this broad discretion, EU law provides exemplary criteria for member states to consider.[126] Besides economic factors such as low income and high energy expenditure, the Directive also lists criteria such as age and health.[127] The Commission Recommendation on energy poverty further includes factors such as climate, gender and ethnic background to consider when assessing the realities of vulnerable citizens.[128]

---

[122] Case C-848/19 P *Germany v. Poland*, ECLI:EU:C:2021:598.

[123] Ibid, para 73.

[124] European Parliament, 'Energy Poverty in the EU' (2023) 1–6 <www.europarl.europa.eu/RegData/etudes/BRIE/2022/733583/EPRS_BRI(2022)733583_EN.pdf>; energy poverty here is defined as 'a household's lack of access to essential energy services that provide basic levels and decent standards of living and health, including adequate heating, hot water, cooling, lighting, and energy to power appliances, in the relevant national context, existing social policy and other relevant policies, caused by a combination of factors, including but not limited to non-affordability, insufficient disposable income, high energy expenditure and poor energy efficiency of homes'.

[125] EU Electricity Directive, Arts 28 and 29.

[126] Ibid.

[127] Ibid.

[128] Commission, 'Recommendation (EU) 2023/2407 of 20 October 2023 on Energy Poverty' 2023/2407, recital 2.

It becomes evident that EU energy regulation recognizes the protection of citizens living in energy poverty as a threshold of injustice. Once this threshold is exceeded, regulatory responses may be triggered. In these instances, EU law also permits public intervention to deviate from market-based mechanisms, such as the price-setting of electricity supply.[129] This approach aligns with the understanding that energy poverty not only equates to economic challenges but also results in social exclusion and the inability to access a minimum standard of living.

Beyond energy poverty, however, EU energy regulation does not foresee interference with market-based mechanisms. On the contrary, the costs associated with the energy transition, including rising costs of passive energy consumption, are seen as an impetus to encourage consumers to resort to energy-efficiency measures or other technological solutions, such as solar panels and EVs, to reduce their energy bills and decarbonize energy services. While safeguarding the most vulnerable from negative market dynamics is crucial, it is noteworthy that the existence of injustices also serves as a pivotal driver for the energy transition, incentivizing its acceleration. In these circumstances, the ambition of consumers to overcome injustices becomes its own catalyst for achieving energy justice.

16.5 Conclusion

This chapter has explored the concept of energy justice against the far-reaching consequences of the Anthropocene and the need to mitigate climate change. It has examined the effectiveness of energy justice in guiding the energy transition while acknowledging the inevitable need to seek trade-offs and navigate unavoidable injustices.

It has become evident that energy justice holds significant promise in steering the energy transition toward more inclusive and equitable legislative and judicial decision-making. Energy justice can function as a complementary framework alongside economic regulation by embracing a whole-of-economy approach. It considers broader aspects, such as procedural and recognition justice, that are not easily addressed through traditional market-based mechanisms. An essential aspect of energy justice also lies in its capability to highlight negative externalities of energy technologies. When these are addressed, they bolster the acceptance of the energy transition by aligning regulation with societal needs. However, it is crucial to recognize that while low-carbon technologies are integral to decarbonization efforts, their roll-out and integration can inadvertently lead to newly arising injustices, particularly if they reinforce existing inequalities or marginalize

[129] EU Electricity Directive, Art 5(3).

vulnerable groups. This underscores the need for energy justice to ensure that the transition remains inclusive and equitable across all levels of society. Ultimately, energy justice creates a dialogue around the risk-risk trade-offs between environmental harm and climate change. It therefore prompts a broader exploration of the risks involved, cultivating a better understanding of the implications of different energy technologies and regulatory strategies.

However, the complexities surrounding energy injustices often obscure their practical resolution, making their regulation challenging and increasing the risk of political instrumentalization in the absence of legal safeguards. This includes the threat of reframing the energy transition primarily as a justice transition. Moreover, while energy justice plays an essential role in exposing injustices, its abstract legal nature significantly limits its justiciability and enforceability. This limits the effectiveness of energy justice as a legal instrument to legally impose significant positive change. In certain instances, tolerating energy injustices becomes an unavoidable necessity and may even act as a catalyst of the energy transition. Nevertheless, protecting the most vulnerable energy consumers from social exclusion must remain an important priority of energy regulation.

Ultimately, energy justice has the potential to serve as an important legal instrument and overarching objective, propelling the acceleration of the energy transition toward socially acceptable energy services. Its practical effectiveness, however, is impacted by a multitude of factors, including political influence, legal and judicial discretion, and the motivation of consumers to move away from injustices.

17

Promethean Dreams: Intellectual Property and Climate Change in the Anthropocene

Matthew Rimmer

17.1 Introduction

Over the past few decades, international climate negotiations have hosted debates over intellectual property, technology transfer and climate change.[1] Developed nations – along with technology developers – have demanded high standards of intellectual property protection and enforcement in respect of clean technologies. Mid-tier countries have called for better intellectual property licensing mechanisms, as well as a system of better climate finance. Developing countries have expressed concerns about the lack of climate finance to secure clean technologies. Such nations have called for wholesale reforms of intellectual property – especially in terms of intellectual property

[1] Matthew Rimmer, *Intellectual Property and Climate Change: Inventing Clean Technologies* (Edward Elgar Publishing, 2011); Mario Cimoli et al (eds), *Intellectual Property Rights: Legal and Economic Challenges for Development* (Oxford University Press, 2014); Matthew Rimmer, 'Intellectual Property and Global Warming: Fossil Fuels and Climate Justice' in Matthew David and Debora Halbert (eds), *The Sage Handbook of Intellectual Property* (Sage Publications, 2014), 727; Peter Menell and Sarah Tran (eds), *Intellectual Property, Innovation and the Environment* (Edward Elgar Publishing, 2016); Joshua Sarnoff (ed), *Research Handbook on Intellectual Property and Climate Change* (Edward Elgar Publishing, 2016); Matthew Rimmer (ed), *Intellectual Property and Clean Energy: The Paris Agreement and Climate Justice* (Springer, 2018); Abbe E.L. Brown, *Intellectual Property, Climate Change, and Technology: Managing National Legal Intersections, Relationships and Conflicts* (Edward Elgar Publishing, 2019); Mahatab Uddin, *Climate Change Law, Technology Transfer and Sustainable Development* (Routledge, 2021); and Ankit Singh and Yogendra Srivastava, *Patent Law, Green Technology and Innovation* (Routledge, 2022).

flexibilities – in order to better achieve climate action and sustainable development. Least Developed Countries (LDCs), Small Island States, and countries vulnerable to climate change have called for clean technologies to be openly shared and dedicated to the public domain. Given such divisions, international climate negotiations have struggled to reach a consensus on intellectual property and technology transfer between developed nations and developing countries, least developed nations, and nations vulnerable to climate change and global warming.

This chapter is part of a collection looking at climate technology and law in the Anthropocene. The Anthropocene is a term used to describe the most recent time period in Earth's history when human activity started to have a significant impact on the planet's climate and ecosystems.[2] The debate over intellectual property and climate change has gone through a number of phases and iterations. There have been historical moments where there has been intense interest in the topic. The academic literature has sought to document some of the key hotspots in this debate. There was a notable debate over technology transfer in the negotiations leading up to the formation of the United Nations Framework Convention on Climate Change 1992.[3] Likewise, there has been a swirling discussion around the TRIPS Agreement 1994, and its implications for intellectual property and climate change.[4] The Kyoto Protocol 1997 featured detailed discussions over technology transfer.[5] There was a fierce contest over intellectual property and climate change in the run-up to the Copenhagen Accord 2009.[6] There was also a major donnybrook over intellectual property and climate change in the discussions

[2] 'Anthropocene', National Geographic, <https://education.nationalgeographic.org/resource/anthropocene/>. See also Noel Castree, 'Anthropocene and Planetary Boundaries' in American Association of Geographies, *International Encyclopedia of Geography: People, the Earth, Environment and Technology* (Wiley, 2018) <https://doi.org/10.1002/9781118786352.wbieg0027.pub2>. All Internet references accessed 1 September 2024.

[3] *United Nations Framework Convention on Climate Change*, opened for signature 4 June 1992, 1771 UNTS 107 (entered into force 21 March 1994). For discussion, see Daniel Bodansky, 'The United Nations Framework Convention on Climate Change: A Commentary' (1993) 18 *Yale Journal of International Law* 451.

[4] *Marrakesh Agreement Establishing the World Trade Organization*, opened for signature 15 April 1994, 1867 UNTS 3 (entered into force 1 January 1995), annex 1C (*Agreement on Trade-Related Aspects of Intellectual Property Rights*) 1867 UNTS 299.

[5] *Kyoto Protocol to the United Nations Framework Convention on Climate Change*, opened for signature 16 March 1998, 2303 UNTS 162 (entered into force 16 February 2005) (*Kyoto Protocol*).

[6] *Copenhagen Accord* 2009 UNFCCC/CP/2009/11/Add.1 (30 March 2010), <http://unfccc.int/resource/docs/2009/cop15/eng/11a01.pdf>. For discussion, see Daniel Bodansky, 'The Copenhagen Climate Change Conference – A Post-Mortem' (2010) 104 *American Journal of International Law* 230.

over the Paris Agreement 2015.[7] There have been ongoing discussions over intellectual property and climate change in the climate conferences following the Paris Agreement 2015.[8] As part of its time-setting, this chapter will focus on recent history – looking at the climate conferences in the early 2020s in Glasgow (COP26),[9] Sharm el Sheikh (COP27),[10] and Dubai (COP28).[11]

This chapter will focus on the debate over intellectual property, technology transfer and climate change in these international climate conferences in the 2020s. It should also be acknowledged that the geopolitical debate over intellectual property, technology transfer and climate change has played out in multiple arenas. Key stakeholders have engaged in a dizzying array of forum-shifting in debates over intellectual property and climate change.[12] The international climate talks have had a long-running thread of discussion about climate law and technology – particularly looking at intellectual property and associated issues of finance.[13] The international trade regime of the World Trade Organization has hosted debates over intellectual property

[7] *Paris Agreement to the United Nations Framework Convention on Climate Change*, opened for signature 12 December 2015 (entered into force 4 November 2016) (in UNFCCC, Report of the Conference of the Parties on its Twenty-First Session, Addendum, UN Doc FCCC/CP/2015/10/Add.1, 29 Jan 2016). For commentary, see Matthew Rimmer, 'Beyond the Paris Agreement: Intellectual Property, Innovation Policy, and Climate Justice' (2019) 8(1) *Laws* 1; and Matthew Rimmer, 'Article 10: Technology Development and Transfer' in Geert Van Calster, and Leonie Reins (eds), *The Paris Agreement on Climate Change: A Commentary* (Edward Elgar Publishing, 2021), 237.

[8] Marrakech Climate Change Conference, COP22, <https://unfccc.int/conference/marrakech-climate-change-conference-november-2016>; UN Climate Change Conference, Bonn, COP23, November 2017, <https://unfccc.int/conference/un-climate-change-conference-november-2017>; Katowice Climate Change Conference, COP24, December 2018, <https://unfccc.int/conference/katowice-climate-change-conference-december-2018>; and UN Climate Change Conference, Madrid, COP25, December 2019, <https://unfccc.int/conference/un-climate-change-conference-december-2019>.

[9] UN Climate Change Conference in Glasgow, COP26, October–November 2021, <www.un.org/en/climatechange/cop26>.

[10] Sharm El-Sheikh Climate Change Conference, COP27, November 2022, <https://unfccc.int/cop27>.

[11] UN Climate Change Conference, United Arab Emirates, COP28, November–December 2023, <https://unfccc.int/cop28>.

[12] On intellectual property forum-shifting, see John Braithwaite and Peter Drahos, *Global Business Regulation* (Cambridge University Press, 2000); and Susan Sell, 'TRIPS Was Never Enough: Vertical Forum Shifting, FTAS, ACTA, and TTP' (2011) 18(2) *Journal of Intellectual Property* 447; and Susan Sell, 'Cat and Mouse: Forum Shifting and the Battle over Intellectual Property Protection and Enforcement', ANU College of Asia and the Pacific, 18 September 2015, <https://papers.ssrn.com/sol3/papers.cfm?abstract_id=1466156>.

[13] Matthew Rimmer, 'Article 10: Technology Development and Transfer' in Geert Van Calster and Leonie Reins (eds), *The Paris Agreement on Climate Change: A Commentary* (Edward Elgar Publishing, 2021), 237.

and climate change in the TRIPS Council.[14] The World Intellectual Property Organization has a longstanding interest in technology transfer – and has increasingly looked at facilitating transactions in respect of environmentally sound technologies.[15] The United Nations Development Programme has an increasing interest in climate law and technology – particularly with the focus on access to clean energy and climate action in the United Nations Sustainable Development Goals 2015.[16] The proliferation of bilateral and regional trade agreements has also led to the creation of new fora in which to consider the intersection of intellectual property and climate change.[17] There has been a similar dynamic in other areas of intellectual property regulation – such as in global issues of access to essential medicines and internet freedom.[18]

This chapter will investigate some key distinctive approaches and discourses to dealing with intellectual property and climate change in modern geopolitics. Section 17.2 explores the growth of litigation over intellectual property and clean technologies. Section 17.3 will focus upon the approach of international institutions to intellectual property and climate technologies. In Section 17.4, we look at the positions of key nation states over intellectual property and climate change. Sectioin 17.5 focuses on non-state actors. It considers industry perspectives in respect of intellectual property – both from renewable energy companies and the fossil fuel industry. It explores the contributions of civil society – particularly in relation to intellectual property and climate change. Section 17.6 highlights UN Secretary-General António Guterres' call for a common agenda – which involves renewable energy technologies being treated as 'global public goods', free of intellectual property restrictions. The conclusion reviews the competing discourses over climate technology at the climate conferences of the early 2020s.

14 World Trade Organization, 'Climate Change and TRIPS', <www.wto.org/english/tratop_e/trips_e/cchange_e.htm> and World Trade Organization, 'The WTO TRIPS Agreement – A Practical Overview for Climate Change Policymakers', 2010, <www.wto.org/english/tratop_e/trips_e/ta_docs_e/8_3_overviewclimatechange_e.pdf>.

15 World Intellectual Property Organization, 'Climate Change and Intellectual Property', <www.wipo.int/policy/en/climate_change/>.

16 UN General Assembly, *Transforming our World: the 2030 Agenda for Sustainable Development*, 21 October 2015, A/RES/70/1.

17 *Trans-Pacific Partnership* 2015; *Comprehensive and Progressive Trans-Pacific Partnership* 2018 ATS 23; and *Regional Comprehensive Economic Partnership* 2022 ATS 1.

18 Carolina Rossini, 'The Unsettling Trend of Forum Shifting in International Intellectual Property' (*Electronic Frontier Foundation*, 21 December 2012), <www.eff.org/deeplinks/2012/12/iip-forum-shifting-trend>.

17.2 Intellectual property litigation over clean technologies

While negotiations have been taking place over intellectual property and climate change at the international climate conferences, there has been major litigation in respect of intellectual property and clean technologies in the courts. In particular, there has been a major dispute over electric batteries between two South Korean companies; a major patent dispute over wind turbines; and a trade secrets misappropriation matter involving Sinovel and the United States government. There has been increasing intellectual property litigation over clean technologies, renewable energy and energy efficiency.[19] In the field of patent law, there have been conflicts over patent ownership of electric batteries, hybrid cars, wind technologies, solar energy and climate-ready crops. There have also been disputes under other regimes of intellectual property. There have been design law disputes over electric trucks. In the field of trademark law, there has been debate over the use of green trademarks, eco-labels and certification systems. Under consumer law, there have been class actions and regulatory intervention over greenwashing by companies. Increasingly, clean technology companies have also taken action to protect confidential information and trade secrets against rivals and competitors. There have been trade disputes between the United States and China over intellectual property and climate change. It is worthwhile discussing some recent examples of intellectual property disputes in depth.

17.2.1 LG Energy Solution Ltd v. SK Innovation Co

In the field of trade secrets, there was a major dispute between two South Korean electric battery companies in the United States courts.[20] In April 2019, LG Chem (a chemical company which holds LG Energy as a subsidiary) filed an EV battery trade secret misappropriation suit against SK Innovation with the International Trade Commission. LG alleged that SK had misappropriated LG's trade secrets by hiring over 100 former LG employees, including 70 engineers.[21] LG alleged that these former employees disclosed

[19] Eric Lane, *Clean Tech Intellectual Property: Eco-Marks. Green Patents, and Green Innovation* (Oxford University Press, 2011).

[20] All News, 'Timeline of EV Battery Dispute Between LG and SK' (*All News*, 11 April 2022), <https://en.yna.co.kr/view/AEN20210411002200320>.

[21] Robert B. Kornweiss, Molly A. Jones and Josh Pond, 'ITC Finds Trade Secret Misappropriation and Bars Electric Vehicle Batteries from SK Innovation—With Exceptions' (*Trade Secrets Trends*, 11 February 2021), <www.crowelltradesecretstrends.com/2021/02/itc-finds-trade-secret-misappropriation-and-bars-electric-vehicle-batteries-from-sk-innovation-with-exceptions/>.

confidential information during the recruitment process and downloaded confidential documents from LG's database.

In September 2019, SK Innovation lodged an EV battery patent infringement suit with the International Trade Commission against LG Chem. LG filed a counter claim against SK Innovation over the EV battery patent technology.

In February 2020, the International Trade Commission accepted LG Chem's demand for a default judgment against SK Innovation over destruction of evidence in the trade secrets case.

In February 2021, the United States International Trade Commission (ITC) issued a ruling, finding that SK Innovation had misappropriated the trade secrets of its competitor LG Chem in violation of Section 336 of the *Tariff Act* of 1930.[22] The ITC determined that the appropriate remedy was: '(1) a limited exclusion order prohibiting the entry of certain lithium ion batteries, battery cells, battery modules, battery packs, and components thereof; and (2) cease and desist orders directed to respondents.'[23] The ITC held that the remedial orders would expire ten years from their issuance, and cover the trade secrets that LG identified.

LG Chem was triumphant at the ruling.[24] Jong Hyun Kim, CEO of LG Energy Solution commented: 'SKI's total disregard of our warnings and intellectual property rights gave us no choice but to file this case and we are grateful to the International Trade Commission for protecting our innovations and significant economic investments in the United States.'[25] The executive added: 'As a global leader and technology innovator, we will further strengthen the protection of intellectual property rights going forward.'[26]

In April 2021, the ITC handed down a preliminary ruling in favour of SK Innovation in a patent suit on lithium-ion battery technology filed by LG.

In April 2021, LG Chem and SK Innovation reached a settlement to resolve the two-year trade dispute – just before a presidential review deadline. According to news reports, SK Innovation agreed to pay LG Energy Solution, a wholly owned subsidiary of LG Chem Ltd, 2 trillion won ($1.8 billion)

---

22 *In the Matter of Certain Lithium Ion Batteries, Battery Cells, Battery Modules, Battery Packs, Components Thereof, and Processes Therefor*, United States International Trade Commission, Inv. No. 337-TA-1159, 10 February 2021.

23 Ibid.

24 Business Wire, 'LG Energy Solution Prevails Over SK Innovation in Final Trade Secret Theft Decision at International Trade Commission' (*Business Wire*, 10 February 2021), <www.businesswire.com/news/home/20210210005988/en/LG-Energy-Solution-Prevails-Over-SK-Innovation-in-Final-Trade-Secret-Theft-Decision-at-International-Trade-Commission>.

25 Ibid.

26 Ibid.

in cash and royalties.[27] The companies agreed to end all litigation in the United States and South Korea and not raise further lawsuits against each other for ten years.

17.2.2 Siemens Gamesa Renewable Energy A/S. v. General Electric (GE)

There has been major litigation over patents and wind technologies between General Electric (GE) and Siemens Gamesa Renewable Energy A/S. GE has been a vocal advocate for strong protection of intellectual property rights in international climate negotiations. So, it is striking to see its positioning as a defendant in a case in which it is being sued for intellectual property infringement.

In 2020, Siemens Gamesa sued GE for patent infringement in Massachusetts over GE's Haliade-X turbines. The judge ruled in September 2022 that Siemens Gamesa's technology was a 'key element' of the turbines' functionality.[28] The judge banned GE from making and selling the turbines in the United States. However, the judge allowed GE to continue using the turbines in existing projects off the coasts of Massachusetts and New Jersey with royalty payments, noting the 'rapidly developing climate crisis', and the 'thousands of jobs' the projects have created.[29] A jury had awarded Siemens Gamesa a royalty rate of US$30,000 per megawatt. Siemens Gamesa asked the court to triple the rate for the New Jersey project to US$90,000 per megawatt for total damages of US$109.3 million.

In the February 2023 case of *Siemens Gamesa Renewable Energy A/S v. General Electric Co*, District Judge Young reflected upon the dispute.[30] The judge observed: 'The issue before this Court requires it to scrutinize one small area of a particular battlefield between Siemens Gamesa Renewable Energy A/S ("SGRE") and General Electric Co. ("GE") who are locked in a global struggle for dominance in the lucrative and vitally important market to develop offshore wind turbines.'[31] The judge chastised the two

27 Hyunjoo Jin, Heekyong Yang and David Shepardson, 'South Korean Battery Makers Agree $1.8 bln Settlement, Aiding Biden's EV push' (*Reuters*, 11 April 2021), <www.reuters.com/technology/skorean-battery-makers-agree-18-bln-settlement-aiding-bidens-ev-push-2021-04-11/>.

28 Blake Brittain, 'Siemens Games Wins U.S. Ban on GE Wind Turbines in Patent Dispute' (*Reuters*, 8 September 2022), <www.reuters.com/legal/litigation/siemens-gamesa-wins-us-ban-ge-wind-turbines-patent-dispute-2022-09-07/>.

29 Ibid.

30 *Siemens Gamesa Renewable Energy A/S v. General Electric Co.* (2022) 617 F. Supp. 3d 55 (US DC Mass.).

31 Ibid.

multinational companies for engaging in this extended patent conflict over essential technologies: 'This struggle does nothing to advance the public interest of the people of the United States or any other nation.'[32] The judge commented: 'Indeed, [this litigation] inhibits efforts to combat climate change world-wide.'[33] These comments are significant in the context of the international debates over intellectual property and climate change – especially given that some nations and companies have denied that intellectual property rights could act as a barrier to climate action. Here is an American judge complaining that intellectual property litigation is an impediment to the public interest in access to clean energy and to international climate action.

The judge called for a cooperative approach to intellectual property between the two companies: 'It would be far, far better for these two corporate behemoths –consistent with antitrust requirements – to cross license their particular technological contributions and forge ahead with production in the global public interest.'[34] The judge was exasperated about the need to adjudicate the ongoing dispute between the two parties: 'In these premises, however, corporate self-interest appears to hold full sway, and this Court must do its duty.'[35]

The judge considered in this matter submissions from the two companies as to damages. The judge noted that the assessment

> should take into account the change in the parties' bargaining positions, and the resulting change in economic circumstances, resulting from the determination of liability—for example, the infringer's likelihood of success on appeal, the infringer's ability to immediately comply with the injunction, the parties' reasonable expectations if the stay was entered by consent or stipulation, and so on—as well as the evidence and arguments found material to the granting of the injunction and the stay.[36]

The judge was frustrated by the submissions of the two parties, noting: 'It is difficult to understand why GE would continue to use the infringing patent if the costs of designing around it are as low as it represents them to be.' The judge observed: 'The parties have made it incredibly difficult to find common ground on whether a design-around would or would not

32 Ibid.
33 Ibid.
34 Ibid.
35 Ibid.
36 Ibid.

best serve the public's interest'.[37] The judge wondered: 'What would this design-around mean for the public interest?'[38] The judge observed: 'These open-ended questions cause the Court to place SGRE and GE on roughly equivalent ground regarding their bargaining positions post-verdict.'[39]

The judge concluded: 'What is indubitably clear is that, by permitting GE to proceed on two offshore wind turbine projects notwithstanding the general injunction earned by SGRE, this Court is conferring on GE, a patent infringer, a significant if incalculable economic advantage.'[40] The judge observed: 'GE will thus be able to build up relationships with reliable vendors and subcontractors, and work out supply chain issues – in short, gain the advantage of learning how to address that myriad of issues that become apparent only by actually building a project out.'[41]

In conclusion, the judge held:

> In light of these considerations, as a result of GE's obligation to meet the public interest as an infringer, joined with the fact that GE is now voluntarily infringing with the Court's permission, this Court decrees it must pay, as to the Ocean Wind I carve out, \$60,000 per M/W to compensate SGRE most fairly for the irreparable harm it is causing.[42]

Moreover, the judge ruled: 'For each infringing Haliade-X wind turbine installed at Ocean Wind 1 Offshore Energy Project, GE shall pay SGRE a royalty consisting of \$60,000 per megawatt of rated capacity within 30 days after the end of the month in which the Haliade-X wind turbine is installed.'[43]

In media reports, a GE Renewable Energy spokesperson lamented that the ruling 'does not follow the jury's verdict or account for today's economic realities'.[44] In March 2023, GE and Siemens Gamesa announced a settlement of the wind technology patent disputes in the United States and Europe.[45]

---

37 Ibid.
38 Ibid.
39 Ibid.
40 Ibid.
41 Ibid.
42 Ibid.
43 Ibid.
44 Blake Brittain, 'GE must Double Payments to Siemens Gamesa in Wind-Turbine Case, U.S. Court Says' (*Reuters*, 3 February 2023), <www.reuters.com/legal/litigation/ge-must-double-payments-siemens-gamesa-wind-turbine-case-us-court-says-2023-02-02/>
45 Andrew Lee, 'GE and Siemens Gamesa End Wind Turbine Patent War that Judge Said Harmed Climate Fight' (*ReCharge*, 31 March 2023), <www.rechargenews.com/wind/ge-and-siemens-gamesa-end-wind-turbine-patent-war-that-judge-said-harmed-climate-fight/2-1-1429547>.

The highly critical comments of the judge most likely encouraged the parties to reach a settlement.

17.2.3 US Government v. Sinovel

In 2018, a United States court found that Sinovel, a manufacturer and exporter of wind turbines based in the People's Republic of China, misappropriated trade secrets from AMSC, a US-based company formerly known as American Superconductor Inc.[46] After an 11-day trial, a jury sitting in Madison, Wisconsin, convicted Sinovel Wind Group Co Ltd, doing business as Sinovel Wind Group (USA) Co Ltd (Sinovel), of conspiracy to commit trade secret theft, theft of trade secrets and wire fraud. Acting Assistant Attorney General Cronan commented: 'Sinovel nearly destroyed an American company by stealing its intellectual property.'[47] He observed that 'this type of conduct, by any corporation – anywhere – is a crime, and won't be tolerated'. US Attorney Blader added: 'Today's verdict sends a strong and clear message that the theft of ideas and ingenuity is not a business dispute; it's a crime and will be prosecuted as such.'[48]

In sentencing, the court held that AMSC's losses from the trade secrets appropriation exceeded US$550 million, and imposed the maximum statutory fine in the amount of US$1.5 million on Sinovel Wind Group LLC.[49] The Court found that the parties settled the restitution amount, and imposed a year of probation until Sinovel paid the full restitution amount. Sinovel paid US$32.5 million to AMSC – of which US$25 million was paid within its year of probation. Sinovel was required to pay US$850,000 to additional victims within its year of probation.

Commenting on the sentencing, Acting Assistant Attorney General Cronan observed: 'Rather than pay AMSC for more than US$800 million in products and services it had agreed to purchase, Sinovel instead hatched a scheme to brazenly steal AMSC's proprietary wind turbine technology, causing the loss of almost 700 jobs and more than $1 billion in shareholder equity at AMSC.'[50] Cronan warned: 'As demonstrated by this prosecution,

---

46 United States Department of Justice, 'Chinese Company Sinovel Wind Group Convicted of Theft of Trade Secrets', Press Release, United States Government (24 January 2018), <www.justice.gov/opa/pr/chinese-company-sinovel-wind-group-convicted-theft-trade-secrets>.

47 Ibid.

48 Ibid.

49 United States Department of Justice, 'Court Imposes Maximum Fine on Sinovel Wind Group for Theft of Trade Secrets', Press Release, United States Government (6 July 2018), <www.justice.gov/opa/pr/court-imposes-maximum-fine-sinovel-wind-group-theft-trade-secrets>.

50 Ibid.

intellectual property theft poses a serious threat to American companies, and the Department of Justice is committed to aggressively investigating and prosecuting individuals and corporations who undermine American competitiveness by stealing what they did not themselves create.'[51]

17.2.4 Summary

For some time, there was debate as to whether there was any evidence of intellectual property rights being relevant to questions of access to clean technologies and renewable energy. These select case studies highlight that the topic of intellectual property and climate change is one in which there are massive conflicts, involving major companies and nation states. There have been serious matters of intellectual property infringements in the field of green technologies. There have also been concerns about intellectual property conflicts creating issues in respect of access to clean technologies and renewable energy. The question is not a simple one of whether intellectual property rights facilitate technology transfer, or create barriers to technology transfer. The relationship between intellectual property and clean technology is a complex and dynamic one, with various effects.

17.3 International organizations

There has been debate as to whether international institutions should develop and elaborate rules and norms on intellectual property and climate change in order to avoid legal conflicts and disputes. The UN Secretary-General António Guterres has encouraged renewable energy technologies to be treated as global public goods. The international climate conferences have hosted rolling debates on intellectual property and climate change – but have only agreed to text on technology transfer and climate finance, not intellectual property. The United Nations Framework Convention on Climate Change (UNFCCC) Climate Technology Centre and Network has been established to help accelerate research, development and deployment of clean technologies. The World Intellectual Property Organization has also held dialogues on intellectual property and climate change, and established the WIPO GREEN mechanism to help facilitate technology transfer. The TRIPS Council of the World Trade Organization has heard a range of views from nation states on the interaction between intellectual property and climate change. The United Nations Development Programme has established a network of accelerator labs to help incubate sustainable innovation, and promote the United Nations Sustainable Development Goals.

[51] Ibid.

17.3.1 The UNFCCC

The debate over intellectual property and climate change during international climate talks has been inconclusive. During the discussions over the Paris Agreement 2015, India urged nation states to agree to language on intellectual property, finance and technology transfer in the agreement.[52]

Draft Article 56.3 of the negotiating text of 12 February 2015 considered a range of options.[53] The first option highlighted means to assist technology transfer. There were four items under it. Item A called for developed countries to 'provide financial resources to address barriers caused by intellectual property rights (IPRs) and facilitate access to and the deployment of technology, including inter alia, by utilizing the Financial Mechanism and/or the establishment of a funding window under the Green Climate Fund/the operating entities of the Financial Mechanism'.[54]

Item B wanted 'an international mechanism on IPRs to be established to facilitate access to and the deployment of technology to [developing country Parties]'.[55] Item C asked for 'collaborative research and development, shareware, commitments related to humanitarian or preferential licensing, fully paid-up or joint licensing schemes, preferential rates and patent pools'.[56] Item D recommended that

> funds from the Green Climate Fund will be utilized to meet the full costs of intellectual property rights (IPRs) of environmentally sound technologies and know-how and such technologies will be provided to developing country Parties free of cost in order to enhance their actions to address climate change and its adverse impacts.[57]

The second option submitted that 'Parties recognize that IPRs create an enabling environment for the promotion of technology innovation in environmentally sound technologies'.[58] The third option, mooted by developed countries, was that 'IPRs are not to be addressed in this agreement'.[59]

---

52 Matthew Rimmer, 'Article 10: Technology Development and Transfer' in Geert Van Calster and Leonie Reins (eds), *The Paris Agreement on Climate Change: A Commentary* (Edward Elgar Publishing, 2021), 237.

53 Ad Hoc Working Group on the Durban Platform for Enhanced Action, 'Work of the Contact Group on Item 3, Negotiating Text', 12 February 2015, <http://unfccc.int/files/bodies/awg/application/pdf/negotiating_text_12022015@2200.pdf>.

54 Ibid.

55 Ibid.

56 Ibid.

57 Ibid.

58 Ibid.

59 Ibid.

The fourth option was for 'Developed country Parties to make available Intellectual Property (IP) through multilateral institutions as [a] public good, through purchase of intellectual property'.[60] In the end, neither the Paris Agreement 2015 nor its Adopting Decision included any text on intellectual property. There was simple, basic text on technology research, development and deployment.

Over the course of the international climate talks, a number of technology mechanisms and networks have been established. The UNFCCC Climate Technology Centre and Network released a paper on Climate Change Strategies in 2020.[61] The executive secretary of the UNFCCC, Patricia Espinosa, considered the question of climate technology in the foreword to the document. She emphasized: 'The full deployment of technologies to reduce greenhouse gas emissions and build resilience to the impacts of climate change is vital if we are to realise the vision of the Paris Agreement.'[62] Espinosa called for a broad understanding of 'climate technology':

> The term 'climate technology' covers a broader scope than many realise. In recent years, climate technologies have been deployed on an unprecedented scale around the globe. Certain renewable energy technologies are now competitive with fossil fuel options for electricity generation, and technology assessments and roadmaps developed by the CTCN and its partners help ensure that appropriate technology choices are made by governments and that enabling environments are in place to support domestic and international investment to scale-up their deployment.[63]

Espinosa noted that 'the Financial and Technology Mechanisms of the Convention are collaborating to support the climate mitigation needs of developing countries'.[64] Moreover, work was also under way on the deployment of climate technologies for adaptation: 'Countries require the full support of the global community to strengthen their resilience to changing rainfall patterns and rising seas, enhance agricultural production and food security, and support the efforts of governments to meet the needs of their populations.'[65] Espinosa called for the work of the Climate Technology Centre and Network to be expanded further: 'We must collectively raise our

---

60 Ibid.

61 Climate Technology Centre and Network, *Climate Change Strategies 2020* (Climate Technology Centre and Network, 2019).

62 Ibid, 6.

63 Ibid, 6.

64 Ibid, 7.

65 Ibid, 7.

ambition if we are to rise to the level of the challenge before us.'[66] Espinosa concluded: 'We need tomorrow's technologies and innovation, but while also scaling up the deployment of those in existence today.'[67]

Dr Rose Mwebaza, the director of the Climate Technology Centre and Network, has discussed the importance of access to climate technology:

> Technology can help us tackle some of the worst impacts of climate change. And innovation is the key to unlocking technology's power to advance adaptation action. Innovation is critical to developed and developing countries alike. It is fundamental to sectors demanding creativity, ingenuity and 'thinking outside the box' to address the increasingly complex issues caused and exacerbated by climate change. [68]

Nonetheless, Mwebaza observes that, although '[t]echnology is a recognized enabler of climate action for adaptation … we have yet to fully maximize its potential'.[69] She emphasized: 'Technology and innovation are needed to address the myriad considerations that go into planning for more sustainable cities, improving agricultural production and transforming food systems, building resilience in coastal communities, and protecting and conserving our dwindling water resources.'[70]

Mwebaza maintained that the Technology Mechanism played an instrumental role in supporting 'all facets of technology innovation through technical assistance, capacity building and knowledge sharing'.[71]

In 2023, the First Global Stocktake at COP28 emphasized that 'finance, capacity-building and technology transfer are critical enablers of climate action'.[72] It noted, with concern 'that current levels of climate finance, technology development and transfer, and capacity-building for adaptation remain insufficient to respond to worsening climate change impacts in developing country Parties, especially those that are particularly vulnerable to the adverse effects of climate change'.[73]

---

[66] Ibid, 7.

[67] Ibid, 7.

[68] World Intellectual Property Organization, *Green Technology Book 2022: Solutions for Climate Change Adaptation* (World Intellectual Property Organization, 2022), 8 <www.wipo.int/en/green-technology-book/>.

[69] Ibid, 8.

[70] Ibid, 8.

[71] Ibid, 8.

[72] *First Global Stocktake*, Conference of the Parties Serving as the Meeting of the Parties to the Paris Agreement, FCCC/PA/CMA/2023/L.17, 13 December 2023.

[73] Ibid.

The First Global Stocktake underlined 'the fundamental role of technology development and transfer, endogenous technologies and innovation in facilitating urgent adaptation and mitigation action aligned with achieving the goals of the Paris Agreement and sustainable development'.[74] The First Global Stocktake highlighted 'the persistent gaps and challenges in technology development and transfer and the uneven pace of adoption of climate technologies around the world'.[75] The First Global Stocktake has established a 'technology implementation programme' to help address some of the outstanding challenges in technology transfer.[76]

17.3.2 The World Intellectual Property Organization

The World Intellectual Property Organization (WIPO) has been an arena of debate between member states over intellectual property, climate change and sustainable development for some time.

The WIPO set up WIPO GREEN to help facilitate technology transactions in respect of clean technologies. This institutional mechanism has seen modest activity in this respect. In 2023, the Director-General of the WIPO, Daren Tang, sought to emphasize the relevance of the institution to climate action. Tang observed: 'In the past 10 years, this platform has grown to cover 130,000 technologies from over 140 countries, becoming the biggest green tech exchange matching platform that the UN offers today.'[77] He recognized that it remains a challenge to ensure that these technologies 'create impact on the ground'.[78] Tang observed: 'WIPO will continue to build innovation and tech transfer capabilities in member states so that tech transfer can lead to actual deployment on the ground, and homegrown innovation solutions can move from mind to market and be deployed and diffused across the world.'[79]

The WIPO released its *Green Technology Book: Solutions for Climate Change Adaptation* at COP27.[80] In the book, Tang stressed: 'As the UN agency for

[74] Ibid.

[75] Ibid.

[76] Ibid.

[77] Kerry Cullinan, 'As Climate Crises Loom, WTO Head Urges Developing Countries to Prepare to Use TRIPS Flexibilities' (*Health Policy Watch*, 15 November 2023), <https://healthpolicy-watch.news/as-climate-crises-loom-wto-head-urges-developing-countries-to-prepare-to-use-trips-flexibilities/>.

[78] Ibid.

[79] Ibid.

[80] World Intellectual Property Organization, 'WIPO Does COP27' (*WIPO GREEN*, 4 November 2022), <www3.wipo.int/wipogreen/en/news/2022/news_0026.html>. See World Intellectual Property Organization, *Green Technology Book 2022: Solutions for Climate Change Adaptation* (World Intellectual Property Organization, 2022), <www.wipo.int/publications/en/details.jsp?id=4633>.

innovation, creativity and intellectual property (IP), WIPO is committed to supporting the development of green technologies and new innovative solutions in the fight against climate change.'[81] Tang insisted that 'the global IP system plays a critical role in ensuring that such solutions create impact'.[82] Tang also argued that the global IP system 'helps translate basic research into tangible climate change solutions, incentivizes public and private R&D investment into environmentally-friendly technologies and generates a vast treasure trove of information that is open and accessible to all'.[83] However, Tang avoided the ideological debates in intellectual property law as to what approach would best help climate action and access to clean energy.

At COP27, the WIPO also participated in an International Trade Centre event on 'Accelerating the Low-carbon Transition through Sustainable Inclusive Trade'.[84] WIPO also co-organized the event 'Advancing Paris Agreement Implementation and Climate Action through Innovation and Technology'.[85]

The WIPO published its *Green Technology Book: Solutions for Climate Change Mitigation* at COP28.[86] The book emphasized: 'Eighty percent of the technologies we need to achieve the 2030 climate goals are already on the market – with many more emerging.'[87] The collection highlighted 'technologies for lowering energy consumption, electrifying transport and enabling material efficiency'.[88] The book maintains that 'a well-functioning innovation ecosystem is underpinned by an efficient and fair intellectual property rights system, which in itself stimulates innovation and disseminates technology to global markets'.[89]

The focus of WIPO GREEN and the green technology books has been very much on voluntary licensing of intellectual property, and use of public domain technologies. Curiously, the WIPO has shied away from politics debates over substantive law reform in respect of intellectual property, sustainable development and climate change.

---

81 World Intellectual Property Organization, *Green Technology Book 2022: Solutions for Climate Change Adaptation* (World Intellectual Property Organization, 2022), 4, <www.wipo.int/publications/en/details.jsp?id=4633>.

82 Ibid.

83 Ibid.

84 Ibid.

85 Ibid.

86 World Intellectual Property Organization, *Green Technology Book 2023: Solutions for Climate Change Mitigation* (World Intellectual Property Organization, 2023), <www.wipo.int/publications/en/details.jsp?id=4696>.

87 Ibid, 11.

88 Ibid, 11.

89 Ibid, 11.

17.3.3 The World Trade Organization

The World Trade Organization (WTO) has been the host of some fierce debates over intellectual property and climate change.

There has been some debate about the application of the TRIPS Agreement to the topic of environmentally sound technologies in an age of climate change.[90]

In 2013, there was a major discussion on the TRIPS Council over intellectual property and climate change.[91] There were two major discussions on the topic of intellectual property and climate change in 2013.[92] The Government of Ecuador argued that intellectual property should help facilitate the transfer of environmentally 'rational' technology.[93] This submission was supported by a number of other countries, such as Chile, Cuba and El Salvador. The BRICS/BASIC countries made submissions in respect of the need to make use of flexibilities under the TRIPS Agreement 1994 to tackle climate change and sustainable development. Developed countries, such as the United States, members of the European Union (EU), Switzerland, Japan, South Korea and Australia, maintained that there was a need for strong protection of intellectual property rights in respect of clean technologies. LDCs and nations vulnerable to climate change have pushed for environmentally sound technologies to be treated as global public goods. In 2014, there was a further debate in the TRIPS Council regarding the topic of intellectual property and climate change.[94] However, there was little

---

90 Wei Zhuang, *Intellectual Property Rights and Climate Change: Interpreting the TRIPS Agreement for Environmentally Sound Technologies* (Cambridge University Press, 2017).

91 Matthew Rimmer, 'Intellectual Property and Global Warming: Fossil Fuels and Climate Justice' in Matthew David and Debora Halbert (eds), *The Sage Handbook of Intellectual Property* (Sage Publications, 2014), 727; and Matthew Rimmer, 'Trade Wars in the TRIPS Council: Intellectual Property, Technology Transfer, and Climate Change' in Panagiotis Delimatsis (ed), *Research Handbook on Climate Change and Trade Law* (Edward Elgar, 2016), 200.

92 World Trade Organization, 'Intellectual Property, Climate Change and Development', TRIPS Council, 5–6 March 2013, <www.wto.org/english/tratop_e/trips_e/march2013_on_climate_e.pdf>; World Trade Organization, 'Intellectual Property, Climate Change and Development', TRIPS Council, 11–12 June 2013, <www.wto.org/english/tratop_e/trips_e/june2013_on_climate_e.pdf> and World Trade Organization, 'Contribution of Intellectual Property to Facilitate the Transfer of Environmentally Rational Technology', 10–11 October 2013, <www.wto.org/english/tratop_e/trips_e/october2013_on_climat e_e.pdf>.

93 Ecuador, 'Contribution of Intellectual Property to Facilitating the Transfer of Environmentally Rational Technology', TRIPS Council, World Trade Organization, 27 February 2013, <www.wtocenter.org.tw/SmartKMS/fileviewer?id=131645>.

94 World Trade Organization, 'Contribution of Intellectual Property to Facilitate the Transfer of Environmentally Rational Technology', TRIPS Council, 25–26 February 2014, <www.wto.org/english/tratop_e/trips_e/february2014_on_climate_e.pdf>.

consensus among the nation states in the TRIPS Council as to an agreed approach to intellectual property, technology transfer and climate change.

In 2022, the WTO published a special world trade report on the topic of climate change and trade.[95] The report asserts: 'The protection and enforcement of intellectual property rights, as provided by WTO rules, is also essential to support innovation in environmental technologies while promoting the transfer of technology.'[96] The report seems to assume that the TRIPS Agreement 1994 helps 'ensure [a] balanced framework for innovation and diffusion of climate adaptation technologies' and 'support[s] technology transfer to LDCs'.[97] The report offers an anodyne account of the role of intellectual property in respect of clean technologies:

> The [TRIPS Agreement] establishes a balanced framework for the innovation and dissemination of climate technologies for the mutual benefit of innovators and technology users, in particular through a range of tailored domestic measures concerning the governance of the intellectual property (IP) system for social and economic welfare. The IP system works in conjunction with international trade to facilitate knowledge transfers and diffusion of critical mitigation technologies, including through the effect of GVCs [global value chains] and knowledge spillovers, and trade in knowledge intensive goods.[98]

There is, however, very little recognition in the TRIPS Council debates between nation states as to whether the TRIPS regime is well adapted to dealing with climate change.[99] The report fails to acknowledge that the technology transfer mechanism under Article 66.2 of the TRIPS Agreement 1994 has not been working very well for LDCs.

The WTO participated in a number of activities at Sharm El Sheikh in Egypt for COP27.[100] WTO Director-General Ngozi Okonjo-Iweala

---

95 World Trade Organization, *World Trade Report 2022: Climate Change and International Trade* (World Trade Organization, 2022), <www.wto.org/english/res_e/publications_e/wtr22_e.htm>.

96 Ibid, 14.

97 Ibid, 44.

98 Ibid, 70.

99 Matthew Rimmer, 'Trade Wars in the TRIPS Council: Intellectual Property, Technology Transfer, and Climate Change' in Panagiotis Delimatsis (ed), *Research Handbook on Climate Change and Trade Law* (Edward Elgar Publishing, 2016), 200.

100 World Trade Organization, 'WTO at Climate Change COP27', <www.wto.org/english/tratop_e/envir_e/wto_cop27_e.htm#day1>.

argued that trade was critical in the global response to climate change.[101] She observed: 'We are advocating under the WTO for a global pricing framework, which will take into account the target we want of limiting the global temperature rise to 1.5 degrees, as agreed under the Paris Agreement.'[102]

In 2023, Okonjo-Iweala argued that developing countries should put in place 'effective mechanisms in their domestic laws' that allow them to use the TRIPS flexibilities to deal with the climate crisis.[103] She commented: 'With the impact of climate change on health becoming more evident, this is the time to get ready.'[104] Okonjo-Iweala recognized: 'As you all know, at the WTO we have also been grappling with sensitive issues around intellectual property (IP) and technology transfer.'[105] She noted: 'To solve problems in public health and the climate, breakthrough technologies must be incentivized, invented, developed and widely diffused.'[106] In her view, '[i]nnovation and access must go together'.[107] Okonjo-Iweala commented: 'Governments have legitimate scope to intervene when necessary to protect the public interest.'[108]

The position of Okonjo-Iweala has been informed by geopolitical struggles in respect of IP, and public health during the COVID-19 crisis. It should be noted that, in the parallel of field of access to medicines, there was a proposal for a TRIPS Waiver from South Africa and India during the COVID-19 crisis.[109] The Biden Administration was willing to support a TRIPS Waiver for vaccines.[110] However, the EU, the United Kingdom and Switzerland opposed a TRIPS Waiver. In the end, the WTO could

---

[101] World Trade Organization, 'DG Okonjo-Iweala: Trade Is Critical in Global Response to Climate Change', Press Release, World Trade Organization (8 November 2022), <www.wto.org/english/news_e/news22_e/cop27_08nov22_e.htm>.

[102] Ibid.

[103] Kerry Cullinan, 'As Climate Crises Loom, WTO Head Urges Developing Countries to Prepare to Use TRIPS Flexibilities' (*Health Policy Watch*, 15 November 2023), <https://healthpolicy-watch.news/as-climate-crises-loom-wto-head-urges-developing-countries-to-prepare-to-use-trips-flexibilities/>.

[104] Ibid.

[105] Ibid.

[106] Ibid.

[107] Ibid.

[108] Ibid.

[109] World Trade Organization, 'Waiver from Certain Provisions of the TRIPS Agreement for the Prevention, Containment and Treatment of COVID-19, Communication from India and South Africa', (2 October 2020) IP/C/W/669.

[110] Katherine Tai, 'Statement on the COVID-19 TRIPS Waiver', United States Trade Representative, 5 March 2021, <https://ustr.gov/about-us/policy-offices/press-office/press-releases/2021/may/statement-ambassador-katherine-tai-covid-19-trips-waiver>.

only agree on Ministerial Decision on the TRIPS Agreement 2022.[111] This Decision allowed for further compulsory licensing of patents for the purposes of the export of COVID-19 vaccines.

17.3.4 The United Nations Development Programme

The United Nations Development Programme (UNDP) has also been interested in promoting sustainable innovation. The UNDP has set up a network of accelerator labs to help generate sustainable innovation, and facilitate the realization of the United Nations Sustainable Development Goals.[112]

The UNDP has been particularly interested in promoting 'grassroots innovation that moves us toward a safer, more inclusive, and sustainable planet'.[113] UNDP Administrator Achim Steiner has been seeking to develop a new policy on intellectual property: 'Looking to the private sector, UNDP is developing a new policy on intellectual property rights to navigate constraints on our ability to engage with small and medium enterprises and start-up firms, and devising a tailored private sector partnership approach for fragile and crisis contexts.'[114] Steiner has been engaging with the larger debate over whether intellectual property rights enhance or constrain sustainable innovation.[115]

Steiner has also advocated for ambitious climate action: 'Global citizens want their leaders to transcend their differences, to act now and to act boldly to fight the climate crisis.'[116] He emphasized: 'We urge leaders and

[111] Ministerial Decision on the TRIPS Agreement 2022, Ministerial Conference, Twelfth Session, Geneva, WT/MIN(22)/39m WT/L/1141, adopted on 17 June 2022.

[112] UNDP Accelerator Labs, <www.undp.org/acceleratorlabs>. For a commentary, see Matthew Rimmer, 'The UNDP Accelerator Lab Network: Intellectual Property, Innovation Policy, Sustainable Development, and the COVID-19 Recovery' in Taina Pihlajarinne, Jukka Mähönen and Pratyush Nath Upreti (eds), *Intellectual Property Rights in the Post Pandemic World: An Integrated Framework of Sustainability, Innovation and Global Justice* (Edward Elgar Publishing, 2023), 246–76.

[113] UNDP Accelerator Labs, 'For Tomorrow: The Documentary', <www.undp.org/acceleratorlabs/our-work/fortomorrow>.

[114] Achim Steiner, 'UNDP Administrator's Speech to the June 2023 Annual Executive Board Meeting', United Nations Development Programme (6 June 2023), <www.undp.org/speeches/undp-administrators-speech-june-2023-annual-executive-board-meeting>.

[115] Radmilla Suleymanova, 'Getting SDGs Back on Track: A Conversation with UNDP's Steiner' (*Al Jazaeera*, 24 June 2021), <www.aljazeera.com/economy/2021/6/24/getting-sdgs-back-on-track-a-conversation-with-undps-steiner>.

[116] UN Development Programme, '80 Percent of People Globally Want Stronger Climate Action by Governments According to UN Development Programme Survey', Press Release (20 June 2024), <www.undp.org/press-releases/80-percent-people-globally-want-stronger-climate-action-governments-according-un-development-programme-survey>.

policymakers to take note, especially as countries develop their next round of climate action pledges – or "nationally determined contributions" under the Paris Agreement.'

It is notable that UNDP administrator Achim Steiner has often been much more radical in his solutions than some of his institutional leaders for the WTO, WIPO and the UNFCCC. As highlighted in the lead quote, Steiner has often advocated for the treatment of clean technologies as global public goods to be shared.

17.4 States

There has been a geopolitical debate as to whether intellectual property rights encourage research and development in relation to clean technologies. There has often been a divide between states on this issue. Developed nations, such as the US, Japan, South Korea and members of the EU, have argued that strong intellectual property rights protection is necessary to stimulate investment in the mid- to long-term development of clean technologies. However, many mid-tier nations have maintained that there should be greater access to clean technologies, subject to intellectual property rights protection. Developing countries have expressed concerns as to whether intellectual property rights create barriers to the dissemination of clean technologies. They have called for the use of intellectual property exceptions to enable the sharing of critical technologies for climate adaptation and mitigation. LDCs, Small Island States and nations vulnerable to global warming have argued that clean technologies should be placed in the public domain. Individual nations have been identifying intellectual property as a critical issue in national reports and assessments in respect of climate action.

17.4.1 The United States

Traditionally, the United States has taken a hard line on intellectual property during international climate talks, demanding that there be strong protection and enforcement of intellectual property rights in respect of clean technologies.[117] The Trump administration asserted in a Special 301 Report in 2020: 'Strong IP protection and enforcement are essential to promoting investment in innovation in the environmental sector.'[118]

[117] Damian Carrington, 'US Goes to Basics over Copenhagen Accord Tactics' (*The Guardian*, 4 December 2010), <www.theguardian.com/environment/2010/dec/03/us-basics-copenhagen-accord-tactics>.

[118] Office of the United States Trade Representative, 'Special 301 Report', April 2020, 36.

The position of the United States has shifted under the Biden administration on the topic of intellectual property, climate change and trade. The Biden administration has taken a new approach to trade and the environment, and has sought to promote 'Green Partnerships', and to diffuse conflicts over intellectual property.[119] Moreover, the United States government has also been willing to support a TRIPS Waiver in respect of vaccines during the COVID-19 crisis – perhaps showing a new approach to the role of intellectual property in respect of global challenges.[120]

In a 2021 speech, the new United States Trade Representative (USTR), Ambassador Katherine Tai, discussed her approach to 'Greening U.S. Trade Policy'.[121] She lamented: 'For too long, the traditional trade community has resisted the view that trade policy is a legitimate tool in helping to solve the climate crisis.'[122] Tai noted that such an approach was unhelpful: 'But that dated line of thinking only perpetuates the chasm that exists between the lived experiences – and expectations – of real people on the one hand, and trade experts on the other.'[123] Tai noted: 'The intersection of environment, climate change, labor, and trade are key to our collective ability to compete, innovate, and create liveable wage jobs that will provide hope and opportunity for future generations and underserved communities.'[124] She highlighted that there should be a strong connection between trade policy and climate action: 'This is why I believe that trade policy is an essential and strategic part of the solution to these huge challenges.'[125]

Tai sought to promote United States clean technologies: 'Together, we can foster U.S. innovation and the production of new technology, while promoting resilient renewable energy supply chains.'[126] She emphasized: 'This work has the potential to unlock new opportunities for U.S. companies who can help reduce emissions and adapt to climate change.'[127] Tai insisted: 'Our

---

119 Katherine Tai, 'Remarks on Trade Policy, the Environment and Climate Change', United States Trade Representative (April 2021), <https://ustr.gov/about-us/policy-offices/press-office/speeches-and-remarks/2021/april/remarks-ambassadaor-katherine-tai-trade-policy-environment-and-climate-change>.

120 Katherine Tai, 'Statement on the COVID-19 TRIPS Waiver', United States Trade Representative (5 March 2021) <https://ustr.gov/about-us/policy-offices/press-office/press-releases/2021/may/statement-ambassador-katherine-tai-covid-19-trips-waiver>.

121 Katherine Tai, 'Remarks on Trade Policy, the Environment and Climate Change', United States Trade Representative (April 2021), <https://ustr.gov/about-us/policy-offices/press-office/speeches-and-remarks/2021/april/remarks-ambassadaor-katherine-tai-trade-policy-environment-and-climate-change>.

122 Ibid.

123 Ibid.

124 Ibid.

125 Ibid.

126 Ibid.

127 Ibid.

bold, collective action can create enormous new economic opportunities and good paying jobs for all Americans while building the industries of the future.'[128]

In April 2021, Tai had personally sought to resolve the intellectual property and trade dispute between SK Innovation and LG Solutions over electric batteries.[129] She praised the parties for resolving their IP conflict: 'I congratulate both companies for working through their significant differences to resolve this dispute, which builds confidence in their reliability and responsibility as suppliers to the U.S. auto industry.'[130]

Tai emphasized that the US would focus on climate justice: 'We expect justice and equity to be on everyone's agenda, and we welcome creative solutions to the massive challenges we face with the environment, with climate change, and trade as a whole.'[131] She emphasized a desire for a cooperative diplomatic approach: 'By engaging new – and all too frequently silenced – voices and encouraging fresh, collaborative thinking, we can forge consensus and find solutions that we never knew existed.'[132]

In 2023, the USTR released a Special 301 Report on Intellectual Property and Enforcement.[133] The report maintained that '[s]trong intellectual property (IP) protection and enforcement are essential to promoting investment in innovation for a sustainable future'.[134] In its view, '[s]uch innovation not only promotes sustainable economic growth and supports jobs, but also is critical to responding to environmental challenges such as climate change'.[135] The report also highlighted the importance of public sector research into climate change and sustainability: 'IP provides incentives for research and development in this important sector, including through university research.'[136]

---

[128] Ibid.

[129] Hyunjoo Jin, David Shepardson and Heekyong Yang, 'S. Korean Battery Makers Agree $1.8 Billion Settlement, Aiding Biden's EV Push' (*Reuters*, 11 April 2021), <www.reuters.com/world/europe/korean-battery-firms-set-announce-deal-resolve-us-trade-dispute-sources-2021-04-10/>.

[130] Ibid.

[131] Katherine Tai, 'Remarks on Trade Policy, the Environment and Climate Change', United States Trade Representative (April 2021), <https://ustr.gov/about-us/policy-offices/press-office/speeches-and-remarks/2021/april/remarks-ambassadaor-katherine-tai-trade-policy-environment-and-climate-change>.

[132] Ibid.

[133] Office of the United States Trade Representative, *2023 Special 301 Report*, Washington, DC: United States Trade Representative, 2023.

[134] Ibid, 35.

[135] Ibid, 35.

[136] Ibid, 35.

In 2024, the USTR issued a communication to WTO members on Climate and Trade.[137] The communication acknowledged: 'Most [WTO members] recognize the urgent need to take action to mitigate greenhouse gas emissions and adapt to climate change in order to help ensure a safe future for families and communities, good paying jobs for workers, and a healthy planet for our and future generations.'[138] The communication recognized that 'significant work remains, and needs to be accelerated, if we are going to meet greenhouse gas (GHG) emission reduction targets and net-zero goals'.[139] The communication commented: 'Trade policy can play an important role in incentivizing decarbonization and reducing GHG emissions in our economies, spreading technological innovations that enable lower- or zero-emission production and transport, and greening international supply chains.'[140]

The United States Patent and Trademark Office (USPTO) has also highlighted the importance of intellectual property in respect of green energy inventions.[141] Secretary of Commerce Gina Raimondo commented: 'The Commerce Department is committed to creating new solutions that we need to overcome climate change and related global energy challenges of the future.'[142] USPTO Director Kathi Vidal observed: 'Our nation's innovation community is playing a crucial role in devising creative solutions to the challenge of climate change.'[143] She stressed: 'Innovation in this field is critical to how we will live and prosper in the future.'[144] The USPTO provided for expedited patent examination for the new categories of Climate Change Mitigation and Green Energy in the Patents for Humanity scheme.

The USPTO has been enthusiastic about the Patents for Humanity scheme. Attorney adviser Edward Elliott of the USPTO has argued: 'Patents for Humanity seeks to recognize innovators of all types by celebrating their varied contributions to our common goal: bringing prosperity to every corner of the globe.'[145] The Patents for Humanity Act of 2022 (US) has

[137] United States Trade Representative, 'Understanding the Opportunities and Challenges of the Green Transition: Coherence and Interoperability of Trade-Related Climate Measures', World Trade Organization (4 April 2024), WT/CTE/W/260, G/V/W/843.

[138] Ibid.

[139] Ibid.

[140] Ibid.

[141] United States Patent and Trademark Office, 'USPTO Announces New Patents for Humanity Green Energy Category', Press Release, United States Patent and Trademark Office (6 March 2023), <www.uspto.gov/about-us/news-updates/uspto-announces-new-patents-humanity-green-energy-category>.

[142] Ibid.

[143] Ibid.

[144] Ibid.

[145] Edward Elliott, 'Patents for Humanity: Improving Lives Across the Globe' (*WIPO Magazine*, April 2017), <www.wipo.int/wipo_magazine/en/2017/02/article_0003.html>.

entrenched the system. However, in the view of this scholar, the Patents for Humanity scheme has not had much more than a symbolic impact thus far.[146] The regime seems to be a public relations exercise rather than a structural, substantive measure to improve green innovation. The regime does little to ultimately change incentives for intellectual property owners to share clean technologies.

In a speech to the APRA-E Energy Innovation Summit, Kathi Vidal maintained: 'Patents are the primary vehicle used by our universities, our government labs, our big companies, and our startups, and our individuals to turn new inventions and technologies into commercial products and entirely new industries.'[147] She stressed: 'We must do this in the clean-energy space.'[148] She maintained that the 'U.S. patent system has been the main driver of our economy for centuries, and at the USPTO we want it to power the new technologies aimed at the reduction of greenhouse gas emissions.'[149] Vidal promised that the USPTO would expedite the examination of clean and green energy technologies: 'If you have a technology that could have a positive impact on the climate, you can file a petition to have an expedited examination of your patent application.'[150] She suggested: 'To all of you involved in the next generation of clean energy technologies, you can look to the U.S. Patent and Trademark Office as your partner in developing your technologies into products and systems that will change the trajectory of our energy future.'[151]

17.4.2 The EU

The European Commission has been pursuing a Green Deal. There has been discussion as to whether the EU Green Deal will involve the development of a Green Intellectual Property Policy.[152] There has long been debate as to whether the EU needs a Clean Technology Directive in intellectual property law, policy and practice.[153]

---

146 Matthew Rimmer, 'Patents for Humanity' (2012) 3(2) *The WIPO Journal* 196.

147 Kathi Vidal, 'Remarks by USPTO Director at the ARPA-E Energy Innovation Summit', United States Patent and Trademark Office (24 May 2022), <www.uspto.gov/about-us/news-updates/remarks-uspto-director-kathi-vidal-arpa-e-energy-innovation-summit>.

148 Ibid.

149 Ibid.

150 Ibid.

151 Ibid.

152 Rouse, 'The New EU Green Deal: Greening IP Rights' (*Rouse*, 4 May 2020), <www.lexology.com/library/detail.aspx?g=542d8840-b0ee-456d-b0ca-3d6c8dcdb5bb>.

153 Matthew Rimmer, 'A Proposal for a Clean Technology Directive: European Patent Law and Climate Change' (2011) 3 *A Journal of Renewable Energy Law and Policy* 195.

The European Patent Office has commissioned extensive research on intellectual property and climate change.[154] The European Patent Office has published detailed intellectual property landscapes in respect of patents, trademarks and other forms of intellectual property, and has sought to provide a new classification for climate mitigation technologies in its patent regime.[155] The European Patent Office has argued that technology transfer need not be obstructed by intellectual property rights.[156]

In the course of climate negotiations, the EU has denied that intellectual property rights serve as a barrier to access to clean technologies.[157] The EU has questioned the need for specific intellectual property flexibilities to tackle the problem of climate change.[158]

The EU opposed the proposal for a broad TRIPS Waiver for access to essential medicines during the COVID-19 crisis.[159] Instead, the EU argued for voluntary licensing and clarified compulsory licensing in respect of essential medicines.[160] In the end, there was a Ministerial Declaration on the TRIPS Agreement 2022, which allowed for the compulsory licensing of patents to facilitate the export of essential medicines.[161] No doubt the

---

[154] Konstantinos Karachalios, Nikolaus Thumm, Ahmed Abdel Latif, Pedro Roffe, Benjamin Simmons and Tahir Amin, *Patents and Clean Energy: Bridging the Gap Between Evidence and Policy* (UNEP, European Patent Office, and International Centre for Trade and Sustainable Development, 2010), <https://seors.unfccc.int/applications/seors/attachments/get_attachment?code=5LZ91UKP0QMRH9POJTXL4Y1EAK5NXHIO>.

[155] European Patent Office, 'Sustainable Technologies', European Patent Office (25 November 2019), <www.epo.org/news-events/in-focus/classification/classification.html>.

[156] European Patent Office and UNEP, 'Under One Per Cent of Clean Energy Technology Patents Filed in Africa, Highlighting Huge Potential for Exploiting Renewable Sources', Press Release, European Patent Office and UNEP (13 May 2013), <www.unep.org/news-and-stories/press-release/under-one-cent-clean-energy-technology-patents-filed-africa>.

[157] Matthew Rimmer, *Intellectual Property and Climate Change: Inventing Clean Technologies* (Edward Elgar Publishing, 2011), 63.

[158] European Union, 'Contribution of Intellectual Property to Facilitate the Transfer of Environmentally Rational Technology', Extract from Minutes of Meeting of the TRIPS Council, World Trade Organization, 11 June 2014, 29, <www.wto.org/english/tratop_e/trips_e/june2014_on_climate_e.pdf>.

[159] Priti Patanik, *The TRIPS Waiver Negotiations at the World Trade Organization*, (Geneva Health Files, 2022); Human Rights Watch, 'Seven Reasons the EU Is Wrong to Oppose the TRIPS Waiver' (*Human Rights Watch*, 3 June 2021), <www.hrw.org/news/2021/06/03/seven-reasons-eu-wrong-oppose-trips-waiver>; and Siva Thambisetty et al, 'The TRIPS Intellectual Property Waiver Proposal: Creating the Right Incentives in Patent Law and Politics to End the COVID-19 Pandemic', LSE Legal Studies Working Paper No. 6, 2021, <https://papers.ssrn.com/sol3/papers.cfm?abstract_id=3851737>.

[160] MSF, 'MSF Analysis of EU Communications to TRIPS Council on COVID-19 IP Waiver Proposal', Technical Brief, 29 June 2021, <https://msfaccess.org/msf-analysis-eu-communications-trips-council-covid-19-ip-waiver-proposal>.

[161] Ministerial Decision on the TRIPS Agreement 2022, Ministerial Conference, Twelfth Session, Geneva, WT/MIN(22)/39m WT/L/1141, adopted on 17 June 2022.

EU would also oppose a similar proposal for a TRIPS Waiver in respect of clean technologies to tackle the climate crisis. The EU has already opposed much more modest proposals for broader intellectual property flexibilities for clean technologies in the TRIPS Council.[162]

It is a paradox that the EU is a great innovator in respect of clean technologies, but it is reluctant to transfer such technologies, without strong intellectual property protection.

In 2023 and 2024, European scholars writing in *GRUR International* have argued that the EU could realign intellectual property law with objectives in respect of the environment, climate change and sustainable development.

At an administrative level, Giorgio Spedicato has suggested that patent offices can play a new and more decisive role in promoting environmentally sustainable innovation.[163] He commented:

> While it is true that there are other public authorities and bodies tasked with steering technological innovation towards environmental sustainability, and perhaps they can prevent unsustainable innovations from reaching the market more effectively than patent offices themselves, it is also true that the challenges posed by climate change and environmental degradation require the active commitment of all parties to be effectively addressed.[164]

At a policy level, Reto Hilty and Pedro Henrique Batista have explored the potential and limits of patent law to address climate change.[165] The scholars conclude: 'If patent legislation needs to be improved, international coordination, involving as many states as possible, including those from which the required technologies originate, is crucial to making a reasonable impact.'[166]

Josef Drexl has further considered the responses of intellectual property and competition law to the challenges of climate change.[167] He observed: 'To

[162] European Union, 'Contribution of Intellectual Property to Facilitate the Transfer of Environmentally Rational Technology', Extract from Minutes of Meeting of the TRIPS Council, World Trade Organization, 11 June 2014, 29, <www.wto.org/english/tratop_e/trips_e/june2014_on_climate_e.pdf>.

[163] Georgio Spedicato, 'The Untapped Potential of Patent Law in Deterring Environmentally Unsustainable Innovation' (2024) 73(8) *GRUR International* 751.

[164] Ibid, 762.

[165] Reto Hilty and Pedro Henrique Batista, 'Potential and Limits of Patent Law to Address Climate Change' (2023) 72(9) *GRUR International* 821.

[166] Ibid, 839.

[167] Josef Drexl, 'Responses of Intellectual Property and Competition Law to the Challenges of Climate Change' (2024) 73(8) *GRUR International* 715.

achieve better results in terms of ecological sustainability, the law should indeed be designed with sufficient flexibility to allow account to be taken of the specific market failures that characterize green innovation.'[168]

Lodewijk Van Dycke has considered whether there is a need to reconsider patent terms for clean technologies: 'Patents typically last for 20 years but, for climate-related technologies, this is arguably too long.'[169] The scholar maintained that there was a need to 'maximise technology spread and climate change mitigation, in order to minimise temperature rises and the catastrophic consequences of unmitigated global warming'.[170]

17.4.3 Other developed countries

A number of other key developed countries – which have holdings in intellectual property – have expressed a reluctance to shift their maximalist position on intellectual property rights. As Enrico Bonadio has observed, 'it is believed that a global protection of IPRs creates a certain and reliable normative context in which private companies are encouraged to transfer their proprietary information and technologies across different countries'.[171]

17.4.3.1 Switzerland

Switzerland has traditionally taken a maximalist stance on intellectual property protection and enforcement. Switzerland has told the TRIPS Council: 'In our view, the IP system plays a key role in the promotion, encouragement and dissemination of green energy technologies and can thus make a positive contribution in the fight against climate change.'[172]

In its fifth national communication under the UNFCCC, Switzerland maintains: 'Alongside public sector activities, the Swiss private sector plays an important role in terms of technology transfer.'[173] Switzerland insists: 'In many fields of environmentally sound technologies, Swiss companies are

---

[168] Ibid.

[169] Lodewijk Van Dycke, 'Why Patents on Climate Technologies Must Be Cut Short' (2024) 73(8) *GRUR International* 717.

[170] Ibid, 741.

[171] Enrico Bonadio, 'Climate Change and Intellectual Property' (2010) 1 *European Journal of Risk Regulation* 72 at 74.

[172] Switzerland, 'Contribution of Intellectual Property to Facilitate the Transfer of Environmentally Rational Technology', Extract from Minutes of Meeting of the TRIPS Council, World Trade Organization, 11 June 2014, 26, <www.wto.org/english/tratop_e/trips_e/june2014_on_climate_e.pdf>.

[173] Federal Office for the Environment, *Switzerland's Fifth National Communication under the UNFCCC: Second National Communication under the Kyoto Protocol to the UNFCCC* (Bern, 2009), 27.

leading in the development, diffusion and implementation of state-of-the-art solutions and pursue climate-related activities in the order of several billion CHF per year, supported by the Swiss export promotion agency Business Network Switzerland.'[174] However, there have been no concessions in respect of intellectual property flexibilities by Switzerland.

17.4.3.2 Japan

Japan has robustly called for strong protection of intellectual property rights at international climate conferences. At the TRIPS Council, the delegation of Japan maintained that intellectual property protection formed a solid and stable foundation for direct investment in environmentally sound technologies.[175] The delegation of Japan maintained that the international frameworks for intellectual property rights were well-balanced. The delegation of Japan disputed the arguments of developing nations that the current intellectual property regime undermined technology transfer – including in respect of green technologies. The Japan Patent Office has collaborated with WIPO GREEN to facilitate technology transfer in respect of clean technologies.[176]

17.4.3.3 Australia

A Labour government replaced a right-of-centre Coalition government in Australia in 2022. At COP27, the new Albanese government vowed to 'demonstrate our potential as a renewable energy superpower and the opportunities of green trade partnerships'.[177] It promised to 'contribute to an effective global response to climate change and support the international community to achieve the goals of the Paris Agreement'.[178] Australia's Minister for Climate Change and Energy, Chris Bowen, maintained at COP27 that there was a need for climate finance: 'We have a moral imperative and driving need for our institutions to work with countries across the developed and developing world not only to reduce emissions but respond to a changing climate and its economic impact on [our]

[174] Ibid, 27.

[175] Japan, 'Extract from Minutes of the TRIPS Council', World Trade Organization, IP/C/M/73/Add.1, 11–12 June 2013, 40.

[176] Japan Patent Office, 'Collaboration with WIPO GREEN' <www.jpo.go.jp/e/news/kokusai/green.html>.

[177] Department of Climate Change, Energy, the Environment and Water, 'Australian Engagement at COP27', Australian Government (2022), <www.dcceew.gov.au/climate-change/international-commitments/cop27>.

[178] Ibid.

nation.'[179] The Australian government has made notifications to the WTO of technology transfer in respect of climate-related technologies. However, it has not introduced specific law reform in respect of intellectual property to facilitate climate action and sustainable development.

17.4.3.4 Canada

The Government of Canada has sought to develop databases of patented inventions related to climate technologies.[180] Canada has developed specialized resources for clean technology businesses on how to leverage intellectual property.[181] Canada has established a patent collective focused on clean technologies called the Innovation Asset Collective.[182] There has been discussion about whether Canada needs to reform its intellectual property laws to promote a circular economy and climate action.[183]

17.4.4 BRICS nations (Brazil, Russia, India, China, South Africa)

Mid-tier nations in the BRICS grouping – such as Brazil, Russia, India, China and South Africa – have played an important role in the debate over intellectual property and climate change. India, in particular, has been vocal in negotiations over intellectual property, climate finance and technology transfer. A BRICS High Level meeting on climate change in 2022 highlighted the need for 'the reinforcement of adaptation and the delivery and urgency of developed countries' commitments to provide credible, adequate, predictable, timely, new and additional financial support and technology transfer to developing countries.'[184]

---

[179] Chris Bowen, 'National Statement to COP27', Australian Government (15 November 2022), <https://minister.dcceew.gov.au/bowen/speeches/national-statement-cop27>.

[180] Canadian Intellectual Property Office, 'Patented Inventions in Climate Change Mitigation Technologies', <https://ised-isde.canada.ca/site/canadian-intellectual-property-office/en/patented-inventions-climate-change-mitigation-technologies>.

[181] Government of Canada, 'Clean Technology and Intellectual Property', <https://ised-isde.canada.ca/site/canadian-intellectual-property-office/en/clean-technology-and-intellectual-property>.

[182] Innovation Asset Collective, <www.ipcollective.ca/> and Isi Caulder and Justin Philpott, 'First Canadian Patent Collective Aimed at Cleantech – Now Open for Business', *Bereskin Parr* (15 April 2021), <https://bereskinparr.com/news-insights/insights/first-canadian-patent-collective-aimed-at-cleantech-now-open-for-business/>.

[183] Graham Reynolds, 'Of Lock-Breaking and Stock Taking: Intellectual Property, Climate Change and the Right to Repair in Canada' (2023) 101 *Canadian Bar Review* 31–60.

[184] Joint Statement issued at the BRICS High-level Meeting on Climate Change, 24 May 2022, <http://brics2022.mfa.gov.cn/eng/hywj/ODMM/202205/t20220529_10694182.html>.

17.4.4.1 India

India's 2022 Long-Term Low-Carbon Development Strategy addresses intellectual property, technology transfer and climate finance.[185] In its executive summary, India emphasized that 'India cannot deploy low-carbon climate technologies at a significant scale unless a facilitative global technology transfer regime is in place, and the incremental and associated costs of these technologies are met by international climate funds'.[186] India argued: 'A collaborative international mechanism needs to ensure that barriers, such as intellectual property rights, are lowered by developed countries to facilitate technology transfer from developed to developing countries.'[187] India is committed to advancing common sustainable development goals: 'However, it also emphasizes that international cooperation is necessary to support developing economies through finance, technology and win-win strategies.'[188]

In the body of the analysis, India elaborates upon the need for progress on research, development and deployment of low-carbon climate technologies:

> Through engagement with multilateral platforms, India will seek to increase access to low-carbon technologies, including solar photovoltaics, offshore wind equipment, LED bulbs, efficient air conditioners, low-carbon iron and steel manufacturing technologies, biofuels and hydrogen technology. India will aim to stimulate markets for clean energy technologies emerging in India, with investment pathways developed in accordance with international best practices for research, innovation, and manufacturing in this field.[189]

India expressed concerns about the lack of adequate and sufficient climate finance: 'In the absence of a roadmap for climate finance, India and other developing countries will face serious constraints in achieving the climate goals in subsequent updates to NDC.'[190] India argued that there was a need to better track climate finance: 'Therefore, it is India's view that the global climate regime should track progress not just on climate mitigation efforts, but on the delivery of international climate finance to developing countries.'[191]

---

[185] Ministry of Environment, Forest and Climate Change, *India's Long-Term Low-Carbon Development Strategy* (Government of India, 2022).

[186] Ibid, 14.

[187] Ibid, 14.

[188] Ibid, 14.

[189] Ibid, 84.

[190] Ibid, 84.

[191] Ibid, 84.

As one of the sponsors of the TRIPS Waiver proposal during the COVID-19 crisis, India suggested that a similar proposal may be needed to address climate change: 'India also notes the example of the COVID-19 pandemic where the advantages of setting aside the intellectual property rights regime, particularly to meet global challenges, became very evident.'[192] India suggested: 'Such an approach if adopted at the global level will promote research and innovation relevant to climate action.'[193]

Krishna Ravi Srinivas has commented that the relationship between intellectual property, sustainable development and climate action is a 'complex conundrum'.[194] He has suggested that like-minded countries could explore options to enable the transfer of clean technologies. He has argued that open sourcing green technologies as 'global public goods' would be an option worth exploring.

17.4.4.2 South Africa

In respect of the implementation of the Paris Agreement 2015, South Africa has argued that 'climate technologies need to flow, without hiding behind the issue of intellectual property rights'.[195] South Africa has insisted: 'The Green Climate Fund needs to make the technology transfer possible and concrete through a dedicated window.'[196] South Africa has also recommended: 'More resources should be dedicated to enhance the efforts and operationalisation of the Climate Technology Centre and Network in accelerating the transfer of environmentally sound technologies for low carbon and climate resilient development in developing countries.'[197] South Africa has stated:

> It is crucial that the operating entities of the financial mechanism support developing countries' actions to tackle climate change, and should ensure predictable, sustainable and adequate resources for the funds that are supporting climate action, in particular the Global

[192] Ibid, 12.

[193] Ibid, 12.

[194] Krishna Ravi Srinivas, 'SDG13 and Intellectual Property Rights: A Complex in Conundrum' in Bita Amani, Caroline Ncube and Matthew Rimmer (eds), *The Elgar Companion to Intellectual Property and the Sustainable Development Goals* (Edward Elgar Publishing, 2024), 415.

[195] Government of South Africa, 'Pre-COP23 Stakeholder Consultation' (2017), <www.dffe.gov.za/event/deptactivity/2017pre_cop23stakeholder_consultationworkshop>. Discussed by Begona Gonzalez Otero, 'IP in Times of Crisis: A Problem or a Solution?' (2022) 53 *IIC – International Review of Intellectual Property and Competition Law* 501.

[196] Government of South Africa, 'Pre-COP23 Stakeholder Consultation' (2017), <www.dffe.gov.za/event/deptactivity/2017pre_cop23stakeholder_consultationworkshop>.

[197] Ibid.

Environment Facility Trust Fund, the Green Climate Fund and the Adaptation Fund.[198]

During the debate over COVID-19, South Africa and India proposed a TRIPS Waiver to help accelerate the provision of essential medicines.[199] There has been parallel discussion by South Africa and India as to whether there should be a TRIPS Waiver to tackle other global issues – such as climate change, biodiversity loss and depletion, and sustainable development. A number of scholars have wonder whether a TRIPS Waiver model would be useful in respect of encouraging climate action through waiving intellectual property rights to clean technologies.[200]

Caroline Ncube from the University of Cape Town contends that 'IP plays a critical role as one of multiple regulatory systems seeking solutions in the highly dynamic and insecure contexts of the climate emergency, uneven global development and pandemics, requiring timely and comprehensive normative development'.[201] She reflected: 'Pandemic prevention and preparedness are critically important, considering the very real possibility of other large-scale health emergencies, endemics and pandemics, which will be accelerated and amplified by the climate crisis and conflict.'[202]

17.4.4.3 China

China has invested heavily in clean technologies, renewable energy and energy efficiency.[203] China has also accumulated intellectual property holdings

---

198 Ibid.

199 India and South Africa, 'Waiver from Certain Provisions of the TRIPS Agreement for the Prevention, Containment and Treatment of COVID-19', World Trade Organization, IP/C/W/669, 2 October 2020.

200 Dalindyebo Shabalala, 'Climate Change, Technology Transfer, and Intellectual Property: A "Modest Proposal" for an IP Enforcement Mechanism' (2020) *Fordham Environmental Law Review* 1; Felicity Deane, 'Climate Change and the WTO: Will a TRIPS Waiver Support Technology Transfer?', Symposium on *The TRIPS Waiver: Intellectual Property, Access to Essential Medicines, and the Coronavirus*, QUT Faculty of Business and Law, 10 December 2021, <https://eprints.qut.edu.au/227139/> and Madeline Thompson, 'A TRIPS Waiver to Combat Climate Change?', *Journal of Technology and Intellectual Property* (10 February 2022), <https://jtip.law.northwestern.edu/2022/02/10/a-trips-waiver-to-combat-climate-change/>.

201 Caroline Ncube, 'Intellectual Property Norms in the Polycrisis: (Still) Omnipresent, Distracting, Irrelevant?' (2024) 19(9) *Journal of Intellectual Property Law & Practice* 717.

202 Ibid, 718.

203 Colleen Howe, 'Explainer: The Numbers behind China's Renewable Energy Boom' (*Reuters*, 16 November 2023), <www.reuters.com/sustainability/climate-energy/numbers-behind-chinas-renewable-energy-boom-2023-11-15/>.

in the field of green technologies.[204] Chen Zhou observes that 'China has a crucial role to play in climate geopolitics'.[205]

China has argued in the TRIPS Council that 'IPRs should contribute to, but not become a barrier to, the transfer and dissemination of environmentally rational technologies'.[206] Moreover, China has observed that 'nothing in the TRIPS Agreement prevents its existing flexibilities from its application to the environmentally rational technologies'.[207]

For its part, China has been embroiled in complex intellectual property and trade disputes with the United States in the field of clean technologies and renewable energy.[208] There has been considerable resentment in the United States government and industry about the Sinovel case – in which the trade secrets of American Superconductor in respect of wind technology were misappropriated. Peter Drahos has reflected: 'US multinationals fear China's stance on intellectual property and continue to partner with the US Government in pressuring China to do more to protect foreign intellectual property monopolies.'[209]

17.4.4.4 Brazil

Historically, Brazil has been a key actor in geopolitics in respect of international intellectual property. Brazil has deployed the threat of compulsory licensing to provide access to essential medicines.[210] Brazil has also been interested in mechanisms to expedite the examination of green patents to promote a circular economy.[211] In the wake of the Bolsonaro era, Brazil is in the process of reorientating its position on intellectual property

---

204 Matthew Glachant and Simon Touboul, 'China's Contribution to Low-Carbon Technology Innovation' (2021) 1 *Geopolitiques*, <https://geopolitique.eu/en/articles/chinas-contribution-to-low-carbon-technology-innovation/>.

205 Chen Zhou, *The Legal Barriers to Technology Transfer under the UN Framework Convention on Climate Change: The Example of China* (Springer, 2019), 243.

206 China, Extract from Minutes of Meeting of the TRIPS Council, World Trade Organization (11 June 2014), 212, <www.wto.org/english/tratop_e/trips_e/june2014_on_climate_e.pdf>.

207 Ibid.

208 Peter Drahos, 'The China-US Relationship on Climate Change, Intellectual Property and CCS: Requiem for a Species?' (2009) 1 *The WIPO Journal* 124–31.

209 Peter Drahos, *Survival Governance: Energy and Climate in the Chinese Century* (Oxford University Press, 2021), 208.

210 Amaka Vanni, *Patent Games in the Global South: Pharmaceutical Patent Law-Making in Brazil, India and Nigeria* (Hart Publishing, 2020).

211 Daniel Chalhub, 'About Circular Economy and Green Patents in Brazil', Executive Agency for Small and Medium-Sized Enterprises, European Commission (3 March 2021), <https://intellectual-property-helpdesk.ec.europa.eu/news-events/news/about-circular-economy-and-green-patents-brazil-2021-03-03_en>.

and climate change. Lula, the new president, told the COP27 summit that Brazil would take concerted climate action.[212] Moreover, he emphasized that Brazil would seek to redress deforestation, and promised better protection of indigenous communities and their lands and territories and associated cultural heritage.

17.4.4.5 Russia

At the BRICS meeting in 2022, Russia maintained that it was adhering to the principles of the UNFCCC and the Paris Agreement 2015.[213] Russia submitted:

> Russia is working on a radical restructurisation of the Russian industry and energy sectors, with 45% of Russia's energy balance already stemming from low-emission energy sources, including nuclear power generation. The Russian Federation intends to increase the utilization of associated gas, implement a large-scale program of ecological modernization and energy efficiency in all sectors of the economy. Russia is creating the infrastructure for the production of hydrogen to be used as a raw material and energy carrier.[214]

But in practice, as recognized by the World Bank, the Russian Federation has been a promoter of fossil fuel industries.[215] Russia has shown a scofflaw approach to intellectual property during the war in Ukraine.[216] As such, Russia has not made a substantive contribution to the debate over intellectual property and climate change.

17.4.5 Developing countries (Group of 77)

The Group of 77 – representing developing countries – has called for further progress in terms of access to clean technologies and sustainable innovation.

---

[212] BBC News, 'COP27: Brazil Is Back on the World Stage, Lula Tells Climate Summit' (*BBC News*, 16 November 2022), <www.bbc.com/news/science-environment-63625698>.

[213] Joint Statement issued at the BRICS High-level Meeting on Climate Change, 24 May 2022, <http://brics2022.mfa.gov.cn/eng/hywj/ODMM/202205/t20220529_10694182.html>.

[214] Ibid.

[215] World Bank Group, 'Climate Change Knowledge Portal: Russian Federation', <https://climateknowledgeportal.worldbank.org/country/russian-federation>.

[216] Brett Doyle and Lili Meunier, 'The War on Intellectual Property in Russia' (*Clayton Utz*, 24 June 2022), <www.claytonutz.com/knowledge/2022/june/the-war-on-intellectual-property-in-russia>.

On behalf of the Group of 77, Senator Sherry Rehman, the Federal Minister for Climate Change of Pakistan, addressed such concerns in a speech at a leadership dialogue for 'achieving a sustainable and inclusive recovery'.[217] She reflected:

> We must re-think our patterns of consumption and production. We can produce with greater respect for nature. We can stop the pollution of our lakes, rivers, and oceans. We can produce more with less water, and without dangerous chemicals. Technology transfers and technical assistance are key to achieve these goals.[218]

Rehman maintained that 'recovery efforts should also address the widening technology and digital gap that exists between the developed and developing countries'.[219] She observed: 'Without equal access to innovative and clean technologies, the current digital divide may become the new face of development divide.'[220] She emphasized that 'we must work towards promoting the alignment of business practices with the SDGs and the UNFCCC and its Paris Agreement'.[221]

The South Centre is an intergovernmental organization which represents developing countries. The South Centre has a long history of engagement with the topic of intellectual property and climate change. In a 2012 research paper,[222] Martin Khor commented that many proposals in respect of intellectual property and climate change had been stymied in international negotiations.

> While the developing countries have advocated new technology mechanisms in the UNFCCC framework and called on 'thinking outside the box' on IPRs, the developed countries inside the UNFCCC by contrast have generally continued to maintain that respect for a strong IPR regime is necessary for innovation and for the transfer of technology to developing countries. The proposals … have not been acceptable to the US or EU, which deny that IPRs constitute a barrier to technology transfer. For the developed countries, which own most

---

[217] Senator Sherry Rehman, 'Statement on Behalf of the Group of 77 and China', Leadership Group 2 of Stockholm +50, 'Achieving a Sustainable and Inclusive Recovery', 3 June 2022, <www.g77.org/statement/getstatement.php?id=220603>.

[218] Ibid.

[219] Ibid.

[220] Ibid.

[221] Ibid.

[222] Martin Khor, *Climate Change, Technology and Intellectual Property Rights: Context and Recent Negotiations* (The Third World Network, 2012).

> of the patents on climate-related technologies, maintaining the normal situation regarding IPRs is a matter of having advantage in economic competitiveness.[223]

As a result, it has been difficult to achieve consensus on the inclusion of decision text dealing with the topic of intellectual property, climate change, and technology transfer.

Carlos Correa – the current executive director of the South Centre – highlighted the importance of intellectual property in the organization's statement at COP27.[224] Correa observed: 'Climate change action is not possible without wide and timely access to technologies.'[225] He provided a reminder that 'an effective fulfillment of the developed countries' obligations in this regard is indispensable'.[226] Deeming current models of technology transfer to be insufficient, Correa recommended that 'a suitable international regime for the appropriation and transfer of cutting-edge technologies needs to be put in place'.[227] Correa maintained that such a regime 'should ensure that green technologies are treated as "public goods" and facilitate patent-free green technology transfers and dissemination through UNFCCC and other mechanisms'.[228] Correa observed: 'Such a regime should move away from a purely market-based approach and be rather based on the principles of cooperation and solidarity.'[229] In such a vision, Correa is calling for intellectual property relating to green technologies to be treated as public goods – which are freely available for sharing.

In a discussion paper for the South Centre, Livia Regina Batista has further explored the inter-relationship between intellectual property, public health and access to climate-related technologies.[230] She observed: 'Fast-tracking programs put in place by national patent offices to accelerate the process of granting a patent on COVID-19-related medicine/vaccines have not assimilated the main criticisms regarding the previous experience on

---

223 Ibid, 25.

224 Carlos Correa, 'Statement of the Executive Director the South Centre', 27th Conference of the Parties (COP27) of the UNFCCC, 16 November 2022, <www.southcentre.int/sc-statement-to-the-cop27-16-november-2022/>.

225 Ibid.

226 Ibid.

227 Ibid.

228 Ibid.

229 Ibid.

230 Livia Regina Batista, *The Intersection Between Intellectual Property, Public Health and Access to Climate-Related Technologies* (Discussion Paper 188, South Centre, 7 December 2023), <www.southcentre.int/wp-content/uploads/2023/12/RP188_The-Intersection-Between-Intellectual-Property-Public-Health-and-Access-to-Climate-Related-Technologies_EN.pdf>.

climate-related technologies and does not represent a solution for a wider dissemination of technologies.'[231] Batista commented: 'Considering broader access to climate-related technologies as a public health issue will link it to human rights rhetoric and allow us to address climate change in a hard law framework.'[232]

The Third World Network have been influential in framing the debate over intellectual property, technology transfer and climate change. The late Martin Khor was a catalyst for raising the profile of this topic.[233] In 2022, the Third World Network emphasized that questions around intellectual property, technology transfer and climate change were unresolved.[234] The Third Network observed that

> promoting economic diversification hinges on the equitable and fair resolution of other issues within the multilateral climate change, trade, and other regimes such as the transfer and development of environmental sound technologies, the provision of intellectual property rights flexibilities with respect to the needed technologies, the reflection of the principles of common but differentiated responsibilities (CBDR) and of Special and Differential Treatment (SDT) for developing countries, the provision of new, additional and adequate finance for mitigation, adaptation, economic diversification, and loss and damage.[235]

The Third World Network observed that 'addressing the twin development and climate change challenge requires an integrated approach to, inter alia, trade policy, economic diversification, the development of productive capacity, environmental policy, investment policy, intellectual property policy, among others'.[236] The Third World Network argued: 'This approach should ensure that the overall development approach is "climate-friendly", alleviate poverty, achieves development, and improves standards of living while adapting to and mitigating climate change.'[237]

---

[231] Ibid, 22.

[232] Ibid, 22.

[233] Sangeeta Shashikant and Martin Khor, *Intellectual Property and Technology Transfer Issues in the Context of Climate Change* (The Third World Network, 2010).

[234] Third World Network, 'Submission on the Assessment of Environmental, Social, and Economic Co-benefits of Climate Change Policies and Actions, Focusing on Economic Diversification' (2022), <https://unfccc.int/sites/default/files/resource/Third%20World%20Network_%20Submission%20to%20KCI%20September%202022f.pdf>.

[235] Ibid.

[236] Ibid.

[237] Ibid.

17.4.6 LDCs, Small Island States, and countries vulnerable to climate change

LDCs have been seeking to establish an agenda at the climate negotiations in respect of the topic of technology transfer.[238] A 2021 briefing paper for negotiators for LDCs highlights a number of key issues. The briefing paper notes that 'the importance of transfer and local adaptation of technologies for sustainable development is widely acknowledged, as is the need for public incentives to support technology transfer for the poorest countries'.[239] The briefing paper laments the lack of progress in this field: 'Despite these commitments, limited progress has been made over the past 10 years in effectively delivering technology transfer to LDCs.'[240] It also highlights the lack of proper debate on access to climate technologies during the international climate discussions: 'Furthermore, there has been little mention of trade aspects of technology transfer in climate negotiations or of climate technologies in trade deliberations and negotiations on technology transfer.'[241] The briefing paper concludes:

> The slow progress on delivering on technology transfer commitments for LDCs in both trade and climate negations means that they have reduced capability to respond to climate change and maintain and expand market access in an era of increasing environmental regulation, such as the imminent introduction by some countries of carbon border adjustment measures.[242]

In the climate arena, the briefing paper comments that 'the delivery of developed countries' commitments on climate finance, including funding for technology transfer, and the flow of funds to LDCs is a major outstanding issue in the negotiations'.[243] It highlights continued problems with LDCs getting sufficient climate finance: 'The link with finance is particularly important for LDCs – especially the need for increased funding to support technology development and transfer, and capacity-building to develop proposals to access funding.'[244] The briefing paper demands better

---

[238] Laura Kelly, Brianna Craft, Fred Onduri Machulu and Maheshwar Dhakal, 'Technology Transfer: A New Agenda for LDC Negotiators', ODI, 10 November 2021, <https://odi.org/en/publications/technology-transfer-a-new-agenda-for-ldc-negotiators/>.

[239] Ibid, 1.

[240] Ibid, 1.

[241] Ibid, 1.

[242] Ibid, 4.

[243] Ibid, 1.

[244] Ibid, 2.

representation of LDCs in the Technology Mechanism: 'The CTCN Advisory Board does not have a specific and consistent slot for LDCs, which the group has argued should change.'[245] Moreover, 'ensuring the representation of technology coordinators, as well as negotiators, on the board (to support information-sharing and any presentation of concerns) will also be discussed'.[246]

In the trade sphere, the briefing paper highlights that the mechanisms for technology transfer under the TRIPS Agreement 1994 have been ineffectual:

> The specific challenges faced by LDCs are recognised in Article 66.2, which creates a legal obligation for developed countries to encourage technology transfer to LDCs through incentivising national enterprises and institutions to transfer technology. However, there is limited evidence of this happening in practice, and the reporting system put in place for Article 66.2 implementation has not been sufficiently rigorous to promote actual transfers.[247]

The briefing paper recommends:

> Discussions on the implementation of TRIPS Article 66.2 and deliberations under the WGTT (the Working Group on Transfer of Technology), the CTE (the Committee on Trade and Environment) and the TESSD (Trade and Environment Sustainability) initiative could provide additional avenues to address concerns related to trade and the transfer of climate change mitigation and adaptation technologies.[248]

Small Island States and countries vulnerable to climate change have highlighted the need for access to climate adaptation and climate mitigation technologies. At the COP26 Glasgow Climate Conference, Prime Minister of Barbados Mia Amor Mottley made an impassioned speech on behalf of Small Island States.[249] She was concerned about the lack of climate finance: 'Failure to provide this critical finance and that of loss and damage is measured in lives and livelihoods being lost in our communities.'[250] She questioned the ethics of the situation: 'It is immoral and unjust.'[251] Mottley's

245 Ibid, 2.
246 Ibid, 2.
247 Ibid, 3–4.
248 Ibid, 8.
249 Mia Amor Mottley, 'Barbados: High-level Segment Statement COP26', Glasgow Climate Change Conference, 3 November 2021, <https://unfccc.int/documents/309276>.
250 Ibid.
251 Ibid.

advocacy was designed to elevate questions of global finance to a higher political level.[252] At the subsequent climate conference at COP27, Mottley criticized industrialized nations for failing to help the developing world deal with the impacts of the climate crisis.[253] Mottley is very much of the view that there should be climate finance to enable research, development and deployment of clean technologies in Small Island States.

Small Island States and associate members were disappointed that COP28 was concluded, without the presence of their representatives.[254] The Alliance of Small Island States observed: 'It is not enough for us to reference the science and then make agreements that ignore what the science is telling us we need to do.'[255]

17.5 Non-state actors

Non-state actors – including business, civil society and indigenous communities – have also been key players in international climate conferences.[256] Such groups have also joined the debate over intellectual property, climate finance and technology transfer.

---

252 Oliver Millman and Peter Walker, '"We are Digging our own Graves": World Leaders' Powerful Words at COP26' (*The Guardian*, 2 November 2021), <www.theguardian.com/environment/2021/nov/01/digging-our-own-graves-world-leaders-speak-cop26>; Climate Home News, 'Mia Mottley: The "Fearless" Leader Pushing a Global Settlement for the Climate Frontlines' (*Climate Home News*,(18 November 2021), <https://climatechangenews.com/2021/11/18/mia-mottley-fearless-leader-pushing-global-settlement-climate-frontlines/ and BBC News, 'COP26: "Two Degrees is a Death Sentence" for Island Nations: Barbados PM Mia Mottley' (*BBC News*, 1 November 2021), <www.bbc.com/news/av/world-59117750>.

253 Patrick Greenfield, Fiona Harvey, Nina Lakhani and Damian Carrington, 'Barbados PM launches Blistering Attack on Rich Nations at COP27 Climate Talks' (*The Guardian*, 8 November 2022), <www.theguardian.com/environment/2022/nov/07/barbados-pm-mia-mottley-launches-blistering-attack-on-rich-nations-at-cop27-climate-talks>.

254 Alliance of Small Island States (AOSIS), 'Statement on the Global Stocktake Text', 28th Session of the Conference of Parties to UNFCCC, Closing Plenary, 13 December 2023, <www.aosis.org/cop28-closing-plenary-aosis-statement-on-gst-decision/>.

255 Ibid.

256 United Nations, 'History of Non-Party Stakeholder Engagement', <https://unfccc.int/climate-action/introduction-climate-action/history-non-party-stakeholder-engagement>; and Thomas Hale, 'The Role of Sub-state and Non-state Actors in International Climate Processes' (*Chatham House*, November 2018), <www.chathamhouse.org/sites/default/files/publications/research/2018-11-28-non-state-sctors-climate-synthesis-hale-final.pdf>.

17.5.1 BINGO – business and industry group organizations

Business and industry groups have often expressed a position, which has been called Prometheanism by the political scientist John Dryzek.[257] Dryzek has noted: 'Prometheans have unlimited confidence in the ability of humans and their technologies to overcome any problem – including environmental problems.'[258]

At the international climate talks, industry groups and associations have often pushed for stronger protection of intellectual property rights associated with clean technologies. Martin Khor commented: 'Business associations in developed countries have demanded that their governments do not make concessions on IPRs in the climate negotiations.'[259] For instance, the International Chamber of Commerce has maintained that 'Intellectual Property Rights (IPR) are vital in the development and diffusion of new and existing technologies for combating climate change'.[260]

Law partner Alison Bryce considered the role of intellectual property in the COP26 discussions at Glasgow.[261] She commented: 'Events like COP26 in Glasgow … are an example of a modern push for collaboration in this field.'[262] Bryce reflected: 'In order to alleviate the impact of modern society on the environment, global IP frameworks will be required to work collaboratively.'[263]

There has been alarm among civil society groups about the infiltration of international climate conferences by fossil fuel industry representatives at COP26[264] and COP27.[265] There has been heavy criticism of the United Arab Emirates using its position as host of the COP28 summit to engage

[257] John S. Dryzek, 'Growth Unlimited: The Promethean Response' in John S. Dryzek, *The Politics of the Earth: Environmental Discourses (3rd edition)* (Oxford University Press, 2012 (print), online 2017), <https://doi.org/10.1093/hepl/9780199696000.003.0003>.

[258] Ibid.

[259] Ibid, 25.

[260] International Chamber of Commerce, 'Intellectual Property Rights Protection Vital in Climate Change Negotiations', International Chamber of Commerce (15 September 2009), <https://iccwbo.org/news-publications/news/intellectual-property-rights-protection-vital-in-climate-change-negotiations/>.

[261] Alison Bryce, 'Intellectual Property: Key Role Against Climate Change', *Journal of the Law Society of Scotland* (14 December 2020), <www.lawscot.org.uk/members/journal/issues/vol-65-issue-12/intellectual-property-key-role-against-climate-change/>.

[262] Ibid.

[263] Ibid.

[264] BBC News, 'COP26: Fossil Fuel Industry Has Largest Delegation at Climate Summit' (*BBC News*, 8 November 2021), <www.bbc.com/news/science-environment-59199484>.

[265] Ruth Michaelson, '"Explosion" in Number of Fossil Fuel Lobbyists at COP27 Climate Summit' (*The Guardian*, 10 November 2022), <www.theguardian.com/environment/2022/nov/10/big-rise-in-number-of-fossil-fuel-lobbyists-at-cop27-climate-summit>.

in greenwashing.[266] There has been a discussion of the need for greater transparency in identifying fossil fuel lobbyists at climate conferences,[267] though Amy Westervelt has observed that it is insufficient to merely identify fossil fuel lobbyists.[268] There have been calls for an exclusion of fossil fuel interests from international climate events – much like there have been rules against tobacco interference in the sphere of public health.[269] There were complaints about the presence of a record number of fossil fuel lobbyists and climate denialists at the COP28 talks,[270] and climate scientists have lamented that COP28 was a 'tragedy for the planet and our future' and a 'dream outcome' for the fossil fuel industry.[271]

17.5.2 Civil society

In international climate talks, there has long been discussion of public policy options of public domain dedications of technology for climate adaptation and climate mitigation. In particular, there has been discussion of options such as technology transfer, public sector licensing, patent pools, compulsory

---

[266] Ben Stockton, 'Cop28 President's Team Accused of Wikipedia "Greenwashing"' (*The Guardian*, 30 May 2023), <www.theguardian.com/environment/2023/may/30/cop28-president-team-accused-of-wikipedia-greenwashing-sultan-al-jaber> and Damian Carrington, 'Leak Reveals "Touchy" Issues for UAE's Presidency of UN Climate Summit' (*The Guardian*, 1 August 2023), <www.theguardian.com/environment/2023/aug/01/leak-uae-presidency-un-climate-summit-oil-gas-emissions-yemen>.

[267] Fiona Harvey, 'Fossil Fuel Lobbyists Will Have to Identify Themselves When Registering for Cop28' (*The Guardian*, 16 June 2023), <www.theguardian.com/environment/2023/jun/15/fossil-fuel-lobbyists-will-have-to-identify-themselves-when-registering-for-cop28>.

[268] Amy Westervelt, 'So What If Fossil Fuel Lobbyists Have to Declare Themselves at Cop28? That Won't Curb Their Power' (*The Guardian*, 28 June 2023), <www.theguardian.com/commentisfree/2023/jun/28/cop-28-fossil-fuel-lobbyists-oil-gas-companies>.

[269] Jennifer Rankin, 'Shun Fossil Fuel Firms by Treating Them Like Tobacco Industry, EU urged' (*The Guardian*, 25 October 2021), <www.theguardian.com/world/2021/oct/25/shun-fossil-fuel-firms-by-treating-them-like-tobacco-industry-eu-urged>.

[270] Nina Lakhani, 'Record Number of Fossil Fuel Lobbyists Get Access to Cop28 Climate Talks' (*The Guardian*, 5 December 2023), <www.theguardian.com/environment/2023/dec/05/record-number-of-fossil-fuel-lobbyists-get-access-to-cop28-climate-talks> and Nina Lakhani, 'Revealed: More than 160 Representatives with Climate-denying Track Records Got Cop28 Access' (*The Guardian*, 12 December 2023), <www.theguardian.com/environment/2023/dec/11/climate-deniers-attend-cop28-talks>.

[271] Damian Carrington, 'Failure of Cop28 on Fossil Fuel Phase-out is "Devastating", Say Scientists' (*The Guardian*, 15 December 2023), <www.theguardian.com/environment/2023/dec/14/failure-cop28-fossil-fuel-phase-out-devastating-say-scientists>.

licensing and public domain dedications.[272] Civil society has played an important role in raising such issues in international climate talks. However, there has been concern about civil society being sidelined and marginalized at COP26,[273] COP27[274] and COP28.[275]

Climate youth leader Greta Thunberg expressed her concerns about whether the international climate conferences were sufficiently inclusive. She observed that the inequitable vaccine distribution during the COVID-19 crisis would disadvantage many from attending the COP26 event.[276] Thunberg expressed her concerns about the COP26 process: 'We need immediate drastic annual emission cuts unlike anything the world has ever seen.'[277] She lamented the technocratic bent of the conference: 'The people in power can continue to live in their bubble filled with their fantasies, like eternal growth on a finite planet and technological solutions that will suddenly appear seemingly out of nowhere and will erase all of these crises just like that.'[278] Thunberg observed: 'All this while the world is literally burning, on fire, and while the people living on the front lines are still bearing the brunt of the climate crisis.'[279] She has since published a collection called

---

[272] Climate Action Network, 'International Submission to UNFCCC Ad Hoc Working Group on Long-Term Cooperative Action Regarding Technology Cooperation and Sharing', 24 April 2009.

[273] Nina Lakhani, 'Cop26 Legitimacy Questioned as Groups Excluded from Crucial Talks' (*The Guardian*, 8 November 2021), <www.theguardian.com/environment/2021/nov/08/cop26-legitimacy-questioned-as-groups-excluded-from-crucial-talks> and Libby Brooks, 'Hundreds of Global Civil Society Representatives Walk Out of Cop26 in Protest' (*The Guardian*, 13 November 2021), <www.theguardian.com/environment/2021/nov/12/global-civil-society-representatives-walkout-cop26-protest>.

[274] UN Experts, 'Egypt: UN Experts Alarmed by Restrictions on Civil Society ahead of Climate Summit', Office for the High Commissioner of Human Rights (7 October 2022), www.ohchr.org/en/press-releases/2022/10/egypt-un-experts-alarmed-restrictions-civil-society-ahead-climate-summit and UN Experts, 'Egypt: UN Experts Alarmed by Harassment of Civil Society Actors at COP27 Climate Summit', Office for the High Commissioner of Human Rights (18 November 2022), www.ohchr.org/en/press-releases/2022/11/egypt-un-experts-alarmed-harassment-civil-society-actors-cop27-climate

[275] Jon Gambrell, 'Protests at COP28 Restricted by "Shocking Level of Censorship" in Host Country UAE' (*PBS*, 9 December 2023), <www.pbs.org/newshour/world/protests-at-cop28-restricted-by-shocking-level-of-censorship-in-host-country-uae>.

[276] Agence France-Presse, 'Greta Thunberg Says She Will Not Attend Cop26 Climate Summit' (*The Guardian*, 9 April 2021), <www.theguardian.com/environment/2021/apr/09/greta-thunberg-says-she-will-not-attend-cop26-climate-summit>.

[277] BBC News, 'COP26: Greta Thunberg Tells Protest that COP26 has been a "Failure"' (*BBC News*, 5 November 2021), <www.bbc.com/news/uk-scotland-glasgow-west-59165781>.

[278] Ibid.

[279] Ibid.

The Climate Book – which collects a variety of climate solutions espoused by climate activists and scholars.[280]

The Climate Action Network (CAN) has long called for a flexible approach to intellectual property rights.[281] CAN has called for a declaration to enable intellectual property sharing:

> Recognizing that climate change is a planetary emergency, the patents and related intellectual property mechanism should be guided by an International Declaration on Climate Technologies and IPRs from the UNFCCC COP or COP/MOP that all possible policy avenues to accelerate research, development, demonstration and diffusion of climate-friendly technology, should be explored, including the use of all flexibilities, exceptions and limitations in international and national patent and related intellectual property rules, as well as innovative uses of intellectual property mechanisms, licensing practices, and alternative modes of innovation such as open source approaches.[282]

CAN maintained: 'The declaration should state that all countries will refrain from using unilateral measures, financial or political (such as trade sanctions or withdrawal of trade preferences), against countries that exercise patent and related intellectual property flexibilities, exceptions and limitations.'[283] They recommended that such a declaration should be jointly adopted by the UNFCCC, the WTO and WIPO.

CAN has canvassed a range of intellectual property flexibilities – including:

a. patent pools and related concepts such as patent libraries;
b. joint research initiatives;
c. compulsory or non-voluntary licensing;
d. patent buy-outs;
e. segmented/parallel markets;
f. parallel imports;
g. patent exclusions;
h. differential patentability requirements;
i. open-source licensing;

[280] Greta Thunberg (ed), *The Climate Book* (Penguin Books, 2022).

[281] Climate Action Network, 'International Submission to UNFCCC Ad Hoc Working Group on Long-Term Cooperative Action Regarding Technology Cooperation and Sharing', 24 April 2009.

[282] Ibid.

[283] Ibid.

j. prize funds;
k. rules on open access to publicly funded technologies.[284]

CAN supported 'the use of alternatives that may require limited or temporary modifications of international intellectual property rules'.[285]

The Open Access movement has also proposed Open Science models for sharing climate innovation, data and knowledge. Catherine Stihler – the CEO of Creative Commons – argued at COP26 that climate research could not be put behind paywalls.[286] She emphasized 'the importance of open data and open science in providing the evidence base and action required to achieve net zero and mitigate the impact of climate change'.[287] Stihler concluded that 'if [the world's leaders] truly want to tackle this emergency, they will recognise the importance of open access to climate research and data so that we can all share the knowledge to take action to save our planet'.[288]

Indigenous communities have also called for stronger recognition of indigenous intellectual property and traditional knowledge in light of the impacts of climate change and global warming upon culture heritage.[289] Indigenous communities and climate justice groups complained that COP28 was 'business as usual'.[290]

17.6 Our common agenda

There has been a growing movement to treat clean technologies as global public goods. Knowledge Ecology International have argued that there

---

[284] Ibid.

[285] Ibid.

[286] Catherine Stihler, 'Climate Change: "Glasgow Agreement" Can Save the Planet but Locking Scientific Research Behind Paywalls is Holding us Back' (*The Scotsman*, 24 August 2021), <www.scotsman.com/news/opinion/columnists/climate-change-glasgow-agreement-can-save-the-planet-but-locking-scientific-research-behind-paywalls-is-holding-us-back-catherine-stihler-3357616>. See also Creative Commons, 'Open Climate Campaign', <https://creativecommons.org/category/science/open-climate-campaign/>.

[287] Ibid.

[288] Ibid.

[289] International Indigenous Peoples' Forum on Climate Change and the Center for International Environmental Law, 'International Indigenous Peoples and Traditional Knowledge in the Context of the UN Framework Convention on Climate Change: Compilation of Decisions and Conclusions Adopted by the Parties to the Convention – 2019 Update', 2019, <www.ciel.org/wp-content/uploads/2019/08/Indigenous-Peoples-and-Traditional-Knowledge-in-the-Context-of-the-UNFCCC-2019-Update.pdf>.

[290] Nina Lakhani, 'Indigenous People and Climate Justice Groups Say COP28 was "Business as Usual"' (*The Guardian*, 14 December 2023), <www.theguardian.com/environment/2023/dec/13/indigenous-people-and-climate-justice-groups-say-cop28-was-business-as-usual>.

should be a treaty on global public goods.[291] UN Secretary-General António Guterres himself has come to the view that intellectual property barriers to access to clean technologies should be removed. He has argued that renewable energy technologies need to be treated like global public goods.

Guterres seems to have become increasingly exasperated by the lack of progress on intellectual property, technology transfer, and climate finance. He has been concerned that stalemates and impasses at international climate conferences have been costly.

In 2021, the Secretary-General released the *Our Common Agenda* report.[292] A key part of the report was a call for nations – large and small – to protect the global commons and deliver global public goods. The report observes: 'Certain public goods have long been acknowledged as being global in nature, in that they cannot be adequately provided by any one State acting alone and they concern the welfare of humanity as a whole.'[293] The report maintains:

> Despite this, the multilateral system is not yet geared for the strategies, investments or solidarity needed, leaving all of us vulnerable to crises, such as in global public health, demonstrated by COVID-19, in the global economy and financial system, as in the 2008 financial crisis and current COVID-19 shock, and in the health of our planet, resulting in the triple planetary crisis of climate change, biodiversity loss and pollution we now face.[294]

Guterres' speeches at COP26 and COP27 could be analysed in terms of their discourse and rhetoric. In 2022, the Secretary-General emphasized: 'Renewable energy technologies, such as battery storage, must be treated as essential and freely-available global public goods.'[295] He commented: 'Removing obstacles to knowledge sharing and technological transfer – including intellectual property constraints – is crucial for a rapid and fair renewable energy transition.'[296]

---

[291] James Love, 'Trade Related Aspects of Public Goods' (*Knowledge Ecology International*, 8 March 2022), <www.keionline.org/37512>.

[292] United Nations, *Our Common Agenda: Report of the Secretary-General*, Geneva: United Nations (2021), <www.un.org/en/common-agenda>.

[293] Ibid, 47.

[294] Ibid, 47.

[295] António Guterres, 'Secretary-General's Video Message on the Launch of the World Meteorological Organization's State of the Global Climate 2021 Report', United Nations (18 May 2022), <www.un.org/sg/en/content/sg/statement/2022-05-18/secretary-generals-video-message-the-launch-of-the-world-meteorological-organization%E2%80%99s-state-of-the-global-climate-2021-report-scroll-down-for-languages>.

[296] Ibid.

At COP27, Guterres warned: 'We are on a highway to climate hell with our foot still on the accelerator.'[297] He called for the development of 'a historic Pact between developed and emerging economies – a Climate Solidarity Pact'.[298] In his view, it would be 'a Pact in which all countries make an extra effort to reduce emissions this decade in line with the 1.5-degree goal'.[299] Guterres envisaged 'a Pact in which wealthier countries and International Financial Institutions provide financial and technical assistance to help emerging economies speed [up] their own renewable energy transition.'[300] He called for 'a Pact to end dependence on fossil fuels and the building of new coal plants – phasing out coal in OECD countries by 2030 and everywhere else by 2040'.[301] Guterres hoped for 'a Pact that will provide universal, affordable, sustainable energy for all'.[302] He asked for 'a Pact in which developed and emerging economies unite around a common strategy and combine capacities and resources for the benefit of humankind'.[303] He declared: 'Humanity has a choice: cooperate or perish.'[304] And he suggested: 'It is either a Climate Solidarity Pact – or a Collective Suicide Pact.'[305]

In 2023, Guterres renewed his call for greater access to renewable technologies.[306] In a video message to a session of the International Renewable Energy Agency, he stressed: 'Only renewables can safeguard our future, close the energy access gap, stabilize prices and ensure energy security.'[307] Guterres articulated a Five-Point Energy Plan for a just transition. First, he emphasized that 'we must remove intellectual property barriers and treat key renewable technologies, including energy storage, as global public goods'.[308] Second, Guterres said that we 'must diversify and increase

[297] António Guterres, 'Secretary-General's Remarks to High-Level Opening of COP27', United Nations (7 November 2022), <www.un.org/sg/en/content/sg/speeches/2022-11-07/secretary-generals-remarks-high-level-opening-of-cop27>.

[298] Ibid.

[299] Ibid.

[300] Ibid.

[301] Ibid.

[302] Ibid.

[303] Ibid.

[304] Ibid.

[305] Ibid.

[306] António Guterres, 'Secretary-General's Video Message to the 13th Session of the IRENA Assembly: "World Energy Transition – The Global Stocktake"', United Nations (14 January 2023), <www.un.org/sg/en/content/sg/statement/2023-01-14/secretary-generals-video-message-the-13th-session-of-the-irena-assembly-%E2%80%9Cworld-energy-transition-%E2%80%93-the-global-stocktake%E2%80%9D> and UN News, 'UN Chief Calls for Renewable Energy "Revolution" for a Brighter Global Future', United Nations (14 January 2023), <https://news.un.org/en/story/2023/01/1132452>.

[307] Ibid.

[308] Ibid.

access to supply chains for raw materials and components for renewables technologies, without degrading our environment'.[309] Third, he said that 'decision-makers must cut the red tape, fast-track approvals for sustainable projects worldwide and modernize grids'.[310] Fourth, Guterres emphasized that 'energy subsidies must shift from fossil fuels to clean and affordable energy' and 'we must support vulnerable groups affected by this transition'.[311] Fifth, Guterres called for climate finance – 'public and private investments in renewables should triple to at least $4 trillion dollars a year'.[312]

At the closing of COP28, Guterres observed: 'COP28 occurred at a decisive moment in the fight against the climate crisis – a moment that demands maximum ambition both in the reduction of greenhouse gas emissions and climate justice.'[313] He noted: 'To those who opposed a clear reference to a phase out of fossil fuels in the COP28 text, I want to say that a fossil fuel phase out is inevitable whether they like it or not.'[314] Guterres concluded: 'The era of fossil fuels must end – and it must end with justice and equity.'[315]

The Summit of the Future in September 2024 is intended to address intellectual property, scientific progress and sustainable development.[316] Guterres is hopeful that the Summit of the Future will help facilitate 'unlocking finance for climate action and development'.[317]

This vision needs to be translated into a policy agenda, and implemented at the international and national levels. The call for a removal of intellectual property barriers will need to be taken up by key international institutions – such as the UNFCCC, the WTO, the WIPO and the UNDP – as well

[309] Ibid.

[310] Ibid.

[311] Ibid.

[312] Ibid.

[313] António Guterres, 'Secretary-General's Statement at the Closing of the UN Climate Change Conference COP28', United Nations (13 December 2023), <www.un.org/sg/en/content/sg/statement/2023-12-13/secretary-generals-statement-the-closing-of-the-un-climate-change-conference-cop28>.

[314] Ibid.

[315] Ibid.

[316] United Nations, Summit of the Future, 20–23 September 2023, <www.un.org/en/summit-of-the-future>; and Ellen t' Hoen, 'UN "Pact for the Future" Intellectual Property Actions Need Revision to Ensure Innovation and Access to Health Products for All' (*Medicines Law & Policy*, 13 September 2024), <https://medicineslawandpolicy.org/2024/09/un-pact-for-the-future-intellectual-property-actions-need-revision-to-ensure-innovation-and-access-to-health-products-for-all/>.

[317] António Guterres, 'Secretary-General's Remarks to the United Nations Civil Society Conference in Support of the Summit of the Future', United Nations (10 May 2024), <www.un.org/sg/en/content/sg/speeches/2024-05-10/secretary-generals-remarks-the-united-nations-civil-society-conference-support-of-the-summit-of-the-future%C2%A0>.

as nation states and key non-state actors. No doubt intellectual property owners will be resistant to any proposal, which impacts upon their intellectual property rights.[318]

17.7 Conclusion

This chapter has explored the raging debate over intellectual property and climate change at recent international climate conferences. In spite of the growing urgency for climate action, the topic of intellectual property and technology transfer has been subject to prevarication and procrastination. There have been deep divisions between key actors as to whether there should even be any text on intellectual property at international climate conferences. In the meantime, there are growing conflicts and disputes between private actors over intellectual property and clean technologies. The climate conferences at COP26, COP27 and COP28 have failed to resolve longstanding differences around intellectual property and climate change. There remain concerns about whether the Technology Mechanism and the Finance Mechanism are adequate to facilitate research, development and diffusion of clean technologies. There have also been some nascent initiatives to encourage cooperative and collaborative approaches to the sharing of intellectual property in respect of clean technologies. Given the difficulties of reaching a consensus on a TRIPS Waiver during the COVID-19 crisis, it is doubtful that there will be a comprehensive agreement on intellectual property and clean technologies. The United Nations Secretary-General António Guterres has called for the treatment of renewable energy and other green technologies as 'global public goods', free from intellectual property restrictions.

[318] For an example of such a viewpoint, Michael Whitt complains: 'There are unintended consequences to monkeying around with IP incentive programs. Just throwing everything out isn't going to work. It may actually [be a disincentive for] innovation, or cause people to keep their patent that's issued for whatever the invention is.' Bennett Jones, 'UN Secretary-General Calls for IP Sharing to Jump Start Renewable Energy Transition' (20 May 2022), <www.bennettjones.com/Publications-Section/In-the-News/UN-Secretary-General-Calls-for-IP-Sharing-to-Jump-Start-Renewable-Energy-Transition>.

18

Climate Technologies and International Trade Law: Tensions and Opportunities

Kateryna Holzer and Yulia Yamineva

18.1 Introduction

The lack of ambition in climate action over the short run and the ease with which countries pledge their carbon-neutral targets for the year of 2050 and even for 2060 can be explained, in our view, by the expectation of today's policy makers that they can escape responsibility in 30 years' time or by their presumption that magical technologies allowing an effective and affordable green transition will be available in due course. This presumption does not look realistic, as, for instance, in the energy sector it can take 30 to 40 years for technologies to become mature or commercially available.[1] It means that for some breakthrough technologies to be available by 2050, there should already be signs of their existence and workability.

The climate goals set by the Paris Agreement clearly require a rapid and transformative change in climate mitigation and adaptation technologies. Technology and access to technology were highlighted in the Sixth Assessment Report of the Intergovernmental Panel on Climate Change (IPCC) as critical enablers for accelerated climate action.[2] Technologies that are in high demand for climate policy are, for instance, those needed

---

1 Robert Gross et al, 'How Long Does Innovation and Commercialisation in the Energy Sectors Take? Historical Case Studies of the Timescale from Invention to Widespread Commercialisation in Energy Supply and End Use Technology' (2018) 123 *Energy Policy* 682, 691.

2 IPCC 2023: Synthesis Report of the IPCC Sixth Assessment Report (AR6). Summary for Policymakers [Core Writing Team, H. Lee et al] Geneva, Switzerland, C.6 and C.7.

for carbon capture and storage and for hydrogen production (electrolysis and fuel cells). Technological solutions for energy storage to make use of electricity generation from renewable-energy sources are also very important. The race for the development of these technologies is under way, and those countries that maintain the lead will benefit the most from future net-zero economic growth.

For climate goals, many necessary technologies are already in place.[3] This is particularly true of the energy sector (for example, photovoltaic panels and wind turbines), road transport (hydrogen fuel cells and e-car batteries), construction (sustainable insulation materials) and agriculture (seeds of well-adapted crops and varieties). To contain climate change before it is too late, there is a need to accelerate the scale-up of technologies which are already available or which are close to being deployed at scale. This means that technology deployment policies are vital.[4] Trade measures, including tariff policies, carbon pricing and regulations, and subsidies, are a critical part of technology deployment policies. They can promote both the generation of technologies by technology-producing countries and the diffusion and adoption of technologies by technology-importing countries.[5]

This chapter examines the adequacy of international trade law for meeting the need for technological transformation necessitated by the climate goals of the Paris Agreement. It explores the current role of trade policy, including the linkages with international trade law, in the diffusion of technologies for climate change abatement and resilience, and identifies related gaps. It also discusses potential ways of strengthening the interface between international trade law and climate technologies for their rapid and wide diffusion.

In discussing climate technologies, this chapter applies the definition used in the United Nations Framework Convention on Climate Change (UNFCCC) regime, where climate technologies are those technologies that help to reduce greenhouse gas emissions, on the one hand, and adapt to the adverse effects of climate change, on the other.[6] Such technologies can consist of hardware (equipment and products), software (processes associated with production and use of hardware) and 'orgware' (the organization needed

[3] United Nations Environment Programme Copenhagen Climate Centre (UNEP-CCC) and United Nations Framework Convention on Climate Change (UNFCCC) Technology Executive Committee (TEC) 2022: Climate Technology Progress Report. Copenhagen, Denmark.

[4] Gross et al (n 1).

[5] Keith E. Maskus, 'Encouraging International Technology Transfer', ICTSD-UNCTAD issue paper no. 7, May 2004.

[6] UNEP-CCC and UNFCCC (n 3), IV.

for the adoption and diffusion of technology).[7] Further, this chapter follows the terminology suggested by the IPCC relating to technology deployment, diffusion and transfer. Accordingly, technology deployment is 'the act of bringing technology into effective application, involving a set of actors and activities to initiate, facilitate and/or support its implementation'.[8] Technology transfer is defined as 'the exchange of knowledge, hardware and associated software, money and goods among stakeholders, which leads to the spread of technology for adaptation or mitigation'.[9] The spread of technology is referred to as technology diffusion. At the same time, in the context of international climate policy, technology transfer is usually understood in a narrower sense – as the purposeful diffusion of climate-related technologies to developing countries.[10]

The remainder of this chapter is organized as follows. Section 18.2 highlights the latest developments in policies promoting climate technologies and the role of trade and trade measures in climate technology generation, diffusion and transfer. Section 18.3 explores the linkages of climate technology policies with the rules of the World Trade Organization (WTO) and assesses if these rules are fit for purpose. Section 18.4 discusses prospects for accelerating the dissemination of climate technologies through the elimination of tariff and non-tariff barriers to trade in environmental goods and services, and cooperation between the international trade and climate regimes.

18.2 Latest developments in policies promoting the development of climate technologies, their diffusion and transfer

18.2.1 Subsidy race for promotion of climate technologies

Realizing the importance of climate technologies for a sustainable future, countries are taking measures to promote their renewable-energy and low-carbon manufacturing. The biggest economic powers, such as the United States (US), European Union (EU), China and India, are spending billions

[7] Jens Müller, 'Perspectives on Technological Transformation: A Conceptual Framework for Technological Analysis' in John Kuada (ed), *Culture and Technological Transformation in the South* (Samfundslitteratur 2003), cited in UNEP-CCC and UNFCCC (n 3), P. IV.

[8] IPCC 2022: Annex I: Glossary [R. van Diemen et al (eds)] in IPCC 2022: Climate Change 2022: Mitigation of Climate Change. Contribution of Working Group III to the Sixth Assessment Report of the Intergovernmental Panel on Climate Change [P.R. Shukla et al (eds)]. Cambridge University Press, Cambridge, UK and New York, NY, USA. doi: 10.1017/9781009157926.020.

[9] Ibid.

[10] G. Blanco et al, 'Innovation, Technology Development and Transfer' in IPCC 2022 (n 8).

of dollars on renewable-energy subsidies and other support for industrial decarbonization. Besides subsidies to its domestic producers, countries also impose local-content requirements[11] and some other conditions for foreign direct investment to support or even enforce technology transfer.[12] These measures are part of industrial policy, which is not new and was used in the past for the industrialization of postcolonial countries. Today, we witness a return to the massive use of industrial policy, but this time for decarbonization.

One of the most prominent examples of green industrial policy measures is the US Inflation Reduction Act (IRA) adopted by the US Congress in August 2022.[13] This law is meant to support the US leadership in clean technology through securing US$370 billion in investment (mainly subsidies) in order to lower renewable-energy costs for families and small businesses and boost development of clean energy solutions in every sector of the economy. In particular, it aims to stimulate R&D and commercialization of leading-edge technologies, such as carbon capture and storage and clean hydrogen. Moreover, buyers of new electric cars will receive a tax break ('the clean vehicle credit') of up to US$7,500.[14] The IRA is expected to reduce US economy-wide greenhouse gas emissions by at least 40 per cent by 2030.

However, as with many other green industrial policy measures, IRA contains elements of trade protectionism in the form of local-content requirements, which are particularly designed to target Chinese products. For instance, to be eligible for the clean vehicle credit, a car must have undergone final assembly in North America, which is defined as the US, Puerto Rico and the US' two closest regional trade agreement partners: Canada and Mexico. There is also a requirement that a certain percentage of the critical minerals used to make batteries are to be sourced in North America or must come from countries with which the US has a free trade agreement, starting with 40 per cent by 2024 and rising to 80 per cent by 2026. This excludes providers not only from China but also

---

11 Local content requirements prescribe a certain amount of the final value of a good or service that needs to be derived from domestic production in exchange for some benefits (for example, tax exemptions).

12 Weihuan Zhou, Qingjiang Kong and Huiqin Jiang, 'Technology Transfer Under China's Foreign Investment Regime: Does the WTO Provide a Solution?' (2020) 54(3) *Journal of World Trade* 455.

13 The White House 2023: Building a Clean Energy Economy: A Guidebook to the Inflation Reduction Act's Investments in Clean Energy and Climate Action, January 2023.

14 US Department of Energy 2023: Federal Tax Credits for Plug-in Electric and Fuel Cell Electric Vehicles Purchased in 2023 or After <https://fueleconomy.gov/feg/tax2023.shtml> accessed 18 December 2023.

from the EU, with which the US does not have a free trade agreement.[15] The US could have provided non-discriminatory subsidies, which would be more effective and less trade distorting. However, subsidies without local-content requirements likely would not garner sufficient domestic political support to be enacted.[16]

The EU is also pursuing a green industrial policy, which has been significantly expanded since the adoption of the European Green Deal in 2019. The 'Fit For 55' legislative package for the implementation of the Green Deal contains a number of green industrial policy elements, including the revision of the EU emissions trading scheme (ETS) with an expanded Innovation Fund. In 2020–22, the EU adopted a post-COVID recovery plan with its Recovery and Resilience Facility, the 'REPowerEU' plan to diversify away from Russian energy supplies as a reaction to Russia's invasion of Ukraine, the 'NextGenerationEU' Green Bonds, and the Green Deal Industrial Plan announced in 2023, with proposals for a Net-Zero Industry Act, Critical Raw Materials Act and European Hydrogen Bank as its key components.

The EU Green Deal Industrial Plan was announced in response to the US IRA and aims to boost Europe's green industry.[17] Like the US, the EU plans to rely on local-content requirements to ensure that at least 40 per cent of the bloc's demand for clean technology is produced domestically by 2030. It also sets targets for technologies deemed necessary to decarbonize the bloc's economy, a move aimed at preventing the EU from deepening its reliance on third countries, such as China. Among the technologies designated as 'strategic' are solar photovoltaic, onshore and offshore wind, battery and storage, heat pumps and geothermal energy, electrolysers and fuel cells, biomethane, carbon capture and storage, and grid technologies. The EU also aims to localize the bloc's supply of critical minerals and boost renewable-hydrogen production and imports. It further plans an adaptation of EU state-aid rules through exemptions from state-aid notification for

---

15 The US is trying to solve this problem with its closest allies through concluding quasi-free trade agreements on trade in critical materials, like the one it recently concluded with Japan, for which it used the existing loophole in the US legislation that does not contain a definition of a free trade agreement. See Inu Manak, 'Biden Is Rewriting the Rules on Trade—and Americans Should Be Worried', *Foreign Policy*, 28 April 2023 <https://foreignpolicy.com/2023/04/28/ira-trade-ustr-tai-biden-congress-fta/> accessed 18 December 2023.

16 Trade and Public Policy Network (TaPP) 2023: The UK in a World of Green Industrial Strategies. Workshop Summary, 6 February 2023.

17 Federica di Sario, 'Commission Releases Net-Zero Industry Act', *Politico*, 16 March 2023 <www.politico.eu/article/commission-releases-net-zero-industry-act/> accessed 18 December 2023.

hydrogen and electric-vehicle sectors and tax breaks for green investments in strategic sectors to counteract the adverse effects arising from the US IRA.

Moreover, driven by the 2050 carbon-neutrality goal that requires tightening the emissions cap under the EU ETS, the EU has adopted legislation on a carbon border adjustment mechanism (CBAM), placing restrictions on embedded emissions in a range of imported products.[18] Under the CBAM, carbon fees paid through the purchase of emission allowances for each tonne of carbon dioxide from the production of imported products are expected to stimulate emission reductions and adoption of climate technologies in affected industrial sectors of EU trading partners. In addition, revenues from the sale of emission allowances can potentially be earmarked to support transfer of technology to third countries in order to accelerate the reduction of carbon emissions abroad.[19]

Promotion of climate technologies has also become a subject of several bilateral and regional initiatives. In December 2022, the G7 countries announced the launch of a climate club.[20] The G7 climate club aims, among other things, to increase climate policy-related cooperation and partnerships, particularly with developing countries on technology transfer and capacity building. How it will be realized in practice remains to be seen. The G7 climate club contrasts with another idea promoted by the US in trade with the EU. In October 2021, the US initiated negotiations for a global arrangement on steel and aluminium aimed at banning dirty and dumped steel and aluminium from the US and the EU markets. While the US proposed this deal to the EU as the next step toward its removal of import duties on EU steel and aluminium, it promised to keep the arrangement open for all countries committed to 'market orientation and reducing trade in high carbon-steel and aluminum products'.[21] It is therefore not clear whether this arrangement will co-exist or merge with the G7 climate club. There is currently a lack of detail on the arrangement itself and the actual state of the bilateral negotiations. However, given the crucial role of

[18] Regulation (EU) 2023/956 of the European Parliament and of the Council of 10 May 2023 establishing a carbon border adjustment mechanism OJ 2023 L 130/52.

[19] Kateryna Holzer, 'Reflections on the Use of Revenues from the EU CBAM', *CCEEL Blog*, 22 May 2024 <https://sites.uef.fi/cceel/reflections-on-the-use-of-revenues-from-the-eu-cbam/> accessed 5 December 2024.

[20] G7 Germany 2022: Terms of Reference for the Climate Club, 12 December 2022 <www.g7germany.de/g7-en 1> accessed 18 December 2023.

[21] The American Presidency Project 2021: Fact Sheet: The United States and European Union to Negotiate World's First Carbon-Based Sectoral Arrangement on Steel and Aluminum Trade, 31 October 2021 <www.presidency.ucsb.edu/documents/fact-sheet-the-united-states-and-european-union-negotiate-worlds-first-carbon-based> accessed 18 December 2023.

production methods and electricity sources in decarbonization of steel and aluminium, technology cooperation and mutual recognition of production standards are likely to play a central role in the implementation of the deal.

Green industrial policy has been crucial for the generation of new clean technologies in some countries, as it helped to drive down the costs of these technologies. The case of China is illustrative. China's industrial policy has not only led to a considerable increase in the deployment of climate technologies (especially renewable-energy technologies) in China itself, but has also brought global benefits through cost reductions that enable other countries to decarbonize more cheaply and quickly.[22] At the same time, green industrial policy is seen as controversial and contrary to a fair-market approach, which may lead to greater efficiency and innovation in the long run.[23] One example of inefficiency is related to export restrictions. Governments impose export restrictions on critical raw materials, such as natural graphite, cobalt and lithium, to be used in the production of batteries to store electricity from renewable-energy sources. These export restrictions then hinder the expansion of climate-friendly production in other countries.[24] Also, it is not clear how successful the imposition of costs on embedded emissions in imports by the EU CBAM will be in terms of creating a stimulus for third countries to increase the deployment of low-carbon technologies. The CBAM may fail to achieve this goal if its implementation causes resource reshuffling, whereby foreign companies deploy emission-reduction technologies only at those installations which export to the EU, while increasing exports from carbon-intensive installations to countries not imposing carbon restrictions.

18.2.2 Challenges of climate technology transfer and diffusion

While some countries have made huge progress in the deployment of climate technologies, many others are lagging behind. Climate technologies are not distributed equally among countries, as countries differ in their investment, policy support and innovation capabilities, and also in the availability of

[22] Allan Bentley, Joanna I. Lewis and Thomas Oatley, 'Green Industrial Policy and the Global Transformation of Climate Politics' (2021) 21(4) *Global Environmental Politics* 1.

[23] Ruchil Agarwal, 'Industrial Policy and the Growth Strategy Trilemma', *IMF Analytical Series*, 21 March 2023 <www.imf.org/en/Publications/fandd/issues/Series/Analytical-Series/industrial-policy-and-the-growth-strategy-trilemma-ruchir-agarwal> accessed 18 December 2023.

[24] Marion Jansen, 'Raw Materials for the Green Transition: What Is the Role for International Trade?' OECD Forum, 21 April 2023 <www.oecd-forum.org/posts/raw-materials-for-the-green-transition-what-is-the-role-for-international-trade> accessed 18 December 2023.

infrastructure to make use of technologies. The latter affects technological diffusion.[25] Access to climate technologies by all requires technology transfer (including legal rights to reproduce technology) from countries where these technologies are available, and large investment in infrastructure for technology adoption.[26]

Enabling access to climate technologies also requires market facilitation and finance. One of the key barriers discussed in the UNFCCC relates to enabling environments for technology diffusion. Enabling environments consist of 'resources and conditions' generated by various structures (for instance, regulatory frameworks) and institutions.[27] Research highlights the importance of appropriate domestic legislation and policy frameworks promoting climate technologies, and specifically the importance of such regulatory incentives as carbon pricing and low-carbon standards.[28] Yet, the debate on the definition and role of enabling environments is far from being settled in the UNFCCC context. For developed countries, enabling environments should focus exclusively on markets, while developing countries consider this approach as too limiting and emphasize local capacities regardless of their relation to markets.[29]

Insufficient financing and investment are also discussed as barriers to clean technology development and diffusion. The picture, however, is more nuanced when considering the allocation of current finance and investment flows. According to the 2021 State of Climate Technologies report, 'the climate tech market is a rapidly maturing asset class'.[30] However,

---

25 While technology transfer depends on supply of technology, technology diffusion depends on demand for technology. Therefore, transfer of technology and legal rights to produce a new product are constrained by the owner's concern for a return on innovation, whereas diffusion of technology is restricted by willingness and ability to adopt it. See Silvia Weko, Andreas Goldthau and Rainer Quitzow, 'Climate Technology Diffusion and Transfer in the International Trade System', White Paper prepared for the Remaking the Global Trading System for a Sustainable Future Project, 2023 <https://remakingtradeproject.org/white-papers> accessed 18 December 2023.

26 Investment needs of developing countries in this respect are estimated to be US$1 trillion in energy infrastructure by 2030 and up to US$6 trillion across all sectors per year by 2050. See Torsten Ehlers et al, 'How to Scale Up Private Climate Finance in Emerging Economies', IMF Blog, 7 October 2022 <www.imf.org/en/Blogs/Articles/2022/10/07/how-to-scale-up-private-climate-finance-in-emerging-economies> accessed 18 December 2023.

27 UNEP-CCC and UNFCCC (n 3), 2.

28 Xiaomei Tan, 'Clean Technology R&D and Innovation in Emerging Countries: Experience from China' (2010) 38 *Energy Policy* 2916.

29 Heleen de Coninck and Ambuj Sagar, 'Technology Development and Transfer (Article 10)', in Daniel Klein et al (eds), *The Paris Agreement on Climate Change: Analysis and Commentary* (Oxford University Press 2017), 262–3.

30 PWC 2021: State of Climate Tech 2021 <www.pwc.com/gx/en/services/sustainability/publications/state-of-climate-tech.html> accessed 18 December 2023.

at present, financial and investment flows are skewed toward the transport and mobility sector, with fewer resources flowing into other sectors.[31] In terms of geographical distribution, the US is leading for venture capital going to climate tech startups, followed by Europe and China.[32]

Intellectual property rights (IPRs) are often seen as a barrier to technology transfer, especially in the UNFCCC context. However, the literature is ambivalent on this question, emphasizing a dual role of IPRs as both incentives and obstacles.[33] For instance, patent systems have been found to have a positive effect on research and development and to facilitate innovation in certain sectors.[34] At the same time, research also alludes to the proliferation of patents as a barrier to technology transfer.[35] In practice, however, suppliers of technology may still be willing to give up on their IPRs if the benefits to a technology supplier with investment in a foreign country outweigh the potential threat of losing competitive advantage provided by IPRs. For instance, access to the large Chinese market has become sufficient motivation for wind-energy companies from the EU to transfer their technologies to the joint ventures they establish with Chinese firms.[36]

In the UNFCCC context, IPRs are one of the most polarized debates, where developed countries consider strong domestic IPR regimes as a precondition for technological innovation, whereas developing countries view them as 'a manifestation of the technological hegemony of the developed world' and the unwillingness of developed countries to share technologies.[37] As is the case with several other items on the UNFCCC agenda, progress in finding a common ground on IPRs has been close to nil, and politics has so far prevailed over evidence-based discussions.

Last, geopolitical tensions and rivalries among major economies can also have a significant impact on clean technology transfer. However, the degree and scope of this impact remain debated, for instance in relation to renewable energy.[38] The US-China trade disagreements are the most illustrative of the nexus between geopolitics (or, rather, geoeconomics[39]) and (climate)

[31] Ibid.

[32] Ibid.

[33] IPCC 2022 (n 8), ch. 16.

[34] Blanco et al (n 10), 1681.

[35] Ibid.

[36] Weko et al (n 24).

[37] de Coninck and Sagar (n 28), 262.

[38] Roman Vakulchuk, Indra Overland and Daniel Scholten, 'Renewable Energy and Geopolitics: A Review' (2020) 122 *Renewable and Sustainable Energy Reviews* 109547.

[39] Henrique Choer Moraes and Mikael Wigell, 'The Emergence of Strategic Capitalism: Geoeconomics, Corporate Statecraft and the Repurposing of the Global Economy', FIIA Working Paper, September 2020.

technology. The competition of those two countries for global technological leadership has been described as 'a bilateral tech war'.[40] Technology transfer is at the heart of this. In constant rivalry on innovation-driven economic development with the US and other developed countries, China has been practising 'forced technology transfer' by forcing foreign firms investing in China to transfer technologies. This is done through various restrictions and discrimination on foreign investment.[41] Partly for this reason, the US started a trade war with China by raising tariff rates on Chinese products in 2018; and initiated an IPR-related dispute against China at the WTO, which was followed by a case against China filed by the EU.[42]

The challenges to climate technology diffusion and transfer are, therefore, numerous. Furthermore, these challenges are a highly politicized subject in the UN climate change negotiations, as developed and developing countries mostly disagree on what hinders climate technology transfer and diffusion, and what support is required for developing countries in this respect.

18.3 Linkages with trade law

As discussed earlier, national policies and legislation play a pivotal role in technology development and transfer. However, access to technology also depends on international cooperation and agreements, including those related to trade. International trade law helps to address regulatory diversity, overcome unnecessary trade barriers and promote investment in climate technologies.[43]

18.3.1 WTO rules applying to green industrial policy measures

Industrial policies aimed at promoting climate technologies are constrained by WTO rules on subsidies. If such policies foresee local-content requirements (LCRs), they also come into tension with WTO non-discrimination rules, particularly the national treatment principle under the General Agreement on Tariffs and Trade (GATT) and the Agreement on Trade-Related Aspects of Investment Measures (TRIMs). That LCRs are illegal under these rules was

40 Zhou et al (n 12), 479.

41 Ibid, 464–73.

42 The US brought a dispute to the WTO over China's inadequate protection of IPRs in *China–IPRs II* (DS542) in March 2018, whereas the EU followed suit by initiating a separate WTO litigation in *China–Transfer of Technology* (DS549) in June 2018. Both disputes are currently suspended.

43 Thomas Cottier, 'Technology and the Law of International Trade Regulation', in Roger Brownsword, Eloise Scotford and Karen Yeung (eds), *The Oxford Handbook of Law, Regulation and Technology* (Oxford University Press 2017), 1017–51.

confirmed in WTO disputes on several occasions.[44] However, it is not clear whether LCRs can be justified under environmental or security exceptions. This issue has never been addressed in WTO jurisprudence. In our view, the mix of economic (promoting domestic industries) and environmental (fighting climate change) motives behind such measures as the US IRA is likely to make it difficult to defend LCRs under environmental exceptions to GATT rules.

At the same time, whether LCRs used for the promotion of climate and other environmental technologies should be allowed under WTO law is not straightforward. Although LCRs can bring short-term benefits for domestic clean technology industries and increase public support for renewable energy, they are not considered to be an effective instrument for the promotion of a green economy.[45] Nevertheless, the calls for the temporal accommodation of LCRs under WTO law through the adoption of waivers (that is, temporal permissions) might be considered in those cases, where these measures are effective in scaling up climate technologies.[46]

TRIMs rules also pose obstacles for the technology-related requirements imposed on investors under some national foreign direct investment regimes. Such requirements often include preferential treatment to foreign investors with advanced technology and know-how, as well as various requirements attaching to imported technologies in the interest of the importing country.[47] Illegality of such requirements under WTO law can potentially become an issue in the context of cooperation on mitigation outcomes under Article 6 of the Paris Agreement. In the past, the issue of inconsistency with TRIMs provisions was raised in connection with Clean Development Mechanism (CDM) projects under the Kyoto Protocol. These projects were supposed to contribute to sustainable development in the host country by, among other things, facilitating the transfer of technology. The adoption of LCRs, therefore, was the perfect solution to achieve the development goals of CDM, but, at the same time, it constituted a prohibited performance

---

44 See Appellate Body Reports in *Canada-Certain Measures Affecting the Renewable Energy Generator Sector* (DS412) and *India-Certain Measures Relating to Solar Cells and Solar Methods* (DS456).

45 Simon Lester, for instance, points out that '[i]f every sub-national entity were trying to promote local production of these products, we would end up with thousands of small, inefficient companies, and we could never make these products in a way that is affordable for the masses'. See Simon Lester, 'Some Comments on a New Foreign Affairs Piece Discussing Trade and Climate Change', IELP Blog, 29 November 2021 <https://ielp.worldtradelaw.net/green-steel-deal/> accessed 19 December 2023.

46 Mahatab Uddin, *Climate Change Law, Technology Transfer and Sustainable Development* (Routledge 2021), 169–70.

47 Zhou et al (n 12).

requirement under TRIMs rules.[48] Negotiations and clarifications on this point are, therefore, urgently needed, including the assessment of the possibility of 'technology transfer' as an essential requirement for investing in green industrial sectors.

But even without LCRs, many industrial policy measures are in tension with WTO subsidy rules. Subsidies provided by governments to producers associated with climate technologies are actionable, that is, they can either be challenged in the WTO dispute settlement or countervailed unilaterally. This is because such subsidies may create adverse effects for the industries of other countries. Moreover, such subsidies are often viewed as 'green protectionism' which is affordable only for rich countries.[49]

Even though support schemes for renewable energy raise questions about fair competition and efficiency, it is subsidies for fossil fuels that create a major obstacle to the deployment of climate technologies.[50] In contrast to renewable-energy subsidies, fossil-fuel subsidies are not captured by WTO subsidy rules. Fossil-fuel subsidies are generally provided to energy consumers across the board, making it difficult to meet the specificity requirement for a subsidy to be captured by WTO disciplines. It is, therefore, widely acknowledged that the rules contained in the WTO Agreement on Subsidies and Countervailing Measures adopted in the era of fossil-fuel promotion cannot be used for the purpose of green transition.[51]

The deficiency of WTO subsidy rules, as well as uncertainty about the outcome of WTO disputes, push other countries to act unilaterally

---

[48] In the past, the issue of inconsistency with TRIMS provisions was raised in connection to CDM projects under the Kyoto Protocol. These projects were supposed to contribute to sustainable development in the host country by increasing the financial returns of local producers, providing a positive impact on the balance of payments of the host country or facilitating the transfer of technology. The adoption of LCRs, therefore, was a perfect solution for the development purposes of CDM but, at the same time, it constituted a prohibited performance requirement in light of the TRIMS Agreement. See Stefan Rechsteiner, Christa Pfister and Fabian Martens, 'TRIMS and the Clean Development Mechanism: Potential Conflicts', in Thomas Cottier, Olga Nartova and Sadeq Z. Bigdeli (eds), *International Trade Regulation and the Mitigation of Climate Change. World Trade Forum* (Cambridge University Press 2010).

[49] Joe Lo, 'IMF Warns Against "Protectionism" in Rich World's Green Subsidies', *Climate Home News*, 1 March 2023 <https://climatechangenews.com/2023/03/01/imf-warns-against-protectionism-in-rich-worlds-green-subsidies/> accessed 19 December 2023.

[50] Harro Van Asselt, 'The SDGs and Fossil Fuel Subsidy Reform' (2023) 23 *International Environmental Agreements: Politics, Law and Economics* 191.

[51] Simon Happersberger, Eleanor Mateo and Selcukhan Ünekbas, 'How to Rein in Fossil Fuel Subsidies? Towards a New WTO Regime' (Geneva Graduate Institute 2021) <www.graduateinstitute.ch/library/publications-institute/how-rein-fossil-fuel-subsidies-towards-new-wto-regime> accessed 19 December 2023.

against foreign subsidies. As a result, subsidies provoke counter-subsidies and countervailing measures, whereby countries are engaged in levelling the playing field for their domestic producers unilaterally, through the imposition of countervailing duties. The US has the most developed practice of countervailing duties, whereas the EU has, until recently, used a multilateral approach to responding to foreign subsidies through complaints in the WTO dispute settlement system. The EU was one of the first countries to bring a dispute over renewable-energy support measures to the WTO.[52] However, the multilateral track has failed to address competitiveness concerns arising from the green subsidy race. The WTO adjudicative bodies have been unable to decide whether feed-in-tariff schemes constitute a subsidy in the first place, to say nothing about their effects on trade. This failure of adjudication coupled with the currently handicapped dispute settlement process due to the dysfunctional appellate review have pushed the EU to follow the US example in taking justice into its own hands by imposing countervailing duties.[53] However, the use of countervailing duties by the EU can, itself, be controversial under WTO rules. Contrary to the usual practice, the EU plans to countervail in cases of subsidies provided by one country to companies registered in another country.[54] Countervailing transnational subsidies is a new phenomenon and it is not entirely clear whether the WTO subsidy rules apply where the subsidizing entity is not within the territory of the member whose goods are allegedly being subsidized.[55]

All these issues highlight the need for reform of WTO subsidy rules, for instance through adding a specific exemption for subsidies supporting climate technologies in the Agreement on Subsidies and Countervailing Measures. However, such an amendment would effectively require the consensus of WTO members, which is unlikely to happen in the foreseeable future due to the current crisis of multilateral negotiations at the WTO and the increasing geopolitical antagonism.

52 *Canada-Feed-in-Tariff Program* (DS426).

53 See Regulation (EU) 2022/2560 of the European Parliament and of the Council of 14 December 2022 on foreign subsidies distorting the internal market.

54 William Alan Reinsch and Sparsha Janardhan, 'Crossing the Line: Transnational Subsidy', CSIS, 14 January 2022 <www.csis.org/analysis/crossing-line-transnational-subsidy> accessed 19 December 2023.

55 This might capture financial assistance of the Chinese government to companies registered in industrial zones established by China on the territory of third countries as part of the Belt and Road Initiative. Export of products from these industrial zones that have benefited from assistance from the Chinese government may be subject to anti-subsidy investigations. See ibid.

18.3.2 WTO rules on protection of intellectual property

Protection of intellectual property that, as mentioned earlier, has an impact on both the development (innovation) and transfer of technology is regulated by WTO rules to the extent it affects trade. These rules are contained in the WTO Agreement on Trade-related Aspects of Intellectual Property Rights (TRIPS). The effects of these rules in the case of technology development and in the case of technology transfer differ. For the former they act as a stimulus, because they impose an obligation to maintain a minimum level of intellectual property protection and enforcement of IPRs,[56] whereas for the latter they can constitute a barrier.

Thus, the system of intellectual property protection is built on a vital trade-off between the need to stimulate innovation by protecting the competitive advantage of the inventor, on the one hand, and the access to inventions in the public interest, on the other. It is true that patent protection and the rights of patent holders restrict a free flow of climate technologies, preventing others from making, using or selling the invention without permission. At the same time, the TRIPS Agreement contains provisions encouraging technology transfer, albeit without enforcement. For instance, in setting the objectives, TRIPS Article 7 states: 'The protection and enforcement of intellectual property rights should contribute to the promotion of technological innovation and to the transfer and dissemination of technology.' Further, in support of technology transfer, TRIPS Article 66(2) imposes an obligation on developed countries to promote and encourage technology transfer to least-developed countries (LDCs). Yet, this obligation is of a vague and recommendatory nature, whereas the rest of TRIPS' rules are not supportive of it.

Attempts have been made by WTO members to work out an implementation mechanism for technology transfer from developed to developing countries, including regular monitoring and reporting requirements.[57] This mechanism has failed to produce tangible results so far.[58] On the contrary, developed countries often use the possibility of including 'grant-back' clauses (which are not prohibited) in licensing agreements, thereby preserving control and

---

[56] For instance, under Art 33 of the TRIPS Agreement, the patent protection is conferred for 20 years from the date of filing a patent application.

[57] In 2001, WTO members established a working group on trade and transfer of technology, to which developed countries were obliged to report on technology transfer issues annually. See WTO 2023: Working Group on Trade and Transfer of Technology <www.wto.org/english/tratop_e/devel_e/dev_wkgp_trade_transfer_technology_e.htm> accessed 19 December 2023.

[58] Maria Anna Corvalglia, 'South-South Technology Transfer Addressing Climate Change and its (Missing) International Regulatory Framework' (2014) 2 *Carbon and Climate Law Review* 125.

avoiding competition with their licensees in case of the development of a superior product at a later stage.[59]

The need to accelerate climate technology transfer calls for introducing flexibilities in the TRIPS Agreement.[60] The main instrument for such flexibility could be compulsory licensing, which can be provided in the case of patent protection under the TRIPS compulsory licence provision (Article 31). Compulsory licensing entitles the government to issue a licence for a certain patented technology without the consent of the holder of the patent but subject to certain conditions and adequate compensation to the patent holder.[61] The problem is that a compulsory licence can only be given for the supply in the domestic market and cannot be used for export.

A solution to this limit could be the adoption of a waiver for exports related to environmental technology under compulsory licensing. Similar waivers have been adopted by WTO members for the exports of medicines and COVID-19 vaccines, allowing exports of these products to developing countries under compulsory licensing.[62] The effect of such a waiver on climate technology transfer will depend on the availability of already-existing infrastructure in technology-receiving countries that will be able to facilitate an increase in manufacturing capacity.[63] The adoption of such a waiver requires political support in WTO member countries producing these technologies, which is very hard to get, as the experience to date with adopted waivers shows.

18.4 Opportunities to promote climate technology transfer within the international trade system

Given that reforms of the international trade regime take time, finding ways to expeditiously facilitate access to climate technologies without changing international trade rules seem more pragmatic.

---

59 Bernard Colas, 'Transfer of Clean Technologies from North to South: Legal Barriers and Mitigations', Policy Brief no. 91, Center for International Governance Innovation, 2016.

60 Uddin (n 45).

61 Nils Meyer-Ohlendorf and Christiane Gerstetter, 'Trade and Climate Change: Triggers or Barriers for Climate Friendly Technology Transfer and Development?' Occasional Paper no. 41, Friedrich Ebert Stiftung, February 2009.

62 The WTO members adopted a waiver for medicines under the Doha Declaration on TRIPS in 2003 and a waiver for COVID-19 vaccines in 2022. See Decision of the General Council of 30 August 2003 'Implementation of paragraph 6 of the Doha Declaration on the TRIPS Agreement and public health' and Ministerial Decision on the TRIPS Agreement of 17 June 2022, WT/MIN(22)/W/15/Rev.2.

63 Sarah Gabrielle, 'What Happened to the COVID-19 Vaccine Patent Waiver?' Bill of Health Blog, 8 November 2022 <https://blog.petrieflom.law.harvard.edu/2022/11/08/what-happened-to-the-covid-19-vaccine-patent-waiver/> accessed 19 December 2023.

18.4.1 Eliminating barriers to trade in environmental goods and services

Removal of trade barriers to environmental goods and services (EGS) is a feasible way forward. Without requiring changes in the text of the WTO Agreement, a deal among countries on elimination of tariff and non-tariff barriers for EGS would support the achievement of the countries' Nationally Determined Contributions under the Paris Agreement and their resilience to climate change. This is because the list of EGS subject to removal of trade barriers might include renewable-energy equipment, energy- and resource-efficient goods and services, drought-resistant crops, early-warning weather systems, water-conservation and storage systems, and other goods and services vital to the development and diffusion of climate technologies. Liberalization of trade in this category of goods and services would facilitate development, transfer and diffusion of climate technologies by reducing their price and hence improving their affordability.

It should be noted, however, that because of a simultaneous increase in the energy consumption resulting from trade, the net effect for climate change mitigation from liberalization of trade in EGS would not be that big. This would enable only a 0.58 per cent reduction in carbon emissions relative to the baseline in 2030.[64] Yet, this strategy is still worth pursuing to facilitate mitigation, and especially resilience to climate change impacts.

The idea of liberalizing trade in EGS is not new. Attempts to negotiate an agreement among WTO members on eliminating tariffs on environmental goods were made in the past but failed.[65] They failed primarily because of a disagreement among negotiating parties over products to be included in the list subject to tariff elimination, as there was no consensus on the definition of an environmental good.[66] Environmental goods do not represent a separate category of products in the Harmonized System of tariff nomenclature used as

---

[64] WTO 2022, 'Leveraging Trade in Environmental Goods and Services to Tackle Climate Change', Policy Brief, World Trade Organization, Geneva <www.wto.org/english/tratop_e/envir_e/policy_brief_environmental_goods_e.pdf> accessed 20 December 2023.

[65] Such negotiations were held at the WTO on a plurilateral basis during 2014–16. China, the US and the EU were part of the negotiating group of WTO members. See Mark Wu, 'The WTO Environmental Goods Agreement: From Multilateralism to Plurilateralism', in Panagiotis Delimatsis, *Research Handbook on Climate Change and Trade Law* (Edward Elgar 2016).

[66] The list first included 54 goods but then was expanded to 300 goods. Whether to include bicycles in the list of environmental products exempted from import duties was a major point of contention between China, which pushed for it, and the US and the EU, which opposed it. See Maureen Hinman, 'Remaking the World Trade Organization: Environmental Goods', White Paper, Remaking Trade for a Sustainable Future Project, September 2022, at 4–5 <https://silverado.org/news/paper-remaking-the-world-trade-organization-environmental-goods/> accessed 20 December 2023, 4–5.

a reference for customs clearance. Moreover, many of these goods are dual-use goods, which means that they can be used for both environmental and non-environmental purposes, resulting finally in free trade for everything.[67] For many WTO members this is unacceptable, as the aim to get access to foreign environmental technologies by some countries does not necessarily match with the aim to get market access for domestic environmental goods by others.[68] Given that mainly rich countries produce technologies, poorer countries are afraid that the liberalization of EGS trade will not deliver any gain to them.[69]

The increasing sense today that the WTO could and should do more to support climate action creates momentum for relaunching the negotiations on liberalization of EGS trade with the extended content of the deal that might also include non-tariff barriers and environmental services, and the extended number of participating WTO members.[70] Once a critical mass of participants in terms of the market share in EGS is achieved, the duty-free commitments of participants could be extended to all WTO members, including those which did not participate in the agreement.[71] A highly contentious debate on the definition of EGS could be avoided by ensuring that EGS in the interest of developed countries and those in the interest of developing countries are included in the list in equal proportions.

A number of recent institutional and political developments seem to be favourable to the revival of negotiations on EGS at the WTO. First, the 2022 edition of the Harmonized System Nomenclature of the World Customs

67 A cast iron pipe, for instance, which is used for wastewater treatment, can also be used for transportation of natural gas. See Wu (n 64), 285.

68 James Bacchus and Inu Manak, 'Free Trade in Environmental Goods Will Increase Access to Green Tech', Cato Institute Free Trade Bulletin no. 80, 8 June 2021, at 4 <www.cato.org/sites/cato.org/files/2021-06/free-trade-bulletin-80.pdf> accessed 20 December 2023, with reference to Maureen Hinman, 'Green Goods Market Access Is Key to Biden's Build Back Better Agenda', Washington International Trade Association, 11 March 2021.

69 This however is not necessarily the case. The liberalization of EGS trade can result in a win-win outcome for both exporting (developed) and importing (developing) countries, if the list of EGS is defined broadly enough to include goods and services of export interest to developing countries (for example, those from the sectors of sustainable agriculture and fisheries, sustainable forest management, biodiversity and sustainable tourism activities). See Meyer-Ohlendorf and Gerstetter (n 60), 10.

70 Debate on addressing pressing environmental issues through trade mechanisms and rules has been intensified at WTO with the launch of three plurilateral groups for discussions on trade and sustainability, plastic pollution and fossil fuel subsidy reform in December 2021. See WTO 2021: New Initiatives Seek to Put Environment at Heart of Trade Discussions <www.wto.org/english/news_e/news21_e/envir_15dec21_e.htm> accessed 20 December 2023.

71 The 2014–16 negotiations of the failed Environmental Goods Agreement included 46 WTO members representing nearly 90 per cent of global trade in environmental goods. See Bacchus and Manak (n 68), 2.

Organization includes new commodity codes specific to technologies that use solar energy and energy-efficient light-emitting diodes. It can be seen as a significant step toward facilitating negotiations on removal of trade barriers for climate technologies.[72] Second, discussions on EGS are currently gaining traction at the WTO within the Trade and Environmental Sustainability Structured Discussions (TESSD) held in a group of 75 WTO members. Although plurilateral, these discussions may lead to a deal that can be extended to all WTO members. Third, progress in liberalization of EGS has been achieved under regional trade agreements (RTAs).[73]

Some RTAs already include commitments to liberalize trade in EGS. An example worth mentioning is the 2013 free trade agreement between New Zealand and Taiwan, which covers the elimination of all tariffs on 132 environmental goods immediately upon the entry into force of the agreement, as well as the movement of businesspersons involved in the supply of environmental services. The 2022 free trade agreement between New Zealand and the UK goes even further by eliminating tariffs on 293 environmental goods, which is the largest list agreed upon so far. However, these commitments to eliminate tariffs on environmental goods are of a non-binding character because they are part of trade and sustainable development chapters (as opposed to chapters on trade in goods and services) that are not subject to dispute settlement mechanisms under the agreements. Discussions to change the traditional practice to make provisions of trade and sustainable development chapters legally enforceable are currently under way, at least in the EU.[74]

As the conclusion of RTAs automatically leads to the removal of tariffs on environmental goods by the very nature of these agreements requiring the elimination of tariffs (at least for industrial goods) in trade among the parties,[75] the main focus of EGS negotiations under RTAs should be on the elimination of non-tariff barriers. Negotiations on non-tariff barriers under RTAs should particularly cover the use of anti-dumping and countervailing

---

[72] WTO (n 63), 11.

[73] Clara Brandi et al, 'Taking Climate Change Seriously in the Design of Trade Agreements', in Kathleen Claussen, Manfred Elsig and Rodrigo Polanco (eds), *The Concept Design of 21 Century Trade Agreements* (Cambridge University Press, 2025).

[74] European Commission 2022: Commission Unveils New Approach to Trade Agreements to Promote Green and Just Growth, Press Release, 22 June 2022 <https://ec.europa.eu/commission/presscorner/detail/en/ip_22_3921> accessed 20 December 2023.

[75] According to GATT Art XXIV.8 (b), a free trade area (a type of RTA referred to in the GATT) 'should be understood to mean a group of two or more customs territories in which the duties and other restrictive regulations of commerce … are eliminated on substantially all the trade between the constituent territories in products originating in such territories'.

duties, which often makes the elimination of tariffs useless. For example, anti-dumping and countervailing duties imposed on imports of solar panels and wind turbines make up 100 per cent of the value of these products.[76] RTA negotiations can also address the problem of diversity of climate technology standards, which is particularly in the interest of developing-country parties to the WTO, as they are most affected by challenges of standard compliance. RTAs can also be useful for the liberalization of trade in environmental services, which are a necessary supplement to the movement of environmental goods and therefore essential for the diffusion of climate technologies.

Finally, to achieve a breakthrough in the liberalization of trade in EGS, EGS negotiations need to include the issues of technology transfer and technical assistance in connection to EGS production. This would make such negotiations more attractive for developing countries that struggle to understand the benefits from promoting trade in EGS. As the experience under RTAs shows, developing countries can be interested in strengthening technology transfer and capacity building by means of grants, development funds or other financial instruments.[77] Financial assistance to developing countries and producers from these countries is also needed to support their compliance with sustainability standards and development of a well-functioning domestic quality infrastructure system.[78]

18.4.2 Increasing cooperation with the international climate regime

Technology development and transfer have been an important element in the UNFCCC regime. Under the Convention's Article 4.1(c), all parties have a commitment to 'promote and cooperate in the development, application and diffusion, including transfer' of climate mitigation technologies. In addition, the Convention mandates Annex II parties to provide new and additional financial resources, 'including for the transfer of technology', as well as technology assistance to developing countries to enable them to implement the Convention: Articles 4(3) and 4(5).

---

76 World Trade Report 2022: Climate Change and International Trade, World Trade Organization, Geneva, 2022.

77 See, for example, chapter 9 on cooperation and capacity building of the 2019 EFTA-Indonesia free trade agreement <www.efta.int/sites/default/files/documents/legal-texts/free-trade-relations/indonesia/efta-indonesia-main-agreement.pdf> accessed 5 December 2024.

78 Aik Hoe Lim and Kateryna Holzer, 'Trading in the Era of Carbon Standards: How Can Trade, Standard Setting and Climate Regimes Cooperate?' (2023) 39(1) *Oxford Review of Economic Policy* 110.

The Kyoto Protocol integrated and enhanced the provisions of the Convention on technology development and transfer. For instance, as a step beyond the Convention, the Protocol covers both mitigation and adaptation technologies: Article 10(c); clarifies and strengthens the scope of the commitment under Article 4(1)(c); and emphasizes 'environmentally sound technologies' as opposed to just any technologies.[79] Technology development and transfer is tied to the provision of financial resources. Yet, these resources, provided through the Global Environmental Facility and the Special Climate Change Fund, are minuscule.[80]

Another element of the Protocol's infrastructure, which was relevant to technology transfer to developing countries while the Kyoto Protocol was still important, was the CDM. According to UNFCCC data for projects registered in the project pipeline by 2010, 30 per cent of all projects, accounting for 48 per cent of estimated emission reductions, involved technology transfer.[81] No official assessments are available for the projects registered after that year.

Technology work under the UNFCCC received a new impetus with the establishment of the Technology Mechanism in 2010.[82] The Mechanism consists of two bodies – the Technology Executive Committee (a policy arm) and the Climate Technology Centre and Network (CTCN) (an implementation arm). The Committee is responsible for providing policy guidance to countries on climate technology actions. Consisting of 20 experts from developed and developing countries, it meets several times a year and prepares regular reports for the COP. The Technology Executive Committee conducts its work through task forces, which include committee members and representatives of civil society, the private sector, international organizations and academia.[83] The CTCN aims at the implementation of climate technology projects through providing expert assistance to developing countries, knowledge exchange and strengthening stakeholder collaboration. It largely relies on donor contributions, limiting the funding available for project implementation.

---

79 Farhana Yamin and Joanna Depledge, *The International Climate Change Regime: A Guide to Rules, Institutions and Procedures* (Cambridge University Press 2004) 306–7.

80 Based on the information provided by the UNFCCC Climate Finance Data Portal <https://unfccc.int/climatefinance?home> accessed 21 December 2023.

81 UNFCCC Report: The Contribution of the Clean Development Mechanism under the Kyoto Protocol to Technology Transfer, 2010.

82 Decision 1/CP.16, The Cancun Agreements, FCCC/CP/2010/7/Add.1 (15 March 2011), paras 113–29.

83 Stephen Minas, 'The Paris Agreement's Technology Framework and the Need for "Transformational Change"' (2021) 14 *Carbon & Climate Law Review* 241, 248.

The Paris Agreement elevated the importance of the technology theme in the overall climate change regime through its creation of the Technology Framework to guide the already established Technology Mechanism (Article 10).[84] Although this Article can be seen as strengthening the workstream on technology development and transfer, its provisions do not contain binding commitments and are mostly worded in a vague language.[85] The Technology Framework was given further substance through subsequent rule development finalized in 2018. It now operates under the principles of 'coherence, inclusiveness, results-oriented approach, transformational approach, and transparency'[86] and focuses on innovation, implementation, enabling environment and capacity building, collaboration and stakeholder engagement, and support.[87] Although not binding on the technology side, the Paris Agreement fares better on the accountability side compared to previous treaties: technology, and in particular 'the effectiveness and adequacy of the support provided to the Technology Mechanism'[88] is part of the Global Stocktake, providing for a regular review of progress and increase of ambition.[89]

An important part of the technology workstream under the UNFCCC is the process of identifying national technology needs through Technology Needs Assessments. More than 90 developing countries have conducted this process between 2009 and 2017.[90] They identified a wide range of climate technologies as essential. On the climate mitigation side, prioritized technologies for instance include: solar PV, hydropower, mass rapid transit road or rail systems, new agricultural practices such as fertilizer dosing and irrigation techniques, as well as technologies related to forest conservation, afforestation, reforestation and forest rehabilitation and restoration. On the adaptation side, countries identified needs in the following technologies: dykes and floodwalls, sprinkler and drip irrigation systems, drought-resistant crop varieties, rainwater harvesting, water storage and catchment, as well as technologies related to coastal protection, wetland restoration, and natural-disaster prevention, such as early-warning systems.[91]

---

84 Ibid, 245.

85 de Coninck and Sagar (n 28), 266.

86 Decision 15/CMA.1, Annex, para 3.

87 Ibid, para 4.

88 Decision 1/CP.21, para 69.

89 de Coninck and Sagar (n 28), 266.

90 Fourth synthesis of technology needs identified by Parties not included in Annex I to the Convention, Report by the Secretariat, FCCC/SBI/2020/INF.1 (3 April 2020).

91 Ibid, pp 17–20.

These technologies are already available; it is their diffusion to developing countries that is missing.

This overview of technology provisions and governance under the UNFCCC and Paris Agreement demonstrates that the international climate regime has developed several important elements to facilitate the transfer of climate technologies to developing countries, including the bottom-up process of identifying technology needs, institutional structures for policy guidance, knowledge exchange and technical assistance, and provisions on financing. However, these can hardly be described as substantiated and impactful. Relating to the topic of the chapter, the discussions under the Technology Mechanism have not so far addressed the role that international trade law plays in the successful development and transfer of climate technologies. More generally, there has been little discussion of the nexus between climate technologies and trade law in the negotiations, except IPRs, owing – no doubt – to underlying political and economic disagreements about many dimensions of technology development, diffusion and transfer. It is fair to say that the international climate regime in its current form is deeply inadequate to the challenge of inducing a large-scale development and transfer of climate technologies.

Against this inadequacy, what are the needed reforms and how feasible are they? The topic could be introduced into the work of the Technology Executive Committee to initiate a series of thematic dialogues or workshops and other events. However, this is only a hypothetical possibility as the charged nature of the surrounding debate would not allow for a productive discussion within the already polarized climate change regime. Given that international trade, including in climate technologies, is under the scope of the WTO and not the UNFCCC, any work in this direction should involve strengthening cooperation between these two international regimes.

The UNFCCC through its Technology Executive Committee and the WTO through its Committee on Trade and Environment (CTE) could establish a cooperation platform. This platform would form a working group responsible for the development of a facilitation mechanism on technology transfer in support of the goals of the Paris Agreement. The CTE is in charge of enhancing policy coordination and understanding of the relationship between trade and environment policies, and cooperation between the WTO and multilateral environmental agreements, including the UNFCCC. CTE staff members have gained valuable expertise from assisting the negotiations on an environmental goods agreement. This expertise can be used for joint work on climate technology transfer with the UNFCCC. Moreover, a UNFCCC-WTO cooperation on technology transfer can be extended to include the World Intellectual Property Organization (WIPO). The model for such cooperation is provided by joint work between the World Health Organization, WIPO and the WTO on public health, intellectual property and trade, which was instrumental in improving access to COVID-19 vaccines, therapeutics, diagnostics and other health technologies.[92]

18.5 Conclusion

Rapid and wide technological transformation is critical to achieve the climate goals set by the Paris Agreement. Given the urgency of climate action and the considerable time lag between invention and commercialization of a technology, the main effort should be directed at the deployment and transfer of already-existing climate technologies for their global diffusion. As technology deployment and transfer policies rely to a large extent on trade measures, in this chapter, we explored the linkages between climate technology deployment and transfer and the international trade regime.

Industrial policy measures aimed at scaling up climate technologies are in tension with WTO non-discrimination and subsidy rules. These measures are either challenged in WTO dispute settlement or counteracted unilaterally through the imposition of countervailing duties. At the same time, support measures that encourage fossil-fuel production, slowing progress with the deployment of climate technologies, are not captured by WTO subsidy rules. Moreover, existing flexibilities in the WTO rules on the protection of IPRs are not sufficient to accelerate transfer of climate technologies to developing countries, where these technologies are unavailable. Overall, our conclusion is that the current state of international trade law is not fit to address the challenge of rapid technological transformation necessary to achieve climate goals. However, changing the regime is deemed politically infeasible, at least for the foreseeable future.

Therefore, we propose ways in which climate technologies can be promoted within the international trade system without amending its rules. This discussion is based on our pragmatic considerations of what is possible to achieve in the context of current geopolitical and geoeconomic rivalries and the state of WTO negotiations. The first option we discuss relates to the elimination of trade barriers for EGS, both in the multilateral and preferential trade. Momentum exists for relaunching the negotiations of a plurilateral agreement that could cover not only a removal of tariffs for environmental goods but also non-tariff barriers to EGS. Liberalization of trade in EGS would reduce the price and increase affordability of climate technologies, speeding up their deployment and diffusion.

---

[92] WTO: Trilateral Cooperation on Public Health, Trade and Intellectual Property <www.wto.org/english/tratop_e/trips_e/who_wipo_wto_e.htm> accessed 21 December 2023.

We also discuss how climate technology transfer can be facilitated through trade- and climate-regime cooperation. This could take the format of a cooperation platform established by the UNFCCC Technology Executive Committee and the WTO CTE, and involve WIPO. Such a platform can be modelled after the example of joint work between the World Health Organization, WIPO and WTO on public health, intellectual property and trade.

19

Debating the Architecture of the International Climate Regime and its Potential for the Regulation of Technology

Alberto Quintavalla and Leonie Reins

19.1 Introduction

The starting point of the present chapter is a technological dilemma that is typical of the Anthropocene: technology is seen as a potential solution to climate change, but it is also associated with greenhouse gas (GHG) emissions, with environmental harm and with danger to human health. Climate change technology is developing rapidly, and its impacts are profound and diffuse, which calls for transboundary regulation. At the same time, regulators are finding it difficult to remain abreast of technological progress.[1] The key regulatory challenges of the Anthropocene become evident from a headline that CNBC ran in February 2023: 'After Mexican government cracked down on solar geoengineering, startup lets off balloons in Nevada.'[2] The US startup had launched balloons that contained 10 grams of sulfur dioxide in order to conduct various solar geoengineering tests. Sulfur dioxide can negatively affect the ozone layer, leading to acid rain and harm to human

---

1 L Reins, 'Legal Principles and Technology at the Intersection of Energy, Climate, and Environmental Law' in O Kanevskaia, P Pałka and B Brożek (eds), *Research Handbook on Law and Technology* (Edward Elgar Publishing 2024).

2 C Clifford, 'After Mexican Government Cracked Down on Solar Geoengineering, Startup Lets Off Balloons in Nevada' (*CNBC*, 22 February 2023) <www.cnbc.com/2023/02/22/solar-geoengineering-startup-make-sunsets-lets-off-balloons-in-nevada.html> accessed 12 January 2024.

health. The Mexican government had cracked down on this use of balloons, and the startup, which at the time was based in Mexico, responded by moving across the border to the US, where it could continue its geoengineering testing activities. Such experiments, if their scale is sufficiently large, can affect several states or even the entire planet. As the CNBC story shows, national regulation is likely to prove impotent in such instances. What of international law? Although the number of regulations and treaties in the field has increased, key legal challenges of the Anthropocene remain: 'It is hard to ignore the fact that IEL [international environmental law] remains unable to achieve deep structural reforms because it lacks normative ambition at a time when precisely as high as possible a level of such ambition is urgently required'.[3] The norms of international environmental law are path dependent,[4] and a normative shift is needed if the law is to prevent humanity from encroaching on planetary boundaries.[5] In 2014, the United Nations admitted that 'environmental legislation has failed to protect the basic structure and integrity of the Earth's ecosystems'.[6] Non-state actors, such as the headlined startup company, are beginning to shape a domain that has traditionally been state-centric, at least insofar as lawmaking is concerned (see Section 19.2.2).

This chapter is an attempt to answer the following question: Is the current architecture of international climate and environmental law suitable for the regulatory problems that climate-related technologies pose? In this research inquiry, 'climate-related technologies' follows and adapts the definition used in the United Nations Framework Convention on Climate Change (UNFCCC) regime which is encompassing of all the various technological applications associated with climate change in both a positive (that is, aimed to the reduction of GHG emissions or to the adaptation of the adverse effects)[7] but also in a negative sense (leading to an increase of GHG emissions). We begin our answer to this question by examining the architecture of international climate and environmental law and its usefulness

---

3 L du Toit and LJ Kotzé, 'Reimagining International Environmental Law for the Anthropocene: An Earth System Law Perspective' (2022) 11 *Earth System Governance* <https://doi.org/10.1016/j.esg.2022.100132> accessed 12 January 2024.

4 Ibid, 3.

5 LJ Kotzé and RE Kim, 'Earth System Law: The Juridical Dimensions of Earth System Governance' (2019) 1 *Earth System Governance* 5 <https://doi.org/10.1016/j.esg.2019.100003> accessed 12 January 2024.

6 Du Toit and Kotzé (n 3) 5.

7 United Nations Environment Programme Copenhagen Climate Centre (UNEP-CCC) and United Nations Framework Convention on Climate Change (UNFCCC) Technology Executive Committee (TEC), 'Climate Technology Progress Report' (Copenhagen, 2022) IV.

for the regulation of the technologies in question. We also evaluate current and mooted initiatives that are aimed at strengthening and adapting that architecture, namely the Global Pact for the Environment, the Earth Charter, Earth System Law and the proposal for a Framework Convention for the Regulation of New Technologies. What emerges from our analysis is that although the general tools of international law should suffice in theory, and although most of the contemporary regulatory attempts have overlooked the need for accommodating climate-related technologies, the Paris Agreement has attempted to fill this gap by covering – to a certain extent – this aspect. Hence, the current architecture of international climate law is mostly reliant on the Paris Agreement when it comes to addressing the issue of climate-related technology and has not yet reached its full potential, including because the architecture is structured around the principle of state sovereignty. This having been said, the initiatives that we examine have yielded valuable proposals for amendments that are tailored to the regulatory problems of the future.

19.2 The architecture of contemporary international climate and environmental law

The main features of international climate and environmental law, its architecture, and its strengths and weaknesses have been covered extensively in the literature.[8] The aim of this section is to examine the suitability of that body of law for the regulation of climate-related technologies. The issues that those technologies pose were outlined in the preceding section. We begin by advancing several arguments for the adequacy of the current architecture of the law; thereafter, we present arguments that point to the opposite conclusion.

19.2.1 Why is the current architecture of international climate and environmental law suitable for the regulation of new climate technologies?

International climate and environmental law has been dubbed the 'laboratory of international law'.[9] In international-law circles, the discipline is seen as flexible, 'new' and 'emerging'.[10] Brown Weiss noted in 1993 that the development of international environmental law entailed the adoption of almost 3,000 multilateral, bilateral and regional treaties. Those treaties,

---

[8] P Sands and J Peels, *Principles of International Environmental Law* (4th edn, Cambridge University Press 2018) 16ff, and also further references there.

[9] JAM Klabbers, *International Law* (2nd edn, Cambridge University Press 2017) 279.

[10] Ibid.

moreover, are underpinned by myriad rules and principles from general public international law. If one were to adopt a purely quantitative perspective on the matter, one would be bound to conclude that the new discipline has met with resounding success. However, in actuality, the proliferation of laws carries the risks of operational inefficiency[11] and fragmentation. However, fragmentation is not necessarily indicative of substantive deficiencies. First, the fragmented framework of international climate and environmental law mirrors the fragmented environmental challenges that institutions had to overcome at the time when most of the relevant treaties were promulgated.[12] Second, fragmented domains are not always more difficult to manage legally,[13] nor does fragmentation necessarily lead to inconsistency;[14] in fact, a harmonized fragmented regulation can be a more effective means of attaining some environmental objectives.[15] The benefits and the drawbacks of fragmentation should, therefore, be identified not in the abstract but in specific contexts (see Section 19.3).

Many treaties refer to the foundational principles of climate and environmental law, which are included in the UNFCCC and the Rio Declaration. Those principles are intimately connected to technological regulation. We cover four such principles here, namely precaution, sustainable development, common concern and equity. First, the precautionary approach is said to have 'grown in profile and authority and in its general implications for the governance not only of science and technology but of wider social issues'.[16] There are numerous approaches to the codification of that principle, and each approach has different implications for its legal standing and weight.[17] Wiener identified three such approaches: '(i) authorization to take

[11] EB Weiss, 'International Environmental Law: Contemporary Issues and the Emergence of a New World Order' (1993) *Georgetown Law Faculty Publications and Other Works* 1628.

[12] PS Berman, 'Global Legal Pluralism' (2007) 80 *Southern California Law Review* 1155; see also WWB White, 'International Legal Pluralism' (2004) 25 *Michigan Journal of International Law* 963.

[13] ILC, 'Fragmentation of International Law: Difficulties Arising from the Diversification and Expansion of International Law' (13 April 2006) UN Doc A/CN.4/L.682, as corrected in UN Doc A/CN.4/L.682/Corr.1 (11 August 2006).

[14] T Stephens, *International Courts and Environmental Protection* (Cambridge University Press 2009).

[15] A Peters, 'The Refinement of International Law: From Fragmentation to Regime Interaction and Politicization' (2017) 15(3) *International Journal of Constitutional Law* 671–704.

[16] A Stirling, 'Precaution in the Governance of Technology' in R Brownsword, E Scotford and K Yeung (eds), *The Oxford Handbook of Law, Regulation and Technology* (Oxford University Press 2017).

[17] J Wiener, 'Precaution and Climate Change' in KR Gray, R Tarasofsky and CP Carlarne (eds), *The Oxford Handbook of International Climate Change Law* (1st edn, Oxford University Press 2016) 167.

policy measures despite uncertain risk, (ii) obligation to take policy measures to address uncertain risk, and (iii) shifting the burden of proof (of safety or of acceptable risk) to the proponent of the activity'. The precautionary principle features in more than 50 international agreements[18] as well as in national[19] and regional[20] regulations. One general criticism that has often been levied at the precautionary principle is that it hampers technological development and innovation. Wiener, conversely, took the view that 'in the case of climate policy, precaution would largely be applied to restrict old technologies, such as fossil fuels and deforestation'.[21] It is at least arguable that the precautionary principle incentivizes innovation by forcing companies to identify new ways of making technology safer.

Second, the principle of sustainable development can also guide the regulation of climate- and environment-related technologies. Similarly to 'precaution' the term 'sustainable development' has many meanings, and its legal status has evolved over time.[22] It was first defined in the Brundtland Report,[23] and it was originally intended to reconcile the notions of economic, social and environmental development. Technology can both imperil sustainable development and accelerate progress toward its attainment, as recognized in, for example, the Sustainable Development Goals.[24] In theory, a sustainable-development approach to regulation would enable the various economic, social and environmental impacts of technology to be harmonized.

Third, 'common concern' has been described as 'the basis for international initiatives to regulate the application of geo-engineering technologies',[25] and it can become relevant to many of the other technologies that this book covers.[26] In brief, the common-concern-of-humankind principle 'encompasses aspects of the global environment that [are worthy of protection] by virtue of their significance and the need for collective action'.[27] The aspects in question cannot be addressed by individual states – collective

[18] A Trouwborst, 'Evolution and Status of the Precautionary Principle in International Law' (2002) 96(4) *American Journal of International Law* 63.

[19] For example, in the Canadian Environmental Protection Act.

[20] For example, in Article 191 of the Treaty on the Functioning of the European Union.

[21] Wiener (n 17) 170.

[22] MC Cordonier Segger and A Khalfan, *Sustainable Development Law: Principles, Practices, and Prospects* (Oxford University Press 2004) 45.

[23] World Commission on Environment and Development, 'Our Common Future' (Oxford University Press 1987).

[24] Reins (n 1).

[25] F Soltau, 'Common Concern of Humankind' in CP Carlarne, KR Gray and R Tarasofsky (eds), *The Oxford Handbook of International Climate Change Law* (1st edn, Oxford University Press 2016) 212.

[26] For example, in Chapters 11 and 13.

[27] Soltau (n 25).

and coordinated responses[28] that account for common but differentiated responsibilities are needed.[29] Although the principle does not give rise to any specific duties to engage in or to refrain from particular courses of conduct, it has been interpreted to '[narrow] the freedom of action of states even in cases where their conduct has not affected the sovereign rights of other states'.[30] Therefore, the scope of the principle of common concern extends beyond the prohibition on cross-border harm.[31] If this reasoning is transposed to the context of the regulation of technology, then, in theory, one could conclude that, even if climate and environment-related technologies are only deployed on the national level and even if their effects are restricted accordingly, there is still an obligation to consider their broader influence on the global environment and the climate and to seek cooperation between states – the 'emphasis is on an equitable approach to cooperation and burden-sharing'.[32] Article 4(1) UNFCCC, which outlines the commitments of the parties to that convention, contains several references to cooperation, which concern, among other things, funding, insurance and technological transfer.

Fourth, the principle of (intergenerational) equity means, in essence, that 'all generations – past, present, and future – are partners in caring for and using the Earth'.[33] That principle is codified in the UNFCCC, the UNESCO World Heritage Convention, and the Convention on Biological Diversity. It is also included as a commitment in various other international treaties.[34] It has been said to inhere in the notion of sustainable development[35] and to be 'one of the tenets of international law on sustainable development'.[36]

---

28 Ibid.

29 On the relationship between Common Concern and Humankind and Common But Differentiated Responsibilities (CBDR): L Rajamani, *Differential Treatment in International Environmental Law* (Oxford University Press 2006).

30 J Brunnée, 'Common Areas, Common Heritage, and Common Concern' in D Bodansky, J Brunnée and E Hey (eds), *The Oxford Handbook of International Environmental Law* (1st edn, Oxford University Press 2008), 566; Soltau (n 25).

31 Soltau (n 25) 212.

32 Ibid 209.

33 EB Weiss, *In Fairness to Future Generations: International Law, Common Patrimony and Intergenerational Equity* (Transnational Publishers and United Nations University Press 1989) 17.

34 MC Cordonier Segger, M Szabó and AR Harrington, 'Introduction' in AR Harrington, M Szabó and MC Cordonier Segger (eds), *Intergenerational Justice in Sustainable Development Treaty Implementation: Advancing Future Generations Rights through National Institutions* (Cambridge University Press 2021).

35 MC Cordonier Segger, 'International Treaty Law and Policy for Future Generations' in AR Harrington, M Szabó and MC Cordonier Segger (eds), *Intergenerational Justice in Sustainable Development Treaty Implementation: Advancing Future Generations Rights through National Institutions* (Cambridge University Press 2021) 47.

36 Ibid 65.

Evidently, the emphasis lies on the implications of sustainable development for intragenerational and intergenerational equity. Brown Weiss wrote of three global contexts that have shaped the principle, which correspond almost exactly to the premises from which this chapter (and even this entire book) proceeds, namely, the '(1) the transition from the Holocene Epoch to the Anthropocene; (2) the exploding effects of measures taken or not taken in one place on other places; and (3) the rapid development of information and communications technology'.[37] The link between the principle and technology is obvious: the choices that the present generation makes about the deployment or non-deployment of certain technologies will shape the lives of future generations. Therefore, a straightforward interpretation of the principle of intergenerational equity would require decision-making to account for the interests of future generations. The focus should not be on 'predicting' the future but on affording future generations the flexibility that they will need 'to achieve their own goals'.[38]

There is more to the architecture of contemporary climate regulation than principles. For example, according to Article 4 of the Paris Agreement, 'In order to achieve the long-term temperature goal … Parties aim … to undertake rapid reductions … so as to achieve a balance between anthropogenic emissions by sources and removals by sinks of greenhouse gases in the second half of this century'. The use of negative emission technologies is thus implicitly recognized.[39] The Paris Agreement does not stipulate particular means or technologies to be employed because that agreement accords with the principle of state sovereignty in international cooperation on climate change (see the Preamble to the UNFCCC).[40] Nevertheless, the treatment of the temperature target in Articles 2 and 4 of the Agreement reflects an 'implicit recognition of the need for negative emission technologies'.[41]

Furthermore, the development of the concepts of technology transfer and development in the history of the UNFCCC and in its architecture holds considerable promise, at least as a matter of theory.[42] Cooperation on

---

37 EB Weiss, 'International Treaty Law and Policy for Future Generations' in AR Harrington, M Szabó and MC Cordonier Segger (eds), *Intergenerational Justice in Sustainable Development Treaty Implementation: Advancing Future Generations Rights through National Institutions* (Cambridge University Press 2021) 17.

38 Ibid 20.

39 B Mayer, 'Article 4 Mitigation' in G van Calster and L Reins (eds), *The Paris Agreement on Climate Change – A Commentary* (Edward Elgar Publishing 2021) para 4.06.

40 Ibid.

41 Ibid.

42 H de Coninck and A Sagar, 'Technology Development and Transfer (Article 10)' in D Klein et al (eds), *The Paris Agreement on Climate Change: Analysis and Commentary* (Oxford University Press 2017).

the governance of technology, as a general matter, has featured in Article 4.5 of the Convention since 1992. The establishment of the Technology Mechanism[43] (TM) in 2010 was a key milestone in the process of its operationalization.[44] Institutionally, the TM is built around two bodies: the Technology Executive Committee (TEC), the policy body which consists of 22 experts on technology who represent both developed and developing countries,[45] and the Climate Technology Centre and Network (CTNC), the implementation arm providing technical assistance, and which enhances access to information and knowledge, and fosters collaboration.[46] Despite being separate bodies, consultation and cooperation is guaranteed through joined activities and areas of work, for example on the role of Artificial Intelligence in climate mitigation and adaptation.[47]

The next milestone was the inclusion of Article 10 in the Paris Agreement of 2015.[48] It provides for 'technology matters in unprecedented detail',[49] and is the first standalone Article on this issue.[50] It has been criticized for its reliance on 'the existing infrastructure' of the TM, which 'has been slow to come into full operation'.[51] Article 10.3 states that '[t]he Technology Mechanism established under the Convention shall serve this Agreement'. This provision entrenched the mechanism and had 'the effect of elevating [its] basis … in the UNFCCC hierarchy'[52] from a COP decision to a

[43] 'Technology Mechanism' (*TT:Clear*) <https://unfccc.int/ttclear/support/technology-mechanism.html> accessed 12 January 2024. Note in this regard that 'the limited nature of the technology provisions [in the Paris Agreement] provides a reflection of a deeper undercurrent – the continuing lack of political will to support climate technology actions adequately and the lack of consensus on how to do it best. It is not enough to insert obligations in to the text – their effectiveness and utility will depend on how seriously the parties translate them into action and build on these in the future'; de Coninck and Sagar (n 42) 47.

[44] S Minas, 'The Paris Agreement's Technology Framework and the Need for "Transformational Change"' (2020) 14 *Carbon & Climate Law Review* 241.

[45] Technology Executive Committee, 'Overview' (*TT:Clear*) <https://unfccc.int/ttclear/tec> accessed 12 January 2024.

[46] Climate Technology Centre & Network, 'UNFCCC Technology Mechanism' <www.ctc-n.org/> accessed 12 January 2024.

[47] For the joint Work Programme of the UNFCCC Technology Mechanism for 2023–2027 refer to TT:Clear, 'Workplan' <https://unfccc.int/ttclear/tec/workplan> accessed 16 January 2024.

[48] Supported by paragraphs 66–71 of COP, Adoption of the Paris Agreement Decision (2015) 1/CP.21.

[49] Minas (n 44) 241 and 245.

[50] Ibid.

[51] M Rimmer, 'Technology Development and Transfer' in G van Calster and L Reins (eds), *The Paris Agreement on Climate Change* (Edward Elgar Publishing 2021) para 10.14.

[52] Minas (n 44) 245.

treaty provision.[53] Furthermore, Articles 10.1 and 10.2, which create 'soft obligation[s]',[54] provide for broader international cooperation on technology. Article 10.4 establishes the Technology Framework, which is intended to supply 'strategic guidance'[55] to those who are involved in the TM, thus improving the alignment between the operations of the TEC and CTNC.[56] As Krause noted, the Framework also 'addresses fragmentation in the UNFCCC technology process'.[57] Article 10.5 reflects an 'explicit focus on the earlier part of the technology cycle', including collaborative approaches to research and development and access to technology for the contracting parties that are developing countries. Articles 10.5 and 10.6 explicitly link the technology mechanism to climate financing and the Financial Mechanism of the Convention. Developing countries had long wished[58] for such a link to be created, but their pleas were resisted by developed states. The text is nevertheless ambiguous, and it 'does not provide clear guidance towards the level of support to be provided'.[59] Article 10.6 specifies that technology development and transfer are included in the global stocktake, which improves accountability.[60] On the whole, Article 10 'provides a sense of direction rather than of action'.[61] Despite its strong language, the commitments that it engenders are not to the attainment of concrete results but only to the observance of certain forms of conduct;[62] 'it does not mention specific countries [and] does not mention numbers or indicators'.[63]

The partial implementation of Article 10 PA through the Technology Framework (TF), which was adopted in 2018 in Katowice,[64] was also an important achievement in the effort to enhance technology transfer and development.[65] The adoption decision stresses that the '[t]he long-term vision for technology development and transfer shared by Parties relates to

---

53 Ibid.

54 L Rajamani, 'The 2015 Paris Agreement: Interplay Between Hard, Soft and Non-Obligations' (2016) 28 *Journal of Environmental Law* 337.

55 K Krause, 'Technology and the Paris Agreement: From Means of Implementation to Climate Innovation to Transformation' in S Minas and V Ntousas (eds), *EU Climate Diplomacy* (1st edn, Routledge 2018).

56 Ibid.

57 Ibid; Minas (n 44) 246.

58 De Coninck and Sagar (n 42) 265.

59 Ibid.

60 Ibid 266.

61 Ibid.

62 Ibid.

63 Ibid.

64 CMA, 'Technology framework under Article 10, paragraph 4, of the Paris Agreement' (2018) Decision 15/CMA.1.

65 For a general overview of the history of the TM refer to Rimmer (n51).

the importance of fully realizing technology development and transfer in order to improve resilience to climate change and reduce greenhouse gas emissions'.[66] The TF is intended to perform a 'strategic role'[67] to that end, and the guiding principles of 'coherence, inclusiveness, results-oriented approach, transformational approach and transparency' are to be embraced.[68] The 'focused areas of action'[69] of the TF are as follows: '(a) Innovation; (b) Implementation; (c) Enabling environment and capacity-building; (d) Collaboration and stakeholder engagement; (e) Support'.[70]

It has been argued that this new mechanism has the potential to facilitate the transformation that is needed if the Paris target is to be met. Obviously, the success of this endeavour will also depend on the extent to which other actors engage with the TM[71] and political will.[72] Rimmer concluded that only a 'modest amount of progress' has been achieved so far. This rate of progress reflects the funding and other resources that have been allocated to the mechanism.[73] According to Minas, the TF has 'refocused the UNFCCC technology bodies on contributing to the achievement of the ambitious Paris Agreement goals',[74] which exemplifies 'the utility of "supreme body" decision making as a flexible and timely means of updating international environmental law'.[75] However 'despite some positive steps, exactly how the TM can contribute to transformational take-up of environmentally sound technologies remains an open question'.[76] The same holds true of the TF:[77] '[its] effectiveness and utility will depend on how seriously the parties translate them, into action and build on these in the future'.[78]

It can be concluded from the foregoing that international climate and environmental law is structured around principles and concrete legal provisions that could, in theory, allow the regulatory challenges of Anthropocene technology to be overcome. However, this theoretical possibility is yet to manifest in legal reality, a point on which we propose to expound in the next section.

66 CMA (n 64).
67 Ibid; CMA (n 64).
68 Ibid; CMA (n 64).
69 Ibid; CMA (n 64).
70 Ibid; CMA (n 64).
71 Minas (n 44) 242 and 244.
72 Coninck and Sagar (n 58) 276.
73 Rimmer (n 51) para 10.12.
74 Minas (n 44) 253.
75 Ibid.
76 Ibid.
77 Ibid 254.
78 Coninck and Sagar (n 42) 276.

19.2.2 Why is the current architecture of international climate and environmental law not *suitable for the regulation of new climate technologies?*

Closer scrutiny of the current international-law framework reveals few causes for optimism. The reasons are manifold, and they have been discussed extensively in the literature.[79] Here, we focus on a single line of argument that is particularly germane to technology. The extent to which international environmental law, as a fragmented and state-centric regime,[80] can meet new challenges promptly is increasingly being called into question.

The current architecture of international environmental and climate law is fragmented. Fragmentation makes it more difficult to manage relationships between different environmental-law regimes, such as the one that governs climate change and the one that regulates the preservation of biodiversity. It also hampers efforts to structure relationships between environmental and non-environmental law, as exemplified by the interplay between the regulation of the climate and that of trade.[81] Inconsistencies and overlaps between norms are thus likely to occur in consequence of fragmentation.[82] The history of environmental law has witnessed several attempts to develop more comprehensive regulation. The first such initiative was the 1972 Conference on the Human Environment.[83] It was followed by similar endeavours,[84] the most recent of which we describe in Section 19.4. For now, it should suffice to say that the difficulties that fragmentation creates can be overcome through the adoption of a more global perspective to regulation.[85] Such a shift would necessitate a departure

---

79 Sands, for example, refers to the adequacy of the legislative process regarding substance and procedure, overlapping environmental issues and a lack of capacity. See Sands and Peels (n 8) 16f. See also KW Abbott, 'An International Framework Agreement on Scientific and Technological Innovation and Regulation' in GE Marchant, BR Allenby and JR Herkert (eds), *The Growing Gap Between Emerging Technologies and Legal-Ethical Oversight: The Pacing Problem* (Springer Netherlands 2011).

80 Kotzé and Kim (n 5) 4.

81 See, for example, S Baughen, *International Trade and the Protection of the Environment* (1st edn, Taylor & Francis 2008).

82 Weiss (n 11) 675, See also: D Bodansky, *The Art and Craft of International Environmental Law* (Harvard University Press 2011) 35, describing 'treaty congestion' as a similar result of the proliferation of MEAs, which 'creates the potential for duplication in effort, lack of coordination, and even conflict between different environmental regimes'); and R Wolfrum and N Matz, *Conflicts in International Environmental Law* (Springer 2003).

83 UNGA Res/2994 (1972) RES/2994 (Stockholm Declaration).

84 References of World Charter for Nature (1982 UNGA RES 37/7); World Commission on Environment and Development (n 23); IUCN, *Draft International Covenant on Environment and Development: implementing sustainability* (2015).

85 For a more in-depth discussion on this aspect, see the following section.

from the state-responsibility approach, that is, from the practice of resolving environmental issues that are of common concern through negotiation or specific agreements. However, such a reorientation either may never take place or would take a lot of time.

Moreover, international environmental law has thus far been of very limited effectiveness in the context of its potential application to relationships between states and individuals. The prominence of non-state actors, such as transnational corporations, has given rise to important concerns about the architecture of the law.[86] For instance, many commentators have inquired whether international corporations should be held accountable in the jurisdictions in which they are based for their actions in host states. The answers to questions of this kind are of fundamental import because it is transnational corporations that play a prominent role in the development and deployment of climate-related technologies, especially when the latter have a negative impact on climate change. In fact, transnational corporations not only have the necessary financial and technical resources for being at the forefront of climate-related technologies but can also optimize technology transfer throughout their supply chain.[87] Hence, transnational corporations have started to be considered as key actors on par with states when it comes to addressing the issues arising from climate-related technologies.[88] However, despite the lively discussions of the current state of the art, concrete solutions are yet to emerge.

Last, the regulation-of-technology problem that international law faces is not merely one of reining in non-state actors – the enforcement of international environmental law in certain jurisdictions is also lacklustre. This deficiency can be traced to the operation of the principles of absolute territorial sovereignty and permanent sovereignty over natural resources – their cumulative operation means that state consent is key for successful interstate cooperation and, ultimately, for the effectiveness of litigation. Moreover, the problem of collective action is particularly pronounced in the context of global environmental issues such as climate change.[89] In principle, all states should cooperate to resolve such issues. In international law, however, there is a tension between the need to ensure that all states are involved and the desire to formulate effective obligations that each state should

[86] A Bianchi (ed), *Non-State Actors and International Law* (Routledge 2017).

[87] S Kalliojärvi, 'Climate Change, Security and the Role of Transnational Corporations' (2020) *Arctic Yearbook* 117.

[88] Ibid 118.

[89] See E Ostrom, *Governing the Commons: The Evolution of Institutions for Collective Action* (Cambridge University Press 1990); see also E Ostrom, 'Polycentric Systems for Coping with Collective Action and Global Environmental Change' (2010) 20 *Global Environmental Change* 550.

bear.[90] The Kyoto Protocol exemplifies this trade-off very clearly. Differentiation is a potential mitigation strategy, in that acknowledging that states may bear different degrees of responsibility would increase the likelihood of agreements being negotiated successfully.[91] However, it should be evident from the foregoing that the accommodation of new climate technologies into the architecture of modern climate law would be neither straightforward nor bound to succeed.

19.3 Potential solutions

The successes of contemporary international law in the regulation of climate-related technologies have been few and far between. Despite the promise of some of its features, the international legal architecture still exhibits several significant deficits, and it does not appear to be especially conducive to the development and deployment of new technologies. Several recent initiatives in international environmental law have highlighted promising approaches to tackling some of these problems. We discuss these regulatory initiatives and evaluate their potential for regulating climate-related technologies in the subsections that follow.

19.3.1 The Earth Charter

The Earth Charter is an international declaration that originated from a large and participatory consultation process. That process was designed to produce a sustainable global society in which environmental protection, human rights, economic justice and peace would be recognized as interdependent objectives.[92] This document, which was published in 2000, can be interpreted as a civil-society response to the various calls for proposals on drafting norms supporting the transition to sustainable development from the preceding decades. Notably, the 1987 United Nations World Commission on Environment and Development urged the development of a new charter for the transition to a more sustainable future,[93] and similar calls were circulated in the run-up to the Earth Summit in Rio de Janeiro in 1992.[94]

---

90 See S Barrett, *Environment and Statecraft: The Strategy of Environmental Treaty-Making* (Oxford University Press 2003); or Young, *International Cooperation: Building Regimes for Natural Resources and the Environment* (Cornell University Press 1989).

91 L Rajamani, *Differential Treatment in International Environmental Law* (Oxford University Press 2006).

92 Independent Earth Charter Commission, 'Earth Charter: Preamble' (2000) <https://earthcharter.org> accessed 16 January 2024.

93 World Commission on Environment and Development (n 23).

94 1992 Earth Summit in Rio de Janeiro.

The preparatory work behind the declaration has attracted considerable scholarly attention.[95] The Charter was the outcome of a five-year (1995–2000) global consultation process that unfolded under the guidance of the Earth Charter Commission.[96] The Commission was tasked with integrating all of the information that had been collected in the course of the consultations into a document that would command global consensus. This decision was driven by the notion that the pursuit of sustainable development in the 21st century must be animated by principles that are genuinely common to all. Likewise, the Earth Charter was intended to become a soft-law instrument that offers an ethical framework for the transition to a just and sustainable world.[97]

The resultant document reflects a non-anthropocentric perspective and integrates concepts of intergenerational, intragenerational and interspecies justice.[98] Its Preamble states, 'it is imperative that we, the peoples of Earth, declare our responsibility to one another, to the greater community of life, and to future generations'.[99] The vision and the principles of the Earth Charter are therefore holistic, in that environmental protection is taken to be tightly intertwined with social justice, institutional transparency and accountability, and active and inclusive public participation. Moreover, there is strong correspondence between the principles and values of the Earth Charter and the principles and values of conventional international law treaties and other (binding) instruments.[100]

It is mostly for these reasons that some commentators have drawn parallels between the Earth Charter and Agenda 21, arguing that the former can follow the trajectory of the latter by becoming soft law.[101] The endorsement

---

95 K Bosselmann, 'Outlook: The Earth Charter: A Model Constitution for the World?' in K Bosselmann and JR Engel (eds) 'The Earth Charter: A Framework for Global Governance' (KIT Publishers 2010) 239–55; K Bosselmann, 'In Search for Global Law: The Significance of the Earth Charter' (2004) 8(1) Worldviews: Global Religions, Culture, and Ecology 62; H Rolston III, 'The Earth Charter Facing the Anthropocene Epoch' in P Burdon, K Bosselmann and K Engel (eds), '*The Crisis in Global Ethics and the Future of Global Governance*' (Edward Elgar Publishing 2019) 72–91.

96 Although the process was formally launched in 1994. this Commission was formed in 1997.

97 Bosselmann and Engel (n 95); LJ Kotzé, 'Human Rights and the Environment in the Anthropocene' (2014) 1(3) *The Anthropocene Review* 252.

98 Bosselman (n 95).

99 Independent Earth Charter Commission (n 92).

100 J Bernstein, 'The Policy Relevance of the Earth Charter for Europe' (2007) Earth Charter International 6.

101 K Bosselmann and P Taylor, 'Promoting Global Ethics: The Earth Trusteeship Initiative' in P Burdon, K Bosselmann and K Engel (eds), *The Crisis in Global Ethics and the Future of Global Governance* (Edward Elgar Publishing 2019) 279–83.

of the Earth Charter has already been discussed at some international-law fora. For instance, the President of South Africa referred to the Earth Charter during the opening session of the 2002 World Summit for Sustainable Development, and some national governments and local authorities have initiated the process of endorsing it.[102] UNESCO also offered its political support at its 32nd General Conference in 2003, where it urged states 'to recognize the Earth Charter as an important ethical framework for sustainable development'.[103] However, despite some official mentions in the early years, the Earth Charter gave way to another regulatory initiative (the Global Pact for the Environment) that the following subsection will discuss in more detail. Hence, the Charter cannot yet be fully considered a soft-law instrument of international law and enjoys only some degree of authority by virtue of its drafting process and the strong grounding of the final text in international law.

Evidently, the Earth Charter can play a (limited) role in the regulation of climate technologies. The text of Principle 7(c), after acknowledging the critical role of technology in tackling anthropogenic environmental impacts, urges the '[p]romot[ion of] the development, adoption, and equitable transfer of environmentally sound technologies'. The focus is thus on transferring technology for the benefit of society. Technological risks, conversely, play a secondary role – at most, they are subsumed into the precautionary principle (Principle 6 in the Charter).

Despite the positives, the usefulness of the Earth Charter as a source of ethical guidance for institutions is circumscribed by its standing in international law. Its main strength is also its main drawback – it has not been endorsed officially by the international community, it is not legally binding and it is not soft law yet. At the time of writing, the United Nations General Assembly had not adopted it formally. Consequently, it is difficult to see a future in which the Earth Charter improves the regulation of climate technology at the international-law level. Furthermore, the Charter does not refer to climate change. The drafters refrained from pinpointing individual environmental problems such as freshwater scarcity or deforestation. Instead, the Charter promotes an ethical framework for transitioning to a sustainable

[102] Earth Charter Steering Committee, 'The Earth Charter at the Johannesburg Summit' (2002) <https://earthcharter.org/wp-content/assets/virtual-library2/images/uploads/2002%2011%20WSSD%20Report.pdf> accessed 12 January 2024. Likewise, the wording used for the Political Declaration can be considered as heavily inspired by the text of the Earth Charter (para 6 of the political declaration as compared to concluding words of the first paragraph of the Earth Charter Preamble).

[103] UNESCO, 'Records of the General Conference, 32nd session, Paris' (2003) 32 C/Resolutions <https://unesdoc.unesco.org/ark:/48223/pf0000133171> accessed 16 January 2024, at 17.

world and for addressing *all* environmental challenges, which makes it difficult to draw on the document in order to formulate *ad hoc* provisions on climate change technologies.

19.3.2 The Global Pact for the Environment

The Global Pact for the Environment is intended to serve as a generic and binding instrument that combines principles of international environmental law and human rights law. Its development has been marred by numerous difficulties, which is unsurprising in the light of its ambitions. The most obdurate obstacle by far has been the need to consolidate the principles of international environmental law and to complement them with references to human rights law.

The drafting process was initiated by environmental-law experts under the aegis of the Environment Commission of the French thinktank *Club des Juristes.*[104] Once it was finalized in 2017, the Global Pact attracted considerable attention at the international level. As a response, the General Assembly adopted the Towards a Global Pact for the Environment Resolution. It called for the preparation of a report by the UN Secretary-General on gaps in international environmental law and on the means of plugging them.[105] This report could have laid the foundation for the adoption of the Global Pact for the Environment. The Resolution stated that the Secretary-General's report would be discussed by an *ad hoc* open-ended Working Group, which would inquire into 'the scope, parameters and feasibility of an international instrument, with a view to making recommendations, which may include the convening of an intergovernmental conference to adopt an international instrument, to the Assembly during the first half of 2019'. The General Assembly published the Secretary-General's report,[106] which noted that collating all principles of international environmental law into a single document would represent genuine progress[107] and thus generated momentum for the adoption of a binding treaty. However, this

[104] Please note that this think tank has previously adopted a report on how to increase the effectiveness of international environmental law. See, for example, Environment Commission of the Club des Juristes, 'Increasing the Effectiveness of International Environmental Law: Duties of States, Rights of Individuals' (November 2015) <https://globalpactenvironment.org/uploads/CDJ_Rapports_Increasing-the-effectiveness_Nov.2015_UK.pdf> accessed 16 January 2024.

[105] UNGA RES 72/277 (2018) A/RES/72/277.

[106] UN Secretary-General, 'Gaps in International Environmental Law and Environment-related Instruments: Towards a Global Pact for the Environment' (30 November 2018) UN Doc A/73/419.

[107] Ibid.

momentum abated after the publication of the recommendations of the Working Group – states preferred a political declaration over a treaty.[108] Accordingly, Resolution 73/333 of the UN General Assembly confirmed that a political declaration would be issued.[109] The Resolution stated that the United Nations Environment Assembly should

> prepare, at its fifth session, in February 2021, a political declaration for a United Nations high-level meeting, subject to voluntary funding, in the context of the commemoration of the creation of the United Nations Environment Programme by the United Nations Conference on the Human Environment, held in Stockholm from 5 to 16 June 1972.

The Global Pact for the Environment did not meet the expectations of its drafters. All of their preparatory work had been directed at the production of a treaty, but it ultimately yielded a political declaration. On a more positive note, that political declaration referred to the right to a healthy environment, which may have contributed to the recognition of that right in Resolution 48/13 of the Human Rights Council and in a Resolution of the UN General Assembly.[110] Therefore, the potential of the Global Pact for the Environment depends mostly on the support of states and on the type of document (treaty, declaration or general framework) that will further build on the Pact.[111]

It is important to note that commentators were highly sceptical of the lofty ambitions of the Pact from the outset.[112] For instance, the integrated approach that it reflects has been questioned because environmental challenges are thought to be too diverse for a one-size-fits-all solution to be

---

[108] L Chabason and E Hege, 'Failure of the Global Pact for the Environment: A Missed Opportunity or a Bullet Dodged? (*IDDRI*, 28 May 2019) <www.iddri.org/en/publications-and-events/blog-post/failure-global-pact-environment-missed-opportunity-or-bullet> accessed 16 January 2024.

[109] UNGA Res 73/333 (2019) A/RES/73/333.

[110] Human Rights Council and UNGA Res 48/13 (2022) A/RES/76/300.

[111] C Voigt, 'How a "Global Pact for the Environment" Could Add Value to International Environmental Law' (2019) 28(1) *Review of European, Comparative & International Environmental Law*,13–24.

[112] See, for example, LJ Kotzé and D French, 'A Critique of the Global Pact for the Environment: A Stillborn Initiative or the Foundation for Lex Anthropocenae?' (2018) 18 *International Environmental Agreements: Politics, Law and Economics* 811; D French and LJ Kotze, ' "Towards a Global Pact for the Environment": International Environmental Law's Factual, Technical and (Unmentionable) Normative Gaps' (2019) 28 *Review of European, Comparative and International Environmental Law*; and S Biniaz, 'The UNGA Resolution on a "Global Pact for the Environment": A Chance to Put the Horse before the Cart' (2019) 28 *Review of European, Comparative and International Environmental Law*.

viable.[113] Moreover, some have argued that the fragmentation of international environmental law may make it more flexible and adaptive and thus more suitable for the management of variegated issues and innovations.[114] These scholarly opinions indicate that the principal aim of the Pact should not have been the production of integrated legislation but improvements to the efficacy of international environmental law. In this context, it has been suggested that integrating the human rights architecture into the principles of international environmental law may be a desirable development.[115] Alternatively, the Global Pact for the Environment can become an umbrella treaty that summarizes the various principles of international environmental law without specifying their content. In that case, the focus would need to shift to the relationships between the various instruments of international law in this domain. As we note in the next subsection, the relationships in question are vaguely defined, and international environmental law remains highly fragmented.[116] Finally, it is important to note that technology was not entirely overlooked in the course of the development of the Global Pact for the Environment – the political declaration of the United Nations Environment Assembly emphasizes the prominent role of technology in the implementation of environmentally friendly policies and in the pursuit of sustainable development.[117]

19.3.3 Earth System Law

Earth System Law (ESL) is a new legal discipline that can complement or even replace international environmental law.[118] Kim and Kotzé defined that discipline as 'an innovative legal imaginary that is rooted in the Anthropocene's planetary context and its perceived socio-ecological crisis. Earth system law is aligned with, and responsive to, the Earth system's functional, spatial and temporal complexities; and the multiple Earth

[113] L Kotzé, 'Fragmentation Revisited in the Context of Global Environmental Law and Governance' (2014) 131(3) *South African Law Journal* 548–83.

[114] S Biniaz, '10 Questions to Ask about the Proposed "Global Pact for the Environment"' (2017) Columbia Law School Scholarship Archive <https://scholarship.law.columbia.edu/cgi/viewcontent.cgi?article=1090&context=sabin_climate_change> accessed 16 January 2024; J Gupta and N Sanchez, 'Global Green Governance: Embedding the Green Economy in a Global Green and Equitable Rule of Law Polity' (2012) 21(1) *Review of European Community & International Environmental Law* 12–22.

[115] Kotzé and French (n 112) 811–38.

[116] Biniaz (n 114) 3.

[117] UN Environment Programme, 'First Special Session: Commemoration of the Fiftieth Anniversary of the Establishment of the United Nations Environment Programme' (Nairobi, 3–4 March 2022) paras 20, 23 and 24.

[118] Kotzé and Kim (n 5) 1–12.

system science and social science-based governance challenges arising from a no-analogue state in which the Earth system currently operates'.[119] The underlying assumption is that a planetary systems-based Earth Law can result in a revision of international law, as traditionally conceived, which would account for fresher perspectives on global and transnational law. At the same time, ESL also calls for a re-examination of both mainstream and critical perspectives on environmental law. In its most mature phase, ESL would produce a conception of law that is 'more responsible for the complexities of the Earth System', that is to say, it would be an 'Earth-centred law for governance by and for all living beings from a planetary systems perspective'[120] and thus also address fragmentation.

Some authors have argued for the constitutionalization of international environmental law[121] or the establishment of a basic norm, along with a governance framework that would support it. Such a development would oblige all the relevant actors across the globe to consider planetary boundaries as limits that must not be exceeded by any polluting activity or by any form of environmental harm.[122] ESL is considered to be a particularly suitable approach to the effectuation of a normative shift that would prevent humanity from encroaching on the planetary boundaries because it addresses the need for 'connectivity'[123] not only between different branches of law but also between the social and the natural sciences; it thus accounts for the Earth-system-thinking perspective on these problems.[124]

---

[119] LJ Kotzé and RE Kim, 'Exploring the Analytical, Normative and Transformative Dimensions of Earth System Law' (2020) 50 *Environmental Policy and Law* 457–70.

[120] Kotzé and Kim (n 5) 6f. See also for a more in-depth discussion, E Cocciolo and L Reins, 'A Critical Review of the Energy Charter Treaty from an Earth System Law Perspective' 2024 *Transnational Environmental Law* 1–27, doi:10.1017/S2047102524000244.

[121] LJ Kotzé and W Muzangaza, 'Constitutional International Environmental Law for the Anthropocene?' (2018) 27(3) *Review of European, Comparative and International Environmental Law* 278–92.

[122] A Muller and M Huppenbauer, 'Sufficiency, Liberal Societies and Environmental Policy in the Face of Planetary Boundaries' (2016) 25(2*) GAIA-Ecological Perspectives for Science and Society* 105–9; P Bridgewater, RE Kim and K Bosselmann, 'Ecological Integrity: A Relevant Concept for International Environmental Law in the Anthropocene?' (2014) 25(1) *Yearbook of International Environmental Law* 61–78; J Ebbesson, 'Planetary Boundaries and the Matching of International Treaty Regimes' (2014) 59 *Scandinavian Studies in Law* 259–84.

[123] This paper transfers the notion of 'connectivity' elaborated by Paul Kjaer in relation to Global Law to ESL. According to Kjaer, from an inter-systemic perspective, connectivity norms function to increase the probability of transfers of condensed social component – *Sinnkomponente* – (as scientific knowledge) from a legally structured context to another within world society; Poul F Kjaer, 'Constitutionalizing Connectivity: The Constitutional Grid of World Society' (2018) 45(S1) *Journal of Law and Society* S114–34.

[124] LJ Kotzé and others, 'Earth System Law: Exploring New Frontiers in Legal Science' (2022) 11 *Earth System Governance*.

The role of technology has received some attention in ESL scholarship, too. Kotzé and Kim note that '[t]he regulation of technological interventions to enable new futures might become a key focus of law, requiring law to also embrace a forward (instead of mostly backward)-looking perspective'.[125] In other words, they acknowledge the role that technology plays in the pursuit of gains in environmental protection and draw attention to the manner in which the law should be drafted if it is to conduce to such gains. Accordingly, ESL is be perceived as a paradigm in which the socio-technical-ecological system can be subsumed.[126] Here, it is important to note that many regard sovereignty as a significant impediment to the practical realization of an ESL. It is for this reason that some scholars of Earth system governance limit themselves to arguing for improvements in the coordination of the current system rather than exhorting states to adopt a new global treaty.[127]

19.3.4 International Framework Agreement on Scientific and Technological Innovation and Regulation

The idea behind the International Framework Agreement on Scientific and Technological Innovation and Regulation (FASTIR) was originally formulated by Kenneth W. Abbott. He proceeded from the premise that some technological innovations have the 'potential to create significant health, safety and environmental risks and/or to pose significant social, cultural and ethical challenges, even as they promise significant benefits'.[128] Several of the technologies that are discussed in this book obviously possess such properties. Abbott recommended a general institutional framework with two main elements: an international framework convention (as in international environmental and climate law, for example, the UNFCCC) and a 'set of international institutions and procedures to coordinate, steer and facilitate national action, which would be established by and operate under the authority of the framework convention'.[129]

Abbott drew on the current practice of accounting for non-state actors in framework conventions. Consequently, he suggested integrating three levels of governance into the new regime. The three are the 'traditional'

[125] Kotzé and Kim (n 5) 1–12.

[126] L Mai and E Boulot, 'Harnessing the Transformative Potential of Earth System Law: From Theory to Practice' (2021) Earth System Governance <https://doi.org/10.1016/j.esg.2021.100103> accessed 16 January 2024.

[127] L Mouysset et al, 'Operationalizing Sustainability as a Safe Policy Space' (2018) 10(10) *Sustainability*; G Chapron et al, 'Bolster Legal Boundaries to Stay within planetary boundaries' (2017) 1(3) *Nature Ecology & Evolution* 1.

[128] Abbott (n 79).

[129] Ibid 129.

international state-to-state level; a 'transgovernmental level', which focuses on the design and implementation of cross-national rules and is populated by 'executive agencies and officials, legislative bodies and legislators, and other units and individuals within national governments';[130] and a 'transnational' level. The last level would leverage the cross-fertilization that results from relationships between societal actors and organizations (private actors and non-governmental organizations) that 'have stakes in the issues of concern and can contribute to effective legal and policy responses' and which operate across borders.[131] The actors at this last level 'produce the innovations to which law must respond, are most knowledgeable about those innovations, and have the authority, access and information to produce meaningful compliance with legal rules on a day-to-day basis in the lab or factory'.[132] At the same time, they represent the constituencies that are affected directly by innovations (workers, consumers and those who suffer the consequences of environmental degradation) and are 'representatives of the larger (transnational) society concerned with the social, cultural and ethical implications of innovations'.[133]

Abbott identified institutional arrangements that reflect the centrality of information problems in the technological domain[134] as well as (concrete) general provisions, possible host institutions, treaty bodies, procedures, means of steering national regulatory systems, good regulatory practices and a mechanism for transgovernmental dialogue that could be included in his proposed convention.[135] The details are beyond the scope of this chapter. More generally, according to Abbott, three of the principal functions of FASTIR would be highly salient to regulation. First, 'it would establish the "pacing law with science and technology" problem as a legitimate issue of international concern, while providing structures for ongoing international cooperation'.[136] Therefore, it would not apply to specific technologies or include substantive provisions. Instead, in the spirit of a framework convention, it would contain overall objectives, principles, general commitments and 'institutions and procedures to coordinate, steer and facilitate national legal and policy actions'.[137] Second, 'protocols', 'annexes' and other subsidiary instruments would be promulgated 'when conditions become ripe for concrete international action'. These instruments would

[130] Ibid.
[131] Ibid 130.
[132] Ibid 79
[133] Ibid.
[134] Ibid 134.
[135] Ibid.
[136] Ibid 136.
[137] Ibid 79

address specific technologies and cross-cutting issues. Therefore, regulation would be incremental.[138] Third, mechanisms for communication and for gathering and sharing knowledge about emerging technologies would be created, as would 'forms of dialogue and deliberation designed to produce deeper understanding of the risks, benefits and social and ethical implications of those innovations'.[139]

Abbott's proposal for an International Framework Agreement on Scientific and Technological Innovation and Regulation is built on the model of existing framework conventions, but some of its features are adapted to the specificities of the regulatory problem at hand. For instance, Abbott did not take the term 'international action' to refer to the regulation of particular technologies at the international level – he took the view that such regulation would not be 'a realistic option' because of the weakness of international law (see also Section 19.3).[140] Furthermore, he recognized that the negotiations for a framework convention would involve 'some of the same problems of delay and cost as any treaty process'.[141] At the same time, he posited that 'the framework convention model is explicitly designed to be initially less demanding and easier to negotiate, while facilitating more detailed incremental and adaptive rule-making over time, as better information on risks and benefits is obtained'.[142] Experience with the UNFCCC and the subsequent Paris Agreement appear to confirm this observation.

The benefits of this approach that Abbott identified generally coincide with the advantages of international cooperation. He argued that 'international arrangements might facilitate speedier and better designed national legal and policy responses, especially in states with limited regulatory capacity'.[143] He also expressed a preference for 'international action to *coordinate national law* and policy'[144] and for 'steering national actions in desirable directions'[145] over regulation and conventional international law. This shift would result in international action becoming more uniform, more efficient, more effective, more legitimate and more closely aligned with the public interest in both form and substance.[146] Addressing 'inconsistent or inappropriate national responses'[147] would boost trade,[148] improve international

[138] Ibid 137.
[139] Ibid.
[140] Ibid 127.
[141] Ibid 129.
[142] Ibid.
[143] Ibid.
[144] Ibid 128.
[145] Ibid.
[146] Ibid 129.
[147] Ibid 130.
[148] Ibid 131.

collaboration[149] and enhance the uniformity of the application of the precautionary principle. The incentives that lead states to over- or under-regulate[150] would be managed more efficiently. The Prisoner's Dilemma[151] and the related concern about a race to the bottom in environmental and climate protection would also be addressed.[152] As Abbott put it, 'The incentives to under-regulate are enhanced when a technology creates negative trans-border externalities, passing some of the costs to others'.[153] According to him, the adoption of a FASTIR-like approach would eliminate these undesirable tendencies.

Since Abbott's proposal operates on three levels of governance, it would be possible for international action to keep pace with scientific and technological innovation:[154]

> legal and policy institutions require early warning of significant innovations; information on new technologies and their potential benefits, risks and other impacts; means of assessing unclear data; even some degree of prediction. In all these areas, more information is better than less, and multiple social, cultural and ethical perspectives are more valuable than unitary, possibly myopic ones.[155]

Due to different 'risk cultures, social relations, legal and regulatory systems, levels of development [and] capacities',[156] the passage of traditional international regulation or the harmonization of controversial technologies or innovations is generally 'very unlikely [or] inappropriate'.[157] A FASTIR-like approach would encourage the exchange of information, assessments and cultural viewpoints when the more aggressive pursuit of uniform regulation is likely to prove counterproductive.[158]

To summarize, the most obvious drawback of the FASTIR approach is that it resembles the UNFCCC and the Paris Agreement too closely – in the last analysis, it is based on the same (flawed) international-law architecture. This is so despite Abbott's emphases on 'steering' rather than 'regulating' and on the potential emergence of new avenues for collaboration and cooperation

149 Ibid.
150 Ibid.
151 Ibid.
152 Ibid.
153 Ibid.
154 Ibid 130.
155 Ibid.
156 Ibid 132.
157 Ibid.
158 Ibid.

between states and other stakeholders – it remains the case that the three layers of governance can easily be replicated under the current structure of international climate and environmental law.

19.4 Concluding remarks: The architecture of international law and the regulation of technology in the Anthropocene

Our objective here was to assess the capacity of international law, with its current architecture, to regulate climate-related technologies. To that end, we explained why the law may be adequate in theory and why it might be inapt in practice. The Paris Agreement has attempted to fill this gap by covering – to a certain extent – this aspect. Hence, the current architecture of international climate law is mostly reliant on the Paris Agreement when it comes to addressing the issue of climate-related technology and has not yet reached its full potential. The proposition that the architecture of international environmental and climate law is fragmented[159] and state-centric commands widespread acceptance. As the balloon example on which we drew initially demonstrates, the structure of the modern law is typified by a plurality of specialized regimes, rules and institutions that govern the relationships between states, international organizations, and the environment.[160] Despite not being a negative feature per se, this structure makes the regulation of climate-related technologies more difficult from an international-law perspective.

The options we discussed in Section 19.3 are useful starting points for prospective solutions. The legal scholars who participated in their formulation assumed roles that were not limited to strengthening and developing the law as presently structured. Some devised wholly novel concepts because they took the view that international environmental law, in its current form, is unlikely to withstand the pressures that are likely to bear on it in the near future. These new concepts reflect the need for the study of law and governance to account for findings from other domains of science. For instance, the literature on planetary boundaries may serve as a new theoretical foundation for a law that protects the environment more effectively – some have argued that ESL should become a new legal

[159] See, for example, H van Asselt, 'Managing the Fragmentation of International Environmental Law: Forests at the Intersection of the Climate and Biodiversity Regimes' (2012) 44 *New York University Journal of International Law and Politics* 1205; F Biermann et al, 'The Fragmentation of Global Governance Architectures: A Framework for Analysis' (2009) 9(4) *Global Environmental Politics* 14–40.

[160] See A Boyle and C Redgwell, *International Law and the Environment* (4th edn, Oxford University Press 2019); D Bodansky (n 82); and Bodansky, Brunnée and Hey (n 30).

discipline. Such a discipline would provide a more holistic perspective on contemporary environmental challenges.[161]

The literature is highly theoretical because the academics in question are aware that their proposals can only bear fruit in the distant future.[162] The principle of sovereignty remains an insurmountable obstacle to such developments because the creation of a supranational institution that protects planetary boundaries and the development of new legal disciplines can only result from exercises of political will on the part of governments, which have thus far hesitated to engage in radical action.[163] For this reason, most modern authors continue to focus on incremental improvements in the coordination of fragmented regimes, rules and institutions.[164]

Acknowledgements

This chapter has been made possible partly thanks to the support of the GroenvermogenNL Hy Success Research Grant.

[161] Kotzé and Kim (n 5) 1.

[162] Ibid 5.

[163] F Biermann, 'Planetary Boundaries and Earth System Governance: Exploring the Links' (2012) 81 *Ecological Economics* 4–9.

[164] Mouysset (n 127) 86.

20

Thinking about Technology Transfer beyond Development

Morag Goodwin

20.1 Introduction

The basic premise of this chapter is that if the Anthropocene is to usher in a revolution in legal relations (or any other kind of relations), we will need to move beyond paradigmatic ways of thinking that have defined the past several centuries. There are by now many works arguing for the need to overcome the binaries of modern European thought (human vs non-human, nature vs culture, and so on).[1] The aim here is to contribute to this literature by drawing attention to one of the key binaries that structures contemporary global relations. The suggestion is that the organizing category of development that separates states into developed and developing plays a profoundly important, but neglected, role in negotiations in relation to climate change. I will attempt to illustrate this by reference to the failure of states to deliver on repeated commitments to international technology transfer.

Technology transfer, if it is to work to enable humankind to stay within two degrees of warming, will necessarily involve the diffusion of 'green'

[1] For example, Donna Haraway, *Staying with the Trouble: Making Kin in the Chthulucene* (Duke University Press, 2016); E. Tendayi Achiume, 'Report of the Special Rapporteur on contemporary forms of racism, racial discrimination, xenophobia and related intolerance: Ecological crisis, climate justice and racial justice', A/77/549, 25 October 2022; Marie-Catherine Petersmann, 'Response-abilities of Care in More-Than-Human Worlds' (2021) 12 *Journal of Human Rights and the Environment* 102–24.

technology[2] to low-income countries of the Global South. Yet technology transfer has a longer history as a battleline between former colonizers and Third World states.[3] This chapter aims to provide some of that context as a necessary frame for reconceptualizing relations. If we want to understand why it is so difficult to achieve our stated goals, we need to be clearer about what we need to change.

This chapter will suggest how development as a structuring global paradigm has determined the legal space within which we understand or facilitate technology transfer. The aim of this chapter is not to speak to legal frameworks directly; this is done in other parts of this book. My purpose instead is to draw attention to the structuring frameworks that underpin the choice for certain legal tools and make them seem natural. There is, of course, nothing natural about the choice to frame technology transfer as individual property rather than to see it as the outcome of a particular Lockean view of human relations. Likewise, the choice to view technology transfer through a development lens individualizes state responsibility and positions Third World states as supplicants rather than partners. We know a lot about how to transfer technology successfully and yet it is not happening, despite the undoubted and increasingly desperate self-interest in doing so. Karen Mickelson, in discussion with Usha Natarajan, suggested that '[a]t its essence, the environmental question challenges the ways we have heretofore understood our relationships with each other and with the world'.[4] I will

---

2 'Green technology' is used here to mean any technologies that can be used to mitigate and adapt to climate change, rather than to a narrower set of, for example, carbon-capture technologies.

3 For what I hope will become obvious reasons, I eschew the categories of developed and developing states, although I do use this language when referring to legal provisions that do so. Instead, I will use the terms 'Third World' and 'Group of 77' and the language of Global North and South to denote a key political and economic dividing line between states. On 'Third World' as a political relationship rather than a geographical or demographic designation, see Vijay Prashad, *The Darker Nations: A Peoples' History of the Third World* (New York/London: The New Press, 2007); also, Sundhya Pahuja, *Decolonising International Law: Development, Economic Growth and the Politics of Universality* (Cambridge University Press, 2011), Annex 2. I have also chosen to include China in the category of Third World states for the purposes of the argument advanced here, despite its current level of emissions; this is both for historical reasons (colonialism), China's own chosen alignment with the Group of 77 and because China, unlike Western nations, has not used more than its fair share of the global carbon budget. On this last point, see Andrew L. Fanning and Jason Hickel, 'Compensation for Atmospheric Appropriation' *Nature Sustainability* (2023); https://doi.org/10.1038/s41893-023-01130-8.

4 Karen Mickelson and Usha Natarajan, 'Reflections on Rhetoric and Rage: Bandung and Environmental Justice' in Luis Eslava, Michael Fakhri and Vasuki Nesiah (eds), *Bandung, Global History and International Law* (Cambridge University Press, 2017), 479. However, it is important to be aware of the dangers of viewing climate change or the Anthropocene as the great leveller. As Chandler and Reid suggest, this risks the '(re)colonisation of

argue that while this insight is undoubtedly true, the practice of international relations instead suggests that we remain stuck in old ways of thinking, of which development is a paradigmatic example.

20.2 Technology transfer and climate change

The core of the United Nations Framework Convention on Climate Change (UNFCCC) framework, adopted by all states in 1992 and reaffirmed on multiple occasions since, is the commitment to limit the amount of greenhouse gases in the atmosphere. The development of this global consensus on the need to prevent 'dangerous anthropogenic interference with the climate system' thus began to take shape in the last decade of the 20th century.[5] The Convention was supplemented in 1997 by the Kyoto Protocol, which imposed legally binding limits on the Global North (Annex I countries) to reduce their greenhouse gas emissions.[6] In 2015, states signed the Paris Agreement. This Agreement marked a new sense of urgency. However, while it signalled the desire to keep global average temperatures at a mere 1.5 °C above pre-industrial levels, this was to be achieved by measures that also enabled states to live up to their promises of sustainable development.[7]

Technology – notably green technology – is at the heart of mainstream mitigation and adaptation moves. As the 2000 Intergovernmental Panel on Climate Change (IPCC) expert report noted, 'the state of the environment today is the result of the technological choices of yesterday. Similarly, the state of the environment in the 21st century will be determined largely by the technologies we choose today.'[8] International technology transfer – the sharing of hardware, software and 'orgware' – is thus widely understood to be crucial to achieving the goals of mitigation and adaptation agreed in the UN Framework.[9] According to the UNFCCC website: 'Developing and transferring technologies to support national action on climate change has

indigeneity': David Chandler and Julian Reid, *Becoming Indigenous: Governing Imaginaries in the Anthropocene* (Rowman & Littlefield, 2019), 4.

5 Art 2 of the UNFCCC 1992. See Md Mahatab Uddin, *Climate Change Law, Technology Transfer and Sustainable Development* (Routledge, 2021).

6 Art 2 and Art 3 of the Kyoto Protocol 1997.

7 Preamble to Paris Agreement 2015, para 8; Art 2 of the Paris Agreement 2015.

8 IPCC, *Methodological and Technological Issues in Technology Transfer* (Cambridge University Press, 2000), 52.

9 While there are many discussions about what precisely is meant by technology transfer, it is broadly accepted that transfer must include the knowledge necessary to make use of the technology by adapting it to local circumstances. See United Nations Conference on Trade and Development, *Transfer of Technology*, United Nations 2001. Also Chapter 34 of Agenda 21 (1992).

been an essential element from the beginning of the UNFCCC process.'[10] Indeed, as the Sixth IPCC Assessment Report, in its headline statements, baldly provides, 'Finance, technology and international cooperation are critical enablers for accelerated climate action.'[11] Yet, most green technologies are protected by the patents of a handful of countries, notably Japan, South Korea, the United States and Germany.[12] Although there is agreement on the fact that technology is essential to meeting the UNFCCC targets, Third World states, on the whole, do not have the resources – financial, technological, institutional – to be able to develop the necessary technologies within the timeframe laid down by planetary boundaries. While the majority of technology transfer takes place between the countries that develop the technology, that is North–North, achieving the goals of mitigation and adaptation demands technology transfer from countries and companies that develop them to those that do not.

This awareness of the key enabling role of technology transfer has been translated into the legal provisions of the UN Framework.[13] Art 4(1)(c) of the UNFCCC requires the Parties to 'Promote and cooperate in the development, application and diffusion, including transfer, of technologies, practices and processes that control, reduce or prevent anthropogenic emissions of greenhouse gases.'[14] Art 4(5) further emphasizes the responsibilities of Annex I and Annex II countries to 'take all practicable steps to promote, facilitate and finance, as appropriate, the transfer of, or access to, environmentally sound technologies and know-how to other Parties, particularly developing country Parties'.[15] Agenda 21, of the same year, set out a vision of sustainable development, including on the importance of technology transfer to realizing it. Chapter 34 of Agenda 21 is directed toward the cooperation necessary to achieve technology transfer, with a focus on partnerships between North and South and between public and private actors.[16]

The 1997 Kyoto Protocol re-stated the centrality of technology transfer and the commitments made in the UNFCCC. Art 10(c) requires all

---

[10] UNFCCC, 'What is technology development and transfer?'; https://unfccc.int/topics/what-is-technology-development-and-transfer (last accessed 25 April 2023).

[11] C.7, Headline Statements, AR6 Synthesis Report: Climate Change 2023; www.ipcc.ch/report/sixth-assessment-report-cycle/ (last accessed 21 March 2023).

[12] Amrita Goldar, Shubham Sharma, Viraj Sawant and Sajal Jain, 'Climate Change & Technology Transfer: Barriers, Technologies and Mechanisms', Working Paper 382, Indian Council for Research on International Economic Relations, July 2019, 2.

[13] For a handy overview, see UNFCCC, *Technology and the UNFCCC: Building the Foundation for Sustainable Development* (2016); https://unfccc.int/ttclear/ (last accessed 20 June 2023).

[14] Art 4(1)(c) of the UNFCCC 1992.

[15] Art 4(5) of the UNFCCC 1992.

[16] United Nations, *Agenda 21: Programme of Action for Sustainable Development*, 1993.

Parties to: 'Cooperate in the promotion of effective modalities for the development, application and diffusion of, and take all practicable steps to promote, facilitate and finance, as appropriate, the transfer of, or access to, environmentally sound technologies, know-how, practices and processes pertinent to climate change, in particular to developing countries'.[17] This chapter of this same Article obliges Parties to take account of 'their common but differentiated responsibilities.' Taken together, Uddin concludes that the Article 'was intended to ensure the transfer of environmentally-sound technologies in a full manner'.[18] The Kyoto Protocol was followed up by a set of recommendations on technology transfer at COP4 in Buenos Aires in November 1998, which laid down criteria for effective technology transfer.

The 2007 Bali Action Plan, ten years after the Kyoto Protocol, introduced four basic pillars to address climate change, the third pillar of which was 'enhanced action on technology development and technology-transfer supporting actions on mitigation and adaptation'.[19] Yet more elements were added to achieve the requirements of technology transfer under the UNFCCC with the Technology Transfer Framework created under the Marrakech Accords of COP7 in 2010.[20] This Framework established the Technology Mechanism, comprised of two bodies: the Technology Executive Committee and the Climate Technology Centre and Network.[21] Parties are given support to identify their technology needs through a 'technology needs assessments' process and, since 2010, countries have been encouraged to develop technology action plans, which are to be concrete implementation plans based on the identified technology priorities.[22] Under Art 10(4) of the Paris Agreement, Parties recommitted themselves to the development of clean technologies and to the international technology transfer of that technology to achieve shared climate goals.

These commitments to support Third World states in achieving technology transfer through financial and technical means are arguably complemented by the principle of common but differentiated responsibility (CBDR). This emphasis on equity in international environmental law has often been understood as a *sui generis* moment in international law, a unique

17 Art 10(c) of the Kyoto Protocol 1997.

18 Uddin (n 5) 119.

19 Decision 2/CP.19, FCCC/CP/2007/6/Add.1, 14 March 2008.

20 UNFCCC, Technology Framework; https://unfccc.int/ttclear/tec/tech-transfer-framework.html (last accessed 25 April 2023).

21 UNFCCC, What is technology development and transfer?; https://unfccc.int/topics/what-is-technology-development-and-transfer (last accessed 25 April 2023).

22 Ibid. See also the role of the Global Environmental Facility (GEF), the operating entity of the UNFCCC Financial Mechanism: www.thegef.org/ (last accessed 25 April 2023).

recognition of our shared fate on this one planet.[23] Included in the Rio Convention under Principle 7, much has been made of CBDR as heralding a paradigm shift and as moving toward a more equitable world order. Sand has described the principle of CBDR as acknowledging 'the breakdown of traditional egalitarian fictions and [as representing] the emergence of a new legal polycentricity'.[24] As Toope notes, while the Rio Declaration is a non-binding instrument, the process of development of the Rio text was itself formal and represented a consensus. As such, Toope sees the principle as generating a set of expectations and as an emerging norm.[25] Winkler and Rajamani suggest that it 'permeates [the] balance of responsibilities between countries' across the UNFCCC as a whole.[26] Shapovalova has gone further, suggesting that it could be seen as the 'cornerstone principle of the international climate change regime'.[27]

Yet, it is not clear what the principle refers to – even more so since Paris.[28] Moreover, CBDR did not satisfy any of the Parties at Rio. For Global North states, for example the US, it was important to emphasize that the new principle did not create any legal obligations. For the states represented by the Group of 77, the principle was a long way from the firm commitment to redistribution that they had sought.[29] Despite this, Shapovalova suggests that

---

[23] Thomas Leclerc, 'The Notion of Common but Differentiated Responsibilities and Respective Capabilities: A Commendable but Failed Effort to Enhance Equity in Climate Law' in Benoit Mayer and Alexander Zahar (eds), *Debating Climate Law* (Cambridge University Press, 2021), 82 (referencing Bartenstein). This claim to uniqueness is set in stark relief against, for example, the privileging of sovereignty over preventing planetary nuclear annihilation in the 1996 ICJ Advisory Opinion on Nuclear Weapons, ICJ Reports 1996, p 226.

[24] Peter H. Sand, 'The Evolution of International Environmental Law' in Daniel Bodansky, Jutta Brunée and Ellen Hey (eds), *The Oxford Handbook of International Environmental Law* (Oxford University Press, 2007), 40.

[25] Stephen J. Toope, 'Formality and Informality' in Daniel Bodansky, Jutta Brunée and Ellen Hey (eds), *The Oxford Handbook of International Environmental Law* (Oxford University Press, 2007), 117.

[26] Harald Winkler and Lavanya Rajamani, 'CBDR&RC in a Regime Applicable to All' (2014) 14(1) *Climate Policy* 102–21, 104.

[27] Daria Shapovalova, 'In Defence of the Principle of Common but Differentiated Responsibilities and Respective Capabilities' in Benoit Mayer and Alexander Zahar (eds), *Debating Climate Law* (Cambridge University Press 2021), 63. As this was a staged debate, the statement may not reflect Shapovalova's own position.

[28] Is the principle referring purely to historical matters or to an ability or greater obligation to address climate change in the present? See Shelton for her consideration of the various ways of approaching equity or CBDR: Dinah Shelton, 'Equity' in Daniel Bodansky, Jutta Brunée and Ellen Hey (eds), *The Oxford Handbook of International Environmental Law* (Oxford University Press 2007), 657 et seq.

[29] The suggested text of the Group of 77 instead provided: 'The major cause of the continuing deterioration of the global environment is the unsustainable patterns of production and

the new incarnation of CBDR agreed at Paris (what she calls CBDR-RC+) has overcome the divisions created by bifurcating states according to their historical responsibility for climate change and now has widespread support across states.[30] Where the Rio Declaration was clear on responsibility – even if it did not lead to any hard legal obligations – the modification of the principle under the Paris Agreement, whereby it is to be implemented 'in the light of different national circumstances', suggests that the notion of differentiated responsibility, albeit duly repeated, means even less than it did before.[31] While Shapovalova sees Paris as a welcome development toward a common responsibility, what it arguably does is shift the emphasis from historical responsibility for climate change to the ability to pay for it now.[32] That is, it shifts the emphasis from differentiated responsibility to a common one. As such, this a-historical move, with its focus on the present, removes the key notion of responsibility from the principle of CBDR.[33]

In sum, states have repeatedly acknowledged the necessity of international technology transfer within the climate change framework, at the same time that they have committed themselves to the idea of CBDR. But the reality has not lived up to the rhetoric. Instead, the actual promises made are very limited and relate largely to creating mechanisms to help states identify suitable green technologies and to support them in negotiating access to it.[34] Each

consumption, particularly in developed countries … In view of their main historical and current responsibility for global environmental degradation and their capability to address the common concern, developed countries *shall* provide adequate, new, and additional financial resources and environmentally sound technologies on preferential and concessional terms to developing countries to enable them to achieve sustainable development.' Rio Conference on Environment and Development, UN Doc./CONF.151/PC/WG.III/L.20/REV.1 (1992); cited in Shelton (n 28) 657 (emphasis mine).

30 Shapovalova (n 27).

31 Paris Agreement 2015. See, too, Vivek Mukherjee and Faizan Mustafa, 'Climate Change and Right to Development' (2019) 5(S3) *Management and Economics Research Journal* 10.

32 Leclerc notes in his riposte to Shapovalova that there has always been ambiguity over whether the basis of the CBDR principle was the contemporary level of economic development or the contribution of GHGs to the atmosphere or both. Leclerc (n 23).

33 This goes beyond the legal ambiguity that Toope is referring to (n 25). While Toope notes the lack of concreteness of the principle, of its legal enforceability, he sees it as having meaning as an emerging norm. Basing financial commitments on the current ability to pay may be practically necessary to get all states on board, but it undoes the principle of CBDR. This is a very different critique to that put forward by Zahar, who – for the sake of argumentation – seeks to undermine the case for historical responsibility. See Alexander Zahar, 'Historical Responsibility for Climate Change Is Political Propaganda' in Benoit Mayer and Alexander Zahar (eds), *Debating Climate Law* (Cambridge University Press 2021).

34 See, for example, Art 10(4) of the Paris Agreement; also, TT: Clear, the home of the UNFCCC for all things climate technology related; https://unfccc.int/ttclear (last accessed 20 June 2023).

new accord or protocol repeats the commitment to international technology transfer and is then followed by expert reports and academic papers that analyse the extent to which, and the reasons why, those commitments have not been met in practice.[35] Similar reports are produced on the nature of the financial commitments actually realized as opposed to promised.[36] Progress is widely acknowledged to be slow and with mixed results.[37] As noted in the introduction, many expected that climate change and the shared threat that it represents would be a game-changer in international relations. Yet it has not proved to be so, despite the overwhelming self-interest of the North in mitigating climate change. The next section considers why this is understood to be the case.

20.3 Barriers to North–South technology transfer

The failure of North–South green technology transfer is, as this chapter is arguing, part of, or bound up with, broader questions of North–South geopolitics. However, much of the literature on climate-related technology transfer chooses to focus on technical barriers or on the moral/political difficulties of cooperation around global public goods, such as the environment.[38] We shall consider them in brief here.

It is difficult to ascertain with any degree of certainty – despite the various mechanisms that have been established – how much climate-relevant technology has been successfully transferred. This was the conclusion of the IPCC Working Group III on mitigating climate change almost 20 years ago and there has been no update.[39] Its report lists many technical barriers to the transfer of environmentally sound technologies, ranging from 'lack of confidence in unproven technologies' to a 'general lack of support for

---

[35] For example, Julian Kirchherr and Frauke Urban, 'Technology Transfer and Cooperation for Low Carbon Energy Technology: Analysing 30 Years of Scholarship and Proposing a Research Agenda' (2018) 119 *Energy Policy* 600.

[36] For example, see the May 2023 report, co-authored by the African Climate Foundation (ACF) and the Firoz Lalji Institute for Africa at the London School of Economics and Political Science (LSE), that predicts that the cost of the EU's Carbon Border Adjustment Mechanism (CBAM) to African countries will be 25 billion dollars a year and will hurt the poorest countries the most. *Implications for African Countries of a Carbon Border Adjustment Mechanism in the EU*; available at https://africanclimatefoundation.org/news_and_analysis/implications-for-african-countries-of-a-carbon-border-adjustment-mechanism-in-africa/ (last accessed 11 May 2023).

[37] See Kirchherr and Urban (n 35).

[38] I am putting aside the question of the lack of hard legal commitments on the basis that this is a result of an unwillingness to commit rather than the cause of it.

[39] IPCC, *Methodological and Technological Issues in Technology Transfer* (Cambridge University Press, 2000), 17.

an open and transparent international banking and trading system'.[40] The absorptive capacity of low-income countries – the ability to adapt and integrate transferred technologies to the technical and social context – is also understood to be a significant technical barrier.[41]

However, the biggest barrier, according to research by the UNFCCC, to the adoption of green technologies in both mitigation and adaptation is financial.[42] Low-income countries cannot pay the full market value for access to these technologies. That this would constitute a significant barrier to achieving the goals of the UNFCCC was clear at the outset: an attempt was made to address this challenge under Art 4(3) of the UNFCCC, which obliges 'the developed country Parties and other developed Parties included in Annex II' to 'provide new and additional financial resources to meet the full costs incurred by developing country Parties in complying with their obligations'.[43] The phrasing 'new and additional' was intended to ensure that wealthier countries did not simply shift funding from existing Official Development Assistance (ODA) budgets.[44] The same Article provides that Annex II countries shall provide such financial resources as are necessary to meet the costs for developing countries of enacting the implementing measures, including for technology transfer. Art 4(5) of the Convention, as noted earlier, also requires Annex II Parties to 'take all practicable steps to promote, facilitate and finance, as appropriate, the transfer of, or access to, environmentally sound technologies and know-how to other Parties, particularly developing country Parties'. Yet, as Uddin notes, all these provisions leave much open to interpretation, of course – for example, over what constitutes an 'incremental cost', what is 'necessary' and, indeed, what is 'practicable' or 'appropriate'.[45]

The financing of technology transfer under the UNFCCC is facilitated by the Global Environmental Facility (GEF), working since 1991, and the Green Climate Fund (GCF), more recently established in 2015.[46] A 2016 UNFCCC report states: 'Since 1991, the GEF has supported developing

---

40 Ibid 19.

41 Goldar et al (n 12).

42 For example, UNFCCC Technology Executive Committee, *Results and Success Factors of TNAs* (2013); https://unfccc.int/ttclear/ (last accessed 20 June 2023). See also Timothy Cadman and Tapan Sarker, 'Climate Finance for Sustainable Development' in Timothy Cadman and Tapan Sarker (eds), *De Gruyter Handbook of Sustainable Development and Finance* (De Gruyter, 2022).

43 Art 4(3) UNFCCC Framework 1992.

44 Yulia Yamineva and Kati Kulovesi, 'The New Framework for Climate Finance Under the United Nations Framework Convention on Climate Change: A Break-through or an Empty Promise?' in Erkki J. Hollo et al (eds), *Climate Change and the Law* (Springer, 2013).

45 For a detailed analysis of these provisions, see Uddin (n 5) chapter 3.

46 See UNFCCC (n 13).

countries to implement more than 800 projects with mitigation technology transfer objectives through over USD 5 billion of funding and USD 40 billion of co-financing.'[47] While this sounds impressive, recent research by Oxfam suggests that the figure of US$100 billion, promised by rich countries in 2009 each year to support adaptation and mitigation, seems only to have been met this year (2023). Moreover, the actual figure is much lower, given – as Oxfam reports – that much of the financing is either not new, that is, it is siphoned off from regular development financing, or it constitutes a loan to be repaid with interest rather than a grant.[48] In addition, it is not clear what part of this general climate financing is reserved for technology transfer, although studies suggest that most climate financed projects claim to involve some element of technology transfer.[49]

That financing is understood to be the key barrier to green technology transfer is, in part, due to the costs created by intellectual property rights (IPR). While IPR are not the only costs – assisting Third World countries in developing the necessary absorptive capacity, for example, also represents a significant cost – IPR does significantly increase the cost of accessing the best green technology.[50] One of the reasons is that the availability and supply of environmentally sound technologies is controlled by private actors, who must be persuaded of the benefit of sharing their assets. It is difficult for states to compel corporations to share their private property; instead, they must pay sufficient financial incentives in order to persuade them to do so.[51]

That wealthy states have shown themselves as yet unwilling to perform this task is ascribed by Kaul to the predominant market-centred policy approach. Such an approach entails that states are not seen (and do not see themselves)

---

47 Ibid 7.

48 Oxfam, *Climate Finance Shadow Report 2023: Assessing the Delivery of the $100 Billion Commitment* (June 2023); https://policy-practice.oxfam.org/resources/climate-finance-shadow-report-2023-621500/ (last accessed 20 June 2023).

49 UNFCCC, 'Mapping Climate Technology Development and Transfer Activities and Initiatives Under and Outside the Convention Relevant to the Implementation of the Paris Agreement', UN Doc FCCC/SBSTA/2016/INF.9 (21 October 2016); also, Chaewoon Oh, 'Contestations Over the Financial Linkages between the UNFCCC's Technology and Financial Mechanism: Using the Lens of Institutional Interaction' (2020) 20 *International Environmental Agreements: Politics, Law and Economics* 559.

50 See Caoimhe Ring, 'Patent Law and Climate Change: Innovation Policy for a Climate in Crisis' (2021) *Harvard Journal of Law and Technology* 375; also Krishna Ravi Srinivas, 'Technology Transfer, IPRs and Climate Change' in D.G. Ockwell and A. Mallett (eds), *Low-Carbon Technology Transfer: From Rhetoric to Reality* (Taylor & Francis, 2012).

51 M. Monirul Azam, 'Climate Change Resilience and Technology Transfer: The Role of Intellectual Property' (2011) 80 *Nordic Journal of International Law* 485–505.

as best placed to solve such problems.[52] However, this claim is arguably belied by the commitments to financing and facilitating technology transfer. If states believe that the market is best placed to respond to climate change, why do they continue to come together to make bold commitments? Another answer is the tragedy of the commons, under which states feel no obligation to act. Yet, they do feel obliged to make commitments, and they appear to accept, in principle at least, the notion of common but differentiated responsibilities for global warming. Similarly, while the coordination challenge is indeed enormous,[53] it is not an insurmountable one, as the 1987 Montreal Protocol on CFCs (chlorofluorocarbons) demonstrated. Here, successful international cooperation on international technology transfer has been able to avert serious stratospheric depletion.[54] The unwillingness on the part of the Global North to make good on commitments to fund and facilitate technology transfer in response to climate change must arguably be located elsewhere.

At the time of the Bandung conference in 1955, and during the high point of the Third World movement in the 1960s and 1970s, the environment was understood as a bargaining chip that the Third World held in their negotiations over the transition to a more equitable international order.[55] While the Third World could not insulate itself from the environmental problems created by the North, the reverse was also true. However, this sense of being trapped on a shared planet has not, despite the optimism, led to real commitment to a sense of being part of a single humanity or to a more just world order. Instead, it has led to a stalemate, which is the status quo. Just as the fact of climate change has not been able to shift the entrenched pattern of relationships, neither has the principle of CBDR.

---

[52] Inge Kaul, 'Global Public Goods and Governance for Addressing Sustainability' in Machiko Nissanke and José Antonio Ocampo (eds), *The Palgrave Handbook of Development Economics* (Palgrave Macmillan, 2019).

[53] Ibid.

[54] The Montreal Protocol on Substances that Deplete the Ozone Layer, adopted 16 September 1987. The ozone layer is recovering and is expected to be back at pre-1980 levels by mid-century – a success widely attributed to the Protocol and its implementation. Former UN Secretary-General Kofi Annan suggested that 'perhaps the single most successful international agreement to date has been the Montreal Protocol'; SG/SM/9471-ENV/DEV/792-OBV/434, 9 September 2004. See Stephen O. Andersen, K. Madhava Sarma and Kristen N. Taddonio, *Technology Transfer for the Ozone Layer. Lessons for Climate Change* (Earthscan, 2007), particularly chapter 14: Lessons. Also, Cass R. Sunstein, 'Of Montreal and Kyoto: A Tale of Two Protocols' (2008) 38 *Harvard Environmental Law Review* 10566.

[55] Mickelson and Natarajan (n 4) 475–7. See also R.P. Anand, 'Development and Environment: The Case of the Developing Countries' (1980) 24 *Indian Journal of International Law* 1 (making a powerful case for the necessity of a relationship between development and respect for the environment).

The next section considers why this might be so by suggesting a basis for the embedded pattern of relationships between North and South that is proving itself to be so enduring – even perhaps to the point of shared disaster.

20.4 The development paradigm

The suggestion here is that 'development' is a fundamental organizing principle of global relations. In order to make good on that claim, it is important to explain what I mean by development.

20.4.1 Development for the developing

Development has many meanings and there are many critical accounts that have done excellent work in unpacking it. The opening page of Escobar's 1995 modern classic *Encountering Development* begins with Harry Truman's 1949 inaugural address as US President, in which he, according to Escobar, 'initiated a new era in the understanding and management of world affairs'.[56] In his address, Truman laid out a division of the world. There were those, more than half of the world's population, that were living in 'misery. Their food is inadequate, they are victims of disease. Their economic life is primitive and stagnant. Their poverty is a handicap and a threat both to them and to more prosperous areas'.[57] The other part of the world, led by 'peace-loving peoples', did not live in desperation because they possessed the modern scientific and technological knowledge that was the key to prosperity and peace. Truman's vision was to see, in the words again of Escobar, 'the American dream of peace and abundance be extended to all peoples of the planet'[58] – and, of course, to see Communism vanquished. Truman's vision rapidly became the orthodoxy, adopted by the institutions of the United Nations and the countries of the West. It remains a powerful organizing concept of contemporary relations.

Truman's notion of development positioned half the world in one fell swoop as under-developed, in contrast to the rest that had already achieved development. This dichotomy of under-developed and developed saw the Third World defined by what it lacked in comparison to the West.[59] The

---

[56] Arturo Escobar, *Encountering Development: The Making and Unmaking of the Third World* (Princeton University Press, reissued 2012).

[57] Inaugural Address of Harry S. Truman, 20 January 1949; available via the digital Truman presidential library at www.trumanlibrary.org/whistlestop/50yr_archive/inagural20jan1949.htm (last accessed 30 April 2023).

[58] Escobar (n 56) 4.

[59] See, here, Escobar's famous analysis: 'The Problematization of Poverty: The Tale of Three Worlds and Development' in *Encountering Development* (n 56) 21–54.

shared understanding of what developing countries lack has shifted over the decades of the 20th century, from the lack of civilization and statehood under colonialism to the lack of industrialization and material goods, education systems and health care, good governance, the rule of law and, more recently, access to justice. The red thread running through these shifts has been the belief that what developing countries lack is the ability to achieve these markers of civilization on their own. While these notions were the product of a particular moment in time – namely the end of the Second World War and the beginning of the Cold War – they drew on previous Western categorizations of the world. The civilized of the colonial era became the developed, the possessor of advanced technological knowledge and the beneficiary of industrially produced abundance. The barbarians and the not-fully-human of the pre- and colonial period became the miserable subjects of 20th-century under-development.[60]

The development discourse that emerged in this period had one crucial difference, however: whereas the colonial subject could only be slightly improved by the benefits of proximity to the enlightened colonizer, the subject of the Third World could be brought into the light. That is, the poverty and backwardness of the Third World subject could be overcome by modernization, that is, technology. This discourse of development was to create consumers and solve the problem of the poor through the work of Western intervention, its largesse and charity, and through the expertise of an industry of development consultants and practitioners that grew up to meet this need.[61] This optimism of the 1950s and 60s gave way to disillusionment in the 70s and 80s, as the promised development failed to materialize. This pattern of energy and optimism followed by disappointment and despair has since been repeated several times,[62] with the discourse of development demonstrating sufficient flexibility so as to be capable of renewal as circumstances change: disillusionment leads to a new wave of optimism, as the next generation redefines development so as to, apparently, address the causes of the failures of the past. Economic development becomes a human-rights-based approach; law and development becomes the rule of

[60] See Peter Fitzpatrick on the 'second imperialism': Peter Fitzpatrick, 'Law's Infamy' in Sammy Adelman and Abdul Paliwala (eds), *Law and Crisis in the Third World* (Hans Zell Publishers, 1993), 30.

[61] On the latter, see Deval Desai, 'Law and Development Experts' in K. de Feyter, G.E. Türkelli and S. de Moerloose (eds), *Encyclopedia of Law and Development* (Edward Elgar, 2021).

[62] For an illustration of this, see David M. Trubek, 'The "Rule of Law" in Development Assistance: Past, Present and Future' in David M. Trubek and Alvaro Santos (eds), *The New Law and Economic Development: A Critical Appraisal* (Cambridge University Press, 2010).

law; industrial development becomes sustainable development; and so on.[63] We are trapped in this pattern of renewal and disappointment because the underlying paradigm of development does not change. Being 'developing' entails constantly trying to attain a moving target, as development is expanded to incorporate new standards of civilization, such as the rule of law, good governance, gender equality, indigenous rights, wildlife protection, and so on. In this way, development has become a key mode of global governance that regulates the behaviour of Third World states.[64] This is the case notwithstanding the decolonization of the Global South in the 1960s and 70s and the regular proclamation of the sovereign equality of all states.

One of the reasons that 'development' has become such a powerful global narrative is its 'colonization of reality'.[65] Development, in the words of Escobar, 'became a powerful instrument for normalizing the world';[66] that is, for attempting to design the world in the image of the West. Development has been so successful as a global discourse because it was adopted and internalized by so many countries. The process of developing has seen decolonized states subject themselves and their populations to a near constant cycle of through-going interventions and austerities in the name of modernization. Integration into the market economy was similarly accepted as the only path available.[67]

Through the crisis of the law-and-development movement,[68] the shifts of economic orthodoxy,[69] the charting of development as a site of social struggle

---

63 See, for example, Mary Robinson, 'What Rights Can Add to Good Development Practice' in Phillip Alston and Mary Robinson (eds), *Human Rights and Development: Towards Mutual Reinforcement* (Oxford University Press, 2005). For a Third World approach to the right to development, see Mickelson's classic analysis of Keba M'Beye's article: Karen Mickelson, 'Rage and Rhetoric: Third World Voices in International Legal Discourse' (1998) 16(2) *Wisconsin International Law Journal* 353.

64 This has been made easier with the rise of global development indicators; see, Morag Goodwin, 'The Poverty of Numbers: Reflecting on the Legitimacy of Global Development Indicators' (2017) 13(4) *International Journal of Law in Context* 485–97.

65 Escobar (n 56) 5.

66 Ibid 26.

67 See Mickelson and Natarajan (n 4).

68 David M. Trubek and Marc Galanter, 'Scholars in Self-Estrangement: Some Reflections on the Crisis in Law and Development' (1974) *Wisconsin Law Review* 1062; Trubek and Santos (n 56); Sammy Adelman and Abdul Paliwala, 'Law and Development in Crisis' in Adelman and Paliwala (n 60).

69 For example, the rise and (maybe) fall of the Washington Consensus and the shift to New Institutional Economics, or the move from free-trade dogmatics and TRIPS to 'Capitalism 3.0': see Dani Rodrik, *The Globalization Paradox* (Oxford University Press, 2011). See also Julio Faundez, 'Rule of Law or Washington Consensus: The Evolution of the World Bank's approach to Legal and Judicial Reform' in Amanda Perry-Kessaris (ed), *Law in the Pursuit of Development: Principles into Practice?* (Routledge, 2010).

rather than of academic or policy theorizing, and the production of critical accounts of development with the rise of new forms of ethnography, the othering of the developed–developing binary has nonetheless remained. The shift to the language of 'sustainable development', which is not explored here, is easily accommodated within the existing paradigm of development and does not challenge it: the burden of delivering development in a sustainable manner falls on individual states, regardless of how broadly one seeks to define sustainability.[70]

20.4.2 Technology transfer in the context of development

The belief in the technological superiority of the West predates Truman's speech. Technology was both a means of colonization and a justification for the oppression of 'backward' societies. However, technological knowledge was at the heart of Truman's vision of development, and he acknowledged the moral imperative to share that knowledge. He stated: 'I believe we should make available to peace-loving peoples the benefits of our store of technical knowledge in order to help them realize their aspirations for a better life.'[71] Since this promise, technology transfer has been a faultline of a development-structured world.

The failure by the 1970s to realize Truman's pledge to share technological know-how saw Third World states begin to use international fora to demand international technology transfer.[72] In the call for a New International Economic Order (NIEO), Third World states put their case for an equitable international system, which necessarily included '[g]iving to the developing countries access to the achievements of modern science and technology, and promoting the transfer of technology and the creation of indigenous technology for the benefit of the developing countries in forms and in accordance with procedures which are suited to their economies.'[73]

---

[70] From among the many efforts to define sustainable development 'holistically' that take into account economic, social and environmental sustainability, see Ramon Pichs et al (eds), *Development, Sustainability, and Equity* (IPCC and WMO, 2000). For sustainable development as both a goal and a process, see Uddin (n 5) 9. See also Marcus Taylor, 'Climate Change and Development' in Henry Veltmeyer and Paul Bowles (eds), *The Essential Guide to Critical Development Studies* (Routledge, 2018).

[71] Truman (n 57).

[72] For a clear overview of international developments in this period, see Pahuja (n 3), particularly chapter 3. Also, Balakrishnan Rajagopal, *International Law from Below: Development, Social Movements and Third World Resistance* (Cambridge University Press, 2003), esp. chapter 4.

[73] Declaration on the Establishment of a New International Economic Order, General Assembly Resolution 3201 (S-VI), A/RES/S-6/3201, 1 May 1974, para 4(p).

This demand was fleshed out in the follow-up General Assembly Resolution that laid down a programme of action for the establishment of the NIEO.[74] The first point of action was to be the negotiation of an international code of conduct for the transfer of technology. Forty years and multiple drafts later, the negotiation process was suspended, and it remains so. According to Blakeney, the main source of disagreement was between the Group of 77 and the Group B countries (the Western, industrialized nations) as to the legal effect of the document: while Group B countries wanted a 'mutually acceptable set of guidelines or standards', the Group of 77 had little faith in the goodwill of these states and would accept only a legally binding agreement.[75] The Group of 77's position is expressed in the twelfth paragraph of the draft document's preamble, which states that 'an international legally binding instrument is the only form capable of effectively regulating the transfer of technology'.[76] This disagreement ensured that the negotiations for a treaty went nowhere, just as the NIEO was stillborn.

In place of a regulatory approach to technology transfer – in which there is an intervention in the market to rectify inequalities between the technology owner and the recipient (such as a legally binding treaty) – instead a free market approach – in which the focus is creating the market conditions for the transfer of technology – triumphed.[77] The principal feature of a free market approach is the reliance on the protection of private rights to technology based on intellectual property laws. Direct intervention in the content or nature of technology transfer transactions is excluded, except where such transactions would violate principles of competition law by reason of their market-distorting effects and/or by their use of unreasonable restrictive trade practices. The suspense over whether or not a code of conduct could be agreed to ended in the negotiation of the Trade-Related Aspects of Intellectual Property Rights (TRIPS) agreement in the Uruguay Round and the pressure on the Global South to adopt its binding rules.[78]

---

74 UN GA Res 3202 (S-VI). Programme of Action on the Establishment of a New International Economic Order, IV. Transfer of technology.

75 Michael Blakeney, *Legal Aspects of the Transfer of Technology to Developing Countries* (ESC Publishing Ltd, 1989), 133–9. See also Pedro Roffe, 'Comment: Technology Transfer on the International Agenda' in Keith E. Markus and Jerome H. Reichman (eds), *International Public Goods and Technology Transfer Under a Globalized Intellectual Property Regime* (Cambridge University Press, 2005), 257–64.

76 UNCTAD doc. TD/CODE TOT/47 (5 June 1985).

77 See United Nations Conference on Trade and Development, *Transfer of Technology* (United Nations, 2001). Also, Kaul (n 52).

78 Agreement on Trade-Related Aspects of Intellectual Property Rights, 15 April 1994. See also William Pretorius, 'TRIPS and Developing Countries: How Level Is the Playing Field?' in Peter Drahos and Ruth Mayne (eds), *Global Intellectual Property Rights* (Palgrave Macmillan, 2002), 183–97.

International technology transfer on a scale large enough to create a more just global economic order remains in the realm of promise.

In place of the rules on technology transfer sought by Third World states, Western powers agreed instead to the 1986 UN Declaration on the Right to Development.[79] This Declaration, in Art 1, proclaimed development to be a human right. While there has been much discussion over the nature of this new human right,[80] one undoubted feature was that it enshrined sovereignty as the basis of development. Whereas Truman had laid out a moral responsibility of the US to provide the means for the development of all states, the Right to Development located that responsibility instead with states, regardless of their economic position. The Declaration grants a right to all individuals and states to development, but the only identifiable duty-holder is states themselves; that is, where an individual has the right to development, it is a right – as with all other human rights – against their own state. The Right to Development states the importance of international cooperation necessary to make a right to development substantively meaningful but provides neither a right to it nor a framework for achieving it. As such, it does not provide the changes to the economic order essential to achieving anything like equity, justice or, indeed, development. Instead, what substituting the Right to Development for an NIEO arguably achieved in practice was to absolve the Global North of any responsibility for suffering caused by global economic practices of those outside their own borders, by limiting obligations to individual states while retaining the right to dictate the terms of existence to Third World states via the conditions under which international credit and access to the global economic system will be granted.[81] Seen in this light, the shift in the Paris Agreement from CBDR as an acknowledgement of historical responsibility for global warming to a responsibility for each state to address the consequences of climate change according to their own national circumstances looks considerably less surprising.

The Truman address and the Declaration on the Right to Development bookend a trajectory of 'development', whereby the underlying impulse as described by Escobar remains unchanged, but where the locus of responsibility for achieving development shifted decisively from the North to

[79] 1986 Declaration on the Right to Development, GA Res. 128, UN GAOR, 41st Sess., Supp. No. 53, UN Doc. A/RES/41/128 (1986) 186.

[80] Among the considerable literature, see Anne Orford, 'Globalization and the Right to Development' in Phillip Alston (ed), *Peoples' Rights* (Oxford University Press, 2001); also, Upendra Baxi, *Human Rights in a Posthuman World* (Oxford University Press, 2009), chapter 4: 'The Development of the Right to Development'.

[81] On this latter point, see Pahuja (n 3), chapter 4. This is not to say that there were not Third World scholars who argued passionately for the Right to Development and had high hopes of what it could achieve. See Mickelson (n 63).

the Global South. It is this development paradigm sketched here – in which Global South states must not only attempt to close the gap between North and South as defined by Northern states and institutions, but in which they must also accept the responsibility for their failure to do so – that forms the context for discussions about actions to address climate change. Global South states have the formal acknowledgement of sovereign equality that came with decolonization and independence, but without the underlying paradigm change that is necessary to lead to any meaningful notion of equality.

20.5 Moving away from 'development': concluding remarks

The development paradigm is one in which the Global South is understood as inferior and thus as in need of Northern interventions; it is a relationship not of equal partners in solving global challenges but of charity. While sovereignty posits a formal equality between states, the development paradigm undermines that equality. However, sovereign equality has not been able to shift the development paradigm described earlier. In the words of Dinah Shelton, '[t]raditional international law largely reflects ideas that protect entitlement'.[82] That is, international law protects gains if they were 'justly' acquired at the time of acquisition, which means that it protects the status quo in terms of the distribution of wealth and privilege under the principle of the equality of states. Similarly, Mason-Case and Dehm note that 'emissions are embedded in, reflect, and reproduce social relations'.[83] They cite Anna Grear as suggesting that climate injustice is a 'manifestation of a structural pathology in which law is central'.[84] Climate change is, they suggest, inextricable from the unfinished business of decolonization, which leads them to call for reparations as the only way to create a bottom-focused redress. I concur with their analysis but would caution that, if development remains a core organizing principle of global governance, there can be little expectation that reparations will move us beyond the entrenched patterns of behaviour. Sovereignty cannot be a means of overcoming 'development'. Rather, the promise of sovereignty – realized formally – has been used to escape responsibility for both past and continuing injustices.

---

82 Dinah Shelton, 'Equity' in Daniel Bodansky, Jutta Brunée and Ellen Hey (eds), *The Oxford Handbook of International Environmental Law* (Oxford University Press, 2007), 654.

83 Sarah Mason-Case and Julia Dehm, 'Redressing Historical Responsibility for the Unjust Precarities of Climate Change in the Present' in Benoit Mayer and Alexander Zahar (eds), *Debating Climate Law* (Cambridge University Press, 2021), 186 (emphasis theirs).

84 Anna Grear, 'Towards "Climate Justice"? A Critical Reflection on Legal Subjectivity and Climate Injustice: Warning Signals, Patterned Hierarchies, Directions for Future Law and Policy' (2014) 5 *Journal of Human Rights and the Environment* 103, 106.

What this means in practice, post-Paris, is that, in place of solidarity, climate change negotiations have fostered new forms of competition between states in the Global South for special vulnerability status and for access to a created scarcity of access to technology. Where some states are positioned as winners that will benefit from recognition of their vulnerability, others are losers, for example 'African' states.[85] In place of the facilitation of technology transfer that the UNFCCC promises, Third World states are instead responsible for ensuring that they drive the process of technology transfer, identifying the best technology for the purpose, acquiring the necessary absorptive capacity to make it work, and seeking the financial support to cover the cost.[86] When they fail, the development doctrine entails that they bear both the material and moral burden of coming up short.

How might we move away from development as a practice of global governance? And what might be the role of (international) law therein? If technology transfer is to work to help all states achieve the goals laid down in Paris, it arguably needs to be based on a spirit of partnership born out of generosity – generosity not in the sense of largesse but in the sense of loving generosity that underpins true friendship.[87] This type of partnership, in my reading, is at the heart of the 2022 Bridgetown Agenda – the latest in a long line of initiatives put forward by Third World states to demand a more equitable world order – which calls, *inter alia*, for new global partnerships to realize technology transfer.[88] Moreover, such partnerships in the area of technology transfer can have knock-on effects, such as enabling the creation of indigenous innovation systems.[89] This would not only contribute to a more equitable global economic system but also be to the benefit of all, as this technology can also be shared, whether South–South or South–North.

---

[85] See the representative body, the African Group of Negotiators (AGN) on Climate Change, founded in 1995: https://africangroupofnegotiators.org/ (last accessed 20 June 2023). Also, Nicholas Chan, '"Special Circumstances" and the Politics of Climate Vulnerability: African Agency in the UN Climate Change Negotiations' (2021) 56(3) *Africa Spectrum* 314–32, 325.

[86] On the need for a push-pull relationship (or partnership), see Chen Zhou, 'Enabling Law and Policy Environment for Climate Technology Transfer: From Perspectives of Host Countries' (2019) 12(1) *Journal of East Asia and International Law* 45–70.

[87] Undoubtedly due to my ignorance of the many other sources that there surely are on this subject, the best account that I have read of such friendship remains that of Michel de Montaigne: 'On Friendship' in *The Complete Essays* (trans. M.A. Screech; Penguin, 1993). On the possibility of partnership in South–North relations, see Marike Botenbal, 'Understanding North-South municipal partnership conditions for capacity development: A Dutch-Peruvian example' (2009) 33 *Habitat International* 100–5.

[88] www.bridgetown-initiative.org (last accessed 23 January 2025). See also Obiora Chinedu Okafor, 'International Solidarity and Climate Change', UN Doc A/HRC/44/44 (2020).

[89] Goldar et al (n 12) 8.

What, then, of the role of (international) law in achieving this? International law can, in principle, be a mechanism through which a more just world order is institutionalized; the Montreal Protocol shows that it is possible for shared goals to be achieved fairly. However, international law is, as Shelton and others note, deeply implicated in the injustices of the current order[90] and, thus, in itself is unlikely to be a vehicle for overcoming those injustices. As such, it cannot be the starting point. Instead, a new way of thinking is necessary – albeit that it may be institutionalized through a new international law. Law, however, is not the answer.

I have suggested here that 'development' is a governance practice, a way of ordering the world, that has fundamental implications for the ability of everyone on the planet to live the best life they can live. It also has serious implications for our ability to achieve the goals of the UNFCCC. In her ground-breaking work on social relations in the Anthropocene, Haraway has described whiteness as a mode of being that has brought the world to the brink of destruction: whiteness as a 'worlding practice premised on the commitment to endless growth and vastly unequal well being'.[91] I am not yet sure what the relationship of this ontological whiteness is to development as a paradigm – whether they are different ways of describing the same phenomenon or distinct manifestations of the same paradigm. Either way, the work of critical Anthropocene scholars, such as Haraway, confirm that it is more important than ever that we excavate and take seriously the underpinning normative structures of this world if we are to create a better one upon a shared and future Earth.

[90] See also, Pahuja (n 3); Antony Anghie, *Imperialism, Sovereignty and the Making of International Law* (Cambridge University Press, 2007). For the harm that international law does to the environment, see Usha Natarajan and Kishan Khoday, 'Locating Nature: Making and Unmaking International Law' (2014) 27(3) *Leiden Journal of International Law* 573–93.

[91] Donna Haraway, 'Making Kin in the Chthulucene: Reproducing Multispecies Justice' in A.E. Clarke and D. Haraway (eds), *Making Kin Not Population* (Prickly Paradigm Press, 2018), 71–2.

Index

J

N

T